Mechanics of Materials

Mechanics of Materials

Paul S. Steif
Carnegie Mellon University

PEARSON

Boston   Columbus   Indianapolis   New York   San Francisco   Upper Saddle River
Amsterdam   Cape Town   Dubai   London   Madrid   Milan   Munich   Paris   Montréal   Toronto
Delhi   Mexico City   São Paulo   Sydney   Hong Kong   Seoul   Singapore   Taipei   Tokyo

Vice President and Editorial Director, ECS: *Marcia J. Horton*
Acquisitions Editor: *Norrin Dias*
Vice-President, Production: *Vince O'Brien*
VP/Director of Marketing: *Patrice Jones*
Executive Marketing Manager: *Tim Galligan*
Marketing Assistant: *Jon Bryant*
Senior Managing Editor: *Scott Disanno*
Production Project Manager: *Clare Romeo*
Operations Specialist: *Lisa McDowell*
Associate Director of Design: *Blair Brown*
Cover Designer: *Blair Brown*
Interior Design: *Blair Brown, Paul S. Steif*
Manager, Rights and Permissions: *Beth Brenzel*
Image Permission Coordinator: *Karen Sanatar*
Composition: *MPS Limited, a Macmillan Company*
Project Management Liaison: *Anoop Chaturvedi*
Printer/Binder: *Courier Kendallville*
Typeface: *9/11 Times Roman*

Credits and acknowledgments borrowed from other sources and reproduced, with permission, in this textbook appear on appropriate page within text.

Pearson Education Ltd., *London*
Pearson Education Singapore, Pte. Ltd
Pearson Education Canada, Inc.
Pearson Education—Japan
Pearson Education Australia PTY, Limited
Pearson Education North Asia, Ltd., *Hong Kong*
Pearson Educación de Mexico, S.A. de C.V.
Pearson Education Malaysia, Pte. Ltd.
Pearson Education, Inc., *Upper Saddle River, New Jersey*

Library of Congress Cataloging-in-Publication Data on File

10   9   8   7   6   5   4   3   2   1

ISBN-13: 978-0-13-220334-0
ISBN-10: 0-13-220334-0

Visual Contents

INTRODUCTION

Chapter 1. Design of products, systems, and structures demands the engineer to consider a broad range of issues. The issues addressed by *Mechanics of Materials* are excessive deformation and material failure. A few general principles enable us to design against excessive deformation and failure for a wide range of part geometries, materials, and loadings. We consider the body to be composed of elements, we study common deformation modes, and we combine contributions of each deformation mode, as needed, to assess deformation and failure.

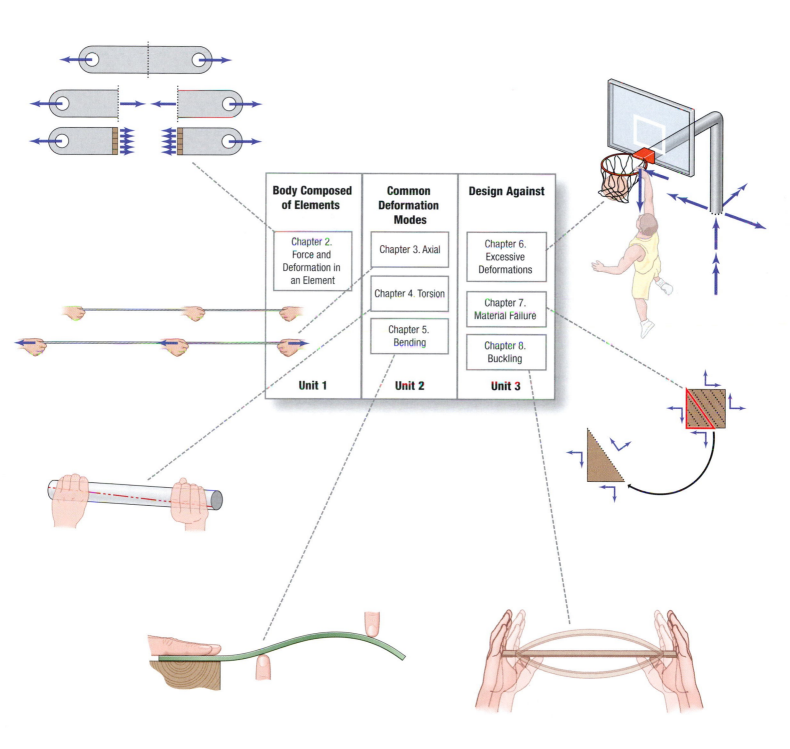

Body Composed of Elements	Common Deformation Modes	Design Against
Chapter 2. Force and Deformation in an Element	Chapter 3. Axial	Chapter 6. Excessive Deformations
	Chapter 4. Torsion	Chapter 7. Material Failure
	Chapter 5. Bending	Chapter 8. Buckling
Unit 1	Unit 2	Unit 3

Contents

1

Introduction 2

Unit 1

Body Composed of Elements

2

Internal Force, Stress, and Strain 18

Unit 2

Common Deformation Modes

3

Axial Loading 84

4

Torsion 136

5

Bending 218

Unit 3

Design Against

6

Combined Loads 364

7

Stress Transformations and Failure 412

8

Buckling 480

Appendices 501

Answers to Selected Problems 552

Key Terms 562

Index 564

Preface

To the Student

This book introduces you to an exciting subject of immense application: how the forces acting on a material relate to its deformation and failure. The range of technologies that rely on insights from *Mechanics of Materials* is vast. They span applications that have seen continual innovation and refinement over many years, such as aerospace structures and propulsion, bridge design, automotive technologies, and prosthetic devices. And, *Mechanics of Materials* underlies applications that were scarcely imaginable a few years ago: atomic force microscopes, micro-scale robotics, wireless sensors for structural monitoring, and engineered biological tissues. *Mechanics of Materials* can be satisfying in another more personal way. It helps us make sense of countless interactions that we have with everyday artifacts: why some are too flimsy, too rigid, or prone to break at certain points.

It is likely you are studying this subject because it is required for your major. But you may have multiple goals: to pass the course or get a good grade, to be intellectually engaged and exercise your mind and curiosity, and to learn something that you can use in later courses or in life outside your courses. Every one of those goals points you in the same direction—to genuinely learn the subject. That means gaining a physical and intuitive feel for its ideas, seeing the big picture, and fitting the ideas together. By just thumbing through this book, you will know it is different from most books you have seen. Let me tell you how the arrangement of this book might help you learn.

We can only communicate the ideas of *Mechanics of Materials* with a *combination* of words, diagrams, and equations. The equation might be necessary to get a quantitative answer or to judge a trend; for example, should a part be thicker or thinner, longer or shorter. But, in real life you are rarely handed the right equations. Someone explains a situation to you with words and diagrams, and you need to make sense of it. Only after you have thought about the words and the diagrams, might you see an equation as useful. For this reason, I have tried to write a book in which words, diagrams, and equations are in balance. In addition, I have laid out this book so the words, diagrams, and equations are near each other on the page to better help you solidify the ideas.

You might also notice a high degree of organization. Each chapter is a series of two-page spreads or sections, with each section dedicated to developing one idea or concept. Further, each two-page spread consists of subsections that break the idea into bite-size pieces. Not only do we break this subject apart for you, we help you put it back together. The Chapter Opener presents the major ideas of the chapter in diagrams and words. At the end of each chapter, we summarize its sections, including the major equations, concepts, and key terms. Finally, Chapters 2 through 8 are grouped into 3 units that capture the overall structure of the subject.

You might also notice many everyday objects depicted on the pages. Familiar, everyday objects can often illustrate the ideas of *Mechanics of Materials*. To genuinely learn this subject, the ideas must ultimately make sense to you. But you are more likely to make sense of new ideas if you see them first in a familiar context. This book tries to take situations that you can already picture, and reframe them in more general, powerful ways. I hope you come to rely on those general ideas and wield them effectively as you explore new applications unimagined today.

To the Instructor

I wrote this book because I love to help other people understand mechanics. I have taught this subject for many years, and I still get excited when I come upon a new way of explaining or illustrating some concept. Often, I bring an object into class—a bungee cord, a pool noodle, a ruler—and I deform it, sometimes with students' help. I point to the deformation, which they can see, and I ask the student helpers what they feel. With this book, I hope to capture some of that classroom experience.

Let me share some of the pedagogic philosophy that informs this book. I think most instructors want students:

1. to understand the concepts in some intuitive way;
2. to grasp the big picture, that is, to see the forest as well as the individual trees;
3. to use the subject to solve problems.

First, to an intuitive understanding of concepts, there are few more important goals than helping students attach physical meaning to the variables and symbols we use, and to their relations with each other. I rarely start with the general case. Instead, I start with a simple situation that exemplifies the idea. This helps to anchor the idea in each students' world. Then, we build a more general mathematical representation, as we need it. Students can picture deformation far better than they can picture forces. So, for most topics, we begin with the deformation, to anchor the topic in reality for the students, and next we deal with the associated forces.

To help students grasp the subject's larger, coherent structure, we have identified the core question that it answers: will a body deform too much or fail (Chapter 1)? And, we have grouped the remaining chapters into three units that delineate how this question is answered. First, we choose to view a body that deforms and may fail as composed of many small, identical pieces or elements (Chapter 2). This step is necessary to address failure, which usually occurs locally, and to separate out the respective contributions of the body's shape and material to the force-deformation relations. Second, we identify three common modes of deformation: stretching, twisting, and bending, which appear repeatedly in engineering and nature (Chapters 3–5). Each mode deserves to be studied independently, considering the deformations and forces overall and within each element. Third, to address deformation and failure in more general situations, we recognize the presence of these common deformation modes, and combine their contributions appropriately (Chapters 6–8). To reinforce the big picture set forth in Chapter 1, the conceptual overview at the start of each chapter features a map that locates the chapter in the overall structure of the subject.

For good reason, the problems in a textbook are very important to most instructors. This book contains problems that illustrate ideas, concepts, and procedures, as well as problems that demonstrate applications to real situations. Studying *Mechanics of Materials* can also offer students a chance to learn about interesting applications. To this end, I have devised a number of problems that highlight selected focused application areas: bicycles, cable-stayed bridges, drilling of wells, exercise equipment, bone fracture fixation, and wind turbines. Focused Application Problems are sprinkled throughout the chapters. The diagram for each such problem references Appendix A, in which that application is described at greater length. An interested student can see how the situation depicted in a single problem fits into the overall application. For different assignments, an instructor can select problems from the same focused application area or problems from a variety of applications.

I hope this book serves your efforts to motivate and teach your students.

Resources for Instructors

- **Instructor's Solutions Manual.** An instructor's solutions manual was prepared by the author. The manual was also checked as part of the Accuracy Checking program.

- **Presentation Resources.** All art from the text is available in PowerPoint slide and JPEG format. These files are available for download from the Instructor Resource Center at www.pearsonhighered.com/steif. If you are in need of a login and password for this site, please contact your local Pearson Prentice Hall representative.

Resources for Students

- **MasteringEngineering.** Tutorial homework problems emulate the instructor's office–hour environment, guiding students through engineering concepts with self-paced individualized coaching. These in-depth tutorial homework problems are designed to coach students with feedback specific to their errors and optional hints that break problems down into simpler steps.

Acknowledgments

Prentice Hall has been a pleasure to work with during the development of this book. I am fortunate to have had continuing guidance and encouragement from three Acquisitions Editors: Eric Svendsen, Tacy Quinn, and Norrin Dias, as well as the insight and enthusiasm throughout from Editorial Director Marcia Horton. This project has benefited greatly from the attention of Marketing Manager Tim Galligan, who helped to shape my appreciation for the multiple audiences this book should seek to satisfy. I am grateful to Senior Managing Editor Scott Disanno, who has both overseen the production of the book and provided the fresh, clear eye that honed the manuscript at its final stages. Designer Blair Brown brought a magical touch and excitement to this unusual project, and I am grateful for his efforts and the fun I had working with him. The expertise of J.C. Morgan and lead artist Matt Harshbarger at Precision Graphics has contributed significantly to the final product, and I am grateful for their patience as the book and artwork evolved. The distinctive integration of text, equations, and artwork in this book could not have been realized without Anoop Chaturvedi and the composition services of MPS Limited. Other than perhaps myself, no one spent more time or agonized more in bringing this project to fruition than Sr. Production Project Manager Clare Romeo. She has been a joy to work with, and I cannot thank her enough for her knowledge, expertise, attention to detail, patience, and humor.

Thank you to the reviewers: Paolo Gardoni, Texas A&M University; Joao Antonio, Colorado State University; Joel J. Schubbe, U.S. Naval Academy; Daniel A. Mendelsohn, Ohio State University; Laurence J. Jacobs, Georgia Tech; Eduard S. Ventsel, Pennsylvania State University; Dashin Liu, Michigan State University; Candace S. Sulzbach, Colorado School of Mines; Amir G. Rezaei, California State Polytechnic University, Pomona; Marck French, Purdue University; Niki Schulz, Oregon State University; Jim Morgan, Texas A&M University; Shane Brown, Washington State University; Christine B. Masters, Pennsylvania State University; Craig Menzemer, University of Akron; Edwin C. Rossow, Northwestern University; Anna Dollár, Miami University; Mark E. Walter, Ohio State University; David Baldwin, University of Oklahoma; Kevin Collins, United States Coast Guard Academy; He Liu, University of Alaska Anchorage; and Anthony J. Paris, University of Alaska Anchorage. At several points during its development, extensive and thoughtful input from the reviewers was critically important in helping the book take shape. Their time and efforts are greatly appreciated.

I am also grateful to faculty members and students who offered ideas for realistic applications and problems, including Jim Papadopoulos, Yoed Rabin, Dustyn Roberts, and Jonathan Wickert. Billy Burkey, Chris D'eramo, Anthony Fazzini, Rob Keelan, Michael Reindl, David Urban, and Derek Wisnieski provided valuable assistance in dimensions and images for a number of application problems. Advice on graphics from Erick Johnson towards the end of project was very helpful. I thank my assistant, Bobbi Kostyak, who provided help with many details that arose. I have relied often, to my great satisfaction, on the design and artistic sense of Ariela Steif, for which I am grateful.

This book has benefited from the many years I have fruitfully and joyfully discussed the learning of mechanics with my long-time friend and collaborator, Anna Dollár. I credit my friend and collaborator, Marina Pantazidou, for giving a pivotal nudge that convinced me to write this book, and for supplying ongoing encouragement in education endeavors generally. I want to thank Robbin Steif for the significant role she played at the start of this project.

My own teachers provided the foundation for my fascination with the subject of mechanics. I have in turn had the pleasure of getting to know many students over the years in my classes. They have helped me recognize the challenges in learning mechanics, and the practical situations in which mechanics comes alive.

During much of the writing of this book, I was fortunate to have the companionship, warmth, and good wishes of many fellow denizens of the Galleria.

My family life provides the perfect counterpoint to my work, and I thank my loved ones, Michelle, Ariela, Talia, and Marigny for making that family life such a desirable distraction to writing this book.

PAUL S. STEIF
Carnegie Mellon University

About the Author

Professor Paul S. Steif has been a faculty member in the Department of Mechanical Engineering at Carnegie Mellon University since 1983. He received a Sc.B. degree in engineering mechanics from Brown University; M.S. and Ph.D. degrees in applied mechanics from Harvard University; and was National Science Foundation NATO Post-doctoral fellow at the University of Cambridge. As a faculty member his research has addressed a variety of problems, including the effects of interfacial properties on fiber-reinforced composites, bifurcation and instabilities in highly deformed layered materials, and stress generation and fracture induced by cryopreservation of biological tissues. Dr. Steif has also contributed to engineering practice through consulting and research on industrial projects, including elastomeric damping devices, blistering of face seals, and fatigue of tube fittings.

Since the mid-1990s, Dr. Steif has focused increasingly on engineering education, performing research on student learning of mechanics concepts, and developing new course materials and classroom approaches. Drawing upon methods of cognitive and learning sciences, Dr. Steif has led the development and psychometric validation of the Statics Concept Inventory—a test of statics conceptual knowledge. He is the co-author of Open Learning Initiative (OLI) Engineering Statics. Dr. Steif is a Fellow of the American Society of Mechanical Engineers and recipient of the Archie Higdon Distinguished Educator Award from the Mechanics Division of the American Society for Engineering Education.

MasteringENGINEERING®

MasteringEngineering is the most technologically advanced online tutorial and homework system. It tutors students individually while providing instructors with rich teaching diagnostics.

Used by over a million students, the Mastering platform is the most effective and widely used online tutorial, homework, and assessment system for the sciences and engineering. A wide variety of published papers based on NSF-sponsored research and tests illustrate the benefits of MasteringEngineering. To read these papers, please visit **www.masteringengineering.com**.

MasteringEngineering for Students

MasteringEngineering improves understanding. As an Instructor-assigned homework and tutorial system, MasteringEngineering is designed to provide students with customized coaching and individualized feedback to help improve problem-solving skills. Students complete homework efficiently and effectively with tutorials that provide targeted help.

▼ **Immediate and specific feedback** shows students their mistakes while they are working on the problem. This allows them to see the explanation behind their misconceptions.

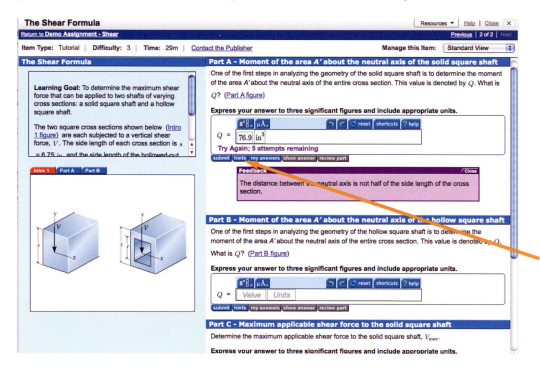

Hints provide individualized coaching and specific feedback on common errors. This helps explain why a particular concept is not correct.

MasteringEngineering for Instructors

Incorporate dynamic homework into your course with automatic grading and adaptive tutoring. Choose from a wide variety of stimulating problems, including Mohr's Circle, Shear and Bending Moment Diagrams, algorithmically-generated problem sets, and more.

MasteringEngineering emulates the instructor's office-hour environment, guiding students through engineering topics with self-paced tutorials that provide individualized coaching.

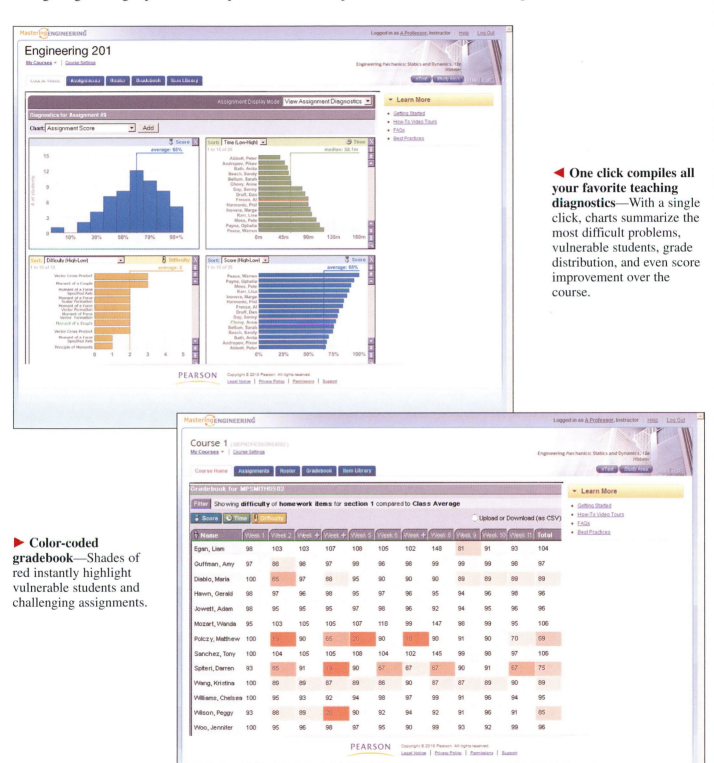

◀ **One click compiles all your favorite teaching diagnostics**—With a single click, charts summarize the most difficult problems, vulnerable students, grade distribution, and even score improvement over the course.

▶ **Color-coded gradebook**—Shades of red instantly highlight vulnerable students and challenging assignments.

Contact your Pearson Prentice Hall representative for more information.

Mechanics of Materials

Introduction to
Mechanics *of*
Materials

Excessive deformation...
Mechanics of Materials
can help with that...

I need to do something
about that sagging shelf...

If I knew the factors that
affect the sagging, I could
redesign the shelf so it
sags less...

How do I model this shelf?

The shelf is bending under the weight of those books, and it's resting on the brackets at the ends. In *Mechanics of Materials* I can represent this shelf approximately as a beam with simple supports. I can approximate the books as applying a uniformly distributed force on the beam.

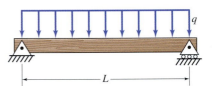

What key result do I need
from analyzing the model?

The maximum deflection v at the center of the shelf is given by this equation:

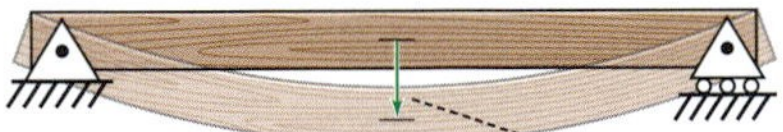

q: the force per length applied by the books

L: the length of the shelf between the brackets

$$v = \frac{5qL^4}{384EI}$$

I: the second moment of inertia — it tells me how the width and the thickness affect the bending

E: the elastic modulus of the shelf — it tells me how the stiffness of the material itself, the wood, affects the bending

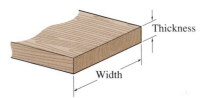

Thickness

Width

From what I just learned, how could I redesign the shelf?

I could make the shelf shorter, or maybe install another bracket under the center — might need some more analysis to see how much that helps . . .

I could use a stiffer material — steel or aluminum, or a carbon-reinforced composite — might be a little overkill for a bookshelf in my apartment . . . The thickness of the shelf has much more effect on the resistance to bending than does its width . . . So it could help a lot to use a thicker board

Or I could put a much thicker reinforcing strip on the front . . . That should help . . . I wonder by how much . . .

. . . Welcome to Mechanics of Materials

CONTENTS

1.1 Why Study *Mechanics of Materials?*

The design of products, systems, and structures demands the engineer to consider a broad range of issues. Here we identify the issues addressed by *Mechanics of Materials.*

1. Account for deformation and the potential for failure when designing systems subjected to forces.

Forces acting on designed artifacts can be significant. All bodies deform under applied forces, and they can fail if the forces are sufficiently large.

Mechanics of Materials addresses two prime questions:

* How much does a body deform when subjected to forces?
* When will forces applied to a body be large enough to cause the body to fail?

Deformation and failure depend on the forces and on the body's material, size, and shape.

2. In most situations, try to avoid failure and keep deformations within acceptable limits.

Usually, the structure or system must remain intact even when subjected to forces. If we know the forces under which failure would occur, we can design to avoid failure. Further, a system often needs to remain close to its original shape to function properly. If we can quantify deformations, we can design the system to avoid undesirably large deformations.

This computerized welding system functions properly only if the deflections of its track are very small.

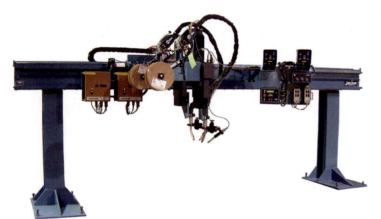

While a structure may still be intact, it could be viewed as having failed if there is a permanent deformation. A bicycle that has deformed this much is unlikely to be useful.

A crack in a structure, such as this support column, is a type of failure. This crack may be repairable. A structure that fractures completely into two parts would clearly be unacceptable.

3. Deformation is desirable in some situations where it depends predictably on the forces.

Some products must deform to carry out their function. They are designed to have a desired relation between the deformation and the acting forces.

For example, such products include pole vaults that flex to temporarily store energy that later propels the vaulter, mountings that accommodate motions of helicopter blades, and support springs that allow for deflection of structural members.

4. Occasionally, failure is desirable, if it occurs at a reproducible level of load.

Although such circumstances are rare, we sometimes deliberately want failure to occur when loads reach a predetermined level. In expensive equipment, failure can be disastrous. So, engineers design into the equipment an inexpensive extra part, which fails at a consistent force that is safely less than the main components can tolerate. For the transmission shaft in a drive train, such a system that protects the shaft is called a torque fuse. Just as an old fashioned electric fuse breaks when the current is too high, the pins in the torque fuse break when the torque is too high.

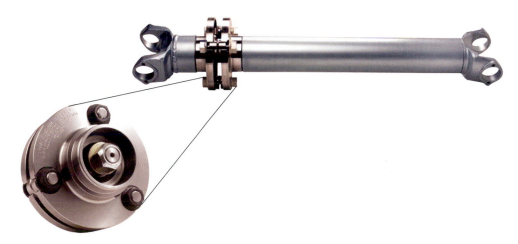

>>End 1.1

1.2 How *Mechanics of Materials* Predicts Deformation and Failure

A few very general scientific principles are needed to predict deformation and failure. With very general principles, we can consider bodies with a wide range of geometries and materials, which are subjected to many types of loads. *Mechanics of Materials* introduces these principles and applies them to bodies and loadings that can be analyzed with relatively simple mathematics.

1. Separate out the effects of material and geometry by viewing a body as composed of many tiny elements.

To predict deformation and failure, mechanics of materials relies on a critical insight: any body can be viewed as *an assemblage of tiny, in fact infinitesimal, cubic elements.* This insight allows us to separate out the effect of the body's *material* from its *shape*. Since a tiny cube is a standard shape, the relations between the cube's deformation and the forces on it depend only on the material, for example, the particular type of ceramic, metal, plastic, or wood. These relations can be measured and described for a given material, and they are relevant to a body of any shape and size composed of that material.

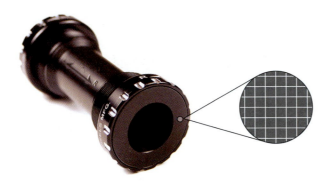

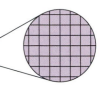

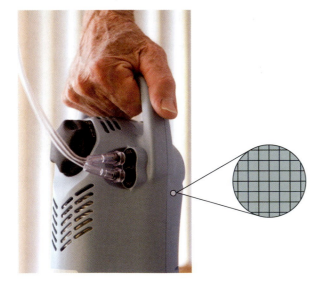

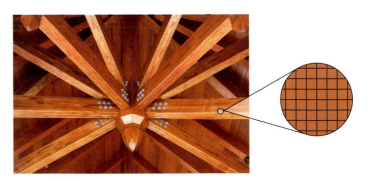

2. Relate forces and deformations at the element level with those at the level of the overall structure.

Mechanics of materials defines stress and strain to describe force and deformation at the level of an elemental cube. To determine a body's overall deformation and potential for failure, we combine (1) the material-specific stress–strain relations for a cubic element, (2) equilibrium relations between forces on the body as a whole and the forces on its elements, and (3) geometric relations between deformations of the whole body and of its elements.

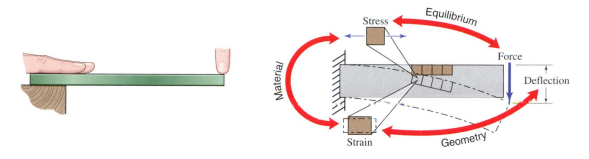

3. Recognize that loaded bodies often deform in simple patterns, namely, stretching, twisting, or bending.

Engineers deal with deformation and failure in structures having a wide variety of shapes, materials, and loadings. However, in mechanics of materials, we study deformation and failure primarily for simple patterns of deformation: stretching, twisting, or bending.

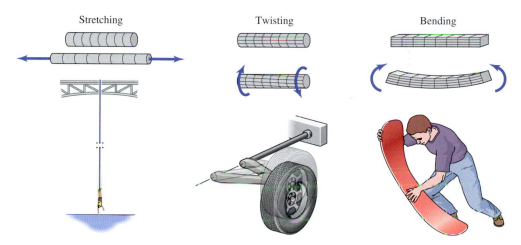

For each pattern, the overall loading is described by equal and opposite forces or moments at the two ends. The overall deformation is described by a single parameter: how much the body stretches, twists, or bends.

4. Study deflection and failure for each pattern individually, and then how they combine.

In mechanics of materials, we learn how the forces and deformations vary from one cubic element to another for each deformation pattern. With that information, we interrelate the overall load and deformation for that pattern, and we find the load at which failure will occur. As a by-product, we gain insight into how the body's geometry (length and cross-section) and the body's material *independently* affect the overall deformation and failure.

Faced with applications that appear complex, we must also learn to detect the presence of these simple deformation patterns, alone or, often, in combination. We typically analyze the deformations and stresses in each pattern and then combine them appropriately to find the total deformation and to determine if failure will occur.

>>End 1.2

1.3 Review of Statics—Forces, Subsystems, and Free Body Diagrams

The forces that we study in *Mechanics of Materials* generally keep the body in equilibrium, even if they also cause the body to deform. For this reason, Statics, which addresses the forces on bodies in equilibrium, is a critical prerequisite to *Mechanics of Materials*. The central ideas of Statics are reviewed here.

1. A force represents a mechanical interaction between two bodies, which often are in contact.

A force describes the equal and opposite mechanical interaction between two bodies, one upon the other. Since a force has a magnitude, direction, and sense, we represent it mathematically by a vector. Two forces applied to a body at the same point have the same effect as their vector sum.

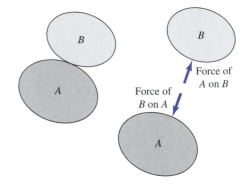

Whenever a force is drawn, it should be clear which body exerts the force on which body.

For example, in this vise-grip forces are exerted between the palm and the upper handle and between the fingers and the lower handle. There are many other forces that one could consider in this example.

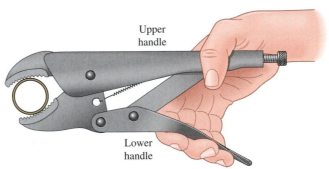

2. Engineering systems of interest may consist of multiple, interconnected parts, which exert forces on each other.

In general, systems studied in engineering are composed of multiple parts. Any pair of contacting parts can exert forces on each other. We must be prepared to consider all such forces and to quantify those deemed necessary.

This vise-grip consists of several connected parts. The clamped object and the lower jaw exert forces on each other. The lower jaw and the lower handle exert forces on each other through the indicated pin.

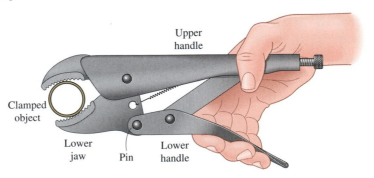

3. All subsystems of a system in equilibrium are also in equilibrium.

A system that is at rest (or at least not accelerating) is in equilibrium. This vise-grip, which is squeezed by the hand and clamps an object, is in equilibrium.

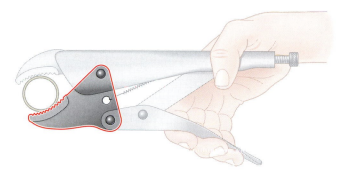

Any part or "subsystem" of a system in equilibrium is also in equilibrium. The lower jaw of the vise-grip, which is highlighted in the figure, must also be in equilibrium.

Because it contacts other parts, each subsystem will typically have multiple forces acting on it. The forces on the subsystem, acting in combination, keep the subsystem in equilibrium. The mathematical conditions for equilibrium are presented later.

4. A free body diagram displays all forces that affect the equilibrium of a subsystem.

In a free body diagram (FBD), we draw a subsystem and all the forces directly exerted on it by bodies external to the subsystem. The FBD is helpful because equilibrium of the subsystem is fully determined by the forces drawn in the diagram.

Here is the lower jaw of the vise-grip. In an FBD of the lower jaw, we would draw forces in the three regions where other bodies touch the lower jaw.

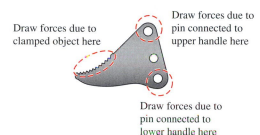

Draw forces due to clamped object here

Draw forces due to pin connected to upper handle here

Draw forces due to pin connected to lower handle here

5. Select subsystems strategically to find forces of interest.

We can choose to focus on any subsystem. We choose particular subsystems because their FBDs contain forces of interest that we wish to determine.

Sometimes, we even consider a portion of a single part as a subsystem. This is important in mechanics of materials, because we often need to find the internal force that acts within a part, between one portion and another.

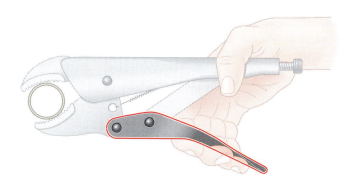

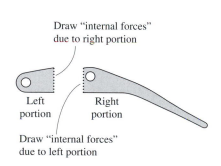

Draw "internal forces" due to right portion

Left portion

Right portion

Draw "internal forces" due to left portion

1.4 Review of Statics—Representing Force Interactions Simply

1. Represent forces of interaction as simply as possible.

In general, two bodies contact at multiple points. Their interaction can be described by a combination of forces. For example, the fingers contact the lower handle at multiple points. Many clamped objects would touch each jaw at multiple points.

The actual, detailed distribution of forces is often impossible to determine. For the purpose of Statics, any complex combination of forces can be represented by relatively simple loads, provided the simpler loads, when acting alone, would resist or cause the same motion as the actual, more complex forces.

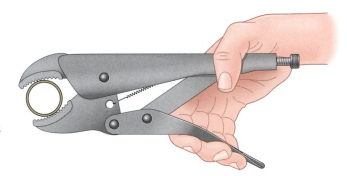

We seek to determine the unknown forces that two bodies exert on each other by representing the unknowns as simply as possible.

2. In simplifying forces, consider their tendencies to cause translational and rotational acceleration.

A single force acting on a body tends to cause translational acceleration. The force also tends to cause rotational acceleration if it produces a non-zero moment about the center of mass, G. The moment due to the force, $M|_G$, is equal to the force, F, times the perpendicular distance, d.

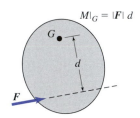

$$M|_G = |F|\, d$$

Two or more forces can still cause a rotational acceleration, even if they sum as vectors to zero net force. We sometimes represent such force combinations with a *couple* (often called a moment), which causes only rotational acceleration. The couple produces the same moment M_O about any point. Here are two combinations of forces that could be represented as a couple, and the drawing of the corresponding couple M_O in 2-D.

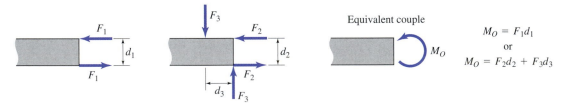

Equivalent couple

$$M_O = F_1 d_1$$
or
$$M_O = F_2 d_2 + F_3 d_3$$

Here a pair of forces that can be represented as a couple are drawn in 3-D, along with three equivalent ways of drawing the couple in 3-D. In any instance, we draw the couple so the figure is clearest and least cluttered. Always note a couple's direction (here, about the z-axis) and its sense (here, about the positive z-axis).

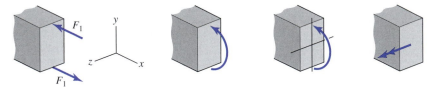

The translational and rotational acceleration produced by any combination of forces can be produced by one force and one couple, provided they are *statically equivalent* to the original combination. Two sets of forces and couples are statically equivalent if they correspond to the same net force and moment. We often represent a combination of forces by just a statically equivalent force and couple.

3. Represent the unknown interaction between two connected bodies by considering which motions of one body the other can resist.

Consider a pair of connected bodies. Hold one body and try to move the other body in various ways. The first body may or may not be able to resist each motion, depending on the connection. If a motion is resisted, the corresponding force or couple can act and must be included in the representation of unknowns.

More specifically,

Ask if a body can resist the connected body's translation in some direction.
If so, then the representation of the interaction must include a force that acts in that direction.

Ask if a body can resist the connected body's rotation about some axis.
If so, then the representation of the interaction must include a couple that acts about that axis.

4. Recall how interactions between bodies joined by common connections are represented.

Here are commonly encountered connections and how the interaction between the connected bodies is represented. We assume there is only loading in the plane (2-D).

A cord that connects two bodies only resists each body's translation parallel to the cord, away from each other. Only the magnitude F is unknown. There can be only a tensile ($F > 0$) force parallel to the cord.

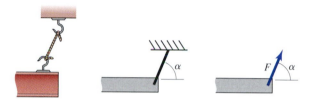

For a roller connection, translation perpendicular to the rolled-on surface is resisted, but not parallel translation or rotation. Only the magnitude F is unknown, and the force acts perpendicularly to the rolled-on surface.

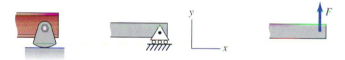

For a pin connection, translation in any direction is resisted, but not rotation. There can be a force in any direction in the plane but no couple. We can represent an arbitrary force in the plane with two independent unknown force components F_x and F_y.

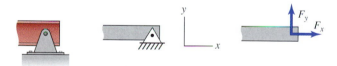

For a fixed or rigid connection, rotation and translation in any direction is resisted. There can be a force in any direction in the plane and a couple. The force magnitudes F_x, F_y, and couple magnitude M are unknown.

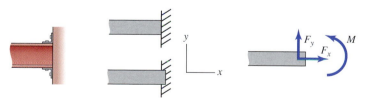

>>End 1.4

1.5 Review of Statics—Conditions of Equilibrium _____

1. A subsystem is in equilibrium if the combined effect of all external forces produces zero translational and rotational acceleration.

If there is to be no translational acceleration, the forces must sum to zero: $\Sigma F = 0$.

If there is to be no rotational acceleration, the moments must sum to zero: $\Sigma M|_O = 0$, where O is any point on or off the body.

Force and moment summation should include all external forces acting directly on the subsystem, namely, the forces drawn in the FBD.

2. Equilibrium can be imposed along coordinate axes.

Vector equilibrium conditions are equivalent to individual force summations along each of three axes:

$$\sum F_x = 0 \qquad \sum F_y = 0 \qquad \sum F_z = 0$$

and moment summations about each of three axes:

$$\sum M|_{O_x} = 0 \qquad \sum M|_{O_y} = 0 \qquad \sum M|_{O_z} = 0$$

The axes for the moment summations can be through different points, not all through the same point O.

If loads can be represented by forces in a single plane, then there are only three independent equilibrium equations:

$$\sum F_x = 0 \qquad \sum F_y = 0 \qquad \sum M|_O = 0$$

where the moment is about the z-axis (perpendicular to the x-y axes) through any point O.

3. Be sure to account for couples correctly when imposing equilibrium conditions.

Do not include couples in the force summations: each couple has zero net force!

In moment summations, include moments due to forces and due to couples. The moment due to a couple which acts on the subsystem is independent of the couple's location or the point about which the moment is taken.

4. Plan out the order in which you impose equations, and consider imposing equilibrium on multiple subsystems.

When the equilibrium equations for a given subsystem are sufficient to find the unknowns, it is often efficient first to sum moments about a carefully chosen point to find one force. Then, use the force summations to find the other forces.

If a desired unknown cannot be found using a single subsystem, draw FBDs of one or more additional subsystems and impose equilibrium on them to find intermediate unknowns.

>>End 1.5

Determine the reactions at the pins B and C. Neglect the weight of the members compared to the distributed load.

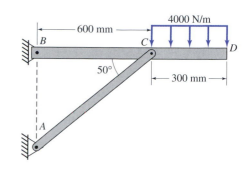

Solution

Choose BCD as a subsystem because the desired unknowns and the external load act on it.

Once we have chosen a subsystem and are about to impose equilibrium on it, we can replace the uniformly distributed force by a statically equivalent concentrated force.

Draw the FBD. In general, the force exerted by a pin has an unknown direction, and hence two unknown components, such as B_x and B_y.

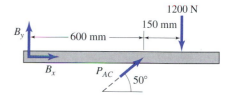

The interaction at C could likewise be represented by C_x and C_y. However, we note that the attached body AC is a two-force member: its weight is neglected and the only forces on it are due to pins at A and C. To maintain equilibrium of AC, these forces must act along the line AC joining the two points of force application. The unknown force, P_{AC}, could act in either sense. It will have the sense that is drawn, if it turns out to be positive.

The summation of moments about point B directly gives the force P_{AC}.

$$\sum M|_B = -(1200\,\text{N})(750\,\text{mm}) + P_{AC}(\sin 50°)(600\,\text{mm}) = 0 \Rightarrow P_{AC} = \frac{(1200\,\text{N})}{\sin 50°}\frac{(750\,\text{mm})}{(600\,\text{mm})} = 1958\,\text{N}$$

Note that $P_{AC} > 0$, so the assumed sense was correct.

Now, sum forces to find the force components due to the pin at B:

$$\sum F_x = B_x + P_{AC}\cos 50° = 0 \Rightarrow B_x = -P_{AC}\cos 50° = -1259\,\text{N}$$

$$\sum F_y = B_y + P_{AC}\sin 50° - 1200\,\text{N} = 0 \Rightarrow B_y = 1200\,\text{N} - P_{AC}\sin 50° = -300\,\text{N}$$

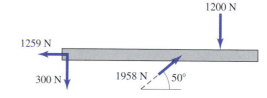

Note that $B_x < 0$, so it is opposite to the sense assumed. $B_y < 0$, so it is also opposite to the sense assumed. Here are the forces in the correct senses.

Note that the force components at B are equivalent to a force of magnitude B in the direction θ, where

$$B = \sqrt{(1259\,\text{N})^2 + (300\,\text{N})^2} = 1294\,\text{N}$$

$$\theta = \arctan\left[\frac{300\,\text{N}}{1259\,\text{N}}\right] = 13.41°$$

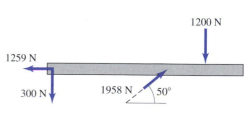

>>End Statics Example 1

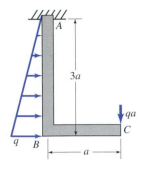

Determine the reactions at the support A, which represents a body to which the structure ABC is rigidly attached. The linearly distributed force has a maximum force per length of q.

Solution

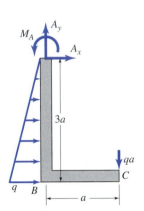

Choose the single member ABC as a subsystem. Draw the FBD.

With a fixed support and loading in the plane, at A there is an unknown couple, and a force of unknown magnitude and direction.

Since we are about to sum forces and moments, the linearly distributed force can be replaced by a statically equivalent force $q(3a)/2$, acting at the point that is 2/3 of the distance from the point of zero distributed force to the point of maximum distributed force: $(2/3)(3a) = 2a$.

Take moments about point A to eliminate the contribution of A_x and A_y and determine M_A.

$$\sum M|_A = \left(\frac{3qa}{2}\right)(2a) - (qa)(a) + M_A = 0 \quad \Rightarrow \quad M_A = -2qa^2$$

Now, sum forces to find the force components A_x and A_y:

$$\sum F_x = A_x + \frac{3qa}{2} = 0 \quad \Rightarrow \quad A_x = -\frac{3qa}{2}$$

$$\sum F_x = A_y - qa = 0 \quad \Rightarrow \quad A_y = qa$$

Note that $A_x < 0$, so it is opposite to the sense assumed. $A_y > 0$, so it is in the sense assumed. $M_A < 0$, so it is opposite to the sense assumed. The reactions at A are re-drawn in the correct senses.

Note that the force components at A are equivalent to a force of magnitude A in the direction θ, where

$$A = \sqrt{\left(\frac{3}{2}qa\right)^2 + (qa)^2} = 1.803qa$$

$$\theta = \arctan\left[\frac{qa}{\frac{3}{2}qa}\right] = 33.69°$$

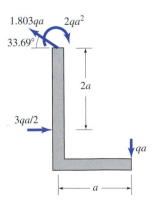

>>End Statics Example 2

Find the reaction at support A.

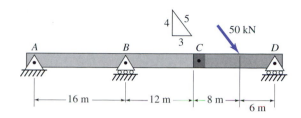

Solution

Choose the entire structure as the subsystem. Draw the FBD. The unknown support reactions are drawn, including the pin at A and rollers at B and D. Notice the pin at C is internal to the subsystem, so no forces at C are drawn.

One cannot solve for all the forces from the subsystem $ABCD$, because there are four unknowns A_x, A_y, B_y, and D_y, but only three equilibrium equations.

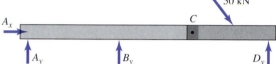

Notice $ABCD$ can be separated into two subsystems without cutting through a part. Here are the FBDs for these two subsystems. Notice that the forces exerted by the pin C now appear, and that they are drawn as equal and opposite on the two subsystems.

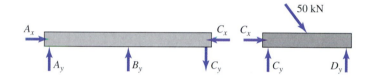

Because it has only three unknowns, use subsystem CD to solve for C_x, C_y, and D_y first.

$$\sum M|_C = -(50 \text{ kN})\left(\frac{4}{5}\right)(8 \text{ m}) + D_y(14 \text{ m}) = 0 \implies D_y = 40 \text{ kN}\frac{(8 \text{ m})}{(14 \text{ m})} = 22.9 \text{ kN}$$

$$\sum F_x = C_x + (50 \text{ kN})\left(\frac{3}{5}\right) = 0 \implies C_x = -30 \text{ kN}$$

$$\sum F_y = C_y - (50 \text{ kN})\left(\frac{4}{5}\right) + 22.9 \text{ kN} = 0 \implies C_y = 17.14 \text{ kN}$$

Now use subsystem ABC to solve for the remaining unknowns.

$$\sum M|_A = B_y(16 \text{ m}) - (17.14 \text{ kN})(28 \text{ m}) = 0 \implies B_y = 17.14 \text{ kN}\frac{(28 \text{ m})}{(16 \text{ m})} = 30 \text{ kN}$$

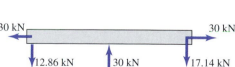

$$\sum F_x = A_x + 30 \text{ kN} = 0 \implies A_x = -30 \text{ kN}$$

$$\sum F_y = A_y + (30 \text{ kN}) - 17.14 \text{ kN} = 0 \implies A_y = -12.86 \text{ kN}$$

Since all components are known, draw all the forces in the correct senses at each point.

>>End Statics Example 3

1.6 Road Map of Book _____

This book has been designed to help you organize the subject of *Mechanics of Materials*. Here, we explain first how each chapter is structured. Then, we explain how the units and chapters of the book form the organization of the subject.

1. Preview the major concepts of the chapter and how they are interrelated.

At the start of each chapter, the major ideas in the chapter are presented in a graphical, integrated way.

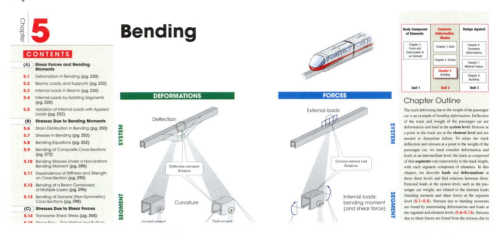

2. Learn the component concepts of the chapter in depth, one at a time.

Each chapter has been broken down into a succession of concepts. Each concept is usually treated in a single two-page spread. This allows all the elements of the concept to be studied at the same time without having to flip pages.

At the start of each two-page spread, the concept is identified. Then, the distinct elements of the component are separated into subsections.

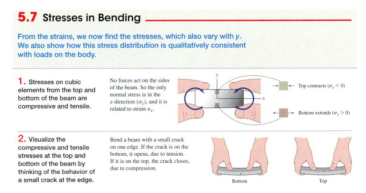

3. Review the concepts in each chapter.

In the Summary at the chapter end, each component (two-page spread) is briefly reviewed. The Summary allows you to appreciate the succession of components at a glance, and, when reviewing, to find the two-page spreads of immediate interest.

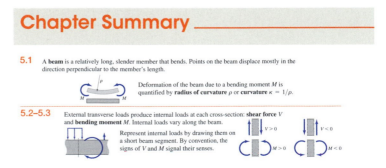

4. Learn the enabling idea of *Mechanics of Materials*: by viewing a body as a collection of tiny cubic elements, we can separate the effects of material and geometry.

In Unit 1 (Chapter 2), the basic idea of viewing the body as composed of tiny elements is explained in depth. We first address the different levels at which force can be viewed. Thinking of force at the level of tiny elements leads to the concept of stress. Thinking of deformation at the level of tiny elements leads to the concept of strain. We then examine the relation between stress and strain, which captures the mechanical response of a material, independent of the size or shape of any particular body composed of that material.

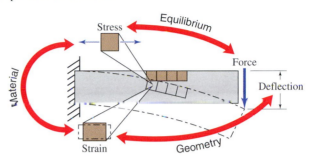

5. Learn the three fundamental patterns of deformation.

In Unit 2, we study three commonly occurring patterns of deformation. Chapter 3 addresses axial loading, that is, bodies that stretch or contract. Chapter 4 addresses torsion, that is, bodies that twist. Chapter 5 addresses bending, that is, bodies which bend or deflect perpendicularly to their length. For each pattern, we quantify the relation between the overall loading and the overall deformation, and we identify the separate roles played by the bodies' material and its geometry. We also address the tendency for failure by finding the distribution of stresses in elements throughout the body.

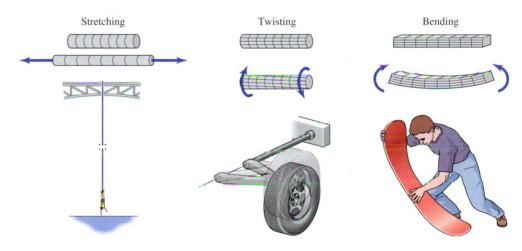

6. Determine deflection and the tendency for failure when multiple fundamental patterns of deformation occur in combination.

In Unit 3, we consider more general situations, including those in which the deformation patterns studied in Unit 2 appear in combination. In Chapter 6 we learn to detect the presence of multiple deformation patterns within a complex situation, to designate their different stresses with a more general notation, and to determine how the resulting strains and deflections combine. In Chapter 7 we learn how failure can be predicted when the combination of stresses becomes too high. In Chapter 8 we study buckling, which is failure that occurs when the body as a whole becomes unstable, rather than stresses becoming high.

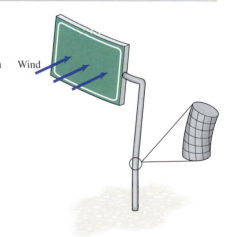

>>End 1.6

Internal Force, Stress, and Strain

1. Distinguish forces acting at different levels

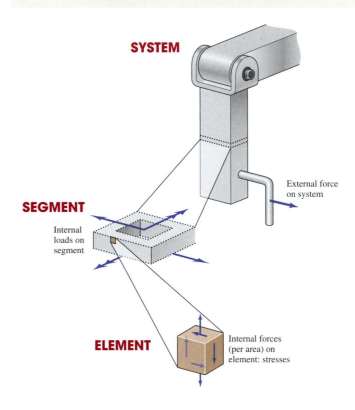

SYSTEM

External force on system

SEGMENT

Internal loads on segment

ELEMENT

Internal forces (per area) on element: stresses

2. Define two types of forces (stress) and deformation (strain) at the element level

	NORMAL	SHEAR
Internal Force (Stress)		
Deformation (Strain)		

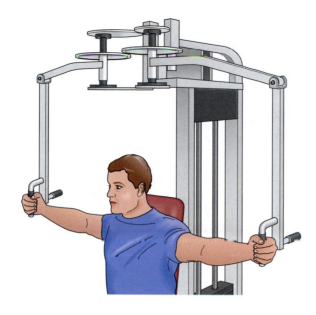

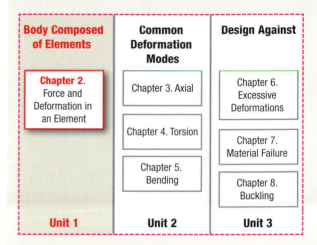

Body Composed of Elements	Common Deformation Modes	Design Against
Chapter 2. Force and Deformation in an Element	Chapter 3. Axial	Chapter 6. Excessive Deformations
	Chapter 4. Torsion	Chapter 7. Material Failure
	Chapter 5. Bending	Chapter 8. Buckling
Unit 1	Unit 2	Unit 3

3. Measure stress and strain in a material, by loading a test specimen

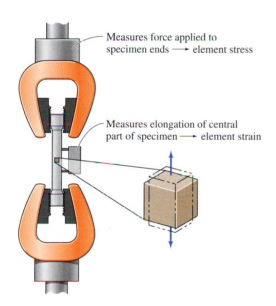

Measures force applied to specimen ends → element stress

Measures elongation of central part of specimen → element strain

4. Interpret plot of measured stress vs. strain to deduce key material properties that are needed for design

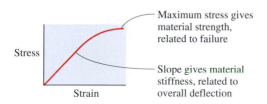

Maximum stress gives material strength, related to failure

Slope gives material stiffness, related to overall deflection

Stress

Strain

Chapter Outline

To predict the **deformation** and **failure** of a mechanical part due to applied forces, we envision the part as composed of many tiny cubic elements (**2.1**). Deformation and failure of the overall part is related to the loads on, and deformations of, individual elements. We usually first determine the forces and moments that act on a thin segment of the part, and then consider individual elements in the segment (**2.2**). Because an element can be arbitrarily small, we quantify the loads on its surface with the *force per area* (**stress**), rather than the total force (**2.3**). Likewise, we quantify the deformation of the element with *elongation per length* (**strain**), rather than the elongation itself (**2.4**). The relation between stress and strain captures the intrinsic stiffness and strength of the material. If we subject a body of a standard shape to known tensile forces and measure its elongation, we can deduce the relation between stress and strain for the body's material (**2.5**). Most materials are *elastic* when the deformation is small: the stress is proportional to the strain, and the original shape is regained if the stress (force) is removed. For many materials, two material-specific constants—the Young's modulus and the Poisson ratio—describe the proportionality between stress and strain (**2.6**). Failure of different materials can take on a variety of forms, but a criterion for failure is usually based on the stress reaching a critical value (**2.7–2.8**). We quantify the deformation of an element that distorts, rather than elongates or contracts, with *shear strain*. The corresponding force per area on the element is the shear stress (**2.9–2.10**).

2.1 Elements

The field of Mechanics has led to the development of methods for determining whether a body fails under given loads and how much it deforms. The methods are powerful: they can be applied in a wide range of situations from simple to complex.

1. Mechanics treats deflection and failure for solid parts of any material or shape.

For example, here are two very different systems: a fuel-efficient vehicle and a hip implant. For each we need to be concerned about failure and deformation. Indeed, virtually any situation in which forces cause parts to deform and potentially fail can be treated with the methods we develop.

2. View any body as composed of small elements.

We view a body as composed of many small (theoretically infinitesimal) cubes of material. An **element** is defined as one such cube of material in the body.

Mechanics treats a wide range of situations by viewing bodies as collections of small **elements**. Generally, failure initiates at some local region in a body because conditions in those elements become critical. By contrast, overall deformation of a body is always due to the cumulative effect of deformation that occurs in elements throughout a body. So the response to forces is related to what happens in the small elements that form the body.

The concepts of force and deformation remain relevant, but they need to be redefined for small elements in a body.

3. Observe the deformed shapes of progressively smaller regions of a body.

Focus on one panel of the vehicle: it is the body of interest. Picture deformation by etching on the body's surface a grid of equally spaced perpendicular lines to form squares. If forces are applied to the body, the etched lines move, and the shapes of the squares change. While bodies under load always deform, those deformations are usually so small as to be invisible. Here, to help visualize, we exaggerate the deformation. Compare the deformed shapes of three progressively smaller regions. Each side of the three squares deforms into a curve.

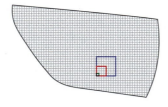
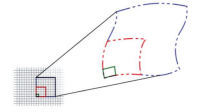

4. A very small portion of any smooth curve appears to be straight.

Consider any smooth function $f(x)$. The plot of $f(x)$ vs. x defines a curve. If one magnifies a smaller-and-smaller portion of the curve, it appears increasingly like a straight segment. So a very small portion of the curve into which each side of an etched square deformed appears to be straight. Therefore, the sides of an infinitesimal square remain straight when the body is deformed.

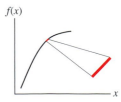

5. An infinitesimal element that was originally a square deforms into a parallelogram.

The two horizontal sides of an infinitesimal square are so close to each other that they deform into lines of the same orientation and length. The vertical sides likewise deform into lines of the same orientation and length. Therefore, the square deforms into a parallelogram. Only three numbers are needed to quantify the shape change from square to parallelogram: the fractional increase in the length of the two sides and the change in the right angle. We describe the deformation of any part by the deformations of each of its square elements.

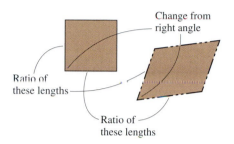

6. Two neighboring infinitesimal elements interact with a uniformly distributed force.

Each of the colored square regions shown before deforms because the immediately surrounding material applies forces to it. Imagine gripping the edges of a stretchy sheet of material that is originally square and applying forces to make it deform in the ways shown.

Here is the force distribution on the lower face of the squares. For the larger square, which deforms in a complex way, the distribution of forces is complex. Just as the shape change is simpler for the smaller square, the distribution of force on it is also simpler. In fact, the force becomes uniformly distributed on the face of each element, and essentially the same force acts on neighboring small elements. So we also consider small elements because the description of forces acting on them is simpler.

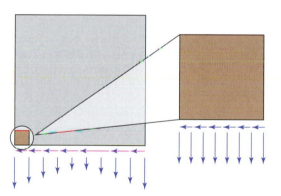

7. To separate the effects of a part's material and geometry on its deformation, consider it to be composed of small elements.

Thinking of a part as composed of small elements has another great benefit. When we quantify the tendency for loads to deform the part, we want to separate out the respective effects of the part's shape (geometry) and material. To do this, we need to describe the *intrinsic* effect of the material of a part, as distinct from its geometry. We do this by considering the deformation of a standard size body, an infinitesimal element, for which the deformations and the forces are simply described. This lets us compare the responses of different materials on an equivalent basis, always for an infinitesimal element.

>>End 2.1

2.2 Internal Force

Force plays a critical role in *Mechanics of Materials*, as it does in Statics. Every force represents a mechanical interaction between two bodies. When we think of or draw a force, we should always have the two interacting bodies in mind. With the exception of gravity, we are primarily interested in bodies that interact by touching each other. In *Mechanics of Materials*, we extend the concept of force by considering a greater variety of contacting bodies.

1. We apply *Mechanics of Materials* to structures, machines, or mechanisms in which forces at many levels are relevant.

We begin to enlarge the concept of force by reconsidering the vise-grip. As the hand squeezes the handles, forces are exerted between the various parts.

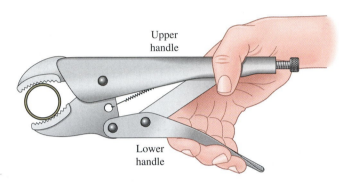

2. In Statics we separate the distinct parts of a system to determine forces between the contacting parts.

In FBDs we display the unknown forces between the distinct parts, as well as the known forces. Here, for example, we might take the force of the hand, F_{hand}, to be known. By applying equilibrium to each part, we can determine each of the unknown forces.

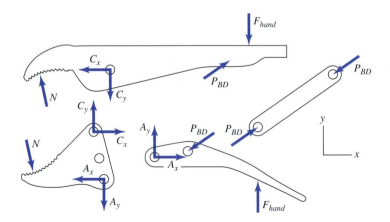

3. *Mechanics of Materials* takes the next step: given the forces on each part determined using Statics, we find the forces and deformations in small elements within a part.

As just shown, Statics is often used to determine the forces acting between individual parts. In *Mechanics of Materials*, we seek to determine if the forces on each part will cause it to fail or to deform excessively. To that end, we will find the forces on and deformations of individual elements of the part. We usually find it convenient to define forces at an intermediate level: the internal force between a group of elements that form an internal surface across a part.

4. Forces between pairs of elements within a part add up to a net force across an internal surface.

Say the forces on a link (similar to BD above, but in tension) have been determined. Consider next the forces exerted between two portions of the link that are separated by a plane that cuts across the link. On each side of the plane, there are individual elements.

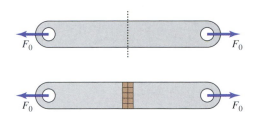

If we view a body as composed of two adjacent parts that meet at an imaginary plane or line passing through the body, then the **internal force** (moment) is defined as the force (moment) exerted by the two adjacent parts on each other.

The individual elements exert forces on each other. We add up the forces across all the elements at the interface. We define the **internal force** P as the net or resultant force transmitted between two adjacent portions of a body over all elements at the separating cross-section.

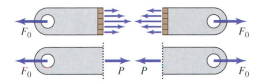

5. Find the force on an internal surface first, and then determine the forces acting between individual elements that add up to the force.

While the force across the internal surface is the resultant of the forces on the individual elements, we do not in practice calculate the internal force by summing the element forces. The forces acting on individual elements are not known in advance—they are the goal of our analysis.

Instead, we find the internal force first by using a FBD in which appears the unknown internal force P, as well as other known forces. Say we isolate the portion of the link to the left of the surface of interest. The only forces on this body are the known force applied to the link, F_0, and the internal force P. Using equilibrium, we determine that P equals F_0 in magnitude, and that it acts in the opposite sense along the same line of action as F_0.

6. Element forces across an internal surface add up, in general, to a net force and moment.

The forces acting between the elements at an internal surface are statically equivalent to a single force in some direction, plus a couple or moment. These *internal loads* (force and/or couple) can be found from applying equilibrium to a portion of the body bounded by the internal surface.

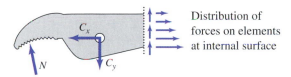

Distribution of forces on elements at internal surface

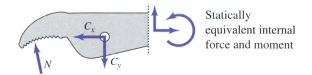

Statically equivalent internal force and moment

In the next section, we define element level forces (stress) and then go on to learn how element level forces are found from internal loads.

A set of planks forms a temporary floor. Each plank is supported by a cable and rests on a pad. Consider the loads induced in the cable and pad, due to an additional load of $F_0 = 400$ lb applied as shown. Do not include the weight of the plank. Determine the internal loads at the two cross-sections shown, S_1 and S_2, draw free body diagrams and use the conditions of equilibrium. Take the force transmitted between the pad and the plank to be uniformly distributed.

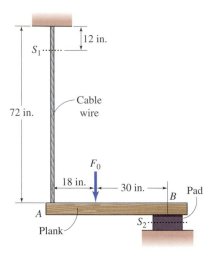

Solution

The unknown loads from the cable and the pad, as well as the given applied force, act on the plank, so we draw a FBD of the plank.

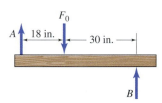

Because it must be in tension, the cable can only exert an upward force on the plank. The pad also exerts an upward force since the plank simply rests on it. Since we have assumed the contact force between the pad and floor is uniform, the unknown force acts through the center of the pad.

If the plank is in equilibrium, then

$$\sum M|_A = -(400 \text{ lb})(18 \text{ in.}) + B(48 \text{ in.}) = 0, \text{ hence } B = 150 \text{ lb.}$$

$$\sum F_y = A - 400 \text{ lb} + B = 0, \text{ hence } A = 250 \text{ lb.}$$

We can draw the forces on the cable as a whole and on the pad as a whole.

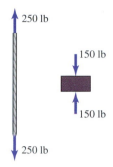

To find the forces that act across an internal cross-section, we must apply equilibrium to a part of a body up to that internal cross-section. Since we know the directions of the forces on the cord and pad, the directions of the internal forces and their magnitudes can be found directly.

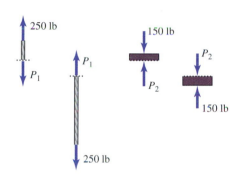

The internal force on the cord has magnitude $P_1 = 250$ lb. The internal force on the pad has magnitude $P_2 = 150$ lb.

The two sides of the cable *pull* on each other across the internal cross-section, so it is in *tension*. The two sides of the pad *push* on each other across the internal cross-section, so it is in *compression*.

>>End Example Problem 2.1

A retractable ladder is supported by two arms and cables in the positions shown. Two people, who have a total mass of 200 kg, step on the ladder. Determine the internal loads in the arm at the indicated cross-section S. Assume that the ladder is centered between the two arms, and that the pins that the arms pivot about exert only forces on the arms.

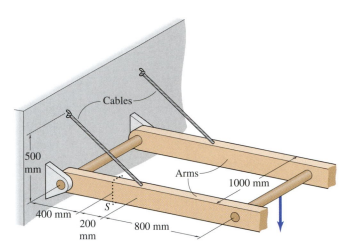

Solution

We recognize that the set of parts (arms, cables, pin, and supports) and the load of the weight are symmetric. So each arm carries loads only in its plane, with half of the weight, or $(100 \text{ kg})(9.81 \text{ m/s}^2) = 981 \text{ N}$, carried by each arm at its end.

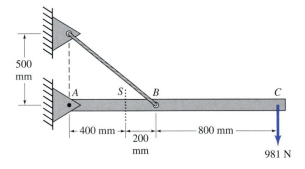

The cable can only exert a force pulling on the arm. The angle of the tension force T is found from trigonometry. The pin support can exert a force in the plane, which we represent with unknown components A_x and A_y.

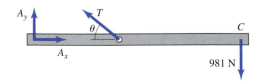

From equilibrium of the arm, we find:

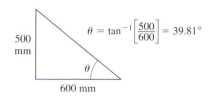

$$\sum M|_A = T (\sin 39.81°) (600 \text{ mm}) - (981 \text{ N})(1400 \text{ mm}) = 0, \text{ hence } T = 3580 \text{ N.}$$

$$\sum F_x = A_x - T \cos 39.81° = 0, \text{ hence } A_x = 2750 \text{ N.}$$

$$\sum F_y = A_y + T \sin 39.81° - (981 \text{ N}) = 0, \text{ hence } A_y = -1308 \text{ N.}$$

We then isolate a part of the arm which terminates at the cross-section of interest. We can either choose to isolate from A to the cross-section or from the cross-section to the right end; they will lead to the same result. (We usually choose to isolate a part which leads to the simplest application of equilibrium. Here they are of comparable difficulty.)

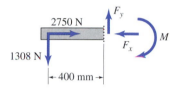

The known forces from the pin are drawn in the correct senses, and the unknown internal forces and moment are drawn at the cross-section. From equilibrium of the chosen part of the arm, we find $F_x = 2750 \text{ N}$, $F_y = 1308 \text{ N}$, $M = (1308 \text{ N})(400 \text{ mm}) = 523 \text{ N-m}$ in the directions shown. These are exerted by the right part of the arm on the left part.

Finally, we show the equal and opposite internal loads that the left part exerts on the right part.

>>End Example Problem 2.2

Additional data on material properties needed to solve problems can be found in Appendix D or inside back cover.

2.1 In an overhead smash, a tennis player strikes the ball with the center of the head of the racquet. Say the contact force is 50 lb. Determine the internal loads in the handle on a plane through the center of the grip. Take $L_1 = 9$ in. and $L_2 = 7$ in.

Prob. 2.1

2.2 A tennis player at the net holds the racquet firmly in a horizontal position. The ball strikes the racquet head below its center, exerting a normal force of 200 N. Determine the internal loads in the handle on a plane through the center of the grip. Take $L_1 = 230$ mm, $L_2 = 180$ mm, $L_3 = 90$ mm.

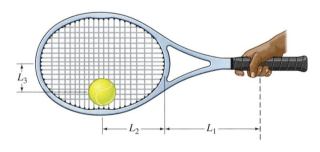

Prob. 2.2

2.3 This 140 lb gymnast supports himself with arms stretched out in the iron-cross position shown. Say the arms each weigh 10% of the body weight (and have approximately uniform distribution of mass), and that each ring exerts only an upward force on each hand. What are the internal loads supported at the shoulders?

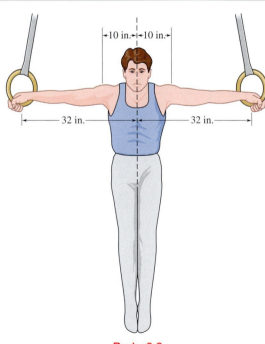

Prob. 2.3

2.4 Consider the loads sustained in an eyeglass frame when the temples of an undersized frame are spread wide enough to fit the face. Say the head exerts an outward force $F_0 = 2$ N perpendicular to the temple. If the contact force acts along the line of the temple, determine the internal loads at A where the temple joins the frame front. Take $L_1 = 105$ mm, $L_2 = 30$ mm, $L_3 = 26$ mm.

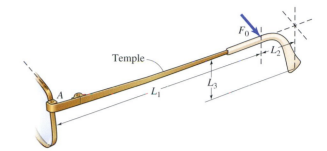

Prob. 2.4

2.5 Consider the loads sustained in an eyeglass frame when the temples of an undersized frame are spread wide enough to fit the face. Say the head exerts an outward $F_0 = 2$ N perpendicular to the plane of the temple. If the head makes contact at the lower tip of the temple, determine the internal loads at A where the temple joins the frame front. Take $L_1 = 105$ mm, $L_2 = 30$ mm, $L_3 = 26$ mm.

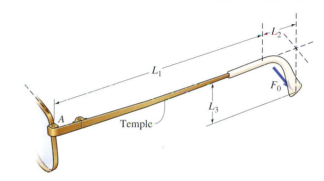

Prob. 2.5

Focused Application Problems

2.6 Say that a tired athlete hangs on the swinging arm shown with his entire 250 lb weight $W = 250$ lb. Determine the internal loads acting on the indicated cross-section. Define normal and shear forces as perpendicular and parallel to the cross-section. Take the dimensions to be $a = 25$ in., $b = 11$ in., and $\theta = 15°$.

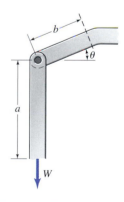

Prob. 2.6 (Appendix A4)

2.7 The external fracture fixation system, with long rod and two pins, maintains the upper and lower fragments of the fractured femur in position. Say no load is transmitted across the fracture plane at the early stage of healing. Consider an upward force of $F_0 = 900$ N exerted by the ground on the foot which is transmitted to the lower fragment as shown.

Determine the internal load at the cross-section A in the pin, where it meets the rod, and at the cross-section B in the rod.

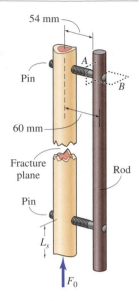

54 mm

A

Pin

B

60 mm

Fracture plane

Rod

Pin

L_s

F_0

Prob. 2.7 (Appendix A4)

2.8 During steady operation, a wind turbine is exposed to a thrust force perpendicular to the blades $F_{overturn} = 60$ kN that tends to overturn the tower. This force acts through a point of the tower that is $h = 65$ m above the base. Determine the resulting internal loads in the cross-section of the tower at the base.

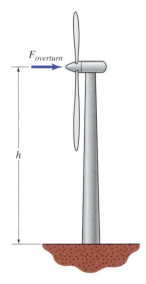

$F_{overturn}$

h

Prob. 2.8 (Appendix A6)

2.9 The sudden loss of a wind turbine blade produces a tangential force $F_{tang} = 50$ kN on the blades. Take the height $h = 70$ m, and the blades to be located $s = 5$ m from the tower axis. Determine the resulting internal loads in the cross-section of the tower at the base.

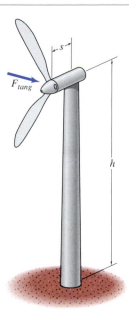

Prob. 2.9 (Appendix A6)

>>End Problems

2.3 Normal Stress

We need a way of describing forces at the level of elements.
Force per area or stress is an effective way of describing forces
on elements, since the force is essentially uniform on neighboring,
identical, infinitesimal elements.

1. Since the forces on small, neighboring elements are equal, force per area describes element forces.

We saw in Section 2.1 that the forces acting on small neighboring elements are equal. Say the force on an element of area ΔA is ΔP. Then, if four of those elements are combined into one with area $4\Delta A$, its force must be $4\Delta P$. Note that the force per area $(4\Delta P)/(4\Delta A) = \Delta P/\Delta A$ is the same. We can summarize all such combinations of element areas and forces by saying that the force per area $\Delta P/\Delta A$ is locally uniform.

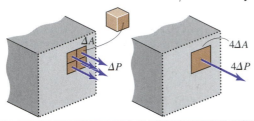

We define stress, σ, as the force per area on an element. Stress alone, without reference to force or area individually, describes forces acting on elements.

$$\sigma = \frac{\Delta P}{\Delta A}$$

Stress has units of $\dfrac{\text{Force}}{\text{Area}} : \dfrac{\text{N}}{\text{m}^2} = \text{Pa}$ or $\dfrac{\text{lb}}{\text{in.}^2} = \text{psi}$

2. Draw stresses on an exposed internal surface of a body or just acting on a single element.

Sometimes we depict stress by showing it acting on the surface of an element still embedded in the rest of the body (as above). Like forces, which obey Newton's 3rd Law, an equal and opposite stress acts on an element in the adjoining body across their common surface.

Sometimes we depict stress by drawing only an element, in 2-D or 3-D, and the stresses on its faces. Usually, the element is aligned with x-y-z axes. Notice that the element is in equilibrium, because the two forces on it, each equal to the stress times the element area, are equal and opposite.

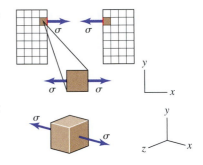

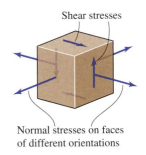

3. Be aware that stress on an element can be very complex.

Normal stress, σ, describes the element-level intensity of internal normal force, and it is defined as the normal force exerted by two adjacent elements on each other, divided by the area on which the force acts.

We will see that stress, even on a single element, can become complicated. The stress considered so far is **_normal stress_**: the force acts perpendicularly (normally) to the element face. We will study *shear stresses* that act parallel to element faces (Section 2.10). Additionally, we will also study normal stresses acting on faces of different orientations (Ch. 6). Although not shown here, we will also study stresses on elements with faces not aligned with x-y-z planes (Ch. 7).

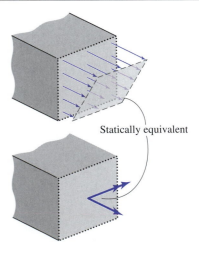

Shear stresses

Normal stresses on faces of different orientations

4. Stresses often vary across a surface cut through a body, but they are statically equivalent to the internal loads acting on the surface.

In general, stress will vary from element to element gradually across a surface. But, the element forces, when combined, are statically equivalent to the internal force and moment. That is why internal loads are sometimes referred to as stress resultants.

Mechanics of Materials addresses commonly occurring situations in which the distribution of stress across an internal surface can be found from the internal load and the shape of the surface.

Statically equivalent

5. In special circumstances, the normal stress may be uniform across an entire surface.

The link shown earlier (Section 2.2.4) has a common loading: two opposite axial forces. Imagine forces of magnitude P were applied to a deformable rectangular sheet of rubber.

Lines have been drawn on the sheet to show the deformation of individual elements when equal and opposite forces are applied at the two dots.

Here the sheet is deformed while forces act. Focus on the elements at two cross-sections: S_1 and S_2.

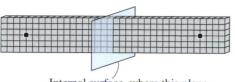

Uniform deformation (S_1) — Nonuniform deformation (S_2)

Internal surface, where this plane cuts through sheet, has area A.

In a region such as S_2, the deformation is nonuniform, so the stress will be nonuniform.

Because the internal force on the sheet is the sum of the forces across all the elements, the average normal stress, σ_{avg}, can be found even if the stress is nonuniform:

$$\sigma_{avg} = \frac{P}{A}$$

We observe that the elements deform identically in the central portion of the sheet (S_1), away from the applied forces. Since the material is uniform, the forces on all elements must be equal.

So, in the central portion of the bar where the element forces are uniform and the stress at each point is also equal to the average stress:

$$\sigma = \frac{P}{A}$$

6. The local normal stress is uniform and equal to P/A under certain conditions.

The normal stress is uniform (so $\sigma = P/A$), under these conditions:

• Member is longer than it is wide, and force acts axially (parallel to the long dimension)

• Material and cross-section are uniform along the length containing the cross-section

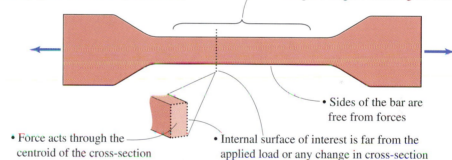

• Sides of the bar are free from forces

• Force acts through the centroid of the cross-section

• Internal surface of interest is far from the applied load or any change in cross-section

St. Venant's principle refers to the observation that the stress distribution becomes uniform as one moves away from the concentrated applied forces, subject to the above conditions.

7. Stresses that act to shorten an element are designated as compressive.

Normal stresses shown so far, which elongate a body, are tensile. Normal stresses may act instead to shorten a body (compressive). Neighboring elements then press on each other, as in this structural post. Besides giving the magnitude of the normal stress, we either specify that the stress is tensile or compressive, or we use a positive sign to denote tension and negative to denote compression.

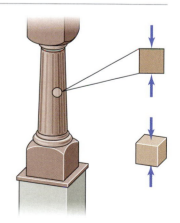

>>End 2.3

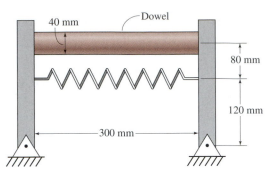

40 mm

Dowel

80 mm

120 mm

300 mm

The dowel 40 mm in diameter and 300 mm long fits into two vertical bars. When a spring is hooked onto the vertical bars, it has a tension of 400 N. Determine the stress in the dowel. Give the magnitude and indicate whether the stress is tensile or compressive. Draw the stress on an element taken from the center of the dowel.

Solution

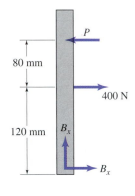

P

80 mm

400 N

120 mm B_x

B_x

With each vertical bar pinned at its bottom, and the spring exerting a tensile force, the dowel can maintain equilibrium by applying an axial force P.

The FBD of the left vertical bar is shown; the force of the spring has been drawn with a sense that is consistent with being in tension. The unknown force components of the pin are also drawn. For equilibrium, we must have:

$$\sum M|_B = P(200 \text{ mm}) - (400 \text{ N})(120 \text{ mm}) = 0, \text{ hence } P = 240 \text{ N}.$$

Using Newton's 3rd law, the force of the vertical bars back on the dowel must be as shown, namely compressive.

240 N   240 N

The dowel cross-section is uniform away from the ends. So, away from where the forces are applied, the normal stress in the dowel will be uniform.

To calculate the stress in SI units, namely Pa (N/m^2), we use forces in N and lengths in m. Then, the magnitude of the stress is found from:

$$\sigma = \frac{P}{A} = \frac{240 \text{ N}}{\frac{\pi}{4}(0.04 \text{ m})^2} = 1.910 \times 10^5 \text{ Pa} = 0.1910 \text{ MPa}$$

The axial stress is in the horizontal direction and it is compressive. The central region where this estimate of the stress is approximately correct is shown. The stress is drawn on the element.

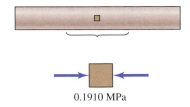

0.1910 MPa

>>End Example Problem 2.3

A simple model of the ski and pivoting arm from an elliptical machine are shown. *ABC* and *CD* are pin-connected at *C*, and *ABC* is supported as shown at *A*. *ABC* has a rectangular cross-section and *CD* is a hollow circular tube. Neglect the weight of the members compared to the applied force corresponding to a 250 lb person pedaling on the ski. Assume that no forces are exerted by the hands on the vertical arm. In which portions, if any, of these members (*AB*, *BC*, and/or *CD*) could there be a uniform normal stress across a cross-section? If so, draw an element showing the direction of the stress. Determine limits on the cross-sectional area of a member, so that the stress in it is less than 10 ksi. Answer the above question for two locations of the pins: (a) C_1 and D_1, and (b) C_2 and D_2.

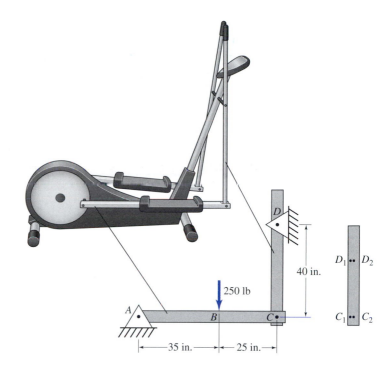

Solution

Note that *CD* must be a two-force member, because it is connected to two pins and no additional forces act on it.

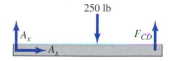

To satisfy $\sum M|_A = -(250 \text{ lb})(35 \text{ in.}) + F_{CD}(60 \text{ in.}) = 0$, we must have $F_{CD} = 145.8$ lb.

$\sum F_x = A_x = 0$ and $\sum F_y = 0$, implies $A_x = 0$ and $A_y = 104.2$ lb.

The forces on *ABC* are vertical, therefore perpendicular to its long direction. The internal forces cannot be axial, so member *ABC* cannot have a uniform normal stress.

The forces on *CD* are axial. For the stress to be uniform under axial loading, the forces need to be directed along the center (more generally centroid) of the cross-section.

(a) For pins attached at points C_1 and D_1, the forces would not act along the bar centerline, so the conditions for uniform stress would not be satisfied (the bar would bend).

(b) For pins attached at points C_2 and D_2, the forces would act along the bar centerline, so the conditions for uniform stress would be satisfied.

The general relation for stress in uniaxial loading is $\sigma = \dfrac{P}{A}$

The area of arm *CD* is unknown (to be determined), but the stress must be less than 10 ksi or 10×10^3 psi.

This implies the inequality: $\sigma = \dfrac{145.8 \text{ lb}}{A} < 10^4 \text{ psi}$

Remember when multiplying or dividing an inequality by a positive number, the inequality remains the same. Therefore one finds:

$$A > \frac{145.8 \text{ lb}}{10^4 \text{ psi}} \quad \text{or} \quad A > 0.01458 \text{ in.}^2$$

>>End Example Problem 2.4

Additional data on material properties needed to solve problems can be found in Appendix D or inside back cover.

2.10 The truss is loaded as shown, with $F_0 = 10$ kN. Determine the stress in each bar, and indicate whether it is tensile or compressive. The bars have areas as follows: $A_{AB} = 200$ mm^2 and $A_{BC} = 250$ mm^2.

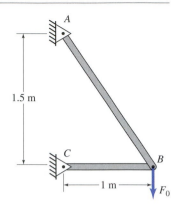

Prob. 2.10

2.11 The frame is loaded by a uniform distributed force q as shown. Consider only the stresses in bars AC and BC; each has a cross-section that is 0.75 in. by 1.5 in. Determine the maximum allowable distributed force q if the stress in each bar is not to exceed 5000 psi.

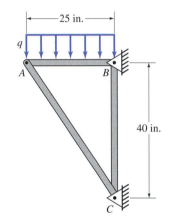

Prob. 2.11

2.12 The truss shown is to be constructed of bars of identical cross-section. Determine the bar cross-sectional area if (a) the stress, tensile or compressive, in each bar is not to exceed 100 MPa; (b) if the tensile stress is not to exceed 50 MPa, and the compressive stress is not to exceed 100 MPa.

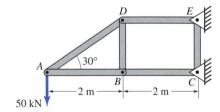

Prob. 2.12

2.13 A steel plate, which is 1.5 m by 1.5 m and 30 mm thick, is lifted by four cables attached to its corners that meet at a point that is 2 m above the plate. Determine the required cross-sectional area of the cables if the stress in them is not to exceed 20 MPa.

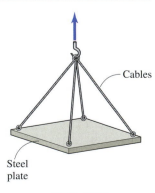

Prob. 2.13

2.14 The required cross-sectional area of cables in a cable-stayed bridge can be estimated by keeping cable stress just at the allowable stress, while supporting the weight of the bridge deck. Say there is a pair of cables, one on each side of the deck attached every 20 m, and that the total loading on the deck is 22 Mg/m. If each cable has an allowable stress of 325 MPa, and is oriented at $\theta = 35°$ from the vertical, determine the minimum acceptable cable cross sectional area. If the cable consists of 10 mm strands, how many strands are needed?

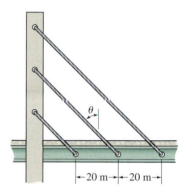

Prob. 2.14 (Appendix A2)

2.15 Collars (and pipes) in drilling consist of 30 ft segments that can be screwed one into the other. One end of each pipe has a so-called pin connection with outer threads and one end has a so-called box connection with internal threads. Once one segment is screwed into the neighboring segment bringing their faces into contact, an additional torque is applied to both segments so that the faces are pressed together with adequate pressure. Neither the average axial stress in the pin connection nor in the box connection should exceed 62500 psi. Determine the contact force that satisfies this condition for the collar geometry shown. Take the parameters to be $D_o = 8$ in., $D_i = 2.8125$ in., and $D_B = 7.375$ in.

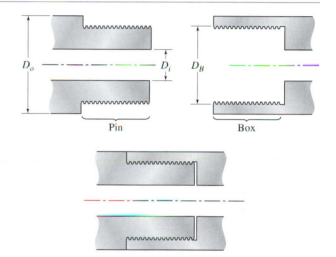

Probs. 2.15–16 (Appendix A3)

2.16 When tightening up two adjacent collar segments of a drill string by the application of a torque, the pin connection is put into tension and the box connection in compression. Say that the force between the two parts due to tightening ("make up") is 470000 lb. For the subsequent design of the drill string, the additional tensile force that the collar can withstand should account for the pre-stress due to make up. Let the allowable tensile stress be 100000 psi. Assume the additional axial force is distributed uniformly over the pin and box connection at the joint. (a) Determine the additional allowable tensile force. (b) Determine the additional allowable tensile force assuming the pre-stress is neglected. Take the parameters to be $D_o = 8$ in., $D_i = 2.8125$ in., and $D_B = 7.375$ in.

2.17 In using this leg curl machine the exertion of the legs results in a torque $M = 120$ lb-ft on the shaft that connects to cord disk. The cord disk has a radius $R = 8$ in. Determine the resulting tensile stress in the 0.125 in. diameter steel cable within the cord that wraps around the cord disk. Assume that frictionless bearings allow the shaft–cord disk assembly to rotate freely in place.

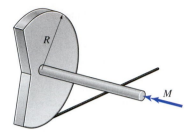

Prob. 2.17 (Appendix A4)

2.18 In weight training machines, the load is varied by passing a pin through one of the weights in the stack. The pin also passes through the pull rod that has a series of holes. In the situation of interest, the pull rod lifts a stack of plates that weigh 195 lb. (a) Determine the normal stress in the pull rod assuming it is solid with $d = 1$ in. diameter. (b) Estimate the average normal stress in the pull rod at a plane through a pin hole. Take the rod to have a 1 in. diameter and each pin hole to have a 0.25 in. diameter. (Note that the sudden change in area means that the assumptions for using P/A to estimate stress at each point are not strictly satisfied.)

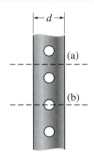

Prob. 2.18 (Appendix A3)

2.19 The wind turbine and tower weigh 650 kip. It is to be bolted to a square concrete slab of depth $H = 4$ ft, which will rest on soil that can withstand pressure of 2000 lb/ft^2. The concrete weighs 150 lb/ft^3. Determine the planar dimension $a \times a$ of the concrete slab if the maximum soil pressure is not to be exceeded.

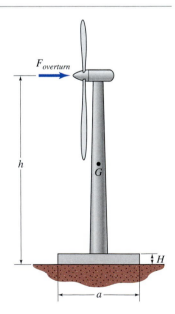

Prob. 2.19 (Appendix A6)

2.20 The wind turbine and tower that weighs 2.868 MN is bolted to a concrete slab of depth $H = 1.5$ m with planar dimensions $a = 10$ m $\times$ $a = 10$ m. The concrete has a mass density of 2400 kg/m³. (a) Determine the average bearing pressure on the soil. (b) Say that the foundation needs to be designed to handle an overturning force $F_{overturn} = 50$ kN that is applied to the blades at a height $h = 70$ m above the ground. Approximate the resulting overturning moment as carried by a linearly varying distribution of normal force per area between the slab and soil. Calculate the maximum force per area at the edge associated with the overturning. (c) Consider two scenarios of failure. In one scenario, the slab lifts off on the tensile side because the maximum force per area due to the overturning moment exceeds the bearing pressure due to weight found in part (a). In a second scenario, the bearing pressure found in part (a), when added to the maximum force per area on the compressive side due to the overturning moment, exceeds the allowable soil pressure of 100 kPa. Determine whether either failure scenario occurs.

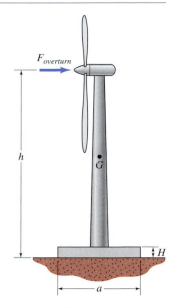

Prob. 2.20 (Appendix A6)

2.21 Small wind turbines can pivot at their base so they can be raised up via a gin pole. The 90 lb turbine is maintained in equilibrium with the 0.25 in. diameter cable in tension as shown. The pole consists of a uniform steel pipe with 2.875 in. outer diameter and a wall thickness of 0.2 in. Determine the normal stress in the cable. Account for both the weight of the pipe and the turbine. The distances are $L_g = 15$ ft, $L_1 = 37$ ft, $L_2 = 5$ ft.

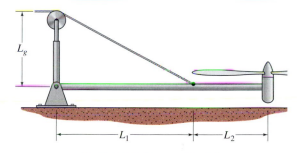

Prob. 2.21 (Appendix A6)

>>End Problems

2.4 Normal Strain

We saw earlier that a small square element changes shape
in a relatively simple way: it distorts into a parallelogram.
We introduce the concept of strain to quantify shape
change in a way that is independent of the element size.

1. The deformation of some portion of a loaded body may be uniform.

Like stress, strain is defined locally. Here again is the sheet on which square elements of equal sizes were drawn. The sheet is elongated by two applied forces. Elements at points B and C have elongated less than and more than point A. But the elongations of all elements around A are equal. The deformation of some length of the bar around A is said to be uniform.

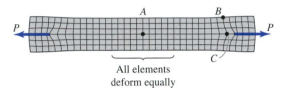

All elements
deform equally

It is easiest to motivate and explain the definition of strain when the deformation is uniform (near A). The definition can be extended to situations (near points B and C) where the deformation is nonuniform.

2. For fixed tensile forces, elongation of a bar increases in proportion to its length.

Consider three bars that have identical cross-sections and material properties. Apply identical tensile forces to each. The bars differ only in their lengths: L, $2L$, and $3L$. Say the deformation is uniform over the whole length of each bar. How do the elongations of the bars differ?

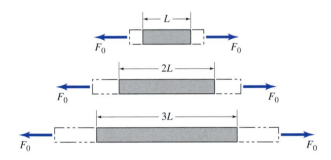

Think of each bar as composed of a series of identically sized segments. If the first bar (L) has N segments, there are $2N$ segments in the second bar, and $3N$ segments in the third bar. Every segment experiences the same force on the same area and material, and so it elongates the same amount. If the elongation of the first bar is δ, the second bar, which has twice as many segments, elongates by 2δ, and the third bar by 3δ. So the elongation is proportional to length.

One segment

3. A segment of a uniformly deforming bar elongates in proportion to the segment's length.

By the same logic, consider a bar of length L that elongates uniformly by δ. A sub-portion of that bar, having length L_1, must elongate by only $(L_1/L)\delta$ or $L_1(\delta/L)$. So the ratio δ/L captures the "intensity" or severity of the deformation. From δ/L, the elongation of any portion of a bar may be found.

For example, this bar was originally 160 mm long and has elongated by 8 mm.

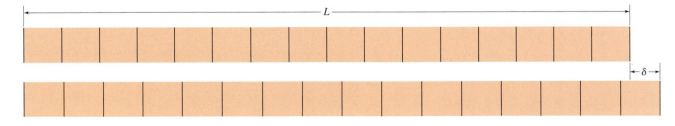

Look at one half of the original bar (any segment 8 units in length). Measure its deformed and initial lengths. Their difference, 84 mm − 80 mm, is the elongation, which is 4 mm or one half of the total elongation. Likewise, one quarter of the original bar (any quarter) elongates by 2 mm.

The ratio of $\delta/L = 8/160 = 0.05$ describes the deformation in any part of this bar, since $4/80 = 2/40 = 0.05$.

4. Normal strain, ε, is defined as the elongation per initial length, and can refer to an infinitesimal element or to a finite segment of a body.

> **Normal strain**, ε, describes the element-level intensity of deformation due to elongation, and it is defined as the increase in length of an element, due to deformation, divided by the element's original length.

For an infinitesimal element of length ΔL that elongates by $\Delta\delta$, we define **normal strain**, ε, by:

$$\varepsilon = \frac{\Delta\delta}{\Delta L}$$

Because the elements in this sheet can be seen to elongate by different amounts, the deformation (strain) in this sheet varies from element to element.

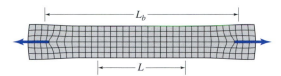

The deformation is uniform over the segment of length L. If that segment elongates by δ, then the strain, ε, at each element in that segment is equal to the segment's average normal strain: $\varepsilon = \dfrac{\delta}{L}$.

We can still define an average normal strain ε_{avg} for the loaded part of the body. If the length L_b elongates by δ_b, then $\varepsilon_{avg} = \dfrac{\delta_b}{L_b}$

Strain is a pure number with no units, since the units of δ and L are both, say, meters. For machines and structures, strains are usually small (10^{-5} to 10^{-3}). To avoid exponents with small numbers, we sometimes use percent strain or micro-strain: 1% strain = 0.01 and 1 micro-strain = 10^{-6}. In some situations, e.g., rubber and some biological tissues, strains can be large ($\sim$1 or more).

---

5. Normal strain describes deformation that is independent of the size of the deformed region and can sometimes be found from measurements.

Like with stress, the definition of strain is particularly helpful for two reasons:

- Strain describes the deformation of an element, without the need to specify the size of the element.
- Provided the deformation in some region is uniform, one can compute the strain (equal in all elements in the region) by $\varepsilon = \delta/L$, where L is the length of the elongating region and δ is its elongation.

---

6. Strains corresponding to shortening of a body are designated as compressive.

Strain is also used to describe element deformation when a body is compressed and shortens. Compressive strain is still change in length (shortening) per initial length.

We can specify the strain with a positive number, and state that it is compressive, or we give the strain a negative sign, provided we agree that negative strain means compression.

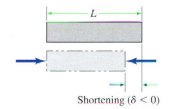

Shortening ($\delta < 0$)

---

7. Don't confuse strain with elongation and don't confuse stress with strain.

Elongate a 100 mm long strip (A) of rubber by 10 mm. Elongate a 20 mm long strip (B) of the same rubber by 5 mm. Compare their strains:

$$\varepsilon_A = \frac{\delta_A}{L_A} = \frac{10\ \text{mm}}{100\ \text{mm}} = 0.1 \qquad \varepsilon_B = \frac{\delta_B}{L_B} = \frac{5\ \text{mm}}{20\ \text{mm}} = 0.25$$

Strip A has a lower strain even though its elongation is greater. Think of each 10 mm segment of B as elongating more than each 10 mm segment of A.

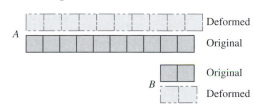

In engineering, stress quantifies the intensity of force per area that seeks to elongate or shorten an element. Strain quantifies the actual elongation or shortening, normalized by length. As we see next, the same stress can produce very different strains in two bodies of identical dimensions, if they are composed of different materials.

>>End 2.4

The Roberto Clemente bridge spanning the Allegheny river in Pittsburgh is a self-anchored suspension bridge. Rather than continuous cables, it has a series of links, connected by pins. Except at its ends, each link has a uniform cross-section. To monitor the effect of a large heavy truck driving over the bridge, extensometers from points B to C and from points D to E are attached to one of the links. One extensometer indicates that the elongation of the link from B to C is 0.024 mm.

(a) If each link acts approximately like a two-force member, with only two equal and opposite forces applied by the pins at A and F, what would be the elongation of the extensometer spanning points D and E?
(b) Determine the strain in the region CD.
(c) Where would it be most difficult to estimate the strain accurately and why?

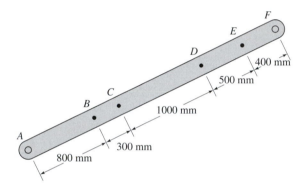

Solution

If the link is approximated as a two-force member, then the stress and strain are uniform. (We are presuming that the other conditions for uniform normal stress are satisfied: long uniform cross-section, uniform material, and the force is applied far from the cross-section of interest and acts through the centroid.)

The average strain in BC is given by $\varepsilon_{BC} = \dfrac{\delta_{BC}}{L_{BC}} = \dfrac{0.024 \text{ mm}}{300 \text{ mm}} = 8 \times 10^{-5}$

The segment DE is also far from the end, so it has the same strain

$$\varepsilon_{DE} = 8 \times 10^{-5} = \frac{\delta_{DE}}{L_{DE}} = \frac{\delta_{DE}}{500 \text{ mm}}$$

Therefore, $\delta_{DE} = 8 \times 10^{-5} (500 \text{ mm}) = 0.040 \text{ mm}$.

The segment CD is also far from the end, so it has the same strain: $\varepsilon_{CD} = 8 \times 10^{-5}$

In the segments AB and EF, at least near the pins, the strain will differ from the values found in $BCDE$. This is because the force is applied nearby (at the pin), and the cross-section of the link is changing near the pin.

>>End Example Problem 2.5

An elastomeric pad is bonded to two plates to isolate the vibrations of one from the other. Say the pad is rated by the manufacturer to withstand strains (in the vertical direction perpendicular to the bond surfaces) up to 20% in compression and up to 9% in tension. The pad is 2 in. wide into the page.

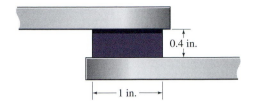

Determine the limits on the (a) downward and (b) upward displacement of the upper plate, given that the lower plate does not displace downward, but could displace upward by any distance from 0 in. to 0.05 in.

Solution

For a part of length L, the general relation between strain ε and change in length δ is $\varepsilon = \dfrac{\delta}{L}$

The relevant length for vertical strain is $L = 0.4$ in.

Tensile strain corresponds to elongation and compressive strain corresponds to shortening.

The maximum tensile strain is 9% or $\varepsilon = 0.09$.

Therefore, the maximum elongation is $\delta = \varepsilon L = (0.09)(0.4 \text{ in.}) = (0.036 \text{ in.})$
The maximum compressive strain is 20% or $\varepsilon = 0.2$.

Therefore, the maximum shortening is $\delta = \varepsilon L = (0.2)(0.4 \text{ in.}) = (0.080 \text{ in.})$

Say the lower plate does not move.

Upward motion of the upper plate causes tensile strain with elongation equal to the displacement. So the upward motion must not exceed 0.036 in.

Downward motion of the upper plate causes compressive strain with shortening equal to the displacement. So the downward motion must not exceed 0.08 in.

Say the lower plate moves up by the maximum amount 0.05 in.

The upper plate may stay fixed, since the compression would be 0.05 in. which is less than the allowable compression of 0.08 in. Upward motion of the upper plate would first cause less compression, which is allowed and eventually causes tension, with elongation equal to the displacement *minus* the lower plate motion of 0.05 in. So the upward motion must not exceed 0.036 + 0.05 = 0.086 in.

Downward motion of the upper plate increases the compression relative to zero displacement. The total compression, which must not exceed 0.08 in., would equal the downward displacement *plus* the lower plate motion of 0.05 in. Thus, the downward motion must not exceed 0.08 − 0.05 = 0.03 in.

Combine the above.

To operate safely, given the possible range of motion of the lower plate, the upper plate cannot move (a) downward more than 0.03 in. and (b) upward more than 0.036 in.

>>End Example Problem 2.6

Additional data on material properties needed to solve problems can be found in Appendix D or inside back cover.

2.22 When magnetic tape is transported it needs to be under tension. Tension is controlled by a set of motors, idlers, and brakes. This tension in the tape results in distortion of the length of the data segments along the tape. Each segment, which is oriented parallel to the long dimension of the tape, has length 0.5 μm when the tape is under zero tension. The length of each segment can be up to 0.5002 μm and still be read properly. Determine the maximum strain that can be allowed in the tape.

Prob. 2.22

2.23 In the exercise machine shown, the flexible rubber belt is fixed to a belt disk of diameter D_b and is guided by pulleys of diameter D_p to the weight stack. If the weight stack is lifted by 200 mm, by how many degrees is the belt disk rotated by the exerciser if (a) the strain in the belt is known to be 0.03; (b) the strain in the belt is neglected (set to zero). Take the dimensions to be L_1 = 800 mm, L_2 = 600 mm, L_3 = 1200 mm, L_4 = 700 mm, D_p = 175 mm, and D_b = 400 mm.

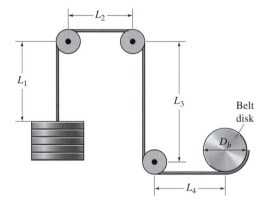

Prob. 2.23

2.24 A rubber bearing pad under the steel girder above is used to produce a relatively uniform bearing stress on the concrete below. The literature from the pad manufacturer suggests that strain in the particular bearing pad should not exceed 12%. Due to the dead loading of the structure's weight, the pad is already strained by 3%. Say the live loads could produce an additional 0.25 in. of vertical motion of the girder, which must be tolerated. Determine the acceptable range of pad thicknesses that keeps strains below the maximum allowable strain.

Prob. 2.24

2.25 Closely spaced lines are etched into a silicon wafer, and an optical scanner is used to count the lines. When the wafer is unloaded, the scanner measures 9512 lines within a 1 mm length. While the same wafer is subject to an axial loading acting perpendicular to the lines, the scanner measures 9525 lines within a 1 mm length. Determine the magnitude of the strain and whether it is tensile or compressive.

Prob. 2.25

2.26 An elastic band, which forms a 2 in. diameter circle when unstretched, is stretched so it wraps around the box shown. Assume the band wraps smoothly around the box so its strain is uniform. (a) Determine the strain in the band. (b) What is the initial length of that portion of the band that wraps over just the top of the box?

Prob. 2.26

2.27 An elastic cord is to be used to secure the stack of concrete pipes to the truck bed as shown. The pipes have an outer diameter of 15 in. If the desired tension in the cord requires a strain of 8%, what should be the initial length of the portion of the cord running from the bed surface on one side of the pipes to the other?

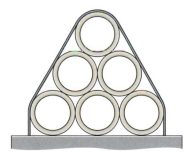

Prob. 2.27

2.28 During construction of the portion of building shown, additional weights are placed on the structure. We want to measure the load taken up by the post and whether the load is primarily axial compression. A strain gage that is attached at the upper left gives a strain reading of 5.3×10^{-4} compression. At the lower right an extensometer with gage length 40 mm has been attached. It indicates a shortening of 0.021 mm. (a) Determine the ratio of the strain difference to the strain measured by the strain gage. (This gives the departure from strain uniformity.) (b) Using the average of the two strain measurements, what would be the vertical shortening of the whole post of height $h = 4$ m under this load?

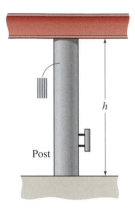

Post

h

Prob. 2.28

2.29 The flexible toothed belt is stretched when it engages the gear shown. Say there can be effective engagement between the belt and the gear even if there is up to a 3% mismatch in the tooth spacing. Determine the range of strains in the belt that is acceptable. Take $s = 10$ mm and $D = 30$ mm.

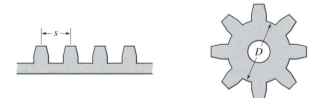

s

D

Prob. 2.29

2.30 A cable is attached as shown to the girder. The member *BCD* rotates rigidly (without deforming) by 1° counter-clockwise around pin *D*. Determine the strain in the cable *AC*. Let the dimensions be $L_1 = 3$ m, $L_2 = 2$ m, and $L_3 = 4$ m.

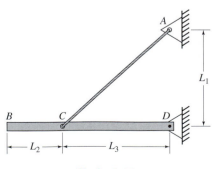

A

*L*₁

B *C* *D*

*L*₂ *L*₃

Prob. 2.30

2.31 The cables in this cable-stayed bridge are limited to a maximum strain of 0.003. Determine the maximum allowable lengthening of the cable shown. The dimensions are $h = 80$ m and $w = 50$ m.

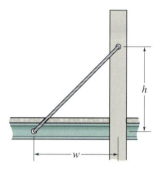

Prob. 2.31 (Appendix A2)

>>End Problems

2.5 Measuring Stress and Strain

The relation between the forces on a body and its deformation depends on the body's dimensions and material. We have defined stress to quantify the internal force acting on a small element and strain to quantify the deformation of an element. So the stress on an element and its strain are also related. However, because stress and strain characterize force and deformation of an element, *independent of its size*, the stress–strain relation depends purely on the material. With the stress–strain relation, we capture the specific role played by a body's material.

1. Measure the relation between stress and strain for a material using a machine that subjects a body of the material to known loads while measuring the deformation.

A common way to measure the stress–strain relation is to load a test specimen in a mechanical testing machine. Both ends of the specimen are gripped. The lower end is prevented from moving, and the upper end moves with the grips attached to the crosshead. The crosshead generally moves at a constant speed, so the specimen's elongation increases steadily with time.

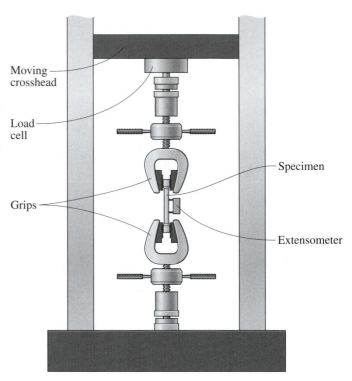

Moving crosshead

Load cell

Grips

Specimen

Extensometer

2. Choose a loading configuration for which the stress and strain can be reliably determined from the force and elongation.

We can determine the relation between stress and strain only if we can determine the stress and the strain in a material simultaneously. Typically, we can directly measure only force and elongation. Stress and strain can be calculated from force and elongation for only a few loading configurations, the simplest one being uniaxial tension. In addition, the test specimen must satisfy these conditions (Section 2.3.6):

*   A portion of the specimen, the gage section, should have a uniform cross-section.
*   The gage section should be relatively long compared to its width.
*   The grips apply axial forces that act through the center of the cross-section.
*   The specimen material should be uniform.

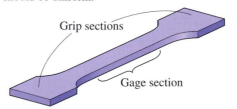

Grip sections

Gage section

Under these conditions, the stress and strain are uniform over the central portion of the gage section. Detailed specifications for tensile tests are put forth by professional societies, such as ASTM International. Likewise, there are specifications for other tests, such as compression tests.

3. Strain is often measured with an extensometer or a strain gage.

The strain may be determined from the elongation of a central portion of the gage section. Elongation is measured, for example, with an extensometer, which attaches to two points in the gage section. The extensometer outputs a voltage proportional to the measured change in separation of attachment points (the elongation). The strain is computed by dividing the elongation by the initial distance between the two points (the gage length).

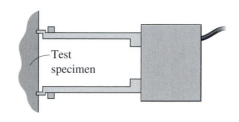

Test specimen

Another way of measuring strain when the strains are not large is with a strain gage. A strain gage consists of a patterned wire that is imprinted on a plastic sheet. The plastic sheet is cemented to the gage section. The electrical resistance of the wire depends on its length. Since the total length of the wire changes with the strain along the length of the wire, the strain can be inferred from measured changes in electrical resistance.

4. Force is often measured with a load cell.

The axial force on the specimen is typically measured with a load cell. The load cell is mounted between the testing machine and the specimen, so it carries the same axial force as the specimen. The load cell outputs a voltage that is proportional to the force. Typically, a load cell contains a member that deforms in response to the applied load. A strain gage is mounted on the member, and the voltage from the strain gage is proportional to the deformation and hence to the force. The correlation between the strain gage voltage and the force is provided by the load cell manufacturer.

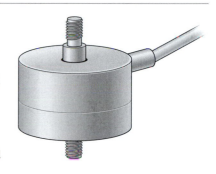

5. Calculate the stress from the force and the strain from the elongation and plot the stress vs. the strain.

The raw data from a tensile test is the axial force, P (measured, for example, by a load cell), and the elongation, δ (measured, for example, by an extensometer).

Since the specimen and loading apparatus are designed so that the stress and strain are nearly uniform, the stress, σ, and strain, ε, are computed as average values from P and δ.

If A is the cross-sectional area on which P acts, and L is the length over which the elongation δ is measured, then the stress and strain are found from:

$$\sigma = \frac{P}{A} \quad \text{and} \quad \varepsilon = \frac{\delta}{L}$$

The tensile strain–strain curve is characteristic of the material, but independent of the precise dimensions of the specimen, provided the conditions for uniform stress and strain are met. In the next sections, we identify key features of stress–strain curves that are important to the design and analysis of machines and structures, and we define parameters that quantify these key features from the curve.

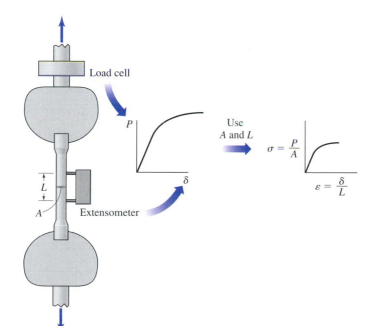

2.6 Elastic Behavior of Materials

To predict deformation and failure we need a mathematical relation between stress and strain. Here we consider the simplest relation, appropriate to materials that exhibit elastic behavior.

1. We usually design parts or structures to have only elastic deformations: when loads are removed, the body returns to its initial size and shape.

A material is said to deform *elastically* if it substantially recovers its initial shape and size when loads are removed. Most materials respond elastically, at least when the deformations, or more precisely the strains, are very small, for example below 0.001 or 0.01, depending on the material.

During elastic deformation, there is usually no permanent internal change in the material. After loads are removed, one usually cannot detect that loads were ever applied. Furthermore, in most engineering applications, we want parts to retain their shape when loads are applied and removed. For these reasons, parts to be used repeatedly are designed to deform elastically under the loads they regularly experience.

2. Elastic deformations are proportional to the load.

In addition, the stress is found to increase in proportion to the strain for most materials that deform elastically, at least at small strains. This is termed *linear elastic* behavior. For most engineering materials, stress is approximately a linear function of strain, starting from zero strain. For some biological materials or fabrics, initial elongation initially occurs with little increase in stress, although eventually the stress increases in proportion to strain.

3. Represent a material that undergoes linear elastic deformation mathematically with a linear stress–strain equation.

Young's modulus (or elastic modulus), E, captures the intrinsic stiffness of a material when it elongates elastically, and it is defined as the proportionality between normal stress and normal strain when the material is subjected to normal stress in one direction.

To make quantitative predictions of the deformation of parts, we need a mathematical equation that captures the stress–strain relation for elastic deformation. The most commonly used equation idealizes the stress to be precisely proportional to the strain, starting from zero strain.

Young's modulus (or elastic modulus), E, is defined as the proportionality between normal stress and strain when the material is linear elastic. E depends on the material. Since strain is unitless, E has the same units as stress (e.g., Pa or psi).

$$\sigma = E\varepsilon$$

4. The stiffer the material, the greater is E.

Stiffness refers to the force necessary to achieve a given deformation. Stiffness of a part depends on its material and geometry.

Because the stress and strain are defined to capture the intensity of force and deformation in an element independent of its size, E quantifies the intrinsic stiffness of a material, irrespective of its size. A material with a larger E is stiffer, because the same strain (ε) requires a larger stress (σ).

5. E varies by a factor of 10^5, even over commonly used materials, and it strongly affects a designer's choice of material.

E has been measured and tabulated for thousands of materials. This table shows typical values for a few materials, and more information can be found in Appendix D and on the back cover. Notice the wide range over which the elastic stiffness can vary. Over commonly-used materials, such as steel and rubber, E is seen to vary by a factor 10^5.

Elastic Moduli (E) of Representative Materials

Material	SI (GPa)	USCS (10^6 psi)
Diamond	1000	150
Steel	200	30
Wood	12	1.8
Plastic	2	0.3
Rubber	0.002	0.0003

6. Materials that are elongated by a tensile stress in one direction, contract in the transverse (perpendicular) directions.

When tensile stress acts in one direction, most materials contract in the perpendicular or transverse direction. This contraction is evident in highly flexible materials, such as rubber.

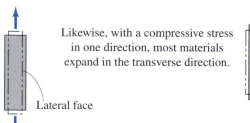

Likewise, with a compressive stress in one direction, most materials expand in the transverse direction.

Lateral face

Lateral face

It is important to note that we are considering uniaxial normal stress, which acts in one direction. There is no tensile or compressive stress acting in the horizontal direction on the lateral faces. But, there is strain in the horizontal direction.

---

7. Normal strain in the transverse direction is similarly defined as change in the transverse length divided by its initial length.

Imagine measuring the change in length parallel to and perpendicular to the tensile axis with extensometers.

Say the initial lengths parallel and perpendicular to the tensile axis are L and L_t, respectively.

Under tension, the axial length L changes by δ (lengthens so $\delta > 0$), and the transverse length L_t changes by δ_t (shortens so $\delta_t < 0$).

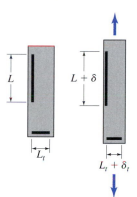

From lengths and elongations we find axial (ε) and transverse (ε_t) strains:

$$\varepsilon = \frac{\delta}{L} \qquad \varepsilon_t = \frac{\delta_t}{L_t}$$

$$\varepsilon > 0 \text{ (tensile)}$$
$$\varepsilon_t < 0 \text{ (compressive)}$$

---

8. For a material that deforms elastically, the transverse strain is proportional to the axial strain.

Poisson ratio, ν, captures the contraction in one direction when a material is elongated in a perpendicular direction, and it is defined as the proportionality between the magnitudes of transverse and longitudinal normal strains when the material is subjected to only normal stress in the longitudinal direction.

If we monitor the changes in both lengths L and L_t during a single tensile test, we can find how the transverse strain changes as the axial strain increases.

When the material is elastic, the transverse strain is found to change essentially in proportion to the axial strain.

Poisson ratio, ν, is defined as the proportionality between the transverse and axial strain magnitudes when a linear elastic material is subjected to uniaxial normal stress. ν depends on the material. Since ε and ε_t are unitless, ν must be unitless:

$$\varepsilon_t = -\nu\varepsilon$$

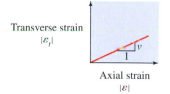

Transverse strain $|\varepsilon_t|$

Axial strain $|\varepsilon|$

---

9. The Poisson ratio varies over a narrow range and rarely affects design.

Though it is more difficult to measure, ν has been tabulated for a number of materials (see Appendix D). Generally, ν has values within a narrow range (generally in $0 < \nu < 0.5$). This table indicates typical values of ν for commonly used types of materials. ν usually has a minor effect on design, and is not as routinely measured as is E.

Typical Poisson ratio (ν) for different types of materials

Material	ν
Rubber	≈ 0.49
Polymers	≈ 0.4
Metals	≈ 0.3
Ceramics	≈ 0.2

>>End 2.6

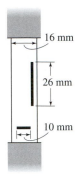

16 mm

26 mm

10 mm

The rectangular specimen, 16 mm wide and 2 mm thick, carries a tensile load of 1200 N. The axially oriented extensometer measures an elongation of 0.014 mm over a gage length of 26 mm. The transversely oriented extensometer measures a contraction of 0.0018 mm over a gage length of 10 mm. From the measured load, the extensometer readings, and the dimensions, determine the elastic modulus and the Poisson ratio, assuming the material is linear elastic.

Solution

The bar is subjected to tensile stress in the vertical direction. The cross-sectional area over which the axial force acts is $(16 \text{ mm})(2 \text{ mm}) = 32 \text{ mm}^2$.

The tensile stress is

$$\sigma = \frac{P}{A} = \frac{1200 \text{ N}}{32 \times 10^{-6} \text{m}^2} = 37.5 \text{ MPa}$$

The axial strain is related to the measured elongation and gage length of the axially oriented extensometer by

$$\varepsilon = \frac{\delta}{L} = \frac{0.014 \text{ mm}}{26 \text{ mm}} = 5.38 \times 10^{-4}$$

The transverse strain is related to the measured elongation and gage length of the transversely oriented extensometer by

$$\varepsilon_t = \frac{\delta_t}{L_t} = \frac{-0.0018 \text{ mm}}{10 \text{ mm}} = -1.8 \times 10^{-4}$$

If the material is linear elastic, the stress and axial strain are related by

$$\sigma = E\varepsilon$$

Since the stress and the axial strain have been determined from measurements, the elastic modulus can be estimated:

$$E = \frac{\sigma}{\varepsilon} = \frac{37.5 \text{ MPa}}{5.38 \times 10^{-4}} = 6.97 \times 10^{10} = 69.7 \text{ GPa}$$

The Poisson ratio ν relates the transverse strain ε_t to the axial strain ε in uniaxial tension according to:

$$\varepsilon_t = -\nu\varepsilon$$

Since both the transverse strain and the axial strain have been determined from measurements, we can estimate the Poisson ratio ν:

$$\nu = \frac{-\varepsilon_t}{\varepsilon} = \frac{-(-0.00018)}{(-0.000538)} = 0.334$$

>>End Example Problem 2.7

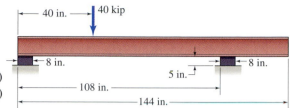

The 12 ft long steel wide-flange I-beam designated W18 × 97 weighs 97 lb/ft. In addition to its own weight, the beam carries a 40 kip load at the point shown from another member resting on it. Bearing pads that are 11 in. by 8 in. and 0.5 in. thick are placed between the beam on the concrete beneath it. (Bearing pads serve to distribute the force more evenly.) The bearing pads have an elastic modulus of 3000 psi. (See Appendix E.)

(a) Estimate the deflection of the beam at each pad, presuming each pad deforms uniformly.
(b) With the deflections computed in part (a), estimate the angle at which the beam tilts.

Solution

The beam weighs (97 lb/ft)(12 ft) = 1164 lb. If we draw a free body diagram of the whole beam to find the forces from the pads, then the weight of the beam (because it is uniformly distributed) acts in the center or 72 in. from the left end. If we assume the pads deform uniformly, the net force due to each pad acts through its center.

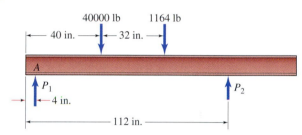

From moment summation about the point A in the center of the left pad, we find:

$$\sum M\,|_A = -(40000\ \text{lb})(36\ \text{in.}) - (1164\ \text{lb})(68\ \text{in.}) + P_2(108\ \text{in.}) = 0,$$
hence $P_2 = 14070$ lb.

$$\sum F_y = P_1 + P_2 - 40000\ \text{lb} - 1164\ \text{lb} = 0, \text{ hence } P_1 = 27100\ \text{lb}.$$

Each pad has an area of (8 in.)(11 in.) = 88 in.

The compressive stresses in the pads are:

$$\sigma_1 = \frac{P_1}{A} = \frac{27100\ \text{lb}}{88\ \text{in.}^2} = 308\ \text{psi} \qquad \sigma_2 = \frac{P_2}{A} = \frac{14070\ \text{lb}}{88\ \text{in.}^2} = 159.9\ \text{psi}$$

From the modulus $E = 3000$ psi, we can determine the strains:

$$\varepsilon_1 = \frac{\sigma_1}{E} = \frac{308\ \text{psi}}{3000\ \text{psi}} = 0.1027 \qquad \varepsilon_2 = \frac{\sigma_2}{E} = \frac{159.9\ \text{psi}}{3000\ \text{psi}} = 0.0533$$

Compressions of the pads are then found to be:

$$\delta_1 = \varepsilon_1(thickness) = 0.1027(0.5\ \text{in.}) = 0.0513\ \text{in.}$$
$$\delta_2 = \varepsilon_2(thickness) = 0.0533(0.5\ \text{in.}) = 0.0266\ \text{in.}$$

The centers of the pads are separated by 108 in., so the beam tilts by

$$\frac{0.0513\ \text{in.} - 0.0266\ \text{in.}}{108\ \text{in.}} = 2.29 \times 10^{-4}\ \text{rad}$$

We could reconsider the assumption that each pad deforms uniformly. The compression varies over each pad by $(2.29 \times 10^{-4})(8\ \text{in.}) = 0.001830$ in. This is 7% of the average deflection of pad 2 and even less of pad 1.

>>End Example Problem 2.8

Additional data on material properties needed to solve problems can be found in Appendix D or inside back cover.

2.32 Tensile testing of human cortical bone is conducted to obtain the elastic moduli. The cross-sectional geometry of bone varies along its length, although the cross-section can be approximated as hollow and circular, with an average outer diameter of 27 mm and inner diameter of 14 mm. Let the gage section over which the elongation is measured be 40 mm. A straight line is fit to the load-elongation data, giving a slope of $1.12 \,(10^8)$ N/m. (a) Estimate the elastic modulus. (b) A gage is also attached which measures the external diameter. The external diameter decreases by 0.0048 mm when the load increases by 2700 N. Estimate the Poisson ratio.

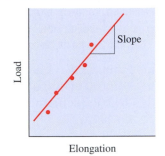

Prob. 2.32

2.33 Properties of frozen tissues have been measured as part of understanding the response of tissue to cryopreservation. In one set of experiments, frozen bovine carotid arteries were tested in uniaxial tension. The cross-sectional geometry was extracted from digital images of slices of the artery obtained after testing. For the specimen tested, the gage length is 30 mm, the outer diameter is 6 mm, and the wall thickness is 1 mm. From the load deflection curve, estimate the elastic modulus of the frozen artery.

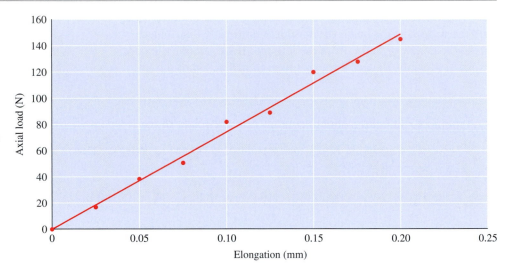

Prob. 2.33

2.34 Properties of frozen tissues have been measured as part of understanding the response of tissue to cryosurgery. In one set of experiments, plugs of frozen rabbit liver were subject to axial compression. The plugs were 10 mm in diameter and 14 mm long. (a) Using the linear portion of the stress–strain curve from points D to E, estimate the elastic modulus. (b) Using the peak value of stress, determine to the nearest 1000 N the maximum load cell force needed if it is to accommodate at least twice the peak stress of the test. (c) Assuming the stress–strain curve is linear from zero stress to failure, with a slope equal to the elastic modulus, what is the shortening of the plug specimen by the end of the test?

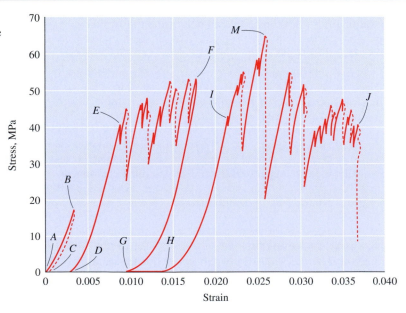

Prob. 2.34

2.35 Thin rubber strips are used to provide increasing resistance to orthopedic patients during large motion exercises. The sheets shown are 0.03125 in. thick and 3 in. wide. Say the rubber properties are $E = 120$ psi and $\nu = 0.47$. One end of the strip is wrapped around the doorknob, and the other end is wrapped around the hand. Say the portion of the strip that elongates is approximately 18 in. (a) What is the increase in the force exerted with every 1 in. increase in length? (b) If a force of 2 lb is applied, what is the change in the strip's width and thickness?

Prob. 2.35

2.36 A solid plastic rod is compressed by an axial force F_0. The rod is 40 mm long and 6 mm in diameter. A second rod of the same material must be designed to undergo the same axial shortening under the axial force F_0 as the first rod. (a) Determine the diameter of second plastic rod, if it is 50 mm long. (b) If the diameter of the first rod expands by 0.005 mm under load, what is the increase in diameter of the second rod under the same load?

2.37 A plastic tube with 20 mm outer diameter and 14 mm inner diameter is compressed by an axial force F_0. The rod must be able to slide without contact within a steel sleeve that has inner diameter 20.03. What is the maximum axial force that the plastic tube can withstand without touching the sleeve? Take $E = 2$ GPa and $\nu = 0.4$.

Prob. 2.37

2.38 In magnetic tape recording, the tape is subjected to tension while it is guided along its path. Tension in the tape causes elastic deformation that results in slight changes in length of the domains on which information is stored. Say the tape has an average modulus of 8 GPa, a thickness of 9 μm, and a width of 12 mm. Let the unit of stored information occupy a length 0.5 μm in a tape that has no tension. Determine the length of the storage unit if the tape is subjected to tension of 0.9 N.

Probs. 2.38–39

2.39 In magnetic tape recording, the tape is subjected to tension while it is guided along its path. The tape consists of many parallel tracks of information that must be individually read. Because of the lateral Poisson contraction, fluctuations in the tension in the tape cause lateral movement of the tracks. Take the tape elastic properties to be $E = 8$ GPa and $\nu = 0.37$ and its thickness and width to be 9 μm and 12 mm, respectively. An individual track is 1 μm wide, with a gap of 0.5 μm between tracks. Motion inward or outward in excess of 0.4 μm cannot be tolerated. Consider the track with inner boundary that is located a distance of 5 mm from the tape center-line when the tape sustains a tension of 1 N. Assuming the center-line does not displace laterally, what increase or decrease in tension from 1 N can be tolerated?

2.40 The truss is composed of steel members that all have a rectangular cross-section 35 mm wide by 6 mm thick. Let the length $L = 1.3$ m, and the angle $\alpha = 40°$. If the force $F_0 = 8$ kN, determine the change in length of each member, and indicate if it elongates or shortens.

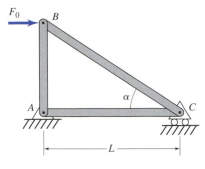

Prob. 2.40

2.41 The cable AC that supports the bar has an effective modulus and area of $E = 140$ GPa and $A = 77$ mm^2, respectively. Let the lengths be $L_1 = 600$ mm and $L_2 = 300$ mm, and the angle $\alpha = 50°$. If the permitted additional elongation of the cable is 0.20 mm, determine the maximum allowed distributed force per length q.

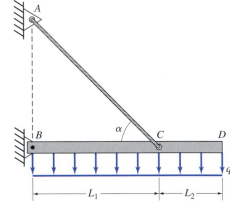

Prob. 2.41

2.42 A cable BD is used to support the post, which is subjected to a lateral force F_0 at its top. The cable has an effective modulus $E = 20(10^6)$ psi and area $A = 0.268$ in.^{2}. Let the lengths be $L_1 = 12$ ft and $L_2 = 6$ ft, and the angle $\alpha = 35°$. What is the load F_0 that brings the cable to its maximum allowable load of 10 kip, and what is the cable elongation under those conditions?

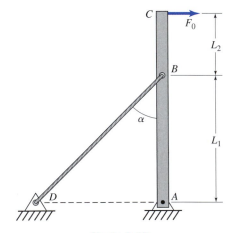

Prob. 2.42

2.43 Cables used in cable-stayed and other bridges consist of many steel wires twisted together. Because the wires partially unwrap, the effective modulus of steel cable is less than that of steel. As a test of its effective modulus, a mass of 500 kg is hung from a cable of original diameter 60 mm and length 50 m. This causes the cable to be 0.52 mm longer. What is the effective modulus of this cable as a fraction of the nominal steel modulus of 200 GPa?

2.44 Tensioning of cables produces compression of the concrete pylons. Consider the shortening of the pylons produced by this loading. Each pylon has only a front and rear stay, and each stay has a tension of 10^7 N. The reinforced concrete pylon is 0.5 m thick and 6 m wide. The modulus of reinforced concrete is 45 GPa. From the attachment points of the cables to its support at the bottom, the pylon is 40 m long. By how much does the column shorten due to the cable tension? Take the dimensions to be $w = 35$ m and $h = 30$ m.

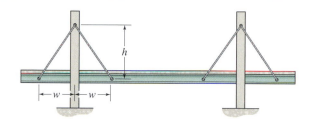

Prob. 2.44 (Appendix A3)

2.45 The forces of the legs on the padded beam are balanced by the two connector links. Assume that the pivot beam is held fixed and that the steel connector links sustain only uniaxial tension or compression. Determine the amount by which the right connector link would shorten, assuming each leg exerts a 50 lb upward force on the padded beam. Take the dimensions to be $a = 4$ in., $b = 16$ in., $c = 4$ in., $d = 3.5$ in., $D_1 = 1.5$ in., $L = 1.75$ in., $m = 5$ in., and $t = 0.25$ in.

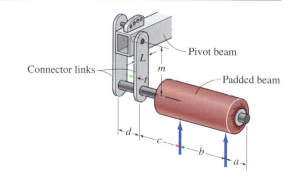

Prob. 2.45 (Appendix A4)

2.46 Wind turbine blades are often made from fiberglass. Tensile testing of a new sample of fiberglass is to be conducted; previous fiberglass samples had an elastic modulus of 40 GPa. Coupon specimens of the material are available that are 2 mm thick, 18 mm wide, and long enough to accommodate an extensometer 35 mm long. The testing machine that covers the right range of load is to be chosen. It is expected to load the material up to strains of 3%. If the material must remain elastic over that range, what would be the maximum load?

>>End Problems

2.7 Failure and Allowable Limit on Stress ⎯⎯⎯⎯⎯⎯⎯⎯

If the strain is increased sufficiently, the deformation is no longer elastic. We usually design so that a material does not experience such high levels of strain. Here we consider how to identify a tolerable level of stress or strain.

1. Materials subjected to high tensile stresses often exhibit a failure that takes one of two general forms.

Many materials can be sorted into one of two major categories: *Ductile* and *Brittle*. These categories are ones that are appreciated by everyday experience.

Ductile materials undergo permanent deformation when the strain reaches some level, but continue to remain intact up to relatively high strains. The stress–strain curve becomes non-linear when permanent deformation begins.

Everyday example: paper clip

Brittle materials do not undergo permanent deformations, but break at very low tensile strains. The stress–strain curve remains linear until failure occurs.

Everyday example: glass

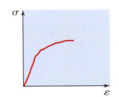

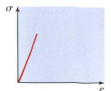

Ductile, and many brittle, materials are often capable of sustaining large compressive strains before failure occurs.

2. Many materials undergo plastic deformation: the stress–strain relation can differ depending on whether the load increases or decreases, and permanent deformation remains after the load is removed.

Consider a thin metal wire (Case I) that is undeformed (State 1). The wire elongates when subjected to low tensile stresses (State 2). The stress is then removed (State 3). Since the stress was low, the wire deformed elastically, and so the length in State 3 equals the initial length. Consider a second thin metal wire (Case II), eventually to be subjected to high stresses. For low stresses, the second wire deforms elastically like the first wire. When the stress exceeds some critical stress level (designated by σ_Y), the stress increases less for the same increase in strain (slope decreases). After the stresses reach the maximum level (State 2), the stresses are removed (State 3). During unloading (State 2 to State 3), the stress decreases in proportion to the strain, essentially following the initial slope E, as in Case I, or for the initial loading, as in Case II. The wire shrinks, but not back to its initial length. The strain that remains is referred to as *plastic* or *permanent* strains, and these are present in State 3 even though the stresses are zero.

1. Initial
2. Loaded
3. Load removed

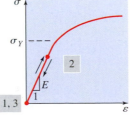
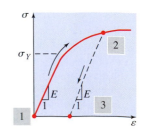

Case I: Lower maximum stress (only elastic deformation)

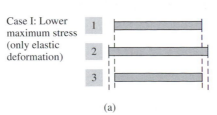

Case II: Higher maximum stress (elastic and plastic deformation)

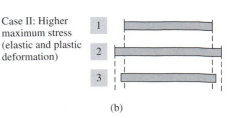

(a)

(b)

3. Design a part to keep the stress below the level at which the material behaves unacceptably either by permanent deformation (ductile) or by complete failure (brittle).

Ductile material

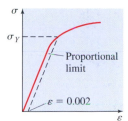

Brittle material

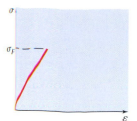

Yield stress, σ_Y, is defined as the uniaxial tensile stress at which a ductile material begins to exhibit noticeable plastic deformation.

The design limit is based on the **yield stress**, σ_Y, at which the material noticeably begins to deform plastically. Since we do not want the loads to produce permanent deformation in material, we usually design so that the stresses stay below σ_Y.

Since the transition from elastic to plastic is often gradual, it is hard to pinpoint the stress marking the first deviation from linearity (the proportional limit). Instead, σ_Y is often defined as the stress at which unloading would produce a plastic or permanent strain of 0.002.

The design limit is based on the failure stress, σ_F, at which the material fractures. We do not want loads to produce complete failure of material, therefore we usually design so the stresses stay below σ_F.

Fracture stress σ_F varies more from one specimen to another of the same brittle material than the yield stress σ_Y varies across specimens of the same ductile material.

Strength is a general term that refers to the maximum load a body can withstand without failure. A body's strength depends on its material and geometry. Because stress and strain pertain to elements, yield stress or fracture stress quantify a material's strength.

---

4. To be safe, given uncertainties, design parts so that stresses are significantly below levels that cause failure.

There are often uncertainties regarding the properties, the geometry, and loading. Because of its capacity to deform plastically without failure beyond σ_Y, a ductile material is relatively tolerant of material flaws and uncertainties. By comparison, a brittle material is relatively intolerant of flaws and uncertainties.

Still, in either case, to account for uncertainties, rather than let the stress reach levels just below the yield or failure stress, one builds in a margin of safety. While there are more advanced design methods, two common methods are:

Allowable Stress Method:

> An allowable stress is defined, typically a fraction (e.g., 1/3, 1/2, 2/3) of the yield or failure stress. Parts are designed so the maximum stress is not greater than this allowable stress.

Safety Factor Method:

> A factor of safety (n) is agreed upon (e.g., 1.3, 1.5, 2). Parts are designed so the maximum stress is not greater than the yield or failure stress divided by n.

Both allowable stress and factors of safety are particular to a material and application. They are arrived at by professional bodies that establish standards, or by company policy, after extensive design and testing experience.

---

5. Take advantage of the superior properties of brittle materials when they are relevant, but be careful using them when tensile stresses are high.

Brittle materials are often used because they have other attractive properties. For example, they may be lightweight, have superior wear or friction properties, or can withstand high temperatures. Often they are subjected to only compressive stresses, and so are less prone to failure. Engineers are also developing new design methods that enable brittle materials to serve in applications with tensile stresses.

>>End 2.7

2.8 Variety of Stress–Strain Response

Different materials, including both man-made and natural, display widely different stress–strain responses. Here we point out some of this variety.

1. Materials that appear to be minor variations within a single class, such as steels, can exhibit widely different stress–strain relations.

Even within what appears to be a narrow class of materials there is in fact great variety. The science and engineering of materials, such as steels, has lead to many technological advances that enable this variety. Types of steel differ in many ways. In particular, there is significant variation in yield strength and ductility, which is quantified as the strain at fracture. Depicted here are two steels; the scales of stress and strain are very different. The steel at the left has low strength and high ductility, while the one at the right has high strength and low ductility. Ductility and strength often need to be traded off one for the other.

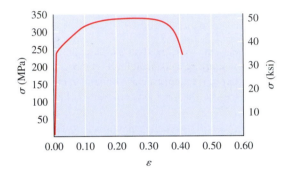

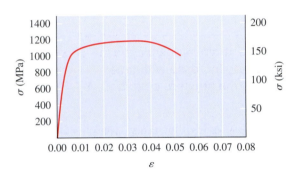

Other metals, for example aluminum, nickel, and titanium have also been extensively developed to produce alloys with different properties suitable for many applications.

2. Many elastomers or rubbery materials are capable of sustaining very large strains while still remaining elastic.

Elastomers are able to withstand very large strains, up to two or three times their initial length. Often their stress–strain response remains nearly linear even for large strains, although the elastic modulus is much less than metals (metals are 10^5 times stiffer than elastomers).

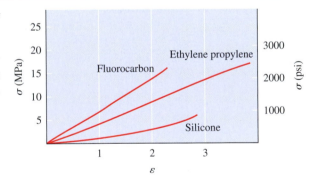

3. Biological materials often exhibit stress–strain curves that differ markedly from engineering materials.

Most biological materials have complex stress–strain curves. For example, ligaments have an initial low stiffness response (called the toe-in region), which corresponds to the uncrimping of fibrils. They then respond linearly until failure occurs at moderate strains. This response further varies with other factors, such as aging and immobility (lack of exercise).

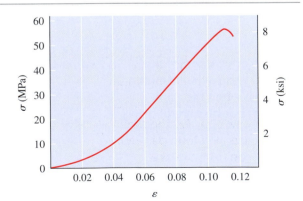

4. Polymeric materials (plastics) play an increasing role in design because of their many favorable properties and the low cost at which they can be mass produced.

Engineers and designers today can take advantage of many materials, including, with increasing frequency, polymers. The mechanical behavior of many types of polymers is roughly similar to that of ductile metals: a linear elastic response is followed by plastic deformation and eventually failure. Extensive effort is devoted to developing polymers that have useful combinations of properties.

Certain general properties of polymers give them design advantages: they are light, inexpensive to fabricate in large numbers, and often have low wear and friction. They are a natural candidate for applications with low stress.

5. Despite their lower strength and stiffness, polymers can substitute for metals in many applications because their mass density is so much lower than metals.

Polymers have a mass density similar to water, so they are roughly 1/7 as dense as steel. They also have lower yield stress σ_Y and stiffness E than do steels. We will see later that a part can be made stiffer or stronger by changing its dimensions or its material. So a polymer part can be as strong as a steel part by making it larger, and it would still not be heavier because its density is so much lower.

Because its E is very much lower than that of metals, a polymer rarely replaces metals in applications that require high stiffness. However, because σ_Y of some polymers is not that much less than some metals, polymers can replace metals in applications that require high strength, such as gears.

6. The strain in polymers can gradually increase with time even if the stress is constant.

The response of polymers can also depend on how long the load is applied. Say we hang a weight from a tensile specimen (applying a constant force). Of course, immediately the specimen elongates elastically. But, in addition, the specimen may continue to elongate slowly with time as long as the weight hangs. The continuing deformation is referred to as creep or viscous strain.

The strain is shown as a function of time for different applied stresses (different weights hung on specimens with the same cross-sectional area). The strain at short times represents the initial elastic response. But, the strain also increases with time. This behavior is referred to as viscoelastic: the response is both elastic (initially) and viscous or liquid-like (strain increasing with time).

If we replot the stress (constant) divided by the strain (increasing), we get a modulus-like quantity that decreases with time.

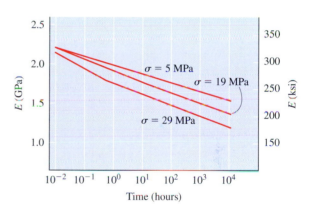

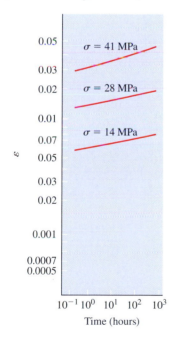

For long time applications, this effect is important. If the load is constant, then the deflection increases with time. If the deflection is fixed, then the load decreases with time. Either way, it can be viewed approximately as elastic, but with a modulus that decreases with time.

7. The tendency for the strain to increase with time depends strongly on temperature.

The tendency for polymers to creep increases with temperature. In fact, metals can also creep when the temperatures become high, for example in jet engines.

Load (kip)	Elongation (in.)
0	0
1.07	0.000375
3.28	0.001125
5.71	0.001875
7.85	0.002625
8.42	0.00375
8.42	0.006
8.57	0.015
11.85	0.03
14.28	0.075
15.35	0.21
13.92	0.3
13.21	0.345

Stress (ksi)	Strain
0	0
7.542512	0.00025
23.12097	0.00075
40.25023	0.00125
55.33525	0.00175
59.35323	0.0025
59.35323	0.004
60.41059	0.01
83.53156	0.02
100.6608	0.05
108.2033	0.14
98.12315	0.2
93.11831	0.23

A tensile test is conducted on a steel specimen with a circular cross-section. The gage length is 1.5 in. and the diameter is 0.425 in. Values of tensile load and elongation are given in the table. (a) Calculate and plot the stress as a function of the strain. (b) Estimate the elastic modulus from the data. (c) Estimate the yield strength using the 0.2% offset method. That is, construct a line with the same slope as the initial, elastic portion of the curve, but starting from strain $= 0.2\% = 0.002$. The 0.2% offset yield strength corresponds to the stress at which the line intersects the stress–strain curve. (Hint: plot just the initial portion of the stress–strain curve with the strain axis spread out.)

Solution

The stress and strain values are derived from load and elongation according to

$$\sigma = \frac{Load}{\frac{\pi}{4}(0.425 \text{ in.})^2} \qquad \varepsilon = \frac{Elongation}{1.5 \text{ in.}}$$

The stress and strain are computed from these formulas, tabulated, and then plotted one against the other for the full range of strain in the left graph. In the right graph stress is plotted for only a small range of strains near zero.

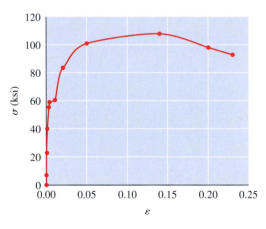
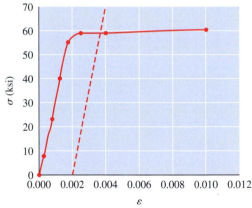

Since the strain scale has been spread out in the right plot, the slope of the initial linear part can be estimated:

$$E = \frac{\sigma}{\varepsilon} \approx \frac{40000 \text{ psi}}{0.00125} = 32.2 \times 10^6 \text{ psi}$$

Also from the right plot, a line parallel to the initial slope has been drawn from strain $= 0.002$. This line intersects the stress–strain curve at 59.35 ksi.

So the yield strength from the 0.2% offset method is 59350 psi.

>>End Example Problem 2.9

A short plastic tube is expected to carry up to a compressive force of 3000 N, and remain below 30% of the yield stress of 50 MPa. Available tubes have a wall thickness of 1.5 mm with outer diameters in increments of 2 mm starting from 10 mm. Find the tube of minimal diameter from among the available sizes that meets the yield requirement and determine the actual factor of safety of the stress relative to the yield stress. Assume the tube is short enough not to buckle under the compressive load.

Solution

The stress in the tube is not to exceed 0.3(50 MPa) = 15 MPa.

The stress is related to the area by $\quad \sigma = \dfrac{P}{A} = \dfrac{(3000 \text{ N})}{A} < 15 \text{ MPa}$

So the area must be at least $\quad A > \dfrac{(3000 \text{ N})}{15 \text{ MPa}} \quad$ or $\quad A > 2 \times 10^{-4} \text{ m}^2 = 200 \text{ mm}^2$

Using a spreadsheet or calculator, determine the area for a range of outer diameters starting from 10 mm. Since the wall thickness is 1.5 mm, the inner diameter is 3 mm less than the outer diameter.

Outer diameter (mm)	Inner diameter (mm)	Area (mm^2)
10	7	160.2212
12	9	197.9203
14	11	235.6194
16	13	273.3186

Since the area must be at least 200 mm^2, we choose a tube with an outer diameter of 14 mm.

The area of the chosen tube is 235.6 mm^2, so the actual stress is

$$\sigma = \frac{(3000 \text{ N})}{235.6 \text{ mm}^2} = 12.73 \text{ MPa}$$

The actual factor of safety, n, is calculated as follows: $\quad n = \dfrac{50 \text{ MPa}}{12.73 \text{ MPa}} = 3.93$

Notice this factor of safety is higher than the minimum required factor of safety of 1/0.3 = 3.33, because we had to choose from a limited number of tube diameters and so had to provide more than enough area.

>>End Example Problem 2.10

Additional data on material properties needed to solve problems can be found in Appendix D or inside back cover.

2.47 A tensile test is conducted on a steel specimen with a diameter of 8.84 mm in the gage section and a gage length of 50 mm. Values of tensile load and elongation are given in the table. (a) Calculate and plot the stress as a function of the strain. (b) Estimate the elastic modulus from the data. (c) Estimate the yield stress using the 0.2% offset method. That is, construct a line with the same slope as the initial, elastic portion of the curve, but starting from strain = 0.2% = 0.002. The 0.2% offset yield stress corresponds to the stress at which the line intersects the stress–strain curve. (Hint: you should also plot just the initial portion of the stress–strain curve with the strain axis spread out.)

Load (kN)	Elongation (mm)
0	0.0000
5.55	0.0180
15.95	0.0610
18.9	0.1030
20.45	0.1650
21.8	0.2500
26.7	1.0150
31.15	3.0500
32.25	6.3500
31.15	8.9000
29.4	11.9400

2.48 A compression test is conducted on a cylindrical concrete sample of 3 in. diameter and 6 in. length. Values of the compressive load and the specimen shortening are given in the table. (a) Calculate and plot the stress as a function of the strain. (b) Estimate the elastic modulus from the data.

Force (lb)	Shortening (in.)
0	0
1500	0.0003
2850	0.0006
4950	0.001
6150	0.0013
7650	0.0017
9000	0.002
10350	0.00225
11550	0.0025
13950	0.0031
15000	0.0035
15900	0.00375

2.49 A tensile test is conducted on an acrylic glass using a specimen with a gage length of 25 mm and a cross-section in the gage section that is 2 mm by 10 mm. Values of tensile load and elongation are given in the table. (a) Calculate and plot the stress as a function of the strain. (b) Estimate the elastic modulus from the data. (c) Estimate the yield stress using the 0.2% offset method. That is, construct a line with the same slope as the initial, elastic portion of the curve, but starting from strain = 0.2% = 0.002. The 0.2% offset yield stress corresponds to the stress at which the line intersects the stress–strain curve. (Hint: plot just the initial portion of the stress–strain curve with the strain axis spread out.)

Load (N)	Elongation (mm)
158	0.08
352	0.182
510	0.277
623	0.322
799	0.407
878	0.462
967	0.522
1075	0.649
1160	0.827
1242	1.072
1244	Fracture

2.50 A tensile test is conducted on a high-strength steel specimen with a diameter of 0.357 in. in the gage section and a gage length of 2 in. Values of tensile load and elongation are given in the table. (a) Calculate and plot the stress as a function of the strain. (b) Estimate the elastic modulus from the data. (c) Estimate the yield stress using the 0.2% offset method. That is, construct a line with the same slope as the initial, elastic portion of the curve, but starting from strain = 0.2% = 0.002. The 0.2% offset yield stress corresponds to the stress at which the line intersects the stress–strain curve. (Hint: plot just the initial portion of the stress–strain curve with the strain axis spread out.)

Load (lb)	Elongation (in.)
500	0.0002
1000	0.0006
3000	0.0019
5000	0.0034
6000	0.0038
6450	0.0043
6700	0.0047
6800	0.0055
6900	0.0063
7000	0.0091
7200	0.0102
7600	0.0129
8400	0.0231
9200	0.0334
10000	0.0505
11200	0.1106

2.51 A tensile test is conducted on a steel specimen, with the stress plotted vs. the strain in the Figure. (a) Estimate the elastic modulus from the plot. (b) Estimate the yield stress using the 0.2% offset method. That is, construct a line with the same slope as the initial, elastic portion of the curve, but starting from strain = 0.2% = 0.002. The 0.2% offset yield stress corresponds to the stress at which the line intersects the stress–strain curve.

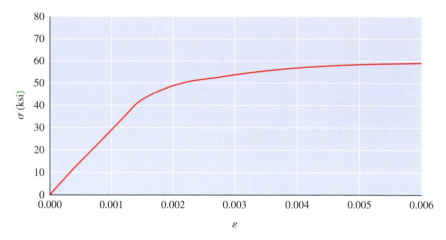

Prob. 2.51

2.52 A tensile test is conducted on an epoxy, with the stress plotted vs. the strain in the Figure. (a) Estimate the elastic modulus from the data. (b) Estimate the yield stress using the 0.2% offset method. That is, construct a line with the same slope as the initial, elastic portion of the curve, but starting from strain = 0.2% = 0.002. The 0.2% offset yield stress corresponds to the stress at which the line intersects the stress–strain curve.

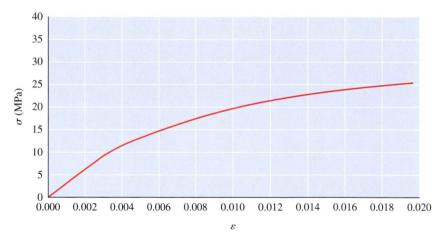

Prob. 2.52

2.53 An aluminum link is subjected to axial tension. The loading is limited by the conditions of not yielding and not elongating excessively. The yield limit is satisfied if the stress remains below 50% of the 325 MPa yield stress. The elongation must remain below 0.1 mm. The link cross-section is rectangular, 3 mm by 7 mm. Assuming elastic behavior, determine the maximum allowable tensile force for two cases: (a) the length of the link is 35 mm and (b) the length of the link is 55 mm.

2.54 A solid nylon cylinder of radius 8 mm and length 30 mm is to be compressed axially and still remain below 30% of the nylon's nominal yield strength of 90 MPa. (a) Determine the maximum allowable compressive force. (b) Determine the resultant shortening of the cylinder if the elastic modulus is 2.4 GPa.

2.55. A structural steel link that is 3 in. wide and 0.5 in. thick is to have a factor of safety of 2 relative to its yield strength of 36 ksi. (a) Determine the maximum allowable tensile force on the bar. (b) If the link is allowed to elongate by no more than 0.05 in. prior to reaching the maximum allowable tensile force, determine the maximum length of the link.

2.56 An aluminum tube is expected to carry up to a compressive force of 2000 lb and remain below 30% of the yield strength of 40 ksi. Commonly available tubes have a wall thickness of 0.058 in. with outer diameters in increments of 0.125 in. starting from 0.375 in. Find the tube of minimal diameter from among the available sizes that meets this requirement and determine the actual factor of safety of the stress relative to the yield strength of 40 ksi. Assume the tube does not buckle.

>>End Problems

2.9 Shear Strain and Shear Stress

We saw at the start of the chapter that a small square element deforms in general into a parallelogram. So far, we have only considered elements with normal strain, those that elongate or shorten, but remain rectangular. Now, we consider shear strain, which quantifies deformation in which element edges no longer form right angles.

1. Shearing corresponds to internal surfaces that slide parallel to themselves and elements that distort.

Take a rectangular block of rubber, attach metal plates to its top and bottom, and etch a grid of square elements on its side face.

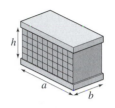

Move the upper plate *relative* to the lower plate in the direction *parallel* to the dimension a, maintaining the plates parallel. The etched grid distorts as shown here.

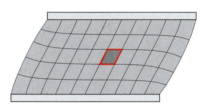

Not all elements deform the same, but elements located away from the left and right edges distort identically. Their lower and upper faces do not lengthen. The side faces are no taller, but they have tilted.

2. Shear strain is quantified by the angle change at an element corner when it is deformed.

Shear strain, γ, describes the element-level intensity of deformation due to shape change, and it is defined as the tangent of the angle change, due to deformation, between two lines that are originally perpendicular.

Here is the key feature of **shear strain**: element edges, which were originally perpendicular, make angles that are greater than and less than 90° or $\pi/2$ radians when deformed. Let β be the *change* from the initial right angle.

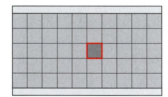

The greater is β, the greater is the distortion or shear of the element. The shear strain, γ, is defined in terms of the angular distortion β.

$$\gamma = \tan \beta$$

If β in radians is small ($\beta \ll 1$), then $\tan \beta \approx \beta$, so $\gamma \approx \beta$. Also, when β is small, the change in length of the vertical elements is negligible.

3. Shear strain can be related to the shear displacement divided by the thickness of the sheared layer.

Using trigonometry, we can find the angle β in terms of the relative shear displacement. Let u be the displacement (motion) of a plate in the shear direction. β is related to the relative shear displacement, Δu_{shear}, (i.e., $u_{top} - u_{bottom}$) and the height, h, perpendicular to the shearing.

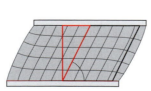

4. Besides two oppositely acting shear forces, there must be additional loads to balance moments.

While one can readily picture the shear strain, it is more difficult to picture the forces on the plates that shear the block. Here are several possibilities (only the second and third are in equilibrium).

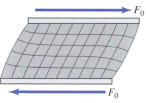

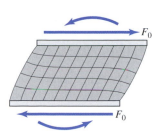

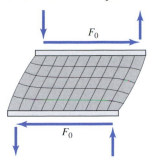

Transverse forces alone cannot be acting: the whole body would spin.

There could be a couple applied to each plate.

In particular, loads might be applied to the plate ends as shown.

5. Bodies that shear must have internal forces (shear stresses) acting on both horizontal and vertical planes.

Assume the forces to be as in the third case above. Consider the internal forces at a surface parallel to F_0.

There must be a horizontal shear force V equal to F_0. We call this a shear force because it acts parallel to, rather than normal to, the face on which it acts.

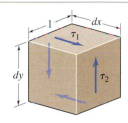

Now cut across the block vertically. There must be a vertical shear force, although its magnitude is unclear because it depends on how the moment is balanced.

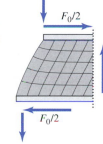

So elements that shear have forces on both horizontal and vertical faces.

Shear stress, τ, describes the element-level intensity of internal shear force, and it is defined as the shear force exerted by two adjacent elements on each other, divided by the area on which the force acts.

Shear stress, τ, is defined as the shear force per unit area.

The shear stress is rarely uniform, but the average value can be found in terms of the shear force, V, and the area, A, on which it acts.

$$\tau = \frac{V}{A}$$

For the sheared rubber block:

$$\tau = \frac{V}{A} = \frac{F_0}{ab}$$

6. Shear stress, or shear force per area on an element, must be equal on horizontal and vertical planes.

The shear stress just found is shown as τ_1 acting on the top and bottom faces of a small element. There are likewise equal and opposite shear stresses, τ_2, on the vertical faces. For this element to be in equilibrium, the moment about the center must be zero. If the element is dx by dy by 1 thick, then $\Sigma M = -\tau_1 (dx)(1)(dy) + \tau_2 (dx)(1)(dy) = 0$. So, the shear stresses must be equal: $\tau_1 = \tau_2$. A single shear stress, τ, describes the shear force per area on horizontal and vertical faces.

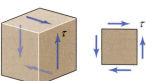

7. Shear stress is proportional to shear strain, if the material is elastic.

Shear modulus, G, captures the intrinsic stiffness of a material when it shears elastically, and it is defined as the proportionality between shear stress and shear strain.

For a linear elastic material, the shear strain is recovered upon unloading and it is proportional to the shear stress. **Shear modulus**, G, is defined as the proportionality between shear stress and shear strain for a linear elastic material.

$$\tau = G\gamma$$

>>End 2.9

2.10 Shear and Bearing Stress in Pin Joints ⎯⎯⎯⎯⎯⎯⎯

You have likely studied pin joints in Statics. Such pins commonly fail in shear, so it is important to be able to calculate the shear stress in pins.

1. Pins often connect two or more members which can exert transverse forces on the pin.

A pin often joins two or more members and allows them to pivot with respect to each other. The forces between members and the pin typically act in the direction perpendicular to the axis of the pin.

The pin joint in this elliptical exercise machine features a typical *symmetric* arrangement: the body that is forked or clevised (ski) straddles the second body (arm) and both bodies contact the pin.

2. Calculate the forces exerted by members on the pin using methods from Statics.

From equilibrium of the system, we would find the magnitude of the force, F_0, which the pin and center body (arm, here) exert on one another. (We would combine x and y components if the statics analysis was done with components.)

The pin is in equilibrium, so each half of the clevis applies equal forces of magnitude $F_0/2$ to the pin. We can draw the joint in the plane of the forces. We are interested in forces between the pin and the connected bodies and within the pin.

With this arrangement of transverse forces, the pin is said to be in "double shear."

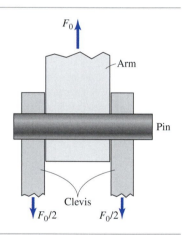

3. Bearing stresses are the forces per area at the surfaces where the members and the pin press on each other.

The members and the pin contact each other on cylindrical surfaces. We say the bodies bear (press) on each other. The forces per area, the contact or bearing stresses, are sometimes computed.

Here we show the contact force, $F_0/2$, between the pin and one side of the clevis as a distributed force. (Analogous contact forces act between the center member and the pin.)

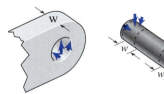

4. Calculation of bearing stress is highly approximate because the area of contact is estimated very roughly.

In general, the bearing stress $\sigma_{bearing}$, bearing force $P_{bearing}$ and bearing area $A_{bearing}$ are related by

$$\sigma_{bearing} = \frac{P_{bearing}}{A_{bearing}}$$

The bodies actually contact over less than the half-cylindrical surface $w\pi d/2$. The area of contact is usually estimated as the projected area wd. So the bearing stress between the clevis and the pin is approximately

$$\sigma_{bearing} = \frac{P_{bearing}}{A_{bearing}} = \frac{F_0/2}{wd}$$

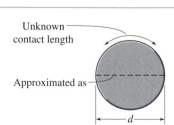

The stresses and strains due to such contact are highly nonuniform, and this estimate is very approximate. If such stresses were considered at all critical, they should be calculated with more accurate methods.

5. Shear stresses act between two portions of the pin across an internal plane perpendicular to the pin axis.

In general, we can look at any cross-section through the pin and determine internal forces and moments. The exaggerated deformation of the pin makes it clear that there is bending. Here we focus on one cross-section and in particular on the shear force at that cross-section. Later we will see why the shear force is maximum at this cross-section.

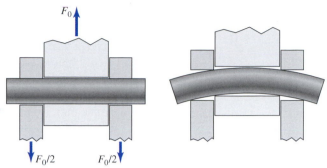

6. Draw a free body diagram of a portion of the pin on which external forces and the shear force act and calculate the shear force and then shear stress.

Consider the cross-section of the pin between the left half of the clevis and the arm. We find the internal loads in the pin at this cross-section by isolating a portion of the pin from one end to the cross-section. The internal force V is a shear force since it acts parallel to the internal cross-section. M is a bending moment, which we do not study further here.

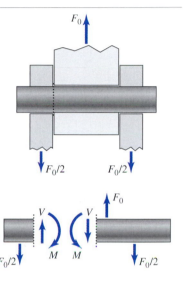

To balance transverse forces, the shear force V at the chosen cross-section must equal $F_0/2$. Again, this situation of a pin symmetrically loaded is referred to as "double shear."

V acts across the circular area $\pi d^2/4$. So the average shear stress is

$$\tau = \frac{V}{A} = \frac{V}{\frac{\pi}{4}d^2} = \frac{F_0/2}{\frac{\pi}{4}d^2}$$

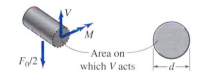

7. A pin on which only two opposing forces acts undergoes "single shear."

Another common configuration of a pin has only two members contacting the pin. The two forces exerted on the pin must be equal and opposite. For equilibrium, one or both members also exert a moment on the pin. To minimize the bending, the members are usually near each other and the pin is short.

By equilibrium, the shear force $V = F_0$. This loading of the pin is referred to as "single shear."

V again acts across the circular area $\pi d^2/4$. So the average shear stress is

$$\tau = \frac{V}{A} = \frac{V}{\frac{\pi}{4}d^2} = \frac{F_0}{\frac{\pi}{4}d^2}$$

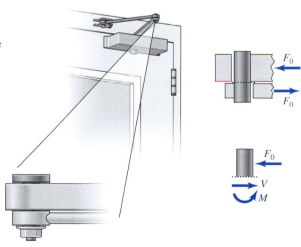

>>End 2.10

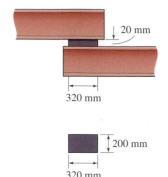

320 mm

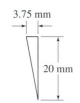

200 mm

320 mm

A bearing pad has been used to distribute more evenly the load between two steel beams. The pad is installed in an unstrained condition at normal temperatures. At a cold temperature, the upper beam displaces to the left by 2.50 mm, and the lower beam displaces to the right by 1.25 mm. (Movements are due to thermal expansion of the beams.) This relative movement of the two beams is resisted by shearing of the pad. Determine the shear force in the pad that resists this motion, assuming the pad has a shear modulus of 7 MPa and the dimensions shown. Draw the resisting forces on the beams.

Solution

The displacements of the steel beams cause displacements of the upper and lower faces of the pad to which the beams are attached. From these displacements, we find the relative shear displacement to be 3.75 mm.

2.50 mm

1.25 mm

3.75 mm

20 mm

The shear strain is then computed from the relative shear displacement and the pad thickness:

$$\gamma = \frac{\Delta u_{shear}}{h} = \frac{3.75 \text{ mm}}{20 \text{ mm}} = 0.1875$$

The shear stress is found from the shear strain and shear modulus:

$$\tau = G\gamma = (7 \text{ MPa})(0.1875) = 1.312 \text{ MPa}$$

The shear stress acts over the cross-sectional area of the pad, so the shear force is:

$$V = \tau A = (1.312 \text{ Pa})(0.2 \text{ m})(0.32 \text{ m}) = 84.0 \text{ kN}$$

The shear forces V on the pad are drawn here, in senses consistent with the shearing.

V

V

These shear forces are due to the steel beams on the pad, so the pad exerts equal and opposite forces on the beams. As expected, the resisting forces of the pad act opposite to the beam displacements.

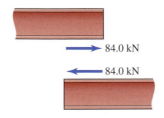

84.0 kN

84.0 kN

>>End Example Problem 2.11

In a leg curl, which exercises the hamstring muscles, the flexing leg lifts the weight arm to the position shown. The pin (actually a bolt) at A permits the rotation of the arm. Say the weight lifted is 100 lb and that the bent legs are oriented vertically and exert a horizontal force. Determine the shear stress in the 0.25 in. bolt at A.

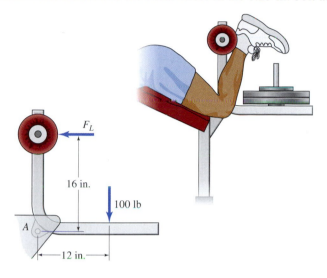

Solution

We apply equilibrium to the FBD of the weight arm, including the force of the *pin on the weight arm*:

$$\sum M|_A = -(100 \text{ lb})(12 \text{ in.}) + (F_L)(16 \text{ in.}) = 0, \text{ hence } F_L = 75 \text{ lb}$$

$$\sum F_x = -F_L + A_x = 0, \text{ hence } A_x = 75 \text{ lb}$$

$$\sum F_y = -100 \text{ lb} + A_y = 0, \text{ hence } A_y = 100 \text{ lb}$$

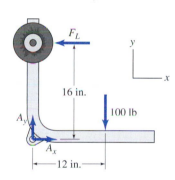

The force of the pin has been found in terms of x-y components for convenience. But, the pin actually exerts on the weight arm a net force of magnitude A.

$$A = \sqrt{A_x^2 + A_y^2} = \sqrt{(75 \text{ lb})^2 + (100 \text{ lb})^2} = 125 \text{ lb}$$

The arm exerts an equal and opposite 125 lb force on the pin. The brackets on the two sides of the arm exert forces F_{b1} and F_{b2} on the pin. For simplicity we have drawn the forces on the pin as acting horizontally, although the 125 lb force has both x- and y-components.

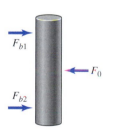

If the arm and bracket forces are applied symmetrically to the pin, then to satisfy equilibrium, $F_{b1} = F_{b2} = (125 \text{ lb})/2 = 62.5 \text{ lb}$.

Consider a portion of the pin from one end to a plane between the bracket that exerts the force F_{b1} and the arm force F_0. From equilibrium, the shear force is found to be $V = 62.5 \text{ lb}$.

The shear force acts over the circular cross-section of the pin. So the average shear stress in the pin at this cross-section is

$$\tau_{avg} = \frac{V}{A} = \frac{62.5 \text{ lb}}{\frac{\pi}{4}(0.25)^2} = 1273 \text{ psi}$$

Additional data on material properties needed to solve problems can be found in Appendix D or inside back cover.

2.57 The body deforms into the shape shown (not drawn to scale). Determine the shear strain.

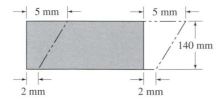

Prob. 2.57

2.58 The shear strain in the concrete slab must remain below 0.4%. The left side displaces downwards by 0.25 in. Determine the range of displacements, up and down, that can be permitted on the right side.

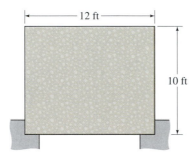

Prob. 2.58

2.59 A compliant layer separates two steel plates, *A* and *B*. Plate *A* is prevented from displacing. (a) Determine the allowable displacement of Plate *B* up or down if the shear strain in the layer is to remain below 12%. (b) Determine the allowable vertical force that can be applied to Plate *B* if the shear stress in the layer is to remain below 40 psi.

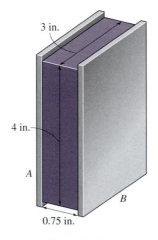

Prob. 2.59

2.60 Three flexible pads are each 10 in. wide into the plane and are sandwiched between four relatively rigid plates. The outer plates are kept from moving horizontally or tilting by rollers. Forces as shown are applied to the plates. Determine the shear stress in Pads *A*, *B*, and *C*. Draw the stresses for each pad on a two-dimensional element in the correct senses.

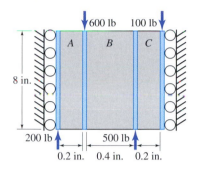

Prob. 2.60

2.61 Three flexible pads are each 10 in. wide into the plane and are sandwiched between four relatively rigid plates. The outer plates are kept from moving horizontally or tilting by rollers. Vertical forces are applied to the plates, causing shear strains in Pads *A*, *B*, and *C* of 10%, 15%, and 5% respectively, in the directions shown. If Plate 4 displaces downward by 0.04 in., determine the displacements of the other three plates.

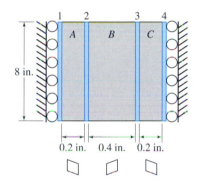

Prob. 2.61

2.62 Three flexible pads (each *G* = 100 psi and 10 in. wide into the plane) are sandwiched between four relatively rigid plates. The outer plates are kept from moving horizontally or tilting by rollers. Forces as shown are applied to the plates. If the plate between Pads *A* and *B* does not displace vertically, determine the magnitude and direction of the displacement of the other three plates.

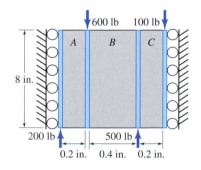

Prob. 2.62

It is often necessary to isolate the vibrations of a truck from delicate goods that it carries. A schematic of one such vibration isolation configuration is shown. Four small rubber pads, two on each side of the crate, connect the crate to its surroundings. The pads are 10 mm thick, have planar dimensions of 40 mm and 60 mm and centers separated by 400 mm, and shear modulus $G = 1.5$ GPa.

2.63 If the crate weighs 200 N (gravity acts along the y-axis), determine the shear stress in each of the four pads.

2.64 What is the permitted motion of the crate up in the direction of the y-axis relative to the zero shear strain position if the strain in each pad is not to exceed 8%?

2.65 If the shear stresses in the pads are to remain below 80 kPa, what is the maximum force that can be applied to the crate along the z-axis? (The 200 N weight of the crate acts as well.)

2.66 The crate is observed to displace by 3 mm along the z-axis when a force is applied in that direction. Determine the magnitude of the force on the crate.

2.67 The crate pivots about the x-axis by 0.3°. Determine the average shear strain in each pad. (Approximate the displacement using the center of each pad.)

2.68 A torque of 50 N-m about the x-axis acts on the crate. Determine the average shear stress in each pad.

2.69 Estimate the torque about the x-axis which is necessary to rotate the crate by 1° about that axis. From this value, determine the rotational stiffness, which is the torque per rotation angle (in radians).

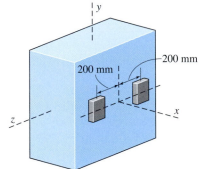

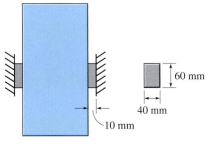

Probs. 2.63–69

2.70 A thin rubber sleeve is used to form a flexible coupling between a shaft and the base to which it is attached. The shaft and base are subjected to torques T as shown. The sleeve has a mean radius R, thickness t, length h, and a shear modulus G.

Treat the rubber sleeve as undergoing a uniform shear across its narrow thickness due to the different rotations of the shaft and the base. Treat the shaft and base as rigid.

Express the rotation of the shaft relative to the base, $\Delta\phi$, as a function of the applied torque T.

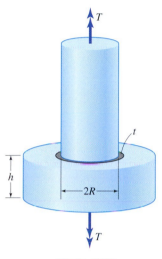

Prob. 2.70

2.71 The horizontal beam is supported by a pin at C and a rod attached to point A. Take $q = 1.1$ kip/ft. Determine the shear stress in (a) the pin at A and (b) the pin at C. Assume the pins have diameter of 0.5 in.

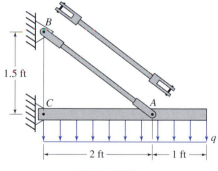

Prob. 2.71

2.72 The beam with hollow box cross-section is supported with a pin at A and a roller at B, where the pin diameter is 1.25 in. The outer dimensions of the box beam are 6 in. high by 8 in. wide and the wall thickness is 0.625 in. The load $F_0 = 3000$ lb. Determine (a) the bearing stress between the pin at A and the box beam and (b) the shear stress in the pin at A.

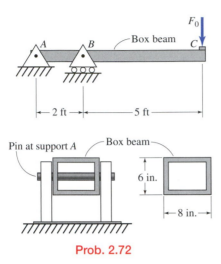

Prob. 2.72

2.73 The shear stress in bolts joining two shafts is not to exceed 4000 psi, under a maximum torque $T = 2 \times 10^4$ lb-in. Assume that bolts are available in diameters of 0.0625 in. increments. Determine the size of available bolts that can safely carry the torque. Assume that the pre-compression of the surfaces due to tightening the bolts has been released, so that the torque is transmitted purely by the bolts (no friction). Take $D = 8$ in.

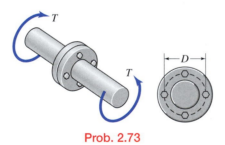

Prob. 2.73

2.74 Two hollow shafts are joined by a pin of diameter 4 mm. The shafts are subjected only to an axial force of P (torque $T = 0$). Find the force P if the shear stress is to remain below 35 MPa. Take $D_1 = 24$ mm, $D_2 = 26$ mm, and $D_3 = 28$ mm.

2.75 Two hollow shafts are joined by a pin. The shafts are subjected only to a torque $T = 50$ N-m (axial force $P = 0$). Determine the minimum diameter of the pin if the shear stress is to remain below 35 MPa. Take $D_1 = 24$ mm, $D_2 = 26$ mm, and $D_3 = 28$ mm.

2.76 Two hollow shafts are joined by a pin of diameter 4 mm. The shafts are subjected simultaneously to a torque $T = 50$ N-m and an axial force $P = 2400$ N. Determine the shear stress in the pin. Take $D_1 = 24$ mm, $D_2 = 26$ mm, and $D_3 = 28$ mm.

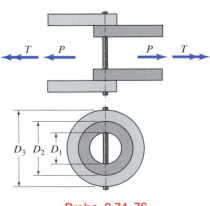

Probs. 2.74–76

2.77 One box beam is rigidly supported at A, and is pinned at B to a second box beam that fits within it. The second beam is also supported by rollers on its sides at C. Find the shear stress in the pins at B and at C. Take the pins to have a diameter 60 mm, $F_0 = 5000$ N, $\alpha = 60°$, $L_1 = 2$ m, $L_2 = 1.2$ m, and $L_3 = 1.6$ m.

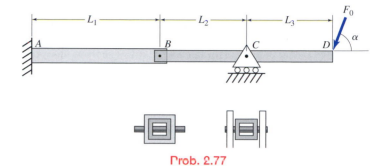

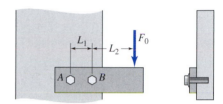

Prob. 2.77

2.78 A bar is bolted at two points to a larger plate, and then subjected to a transverse force F_0. Assume the bolts experience only vertical forces. Determine the shear stress in bolts A and B. Take the bolts to have a diameter 12 mm, $F_0 = 500$ N, $L_1 = 50$ mm, and $L_2 = 120$ mm.

Prob. 2.78

2.79 Consider two configurations for a long duct: (a) a duct of circular cross-section that is formed by screwing together two semicircular halves with flanges as shown and (b) a duct of square cross-section that is formed by two flat sheets that are screwed to two C-shaped parts as shown. Take the screws to be 8 mm in diameter, the ducts to have the dimensions as shown. The ducts handle an internal pressure of 100 kPa (over atmospheric outside). Screws are spaced 200 mm apart along the duct into the page. For (a) and (b) determine the type of load taken by the screw (shear or normal) and the stress magnitude. Take the pressure to load the screws in (a) by pushing apart the two halves along the dashed line and in (b) by pressing against the C-shaped parts.

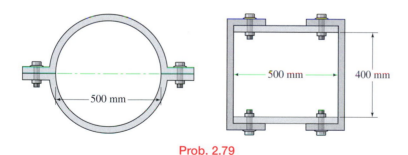

Prob. 2.79

2.80 A connection is formed as shown in which a bolt joins two plates to three plates. The plates are 6 mm thick and the bolt is 8 mm in diameter. Considering different cross-sections in the bolt, (a) determine the maximum shear stress in the bolt and (b) the maximum bearing stress against the bolt.

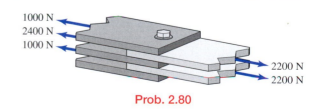

Prob. 2.80

2.81 Two plates are bolted together with the joining bracket. All bolts are 20 mm in diameter. Let the force $F_0 = 5000$ N. Describe the type of loading each bolt experiences. For bolts in tension, compute normal stress, and for bolts in shear, calculate the shear stress and the bearing stress.

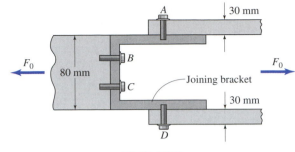

Prob. 2.81

Focused Application Problems

2.82 In this pectoral fly machine load is transmitted from the pivoting arm to the rotating plate via a pin. The relevant part of the force applied to the moving arm is drawn. Let this force $F_0 = 40$ lb. If the average shear stress in the pin is not to exceed 4000 psi, what must be the minimum diameter of the pin? Take the dimensions to be $L = 4$ in. and $w = 20$ in.

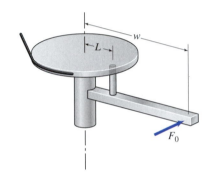

Prob. 2.82 (Appendix A4)

2.83 Occasionally, tired athletes will just hang from the swinging arm of this pectoral fly machine load. Say this results in a vertical F_0 250 lb force through the center of the swinging arm. Determine the shear stress in the 0.375 in. diameter pin that joins the swinging arm to the pivoting arm. Take the dimensions to be $L = 25$ in., $a = 1.5$ in., $b = 2$ in., and $t = 0.25$ in.

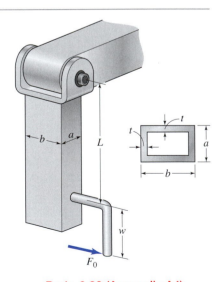

Prob. 2.83 (Appendix A4)

2.84 Pulleys are common in exercise equipment, where they redirect cords carrying tension. For the pulley shown that carries a tension $T = 120$ lb, determine the shear stress in the pin on which the pulley rotates. Take the dimensions to be $d = 0.375$ in., $R = 2$ in., and $w = 2.5$ in.

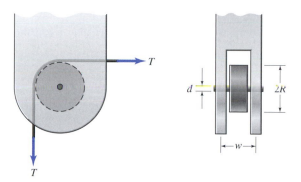

Prob. 2.84 (Appendix A4)

2.85 Occasionally, tired athletes will just hang from the peg on the swinging arm of this pectoral fly machine load. Say this results in a vertical 250 lb force on the swinging arm, but offset by the distance $L = 5$ in. Determine the shear stress in the 0.375 in. diameter pin that joins the swinging arm to the rotating arm. Take $c = 2.5$ in.

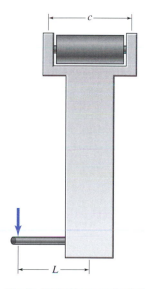

Prob. 2.85 (Appendix A4)

>>End Problems

Chapter Summary

2.1 Study tendency for forces to cause failure or excessive deformation in a body of any shape and material.

View body as composed of identical, infinitesimal cubic regions termed **elements**.

Elements enable us to separate out the effect of material from shape and to predict failure, which usually occurs locally.

Describe force and deformation at the level of an element.

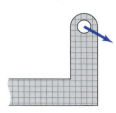

2.2 Common deformations studied in *Mechanics of Materials*—stretching, twisting, and bending—typically occur in elongated, prismatic members with an axis and a cross-section.

To find element-level forces and deformations in prismatic members, consider the member as separated into two parts by a plane perpendicular to the axis.

Find **internal forces/moments** (P, V, M) acting between the two parts separated by the plane.

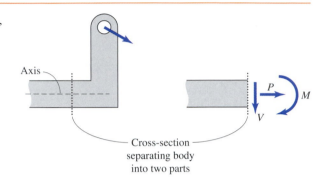

Axis

Cross-section separating body into two parts

2.3 **Normal Force P:** Internal force acting perpendicularly to cross-section with area A

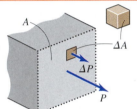

Normal Stress σ: Quantifies force at level of element

$$\sigma = \frac{\text{Force on element face}}{\text{Area of element face}} = \frac{\Delta P}{\Delta A}$$

2-D view (cross-section of area A only seen edge-on)

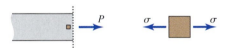

Under certain conditions, stress uniform over cross-section:

$$\sigma = \frac{P}{A}$$

2.4 **Elongation δ:** Change of length of a segment originally of length L

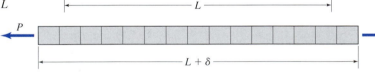

Normal Strain ε: Quantifies deformation at level of element

$$\varepsilon = \frac{\text{Elongation of element}}{\text{Length of element}}$$

Under conditions in which σ and material are uniform, δ is proportional to L, and so ε is uniform.

$$\varepsilon = \frac{\delta}{L}$$

2.5 Relation between stress and strain depends only on the type of material, not on the size of the body.

Testing machine pulls material specimen in tension and measures **P and δ**.

In central region of specimen, stress and strain are uniform.

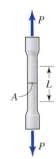

Compute σ from P and cross-sectional area A.

Compute ε from δ and length L.

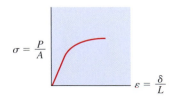

$$\sigma = \frac{P}{A}$$

$$\varepsilon = \frac{\delta}{L}$$

2.6   **Linear Elastic Deformation:** Deformation increases in proportion to load and returns to zero when the load is removed.

When strains are very small, deformation is linear elastic for most materials.

Young's Modulus E: Proportionality between normal stress and normal strain when deformation is linear elastic (slope of σ–ε curve)

Material also strains in direction perpendicular to load. Define each strain using length of body parallel to change in length.

Poisson Ratio (ν): Ratio of transverse strain, ε_t (perpendicular to load) to longitudinal strain, ε (parallel to load).

$$\nu = \frac{|\varepsilon_t|}{|\varepsilon|}$$

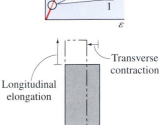

Longitudinal elongation

Transverse contraction

2.7–2.8   There are two common types of failure due to tensile stress: ductile and brittle. We usually design parts so stress remains well below failure stress.

Ductile Failure

Material does not return to initial shape upon unloading, if stress exceeds the **Yield stress**, which is reached just above point where σ–ε curve becomes nonlinear.

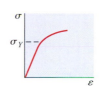

Brittle Failure

Material remains elastic up to **Fracture stress** at which material breaks.

2.9–2.10   Stress and strain can also capture distortion of element rather than elongation or shortening.

Shear Strain γ: Distortional deformation of element that depends on change, β, from right angle formed by element edges
Shear Stress τ: Force per area acting parallel to internal surface, acts on four faces of element
Shear Modulus G: Proportionality between shear stress and shear strain when material is linear elastic

$$\gamma = \tan\beta = \frac{\Delta u_{shear}}{h} \qquad \tau = \frac{V}{A} = \frac{F_0}{ab} \qquad \tau = G\gamma$$

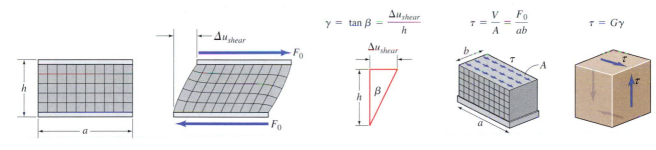

Transverse forces on a pin that joins bodies cause bearing stress between the pin and the bodies and shear stress in the pin.

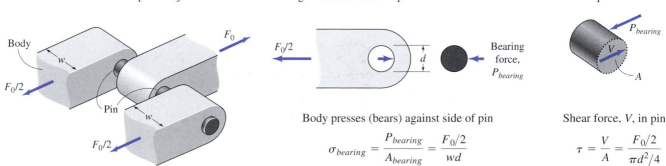

Body presses (bears) against side of pin

$$\sigma_{bearing} = \frac{P_{bearing}}{A_{bearing}} = \frac{F_0/2}{wd}$$

Shear force, V, in pin

$$\tau = \frac{V}{A} = \frac{F_0/2}{\pi d^2/4}$$

Axial Loading

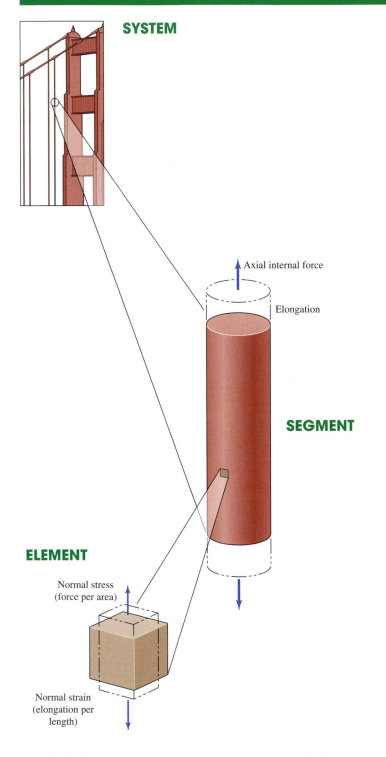

LEVELS OF VIEWING DEFORMATION AND FORCE

SYSTEM

Axial internal force

Elongation

SEGMENT

ELEMENT

Normal stress (force per area)

Normal strain (elongation per length)

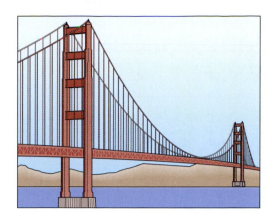

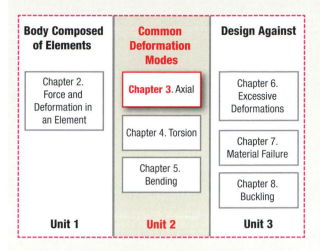

Body Composed of Elements	Common Deformation Modes	Design Against
Chapter 2. Force and Deformation in an Element	**Chapter 3**. Axial	Chapter 6. Excessive Deformations
	Chapter 4. Torsion	Chapter 7. Material Failure
	Chapter 5. Bending	Chapter 8. Buckling
Unit 1	Unit 2	Unit 3

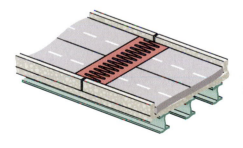

Multiple forces due to players result in different internal axial forces in successive segments of the rope.

Displacements may not be directed along a truss member even though it is in tension or compression.

Temperatures changes cause elongation and contraction of roadway, and expansion joints allow displacements to occur.

Closed loop or band is placed in tension and elongated due to outward pressure from arm.

Chapter Outline

A cable deforming due to the weight of a bridge deck is an example of *axial deformation*. An axially loaded member can be viewed as a series of segments along the length of the member. External forces result in an axial internal force, tension or compression, acting oppositely on the two faces of a segment. Axial internal force results in normal stress that is equal in all elements of the segment. The associated normal strain is also uniform in all elements of the segment (**3.1**). Internal forces can vary from one segment of the member to the next, and the elongation of any segment depends on its length, elastic modulus, and cross-sectional area (**3.2**). Axial deformations can occur under more complex situations, such as a truss or collection of connected, axially deforming members. Such systems can be analyzed by inter-relating the deflections of members to reflect their connections and by accounting for the constraints placed by supports on deflections (**3.3–3.4**). Besides being driven by forces, axial deformations are also caused by temperature changes (**3.5**). Axial deformations also occur in bands and rings wrapping around other bodies that exert radial outward forces (**3.6**).

3.1 Internal Force-Deformation-Displacement ─────

Members in simple axial loading are either stretched or compressed in their long direction. The forces at the ends can be related to the displacements of points on the member.

1. A cord that elongates due to tensile forces at its ends is one simple case of axial loading.

Strings in musical instruments display the principal quantities in axial loading. The internal force in the string, the tension, affects the pitch of the tone. Raise the tension, and hence the tone, by turning the tuning keys and making the string longer.

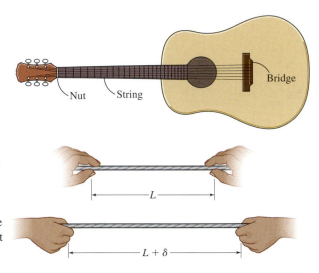

The fingers keep this bungee cord straight at length L, without applying any force.

Now the fingers apply forces that keep the cord stretched at a greater length, $L + \delta$. We seek to determine how the properties of the cord affect the relation between the force and the elongation δ.

2. A long straight member, which could withstand more complex loadings, is in axial loading if the forces act along its length.

Cords or cables can only be in equilibrium if the forces act along their lengths and they produce tension. For members that can handle other loads as well, statics or sometimes other arguments are needed to determine if the loading is axial. Each link in this wheel puller sustains axial loading because it is a two-force member; the line joining the two pins, which are modeled as frictionless, runs through the center of the link.

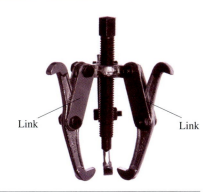

Link   Link

3. Failure and deformation are tied to the internal force.

Stress and deformation are determined by forces acting internally. We find the internal force P from equilibrium of a portion of the body.

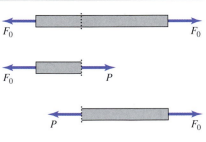

For this simple case, the internal force P is equal in magnitude to F_0. So if the force on each end is $F_0 = 100$ N, the internal force is $P = 100$ N (not 200 N). In more complex situations in which several external forces act, one must take care in relating internal and external forces.

4. Combine relations between force, stress, strain, and elongation to relate the internal force to the elongation.

As we saw in Chapter 2, the stress σ and strain ε are uniform in a bar of uniform material (elastic modulus E) and cross-section (area A) that is subjected to axial forces. We studied these relations:

Stress is related to internal force	Strain is related to elongation and bar length L	Stress is related to strain if the material is elastic
$\sigma = \dfrac{P}{A}$	$\varepsilon = \dfrac{\delta}{L}$	$\sigma = E\varepsilon$

Combine these equations to relate the elongation, δ, to the internal force, P: $\quad \delta = \varepsilon L = \dfrac{\sigma}{E}L = \dfrac{PL}{EA}$

$$\delta = \frac{PL}{EA} \quad \text{or} \quad P = \frac{EA}{L}\delta$$

5. A bar under axial loading is analogous to an elastic spring.

The force-elongation relation of a bar, $P = \left(\dfrac{EA}{L}\right)\delta$, is analogous to the spring law studied in physics:

$$F = kx \qquad F = \text{force}, \ k = \text{spring constant or stiffness}, \ x = \text{displacement}$$

EA/L is analogous to the spring constant k. We call EA/L the axial stiffness of a bar of length L.

We apply $P = \dfrac{EA}{L}\delta$ carefully:
$\quad P = $ internal force in a segment, not the external force
$\quad \delta = $ elongation of a segment, not the displacement of one point

To use this relation, E, A, and P must be uniform over the segment of length L.

---

6. Define the displacement of a point on a body.

Displacement, u or v, is defined as the distance through which a point on a body moves in some direction, due to applied loads.

Individual points on a body move possibly by different amounts in response to forces. We define the **displacement** u as the horizontal distance moved by a point. We define v as the vertical displacement. The sign of u signifies whether the displacement is left or right (v, up or down).

---

7. Relate the elongation or change in length δ of a segment to the difference in displacements at its ends.

A bar with end points A and B is shown first in its unloaded configuration. The bar is uniformly strained by equal and opposite forces (not drawn). Displacements of the end points are given in the table at the right. Consider the relation between u at the two ends and the change in length, δ. Here choose the signs to mean: $u > 0$ to right, $u < 0$ to left, $\delta > 0$ elongates, $\delta < 0$ contracts.

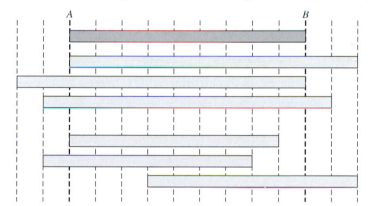

u_A	u_B	δ
0	2	2
-2	0	2
-1	1	2
0	-1	-1
-1	-2	-1
3	2	-1

One can see that the displacement at one point and δ are not uniquely related to each other. Instead, δ is equal to the difference in u at the two end points. For the signs used above: $\delta = u_B - u_A$.

We choose the meanings of the signs for displacement and elongation as convenient for a particular problem. Depending on the signs, the relation $\delta = u_B - u_A$ or $\delta = u_A - u_B$ will be correct.

---

8. Use the relations between elongation and displacements to find displacement at any point along a bar.

Use the same relation between δ and u to find the displacement at any point along the length of a uniformly strained bar. Given displacements of points A and B, what is the displacement of point C?

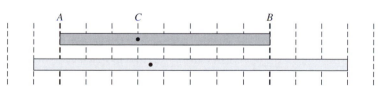

Strain is uniform: $\quad \varepsilon = \delta_{AB}/L = (u_B - u_A)/L = [3 - (-1)]/8 = 4/8 = 0.5$ (large strain is used for illustration)

Elongation of AC: $\quad \delta_{AC} = (L_{AC})(\varepsilon) = (3)(0.5) = 1.5$

Displacement of C: $\delta_{AC} = u_C - u_A \Rightarrow u_C = \delta_{AC} + u_A = 1.5 + (-1) = 0.5$

>>End 3.1

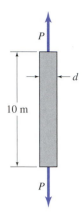

P

d

10 m

P

A length of cable that is 10 m long must withstand a maximum load of 400 N without reaching 25% of the yield stress or stretching by more than 10 mm. Determine the allowable diameter if these conditions are both to be satisfied. Consider two cases: (a) steel with $\sigma_Y = 550$ MPa and $E = 165$ GPa; (b) nylon with $\sigma_Y = 40$ MPa and $E = 2.1$ GPa.

Solution

The axial stress in the cable satisfies: $\sigma = \dfrac{P}{A}$

The elongation in the cable satisfies: $\delta = \dfrac{PL}{EA}$

Steel

At the maximum load of $P = 400$ N, the yield stress is reached when the area satisfies

$$A = \frac{P}{\sigma_Y/4} = \frac{(400 \text{ N})}{(550 \times 10^6 \text{ Pa})/4} = 2.91 \times 10^{-6} \text{ m}^2$$

At the maximum load of $P = 400$ N, the elongation of 10 mm is reached when the area satisfies

$$A = \frac{PL}{E\delta} = \frac{(400 \text{ N})(10 \text{ m})}{(165 \times 10^9 \text{ Pa})(0.01 \text{ m})} = 2.42 \times 10^{-6} \text{ m}^2$$

Notice that for a given axial force P, the stress and the elongation both increase as the area decreases. So A must exceed the values calculated.

$\Rightarrow$ With an area of less than 2.42×10^{-6} m^2, the elongation condition would be violated. With an area of less than 2.91×10^{-6} m^2, the yield condition would be violated.

$\Rightarrow$ If $A = 2.91 \times 10^{-6}$ m^2, both conditions are satisfied.
   (Note that the yield condition, not the elongation condition, limits the area.)

$$d = \sqrt{\frac{4A}{\pi}} = 1.925 \text{ mm}$$

Nylon

At the maximum load of $P = 400$ N, the yield stress is reached when the area satisfies

$$A = \frac{P}{\sigma_Y/4} = \frac{400 \text{ N}}{(40 \times 10^6 \text{ Pa})/4} = 4 \times 10^{-5} \text{ m}^2$$

At the maximum load of $P = 400$ N, the elongation of 10 mm is reached when the area satisfies

$$A = \frac{PL}{E\delta} = \frac{(400 \text{ N})(10 \text{ m})}{(2.1 \times 10^9 \text{ Pa})(0.01 \text{ m})} = 1.905 \times 10^{-4} \text{ m}^2$$

$\Rightarrow$ In this case, the elongation limits the area

$\Rightarrow$ If $A = 1.905 \times 10^{-4}$ m^2 $\Rightarrow d = \sqrt{\dfrac{4A}{\pi}} = 15.57$ mm

>>End Example Problem 3.1

Rubber disks are used to provide cushioning for equipment subjected to vibrations. The vibration analysis indicates that the rubber disk should provide an effective spring constant of 600 lb/in. Spacing considerations dictate that the disk should be 0.5 in. tall. Take the rubber to have a Young's modulus of 400 psi. (a) Determine the desired diameter of the disk. (b) If the rubber strain is not to exceed 0.2, determine the maximum allowable compressive force.

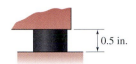

Solution

The axial stiffness of member is given by $\quad k = \dfrac{P}{\delta} = \dfrac{E A}{L}$

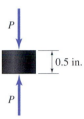

The modulus E and the length of the compressed member are given, so we can solve for the disk area and hence diameter:

$$A = \frac{kL}{E} = \frac{(600 \text{ lb/in.})(0.5 \text{ in.})}{(400 \text{ lb/in.}^2)} = 0.75 \text{ in.}^2$$

So the diameter d is:

$$d = \sqrt{\frac{4A}{\pi}} = \sqrt{\frac{4(0.75 \text{ in.}^2)}{\pi}} = 0.977 \text{ in.}$$

The allowable force can be found in the following two ways:

From the maximum rubber strain (compressive) of 0.2, the maximum stress is

$$\sigma = E\varepsilon = (400 \text{ psi})(0.2) = 80 \text{ psi}$$

Then the allowable force is given by:

$$P = \sigma A = (80 \text{ psi})(0.75 \text{ in.}^2) = 60 \text{ lb}$$

Alternatively, the allowable compression is

$$\delta = \varepsilon L = (0.2)(0.5 \text{ in.}) = 0.1 \text{ in.}$$

Then the allowable force is given by:

$$P = k\delta = (60 \text{ lb/in.})(0.1 \text{ in.}) = 60 \text{ lb}$$

>>End Example Problem 3.2

Additional data on material properties needed to solve problems can be found in Appendix D or inside back cover.

3.1 Piles are used to support structures on soft soil. In some cases, the lower ends of piles rest on firm ground (end-bearing piles). Load from the soil to the side of the pile is then often neglected, and the pile is just compressed by forces at its two ends. The stiffness of a pile is important since it affects the deflection of the structure. A particular site allows for piles of length $L = 30$ ft. The loads on the structure and the arrangement of piles call for the stiffness of individual piles to be at least 5.5×10^5 lb/in. Steel pipes with 0.25 in. wall thickness are available in outer diameters in increments of 0.5 in. Determine the minimum pipe outer diameter that will lead to sufficient stiffness.

Prob. 3.1

3.2 The frequency f in Hz (s^{-1}) at which a string vibrates is related to the tension T, the length L, and the mass per length μ, according to:

$$f = \frac{1}{2L}\sqrt{\frac{T}{\mu}}$$

Consider the "regular light" steel high E string on a guitar, which is 0.254 mm in diameter. The guitar is tuned to 329.6 Hz. (a) If the guitar has a base length of 0.65 m, determine the necessary tension in the string. (b) Let the machine heads that tune the guitar have a 14:1 turning ratio (14 turns of the head produces one rotation of the pin around which the string wraps), and a pin diameter of 4 mm. Say that the string is already wrapped around the pin five times, and make the simplifying assumption that the tension is carried equally along the length of the string (0.65 m plus the wrapped part). How many turns will be necessary to raise the string to a pitch of F at 349.2 Hz?

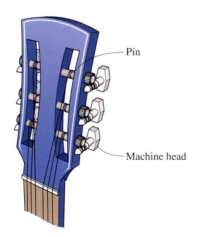

Prob. 3.2

3.3 High internal loads occur in the deck of a cable-stayed bridge near the pylons if the deck
rests directly on the pylons. Instead, a relatively soft bearing may be placed between the deck
and the pylon. For a bridge with a single row of cables supporting the center of the deck, say
the bearing is to have the same stiffness as one of the nearest cables, which are 0.01 m in
diameter, 60 m long, and approximately vertical. The effective modulus of steel in cables
is 165 GPa. Determine the necessary bearing stiffness in kN/m.

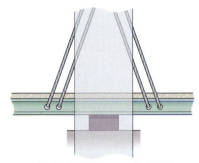

Prob. 3.3 (Appendix A3)

3.4 Unlike cables attached to the deck center, cables attached to the sides of the deck also serve
to resist rotation of the deck about its long axis. We simplify the problem and approximate
the cables as vertical, with length $L = 200$ m and cross-sectional area 0.01 m^2. The cables
are attached at the edges of the 35 m wide deck. Let the portion of the deck in question rotate
by $\phi = 2°$. Determine the resisting moment due to the pair of cables. The effective modulus
of steel cables is 165 GPa. Hint: the cables are already in tension, but the additional force
due to deck rotation on one cable is compressive and the other is tensile, with the pair
of additional forces causing the moment.

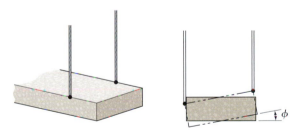

Prob. 3.4 (Appendix A3)

>>End Problems

3.2 Varying Internal Force ———————————

So far, the body has been loaded axially with two equal and opposite forces at its ends. We now consider situations in which a bar is loaded axially at more than two points.

1. Several axial forces act at different points along a member in some applications.

A member in axial loading may not simply be loaded by one force at each end. Rather, there can be loads acting at intermediate points along the length. For example, this vertical post has axial loads due to the weight of each floor beam, and a support reaction from the ground.

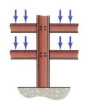

2. When multiple axial forces act on a body, the deformation can be different between each successive pair of forces.

Hands that grip this bungee cord at points A, C, and B apply forces that deform the cord.

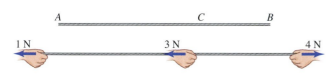

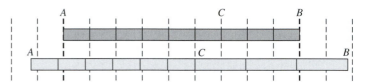

To illustrate the deformation of a cord, we have drawn lines that divide the cord into segments that are initially of equal length. The five segments between A and C elongate equally, and the three segments between C and B elongate equally. Between A and C, the strain, stress, and internal force are uniform, and they differ from the uniform values between C and B.

3. Find the internal force at any cross-section by imposing equilibrium on an FBD that includes a portion of the body ending at the cross-section.

Each portion of the cord must be in equilibrium. We can relate the internal forces to the external forces by imposing equilibrium on a portion of the cord.

Think of the cord as composed of two portions: from A to some point between A and C and from that point to B. The two portions exert equal and opposite internal forces on each other where they join. For example, P_{AC} denotes the internal force between two portions that meet at a cross-section between A and C.

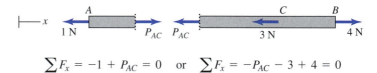

$$\sum F_x = -1 + P_{AC} = 0 \quad \text{or} \quad \sum F_x = -P_{AC} - 3 + 4 = 0$$

Now think of the two portions joining at some point between C and B.

$$\sum F_x = -1 - 3 + P_{CB} = 0 \quad \text{or} \quad \sum F_x = -P_{CB} + 4 = 0$$

The internal force, $P_{AC} = 1$ N, is uniform from A to C, and $P_{CB} = 4$ N is uniform from B to C.

4. Because the internal force is uniform, but different in each of the segments AC and CB, we calculate the stress and elongation for each of those two segments separately.

Since the internal force, area, and material are uniform within the segments AC and CB, we can apply relations for uniformly axially loaded bodies to each separately:

$$AC: \quad \sigma_{AC} = \frac{P_{AC}}{A}, \delta_{AC} = \frac{P_{AC} L_{AC}}{EA} \qquad CB: \quad \sigma_{CB} = \frac{P_{CB}}{A}, \delta_{CB} = \frac{P_{CB} L_{CB}}{EA}$$

5. Combine successive relations between external and internal forces, stress, strain, elongation, and displacement as needed.

When there are several axial forces along the length, our goal is often to find stress and displacements everywhere. The basic formulas that relate internal force, stress, strain, and elongation are applicable when P, A, and E are constant along a segment.

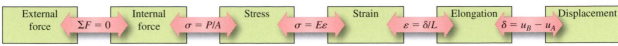

| External force | $\Sigma F = 0$ | Internal force | $\sigma = P/A$ | Stress | $\sigma = E\varepsilon$ | Strain | $\varepsilon = \delta/L$ | Elongation | $\delta = u_B - u_A$ | Displacement |

6. Recognize tension vs. compression, and use the sign of P to distinguish them.

Tension: all forces in outward sense on face

Compression: all forces in outward sense on face

Use $P > 0$ for tension and $P < 0$ for compression, e.g., $P = -2$ kN means 2 kN in compression.

The direction of the internal force in a segment may be hard to guess in advance. To simplify our book-keeping, we usually draw the senses of unknown internal forces assuming they are in tension. When we eventually solve, if $P > 0$, the internal force is tensile, and if $P < 0$, the internal force is compressive.

7. A bar with multiple forces is analyzed by decomposing it into segments in which P, E, and A are uniform.

To illustrate the procedure: consider one bar with a change in cross-section that is attached rigidly to a base. A bar of different material is screwed into the first. External forces are applied to the bars at the points shown.

Identify segments in which internal force P, E, and A are all constant.
Do this by recognizing and labeling points where something changes.

Point	B	C	D	E
Change	Int. Force	Modulus	Area	Int. Force

Label segments as 1, 2, 3, 4, and 5 (could use AB, BC, CD, DE and EF instead).

For each segment, find P from an FBD extending from one end to a cross-section in that segment. Unknown P is drawn so it is in tension, if positive.

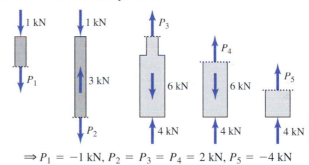

$$\Rightarrow P_1 = -1 \text{ kN}, P_2 = P_3 = P_4 = 2 \text{ kN}, P_5 = -4 \text{ kN}$$

Compute the stretch of each segment using P, L, E, and A appropriate to each:

Stretches: $\quad \delta_1 = \dfrac{P_1 L_1}{E_1 A_1} \quad \delta_2 = \dfrac{P_2 L_2}{E_2 A_2} \quad \cdots \quad \delta_5 = \dfrac{P_5 L_5}{E_5 A_5}$

Relate displacements (take $v > 0$ upward) to stretches

$$\delta_5 = v_E - v_F, v_F = 0 \Rightarrow v_E = \delta_5,$$
$$\delta_4 = v_D - v_E \Rightarrow v_D = \delta_4 + v_E = \delta_4 + \delta_5, \text{ and so forth.}$$

8. Integrate to find elongation when P, E, or A vary continuously along the bar.

The area or modulus may vary continuously along a bar (with x). Also, when external forces are applied in a distributed manner, the internal force varies with x: $P(x)$. We assume the same relation between force, stress, and elongation still applies to a slice of length dx. So the elongation of a segment from cross-section, x_A, to another, x_B, is found from integration:

$$\sigma(x) = \frac{P(x)}{A(x)} \qquad d\delta = \frac{P(x)dx}{E(x)A(x)} \qquad \delta = \int_{x_A}^{x_B} \frac{P(x)}{E(x)A(x)}dx$$

>>End 3.2

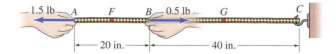

The bungee cord consists of rubber with a modulus of 700 psi and a diameter of 0.48 in. The cord is fixed at the right end C, and forces are applied to the cord as shown at points A and B. Determine (a) how the stress varies in the cord, and (b) the motion of the points F and G located midway from A to B and B to C.

Solution

From summation of horizontal forces, the support at C exerts a 1 lb force to the right. The internal force is uniform in the cord between A and B, and between B and C. The force in the two segments can be found by isolating a portion of the cord as follows:

These free body diagrams can be used to find the internal force between A and B:

$$\sum F_x = -1.5 \text{ lb} + P_{AB} = 0 \quad \sum F_x = -P_{AB} + 0.5 \text{ lb} + 1 \text{ lb} = 0 \Rightarrow P_{AB} = 1.5 \text{ lb}$$

These free body diagrams can be used to find the internal force between B and C:

$$\sum F_x = -1.5 \text{ lb} + 0.5 \text{ lb} + P_{BC} = 0 \quad \sum F_x = -P_{BC} + 1 \text{ lb} = 0 \Rightarrow P_{BC} = 1 \text{ lb}$$

$$A = \frac{\pi}{4}d^2 = \frac{\pi}{4}(0.25 \text{ in.})^2 = 0.0491 \text{ in.}^2 \quad EA = (700 \text{ in.})(0.0491 \text{ in.}^2) = 34.4 \text{ lb-in.}^2$$

$$\sigma_{AB} = \frac{P_{AB}}{A} = \frac{1.5 \text{ lb}}{0.0491 \text{ in.}^2} = 30.6 \text{ psi} \quad \sigma_{BC} = \frac{P_{BC}}{A} = \frac{1 \text{ lb}}{0.0491 \text{ in.}^2} = 20.4 \text{ psi}$$

Because P, E, and A are uniform in each segment, to find stretches we use: $\delta = \dfrac{PL}{EA}$

$$\delta_{AB} = \left(\frac{PL}{EA}\right)_{AB} = \frac{(1.5 \text{ lb})(20 \text{ in.})}{(34.4 \text{ lb-in.}^2)} = 0.873 \text{ in.} \quad \delta_{BC} = \left(\frac{PL}{EA}\right)_{BC} = \frac{(1 \text{ lb})(40 \text{ in.})}{(34.4 \text{ lb-in.}^2)} = 1.164 \text{ in.}$$

The internal force is uniform in AB and $L_{AF} = 0.5 \, L_{AB}$, so $\delta_{FB} = \delta_{AB}/2 = 0.436 \text{ in.}$; likewise $\delta_{GC} = \delta_{BC}/2 = 0.582 \text{ in.}$

$\delta_{GC} = u_C - u_G$ and $u_C = 0 \Rightarrow u_G = -0.582 \text{ in.}$

So G moves to the left by 0.582 in.

$\delta_{BC} = u_C - u_B$ and $u_C = 0 \Rightarrow u_B = -1.164 \text{ in.}$

$\delta_{FB} = u_B - u_F \Rightarrow u_F = -1.164 \text{ in.} - 0.436 \text{ in.} = -1.600 \text{ in.}$

So F moves to the left by 1.600 in.

>>End Example Problem 3.3

For a tall column resting on the ground, axial stress develops due to the column's weight. (a) Derive the stress as a function of position z measured from the bottom of the column, and the total contraction of the column. Express the results in terms of the column area A, mass density ρ, modulus E, and height H. (b) Evaluate the maximum stress for a case of a steel tower that is 70 m tall and has a 3.4 m outer diameter and wall thickness of 25 mm. (c) Redo the calculation of stress at the bottom of the tower for the case of the tower with wall thickness 25 mm, but with outer diameter that varies linearly from 4.4 m at the bottom to 2.4 m at the top.

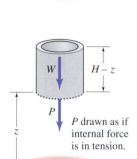

Solution

We are concerned with axial stress in the column due to its weight. External forces are due to the weight and the ground. We isolate a portion of the tower from a height z to the top. The forces on this segment are its weight W and the internal force P.

$$W = \rho g \text{ (Volume)} = \rho g A(H - z)$$

From $\sum F_z = 0$, $P = -W \Rightarrow \qquad \sigma = \dfrac{P}{A} = \dfrac{-W}{A} = -\rho g(H - z)$

The stress is compressive and independent of the area. The stress approaches 0 at $z = H$ (at top). The maximum stress is at $z = 0$: $|\sigma_{\max}| = \rho g H$

We can find the contraction of the whole tower by integration:

$$\delta = \int_0^H \frac{P(z)}{EA} dz = \int_0^H \frac{-\rho g A(H - z)}{EA} dz = \frac{\rho g (H - z)^2}{2E}\bigg|_0^H = \frac{-\rho g H^2}{2E}$$

For the given dimensions, we find:

$$|\sigma_{\max}| = \rho g H = \left(7850\frac{\text{kg}}{\text{m}^3}\right)\left(9.81\frac{\text{m}}{\text{s}^2}\right)(70\,\text{m}) = 5.39 \text{ MPa}$$

$$|\delta| = \frac{\rho g H^2}{2E} = \frac{\left(7850\dfrac{\text{kg}}{\text{m}^3}\right)\left(9.81\dfrac{\text{m}}{\text{s}^2}\right)(70\,\text{m})^2}{2(200 \text{ GPa})} = 0.943 \text{ mm}$$

For a tower with varying cross-section (typical of a wind turbine), we reconsider the force:

At bottom $\qquad P = -W = \rho g \text{ (Volume)} \qquad \sigma_{bottom} = \dfrac{P}{A_{bottom}} = \dfrac{-W}{A_{bottom}}$

To find the weight, we approximate the wall as thin compared to the diameter:

$Area = \pi(\text{diameter})(\text{wall thickness})$

$$\text{Area} = \pi\left[4.4 \text{ m} - 2 \text{ m}\left(\frac{z}{70 \text{ m}}\right)\right](0.025 \text{ m}) \quad \text{(Area varies linearly)}$$

$$Volume = \int (\text{Area})dz = \frac{1}{2}[\text{Area}|_{top} + \text{Area}|_{bottom}](H) = \frac{\pi}{2}[4.4 \text{ m} + 2.4 \text{ m}](0.025 \text{ m})(70 \text{ m}) = 18.69 \text{ m}^3$$

$$|\sigma_{bottom}| = \frac{|P|}{A_{bottom}} = \frac{W}{A_{bottom}} = \frac{\left(7800\dfrac{\text{kg}}{\text{m}^3}\right)\left(9.81\dfrac{\text{m}}{\text{s}^2}\right)(18.69 \text{ m}^3)}{\pi(4.4 \text{ m})(0.025 \text{ m})} = 4.16 \text{ MPa}$$

Note: Axial stresses due to weight are small compared to yield stress of 210 MPa for structural steel. Even for a skyscraper, H ~ 300 m, stress is still reasonably small (~24 MPa).

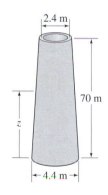

Weight is equal to that of the uniform tower with the same mean diameter (3.4 m), but the stress at the bottom is smaller, since the area there is larger.

>>End Example Problem 3.4

Additional data on material properties needed to solve problems can be found in Appendix D or inside back cover.

3.5 A steel pipe with 200 mm outer diameter and 10 mm wall thickness is buried in the ground, and is to be lifted out. The ground pressure varies linearly from the surface to the bottom, and so the frictional resistance to sliding varies as well. Say the frictional shear stress varies from 0 at the top to 40 kPa at 12 m. Determine the tensile stress in the pipe when the lifting force is just enough to overcome the frictional resistance.

3.6 A steel pipe with 200 mm outer diameter and 10 mm wall thickness is buried in the ground, and is to be lifted out. The ground pressure varies linearly from the surface to the bottom, and so the frictional resistance to sliding varies as well. Say the frictional shear stress varies from 0 at the top to 40 kPa at 12 m. Assuming the bottom does not move, how much does the upper end of the pipe displace when the upward force lifting the pipe is just enough to overcome the frictional resistance?

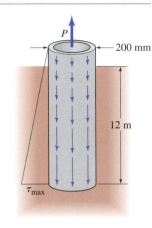

Probs. 3.5–6

Two gears are mounted on a steel shaft as shown. Helical gears can result in axial forces on shafts (as well as transverse forces and torques). Only the axial forces are depicted, and they have been modified to produce only axial loading of the shaft. (Take the distances to be $L_1 = 200$ mm, $L_2 = 300$ mm, and $L_3 = 200$ mm, and the diameters of the three segments to be $d_{AB} = 16$ mm, $d_{BC} = 20$ mm, and $d_{CD} = 18$ mm.)

3.7 Determine the maximum tensile and compressive stress in the shaft and where in the shaft they are reached.

3.8 Axial motions of the gears can affect the meshing of the gears. Determine whether the gears move together or separately and by how much.

3.9 If the point D of the shaft does not displace, determine the displacements of each of the gears, and the displacement of point A.

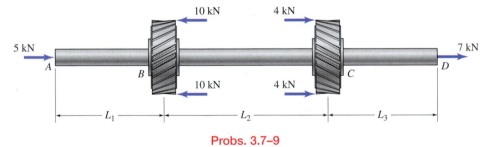

Probs. 3.7–9

The aluminum rod is secured at the top and loaded as shown. The rod has an outer diameter of 1.25 in. and a wall thickness of 0.125 in.

3.10 Determine the maximum axial stress in the rod and indicate if it is tensile or compressive.

3.11 Determine the axial stress in the rod midway between points B and C and indicate if it is tensile or compressive.

3.12 Determine the displacement of the rod midway between points B and C due to the loading.

3.13 Determine whether the collars at B and C move toward or away from each other when they are loaded, and by how much.

3.14 A strain gage is attached to the rod midway between points B and C to measure the axial strain on the surface of the rod. Say all the applied forces are increased by a factor of 2 (twice the values shown). By how much does the strain measured by the strain gage increase?

3.15 The rod is subjected to the loads shown. By how much does the point A further displace if the 1500 lb forces at B are reduced to zero, and does it move up or down?

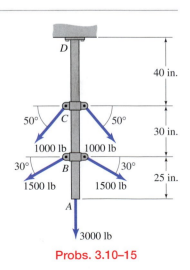

Probs. 3.10–15

A steel column in the building supports downward forces F_1 and F_2 from the respective floors. The upper column has a cross-sectional area of 7 in.^{2}, and the lower column has a cross-sectional area of 17 in.^{2} (1 kip = 1000 lb).

3.16 If F_1 = 100 kip and F_2 = 200 kip, determine the stresses in the upper and lower columns.

3.17 The positions of the columns are monitored, while loads change. Say point A is observed to move down by 0.1 in. while only one of the floor forces changes. (a) If only force F_1 changed, what would be its change? (b) If only force F_2 changed, what would be its change?

3.18 Say the stresses in both posts are to be 6000 psi. Determine the values for the floor forces F_1 and F_2.

3.19 Strain gages are installed on each of the columns to measure the vertical strain. They are zeroed under an initial set of loads. Determine the change in the strain gage readings if (a) only load F_1 increases by 10 kip and (b) only load F_2 increases by 10 kip.

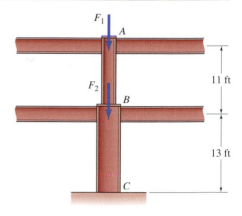

Probs. 3.16–19

An observation compartment can move up and down the tower. As a simplified model of the steel tower, take it to be 300 m tall with an outer diameter of 10 m and wall thickness of 50 mm. The observation compartment is 4 m high and has a mass of 4000 kg.

3.20 As the compartment moves from the top of the tower to halfway down, by how much does the top of the tower displace and does it displace up or down?

3.21 The axial stress in the tower fluctuates as the tower moves up and down. Consider a point 60 m from the bottom of the tower. Determine the maximum and minimum values between which stress at this point fluctuates as the compartment moves up and down.

3.22 A strain gage is attached to a point 100 m from the bottom of the tower. As the compartment moves up past the gage, by how much does the strain change? (Neglect any dynamic effects of motion or braking.)

3.23 Determine the maximum possible height of the tower (rather than 300 m) if the maximum axial stress in the tower (including the effect of the compartment) is to remain below 40 MPa.

Probs. 3.20–23

3.24 A magnetic tape is drawn by a drive at the left and passes through rolls. The rolls have frozen and so provide frictional resistance to the tape motion. The variation in the elongation along the tape is of interest. For simplicity, take the tape at the right end to be fixed. The drive at the left applies a torque M_0 = 9 N-mm to the tape. The rolls squeeze with normal forces N_0 = 0.6 N, and the friction coefficient is estimated to be 0.2. Estimate the strain in the tape between (a) the fixed end and the rolls and (b) the rolls and the drive. The tape has an average modulus of 8 GPa, a thickness of 9 μm, and a width (into the page) of 12 mm. Take the dimensions to be L_1 = 300 mm, L_2 = 200 mm, and D = 20 mm.

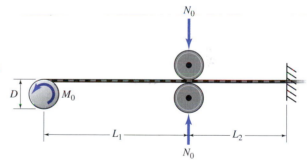

Prob. 3.24

3.25 A magnetic tape is transported from one spool to another over an idler roller. Say that magnetic information is read from the tape on both sides of the idler roller. There is a minor amount of friction between the idler and its axle. In particular, the resisting torque is proportional to the rotational speed: 5×10^{-4} N-mm per rpm. Say the tape is transported at a speed of 3 m/s and that the drive supplies a torque of M_0 = 12 N-mm to the tape. Data are stored in the form of segments along the tape that are each 0.5 μm in length when the tape is unstressed. (a) Determine the torque M_b applied by the brake under steady conditions. (b) Determine the deformed length of the data segment during transport between the drive and the idler. (c) Determine the deformed length of the data segment during transport between the brake and the idler. The tape has an average modulus of 8 GPa, a thickness of 9 μm, and a width (into the page) of 12 mm. Take the dimensions to be D_1 = 25 mm, D_2 = 20 mm, and D_3 = 18 mm.

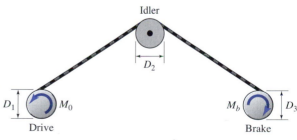

Prob. 3.25

3.26 A long drill used for deep exploration consists of a series of N segments, each of length L_0. Each neighboring pair of segments is coupled by a connector that weighs W_0. We want to determine the deflection at the end due only to the weight of the connectors. As an aside, the deflection due to the weight of the segments themselves could be determined from the continuous gravitational load acting along the length of the member. Let the segments have modulus E and cross-sectional area A, and let the deformation of the connectors themselves be neglected.

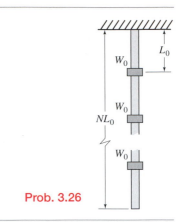

(a) Determine the deflection of the lower end of the drill. Hint: you will need the finite

series summation: $\displaystyle\sum_{n=1}^{N} n = \frac{N(N+1)}{2}$

(b) Compute the end deflection a second way, by treating the weight of the connectors as uniformly along the length. Discuss why the two formulas are slightly different.

Prob. 3.26

3.27 Many materials undergo creep or viscous deformation over a range of temperatures. This is most commonly the case with plastics, but it is also relevant to glass during processing and even to metals when used in high temperature applications, such as turbine blades in aircraft engines. A very simple model for the deformation of plastics treats the deformation as the sum of an elastic part and a viscous part. The elastic and viscous elements are depicted in the diagram at right.

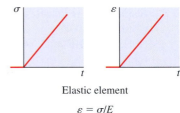

When stress is applied to a viscous element, it strains (creeps) at a rate that is proportional to the stress and inversely proportional to the viscosity η (which is sensitive to temperature). The strain in a viscous element cannot suddenly be changed. By contrast, the strain in the elastic element is proportional to the stress. So, sudden changes in stress produce sudden changes in strain in an elastic element. The time responses of the elements to stress are shown here.

When stress is applied to the material, that stress is felt equally by both elements (they are in series so they feel the same force). Each element contributes to the strain of the material, which equals the sum of the strains due to the two elements. The time history of the stress can be found in terms of the time history of strain or vice versa.

Viscous element

$\dfrac{d\varepsilon}{dt} = \sigma/\eta$

Elastic element

$\varepsilon = \sigma/E$

Prob. 3.27

Let a strain ε_0 be suddenly applied to the material (ε_0 is the total strain contributed by the two elements). Derive and solve a differential equation for how the stress varies with time. Use the fact that the strain of the viscous element cannot be suddenly changed.

Focused Application Problems

3.28 The stress in the steel tower (pylon) of a cable-stayed bridge is due to many sources, including its own weight. Let the tower consist of two box sections, each formed with 40 mm thick plate and with cross-sectional dimensions 10 m by 12 m. If the tower is 330 meters high, determine the axial compressive stress at the base.

Prob. 3.28 (Appendix A-2)

3.29 The pedestrian cable-stayed bridge has a symmetric pair of pylons with a single row of cables in a fan pattern attached to points on the walkway. Let the reinforced concrete walkway have a cross-sectional area of 2.4 m². Take the density of reinforced concrete to be 2400 kg/m³. (a) Assuming the walkway does not rest on the pylons, compute the tensile force in each cable assuming each exerts a force with upward component equal to the weight of the walkway between the points at which the cable connects to the walkway. (b) Consider the axial loading of the walkway induced by the cable tensions found in (a). Determine where along the length of the walkway the average axial stress induced by the cables would be maximum, and determine the value of that average axial stress. Take the dimensions to be $w_1 = 10$ m, $w_2 = 30$ m, $w_3 = 50$ m, and $h = 50$ m.

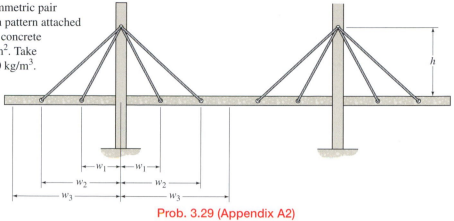

Prob. 3.29 (Appendix A2)

3.30 One constraint in designing a drill string is that the full string needs to be lifted without causing yield. The bore is 10000 ft deep. At the bottom of the string are 21 joints of drill collars each 30 ft long weighing 139 lb/ft. Above the collars, it is proposed to use steel drill pipe with 5 in. outer diameter and 4.408 in. inner diameter. Neglect the weight of the drill bit. If the allowable tensile stress of the steel is 40000 psi, can this size pipe be used?

3.31 Collars on a drill string are used to create sufficient force on the drill bit. At the bottom of a steel drill string is a set of 20 joints of drill collars each 30 ft long. The collar has 7.75 in. outer diameter and 2.8125 in. inner diameter. Above the collars are 300 steel drill pipe segments each 30 ft in length with 5 in. outer diameter and 4.408 in. inner diameter. Neglect the weight of the drill bit. (a) Determine the tensile force that should be applied to the top of the string so that the force on the drill bit is 40000 lb. Determine (b) the maximum tensile stress in the pipe, (c) the maximum tensile stress in the collar, and (d) the maximum compressive stress in the collar.

Probs. 3.30–31 (Appendix A3)

3.32 A rotating wind turbine blade develops stress due to centripetal acceleration. The tensile force at the base of the blade must be sufficient to keep the blade moving in a circular path (like the tension in a string attached to a ball following a circular path). Approximate the blade as having a uniform cross-section along its length, so its center of mass is halfway along its length. Recall that the centripetal acceleration of a steadily rotating body is $r\omega^2$, where r is the radius of its path. Estimate the tensile stress at the base of a solid glass-fiber reinforced epoxy blade that is $L = 30$ m long, has a density of $\rho = 1700$ kg/m³, and rotates at 20 rpm. Show that the result is independent of the cross-section, provided it is uniform along the length.

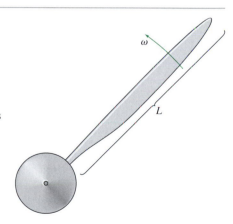

Prob. 3.32 (Appendix A6)

3.3 Systems of Axially Loaded Members

Some axially loaded systems consist of two or more members. To analyze such systems, one must translate their connections into mathematical relations between displacements.

1. Consider systems in which axially deforming members are connected to other members that have negligible deformation.

We have so far only learned to analyze axial deformation. So, we can consider systems with multiple members, but all deformations other than axial (for example, bending) must be neglected. As an example, consider this problem in which two cords with stiffness k are connected to member BD. The force F_0 causes axial deformation of the cords. Determine the displacement at E where load F_0 is applied.

The axial forces in the cords can be found from equilibrium.

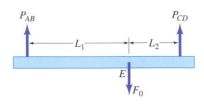

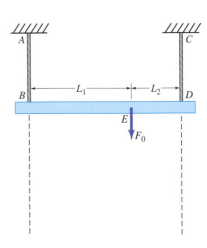

$$P_{AB} = \frac{L_2}{L_1 + L_2}F_0 \quad P_{CD} = \frac{L_1}{L_1 + L_2}F_0$$

$$\delta_{AB} = \frac{P_{AB}}{k} \qquad \delta_{CD} = \frac{P_{CD}}{k}$$

The cord forces and F_0 act perpendicularly to member BED, and so would tend to bend it. Let us neglect bending and treat BED as rigid.

2. View the general displacement of a rigid member as a uniform displacement (no rotation), plus a rigid rotation about some point.

The top of each cord is fixed. The downward displacements v of the ends of the cords at the bar are equal to the stretches: $v_B = \delta_{AB}$ and $v_D = \delta_{CD}$.

Since the member BED is idealized as rigid, it stays straight. It can rotate and displace horizontally and vertically.

To be specific, take point B to move only downward, so BED displaces uniformly down and then rotates about point B. Point D will lie along a circular arc centered at the displaced position of B.

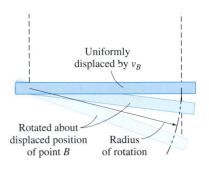

Uniformly displaced by v_B

Rotated about displaced position of point B

Radius of rotation

3. For a body that rotates rigidly by a small angle, the displacement of any point is perpendicular to the line from the center of rotation and is proportional to the distance to that point.

Often, the rotation angle α in radians is small ($\ll 1$). Then, the rotation of a body about point P causes Q to displace to Q', with the displacement perpendicular to PQ and proportional to the length PQ: $u_{QQ'} = (L_{PQ})\,\alpha$.

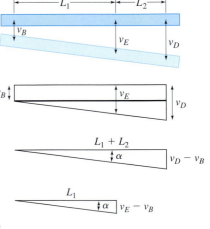

If the bar BED rotates through a small angle, point D of the bar moves an additional distance only downward $v_D - v_B$ (perpendicularly to BED).

For a small rotation, $\tan \alpha \approx \alpha$, and the horizontal displacement is negligible. The rotation angle, α, satisfies $\tan \alpha = (v_D - v_B)/(L_1 + L_2)$.

To find the displacement of E, note that the segment from B to E, or to any point on the bar, rotates about B through the same angle α. So

$$\frac{v_E - v_B}{L_1} = \tan \alpha = \frac{v_D - v_B}{L_1 + L_2} \Rightarrow v_E = v_B + \frac{L_1}{L_1 + L_2}(v_D - v_B) = \delta_{AB} + \frac{L_1}{L_1 + L_2}(\delta_{CD} - \delta_{AB})$$

4. Displacements of the ends of an axially loaded member may not be along the member.

So far, all displacements of an axially loaded member have been parallel to its length. Often though, the displacement will be at an angle to the member. For example, truss members are two force members, and so they are axially loaded. But, since members are connected at various angles, displacements cannot act along the length of all members.

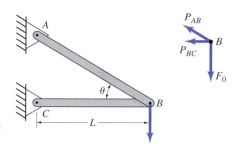

The axial internal forces in the two bars of this truss, P_{AB} and P_{BC}, can be found from statics. From the internal forces, we can determine the elongation of the two bars (assume same EA).

$$P_{AB} = \frac{F_0}{\sin\theta} \qquad \delta_{AB} = \frac{P_{AB}L_{AB}}{EA} = \frac{F_0(L/\cos\theta)}{EA\sin\theta} \quad \text{(lengthens)}$$

$$P_{BC} = -\frac{\cos\theta}{\sin\theta}F_0 \qquad \delta_{BC} = \frac{P_{BC}L_{BC}}{EA} = \frac{-(\cos\theta)F_0(L)}{EA(\sin\theta)} \quad \text{(shortens)}$$

The support points A and C do not displace. If the bars change length without rotating, they would no longer be connected at B. So the bars must rotate and change length, keeping ends B together.

If bars change length, but do not rotate, the ends will not coincide.

Bars must rotate and change length so their ends still coincide.

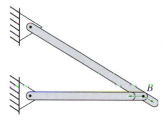

Determine the horizontal u and vertical displacement v that take point B to B'.

5. Even when a bar elongates, the displacement due to an additional rotation by a small angle is perpendicular to the bar's initial direction.

Resolve the displacement at B of each bar into components parallel and perpendicular to its length. If the rotation is small, we interpret the displacement components as follows:

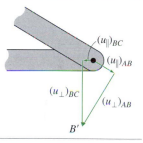

- Displacement component parallel to bar ($u_{\parallel}$) is the elongation (related to δ)
- Displacement component perpendicular to bar ($u_{\perp}$) corresponds to the rotation

6. Define u and v as the displacements of point B (equal for the two bars) along x-y axes, and relate them to the displacements parallel and perpendicular to the bars.

Define the x-y displacements, u and v, at B, resolve u and v relative to each bar, and add their parallel and perpendicular components.

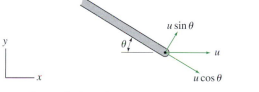 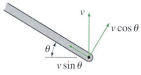

AB: $(u_{\parallel})_{AB} = u\cos\theta - v\sin\theta = $ elongation of $AB = \delta_{AB}$

        $(u_{\perp})_{AB} = u\sin\theta + v\cos\theta$ gives rotation of AB, zero elongation

BC: $(u_{\parallel})_{BC} = u = $ elongation $= \delta_{BC}$

        $(u_{\perp})_{BC} = v$ gives rotation of BC, zero elongation

$$u = \delta_{BC} \Rightarrow u = \frac{-F_0 L\cos\theta}{EA\sin\theta} \qquad -v\sin\theta = \delta_{AB} - u\cos\theta$$

$$\Rightarrow v = \frac{1}{\sin\theta}\left[\frac{-F_0 L}{EA\sin\theta\cos\theta} - \frac{F_0 L\cos^2\theta}{EA\sin\theta}\right] = \frac{-F_0 L}{EA}\left[\frac{1 + \cos^3\theta}{\sin^2\theta\cos\theta}\right]$$

>>End 3.3

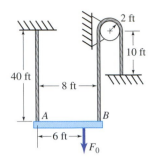

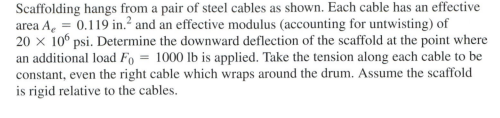

Scaffolding hangs from a pair of steel cables as shown. Each cable has an effective area $A_e = 0.119$ in.^{2} and an effective modulus (accounting for untwisting) of 20×10^6 psi. Determine the downward deflection of the scaffold at the point where an additional load $F_0 = 1000$ lb is applied. Take the tension along each cable to be constant, even the right cable which wraps around the drum. Assume the scaffold is rigid relative to the cables.

Solution

The tensions in the cables can be found from equilibrium of the scaffold as follows:

$$\sum F_y = T_1 + T_2 - F_0 = 0 \qquad \sum M|_A = T_2(8 \text{ ft}) - (1000 \text{ lb})(6 \text{ ft}) = 0$$

$$\Rightarrow T_1 = 250 \text{ lb} \qquad T_2 = 750 \text{ lb}$$

We can then find the elongation of each of the cables:

$$\delta_1 = \frac{T_1 L_1}{E_e A_e} = \frac{(250 \text{ lb})(40 \text{ ft})(12 \text{ in./ft})}{(20 \times 10^6 \text{ psi})(0.119 \text{ in.}^2)} = 0.0504 \text{ in.}$$

$$\delta_2 = \frac{T_2 L_2}{E_e A_e} = \frac{(750 \text{ lb})[(40 + \pi(2) + 10) \text{ ft}](12 \text{ in./ft})}{(20 \times 10^6 \text{ psi})(0.119 \text{ in.}^2)} = 0.213 \text{ in.}$$

The extensions of the cables cause the scaffold to deflect downward. Because the extensions are different, the scaffold tilts by angle α.

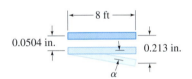

$$\tan \alpha = \frac{(0.213 - 0.0504) \text{ in.}}{8 \text{ ft} \left(\dfrac{12 \text{ in.}}{1 \text{ ft}} \right)} = 1.692 \times 10^{-3}$$

Note $\tan \alpha$ is small, so the rotation α is also small.

Let v be the deflection at the point of interest (where F_0 is applied).

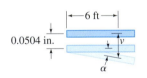

$$\tan \alpha = \frac{(v - 0.0504) \text{ in.}}{6 \text{ ft} \left(\dfrac{12 \text{ in.}}{1 \text{ ft}} \right)} = 1.692 \times (10^{-3})$$

$$\Rightarrow v = 0.1722 \text{ in.}$$

>>End Example Problem 3.5

Two plastic links are pin connected at C where a force is applied. The elastic modulus of the links is 2 GPa, and their cross-sections are identical with dimensions 2 mm × 3 mm shown. Determine the deflection of the point C under the given applied force.

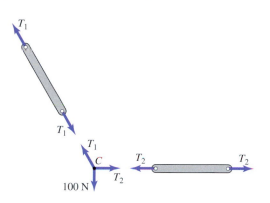

Solution

The links are in uniaxial loading, with tensile forces T_1 and T_2.

From equilibrium of the pin at C, we find:

$$\sum F_x = T_2 - T_1 \cos 60° = 0 \qquad \sum F_y = T_1 \sin 60° - 100 \text{ N} = 0$$

Tensions can be found: $T_1 = 115.5$ N $\quad T_2 = 57.7$ N

From the dimensions we find the cross-sectional area:
$A = (3 \text{ mm})(2 \text{ mm}) = 6 \times 10^{-6} \text{ m}^2$

Next, we obtain the elongations:

$$\delta_1 = \frac{T_1 L}{EA} = \frac{(115.5 \text{ N})(0.03 \text{ m})}{(2 \times 10^9 \text{ Pa})(6 \times 10^{-6} \text{ m}^2)} = 0.289 \text{ mm}$$

$$\delta_2 = \frac{T_2 L}{EA} = \frac{(57.7 \text{ N})(0.03 \text{ m})}{(2 \times 10^9 \text{ Pa})(6 \times 10^{-6} \text{ m}^2)} = 0.1443 \text{ mm}$$

Both members become longer under load. If the links lengthened and remained in their original orientations, they could not remain pinned together.

Now let the members pivot while they elongate. If they pivot the right amounts, they can remain pinned together.

Define the displacements of point C as u_c and v_c, positive as shown:

Members have zero displacements at the ends A and B. The displacement at C must be consistent with elongations of both bars. Resolve displacements of point C parallel to each bar, and equate to the elongation of that bar.

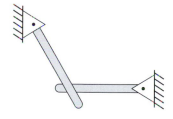

$$\delta_1 = 0.289 \text{ mm} = u_c \cos 60° - v_c \sin 60°$$

$$\delta_2 = 0.1443 \text{ mm} = -u_c$$

Solve these two equations for u_c and v_c:

$u_c = -0.1443$ mm (to the left) $\qquad v_c = -0.395$ mm (downward)

>>End Example Problem 3.6

Additional data on material properties needed to solve problems can be found in Appendix D or inside back cover.

Two cables are used to suspend a bar that is approximated as rigid. A force is applied at the point shown.

3.33 Consider the case of $A_1 = A_2 = 0.1$ in.^{2}, $E_1 = E_2 = E = 20 \times 10^6$ psi, $L_1 = 12$ ft, $L_2 = 7$ ft, $L_3 = 30$ in., and $L_4 = 20$ in. Take $F_0 = 2000$ lb. Determine the deflection of the bar (a) where cable 1 is attached, (b) where cable 2 is attached, and (c) where F_0 is applied.

3.34 Consider the case of $A_1 = A_2 = A$ and $E_1 = E_2 = E$. Allow the cable lengths to differ by retaining the notation L_1 and L_2. Find the ratio of L_3/L_4 at which the bar stays horizontal (the result will depend on variables L_1 and L_2).

3.35 Consider the case of $L_1 = L_2 = L$ and $E_1 = E_2 = E$. Allow the cable areas to differ by retaining the notation A_1 and A_2. Find the ratio of L_3/L_4 at which the stresses in the cables are equal (the result will depend on variables A_1 and A_2).

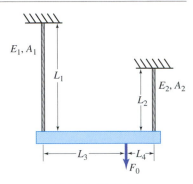

Probs. 3.33–35

The link AC is connected to the bar BCD, which is taken to be rigid relative to AC.

3.36 Let the link AC have Young's modulus $E = 200$ GPa and area $A = 25$ mm^2. Let the lengths be $L_1 = 600$ mm and $L_2 = 200$ mm, and the angle $\alpha = 40°$. A distributed force $q = 4$ kN/m is applied to the bar BCD. Determine the deflection of the end D.

3.37 Let the link AC have Young's modulus $E = 200$ GPa and area $A = 25$ mm^2. Let the lengths be $L_1 = 600$ mm and $L_2 = 200$ mm, and the angle $\alpha = 40°$. Determine the distributed force per length q that results in the end D moving down by 2 mm.

3.38 Let the link AC have Young's modulus $E = 200$ GPa and area $A = 25$ mm^2. Let the lengths be $L_1 = 600$ mm and $L_2 = 200$ mm, and the angle $\alpha = 40°$. Let there be a strain gage mounted on link AC. Express the downward deflection of point D in millimeters as a function of the strain gage reading.

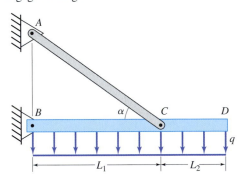

Probs. 3.36–38

Two links that can deform elastically are pinned together. A force can be applied to the pin at B.

3.39 Consider the case of $A_1 = A_2 = 30$ mm^2 and $E_1 = E_2 = 200$ GPa, $L_1 = 300$ mm, $L_2 = 500$ mm, and $\alpha = 60°$. Let there be a downward force $F_0 = 2000$ N applied to the pin at B. Determine the horizontal and vertical deflection of the pin at B.

3.40 Consider the case of $A_1 = A_2 = 30$ mm^2 and $E_1 = E_2 = 200$ GPa, $L_1 = 300$ mm, $L_2 = 500$ mm, and $\alpha = 60°$. Let there be a leftward force $F_0 = 3000$ N applied to the pin at B. Determine the horizontal and vertical deflection of the pin at B.

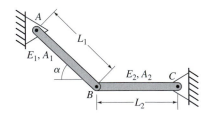

Probs. 3.39–42

3.41 Determine direction and magnitude of the force F_0 applied to the pin B that would result in point B moving straight down by a distance v_0 (zero horizontal displacement). Express the results in terms of the variables in the diagram (E_1, L_1, A_1, E_2, L_2, A_2, and α, not numerical values). Take the displacements and rotations to be small.

3.42 Determine direction and magnitude of the force F_0 applied to the pin B that would result in point B moving directly to the left by a distance u_0 (zero vertical displacement). Express the results in terms of the variables in the diagram (E_1, L_1, A_1, E_2, L_2, A_2, and α, not numerical values). Take the displacements and rotations to be small.

3.43 Consider the case of $P_F = P$, $P_G = 0$, $k_1 = k_2 = k$, and $L_1 = L_2 = L_3 = L$. In terms of the given variables, determine the forces in the two springs and the displacement of point E.

3.44 Consider the case of $P_F = 0$, $P_G = P$, $k_1 = k_2 = k$, and $L_1 = L_2 = L_3 = L$. In terms of the given variables, determine the forces in the two springs and the displacement of point E.

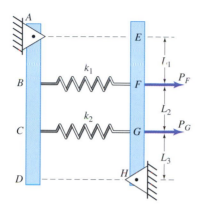

Probs. 3.43–44

Links AC and BC of identical length are pinned together to a third link as shown. All links can deform elastically. Forces could be applied at points C and D.

3.45 Take all bars to have a Young's modulus of $E = 30 \times 10^6$ psi and an area of 0.75 in.^{2}. Let the angle $\alpha = 40°$ and the links have lengths $L_1 = 40$ in. and $L_2 = 60$ in. Find the downward deflection of point D if the forces are $F_1 = 5000$ lb and $F_2 = 4000$ lb.

3.46 Take all bars to have a Young's modulus of $E = 30 \times 10^6$ psi and an area of 0.75 in.^{2}. Let the angle $\alpha = 40°$ and the links have lengths $L_1 = 40$ in. and $L_2 = 60$ in. What must be the forces F_1 and F_2 if point C is to displace downward by 0.02 and point D by 0.036 in.?

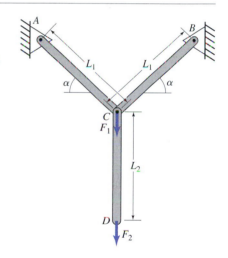

Probs. 3.45–46

Two links that can deform elastically are pinned together. A force can be applied to the pin at C.

3.47 Take both bars to have a Young's modulus of $E = 70$ GPa. AC has an area of 25 mm^2, and BC has an area of 35 mm^2. Let the angle $\alpha = 50°$ and the length $L_1 = 700$ mm. Say the forces are $F_1 = 4$ kN and $F_2 = 0$. Determine the horizontal and vertical deflection of the pin at C.

3.48 Take both bars to have a Young's modulus of $E = 70$ GPa. AC has an area of 35 mm^2, and BC has an area of 25 mm^2. Let the angle $\alpha = 50°$. The length $L_1 = 800$ mm. Say the point C moves downward by 1 mm and not horizontally. Determine the horizontal and vertical force components F_1 and F_2.

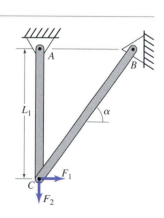

Probs. 3.47–48

3.49 The tension in a guitar string affects the force necessary to deflect the string. Let the string have a tension T_0 and length L. (a) Derive an expression for the lateral force necessary to deflect the string in its middle by a distance d, neglecting any change in string tension. (b) Assuming the string has a stiffness of k, derive an expression for the change in tension associated with deflecting the string in the middle by a distance d.

Prob. 3.49

3.50 The frequency f in Hz (s^{-1}) at which a string vibrates is related to the tension T, the base length L, and mass per length μ. To raise the pitch by one octave (that is, to double the frequency), one needs to halve the vibrating length. This is done by holding down the string at the 12th fret. However, this can produce a frequency which is not precisely doubled if the strings are located far from the fretboard. In pressing down the string to the fretboard, one is elongating the string and raising the tension. Say that the steel E string, which is 0.254 mm in diameter and tuned to 329.6 Hz, is located 8 mm above the fretboard. Let the free length of the string (from nut to saddle) be 0.65 m. If the altered tension is based on the string elongation distributed over 0.75 m, determine deviation in the frequency from exactly one octave 659.2 Hz by

$$f = \frac{1}{2L}\sqrt{\frac{T}{\mu}}$$

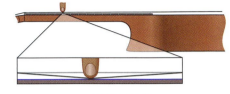

Prob. 3.50

Focused Application Problems

3.51 Due to a severe loading, the pylon deflects rightward by 0.7 m where the cable attaches to it, and the deck deflects downward by 1.1 m where the cable attaches to it. Determine the additional strain induced in the cable due to these deflections. Take the dimensions to be $h = 220$ m and $w = 180$ m.

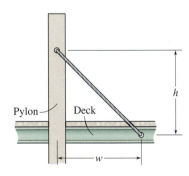

Prob. 3.51 (Appendix A2)

3.52 Cables attached to the sides of the deck serve to resist torsion of the deck (rotation of the deck about its long axis). If the deck were to rotate too much while twisting, a cable could go slack. Let the cables be attached at the edges of the 18 m wide deck. Consider the pair of cables shown: each has a cross-sectional area of 0.001 m^2 and is under a tension of 400 kN. Determine the angular rotation of the deck at which the cable would go slack. Take the dimensions to be $h = 180$ m and $w = 150$ m. The effective modulus of steel cables is 165 GPa. During rotation, take one edge of the deck to displace upward and the other downward by equal amounts.

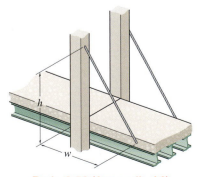

Prob. 3.52 (Appendix A2)

3.53 For many cable-stayed bridges, vertical deflection of the deck of a long span is primarily resisted by changing tension in the cables, rather than by the bending resistance of the girder supporting the deck. Consider the cable shown, which has an area of 0.015 m². Determine the vertical stiffness experienced by the deck due to this cable, where stiffness is defined as the vertical component of cable force induced for every 1 m deflection of deck. The effective modulus of steel cables is 165 GPa. Take the dimensions to be $h = 100$ m and $w = 70$ m.

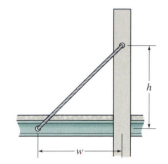

Prob. 3.53 (Appendix A2)

3.54 Because cables are subjected to substantial pre-tension due to the weight of the deck, they resist the deflection of the tower in the transverse direction due to a side wind. Say the tower deflects transversely by 2 m. Determine the restoring transverse force of the cable shown, which carries a tension of 90 kN. Assume that the deflection perpendicular to the plane of the cable induces no additional elongation of the cable. Take the dimensions to be $h = 70$ m and $w = 110$ m. Hint: this deflection does not change the strain, but redirects the force.

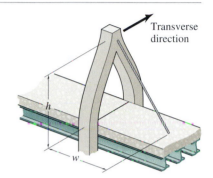

Prob. 3.54 (Appendix A2)

3.55 A small scale wind turbine tower needs to pivot about its base because it is raised up from the ground by a gin pole. The tower is secured in the upright position by a pair of 0.25 in. diameter cables, each attached to an anchorage, as shown. The cable has a working load limit of 1200 lb (typically 1/5 of the breaking strength). The cables are pre-tensioned by equal amounts. The thrust force from the turbine's operation is balanced by an increase in the tension in the front cable and an equal decrease in the rear cable. The rightward deflection of the tower due to the thrust force causes an elongation of the front cable that is equal to the shortening of the rear cable, which is why the magnitudes of the tension changes are equal. The pre-tension in the cables must be chosen so that neither side goes slack nor does the maximum tension exceed the load limit. Assume a steady thrust force of 200 lb. (a) Determine the rearward tilting of the tower at the turbine. (b) Determine the minimum and maximum levels for the cable pre-tensions. Take the effective modulus of the steel cable to be 24×10^6 psi. The distances are $w = 30$ ft, $h_0 = 45$ ft, and $h_1 = 37$ ft. Neglect the re-orientation of the cable force.

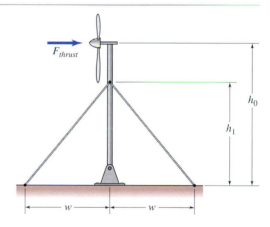

Prob. 3.55 (Appendix A6)

>>End Problems

3.4 Statically Indeterminate Structures

Systems with elastic bars may have more constraints or support reactions than can be solved for with statics alone. To solve such problems, we apply the same principles involving internal force, deformation, and displacement. However, rather than solving the equations sequentially, we solve them simultaneously.

1. A statically indeterminate problem has more support reactions than can be determined by statics (equilibrium) alone.

Consider this problem: a single bar is prevented from displacing at both ends A and B. A known force F_0 is applied at the cross-section C. We want the displacements and stresses in the bar.

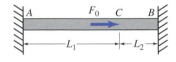

If we assume the end supports apply only axial forces, here is the free body diagram of the whole bar.

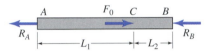

The senses of the unknown reactions R_A and R_B could each have been drawn in either sense. If R_A or R_B turns out to be positive, then the originally drawn sense was correct. If negative, then the sense is opposite.

From the single equilibrium equation, $\Sigma F = 0 \Rightarrow R_A + R_B = F_0$, we cannot find the two unknowns.

One or more reactions in a statically indeterminate problem might be called redundant (extra), since the bar would be in equilibrium with them.

2. Use the same principles and equations involving equilibrium and deformation to solve statically indeterminate problems, but combine the equations rather than solve them sequentially.

In Mechanics of Materials, in general, we relate four types of **quantities** using three types of **relations**, as depicted in this diagram.

Here are those **quantities** and **relations** for the case of axial loading.

For the statically determinate problems that we have studied so far, we applied these relations from left to right: first finding internal loads from external loads, then elongations, and then displacements.

For statically indeterminate problems, there are not enough equilibrium equations to solve for all the loads. But we can still write down all the same equations symbolically and solve them simultaneously.

There are enough total equations to solve statically indeterminate problems: extra relations between the displacements compensate for the extra loads that cannot be found from equilibrium.

3. Use equilibrium to relate the unknown internal axial forces to applied loads and unknown reactions (equilibrium).

Draw FBD's to find internal forces. Use the fact that the internal force is constant in AC (P_{AC}) and in CB (P_{CB}).

As usual, to keep track of signs, it is helpful to draw and label unknown internal forces assuming that tension is positive.

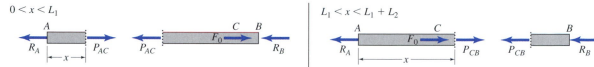

$0 < x < L_1$

$L_1 < x < L_1 + L_2$

$P_{AC} = R_A$ or $P_{AC} = F_0 - R_B$ agree since $R_A + R_B = F_0$ $\quad$ $P_{CB} = R_A - F_0$ or $P_{CB} = -R_B$ agree since $R_A + R_B = F_0$

We can express both P_{AC} and P_{CB} in terms of one reaction (e.g., R_A), but we still don't know R_A in terms of F_0. We proceed as if the internal forces are known, and we just use the symbols P_{AC} and P_{CB}.

4. For each segment in which the internal force, elastic modulus, and area are uniform, relate the elongation to the unknown internal forces (Material law).

Within each segment AC and CB, properties (E and A) and internal force (P) are uniform. Find elongations from $\delta = PL/EA$.

$$\delta_{AC} = \frac{P_{AC}L_1}{EA} \qquad \delta_{CB} = \frac{P_{CB}L_2}{EA}$$

Often, we don't know if a segment is in tension or compression. Because we assumed the unknown P as positive in tension, the relation $\delta = PL/EA$ will still be valid and have consistent signs:

$P > 0$ means tension $\qquad \leftrightarrow \quad \delta > 0$ means lengthening $\qquad (\delta > 0$ if $P > 0)$

$P < 0$ means compression $\quad \leftrightarrow \quad \delta < 0$ means shortening $\qquad (\delta < 0$ if $P < 0)$

5. Relate elongations to displacements and make sure they reflect the support constraints (geometric compatibility).

Displacements must be consistent with the elongations and with constraints imposed by the supports.

Original

Deformed

The bar is shown divided into several segments initially of equal length, and for this problem two approaches lead to the same equation that relates elongations of AC and CB.

1. Combine $\delta_{AC} = u_C - u_A$ and $\delta_{CB} = u_B - u_C$ with fixed supports ($u_A = u_B = 0$) $\Rightarrow \delta_{AC} = -\delta_{CB}$
2. Ends are fixed, so even when deformed, the bar has length $L_1 + L_2$. So the total elongation of ACB must be zero $\Rightarrow \delta_{AC} + \delta_{CB} = 0 \Rightarrow \delta_{AC} = -\delta_{CB}$

So either way, AC stretches by same amount that CB contracts.

6. Combine all relations to find all unknowns.

There is no single method of combining equations, but often we express elongations in terms of reactions (eliminating internal loads) and then insert those expressions for elongations into relations of geometric compatibility.

Combine relations: $\delta_{AC} = -\delta_{CB} \Rightarrow \dfrac{P_{AC}L_1}{EA} = \dfrac{P_{CB}L_2}{EA} \Rightarrow \dfrac{R_A L_1}{EA} = \dfrac{-(R_A - F_0)L_2}{EA} \Rightarrow$ Solve for R_A

Important results: $\quad R_A = \dfrac{F_0 L_2}{(L_1 + L_2)} \qquad R_B = \dfrac{F_0 L_1}{(L_1 + L_2)} \qquad \sigma_{AC} = \dfrac{P_{AC}}{A} = \dfrac{F_0 L_2}{A(L_1 + L_2)} \qquad \sigma_{BC} = \dfrac{P_{BC}}{A} = \dfrac{-F_0 L_1}{A(L_1 + L_2)}$

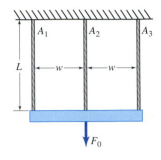

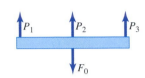

Three cables are connected to a common support, and to a horizontal bar, which is approximated as rigid. The cables have equal Young's modulus E, but one of the cables has a slightly different cross-sectional area: $A_1 = A_0$, $A_2 = A_0$, and $A_3 = 0.9A_0$. Determine the tensions in the cables and the deflection of the point on the bar where the load F_0 is applied.

Solution

The tensions in the cables must satisfy equilibrium:

$$\sum F_y = P_1 + P_2 + P_3 - F_0 = 0 \qquad \sum M|_{center} = -P_1(w) + P_3(w) = 0$$

Two equations cannot be solved for three unknown forces, so the problem is statically indeterminate. In addition to equilibrium, we need to use relations that account for the material properties and geometric compatibility (geometric relations between displacements).

Material properties:

$$\delta_1 = \frac{P_1 L}{E A_1} = \frac{P_1 L}{E(A_0)} \qquad \delta_2 = \frac{P_2 L}{E A_2} = \frac{P_2 L}{E(A_0)} \qquad \delta_3 = \frac{P_3 L}{E A_3} = \frac{P_3 L}{E(0.9A_0)}$$

Notice: there are now five equations, but six unknowns.

All cables are attached at the top, so the elongations are equal to the downward displacement where the cables attach to the bar. If vertical displacements v_1, v_2, and v_3 are defined as positive downward, then

$$\delta_1 = v_1, \delta_2 = v_2, \text{ and } \delta_3 = v_3$$

We have three more equations, but three more unknowns.

These three displacements cannot be independent. Although the bar deflects and tilts, it remains straight because it has been assumed to be rigid (does not bend).

The original line through the center of the bar and the line through the center of the tilted bar create a triangle.

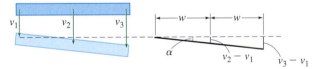

There are similar triangles with a common angle α.

$$\tan \alpha = \frac{v_2 - v_1}{w} = \frac{v_3 - v_1}{2w} \Rightarrow v_2 = \frac{v_1 + v_3}{2}$$

Combine this relation with the earlier relations between elongations and displacements and the relations between tensions and elongations:

$$\delta_2 = \frac{P_2 L}{E A_0} = \frac{1}{2}[\delta_1 + \delta_3] = \frac{1}{2}\left[\frac{P_1 L}{E(A_0)} + \frac{P_3 L}{E(0.9A_0)}\right] \Rightarrow P_2 = \frac{1}{2}\left[P_1 + \frac{1}{0.9}P_3\right]$$

Now combine with the two equilibrium equations to find:

$$P_1 = P_3 = 0.327F_0 \qquad P_2 = 0.345\,F_0 \qquad v_2 = \delta_2 = \frac{P_2 L}{E A_0} = \frac{0.345F_0}{E A_0}$$

>>End Example Problem 3.7

A pair of springs with different spring constants, $k_1 = 20$ lb/in. and $k_2 = 40$ lb/in., are each fixed at one end. They are to be hooked to a rigid plate as shown. (a) Determine the tensions, T_1 and T_2, in the two springs. (b) Once the springs are hooked onto the plate, a net force $F_0 = 12$ lb is applied to the plate as shown. Find the new tensions in the springs.

Solution

When the springs are hooked to the plate, the force on the springs and the plate are:

The plate must be in equilibrium, so $T_1 = T_2 = T$.

Since the spring constants differ ($k_1 \neq k_2$) and the tensions are equal, the spring stretches must differ.

The rigid plate does not deform, but it can displace left or right. In total, the springs must extend by a length s to connect to the rigid plate.

$$\delta_1 + \delta_2 = s$$

$$\frac{T}{k_1} + \frac{T}{k_2} = s \qquad T = s\frac{k_1 k_2}{k_1 + k_2} = 0.5 \text{ in.}\frac{(20 \text{ lb/in.})(40 \text{ lb/in.})}{(20 + 40) \text{ lb/in.}} = 6.67 \text{ lb}$$

Note that a rigid displacement of the plate allows the spring elongations to differ.

The springs have the tension just found. Now consider the additional forces that develop in the springs due to the application of the force F_0. The springs develop additional forces T_1 and T_2. Now the FBD is:

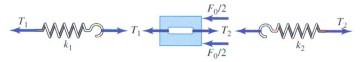

The plate must be in equilibrium, so $\Sigma F_x = -T_1 + T_2 - F_0 = 0 \Rightarrow -T_1 + T_2 = F_0$.

Under the action of the force F_0, the plate moves to the left by an amount u and does not deform.

The contraction of spring 1 = extension of spring $2 \Rightarrow -\delta_1 = \delta_2$

$$-\frac{T_1}{k_1} = \frac{T_2}{k_2} \quad \Rightarrow \quad T_1 = -\frac{k_1}{k_2}T_2$$

Combine with $-T_1 + T_2 = F_0$ to find

$$T_2 = \frac{F_0}{1 + k_1/k_2} = \frac{12 \text{ lb}}{1 + 1/2} = 8 \text{ lb} \qquad T_1 = \frac{-F_0}{1 + k_2/k_1} = \frac{-12 \text{ lb}}{1 + 2/1} = -4 \text{ lb}$$

Now, add the additional spring forces just found, T_1 and T_2, to the pre-tension, $T = 6.67$ lb, associated with first attaching the springs to the plate. The final tensions in springs are:

$$T_1 = 6.67 \text{ lb} - 4 \text{ lb} = 2.67 \text{ lb} \qquad T_2 = 6.67 \text{ lb} + 8 \text{ lb} = 14.67 \text{ lb}$$

>>End Example Problem 3.8

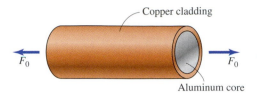

Copper cladding

F_0 F_0

Aluminum core

A copper clad aluminum wire is subjected to axial tension. The loading of the wire must satisfy these conditions: neither the copper nor the aluminum are to exceed 50% of their yield stress, nor is the elongation of the wire to exceed 0.4 mm. Determine the maximum allowable tensile force the wire can withstand. The wire is 700 mm long, the aluminum core is 5 mm in diameter, and the copper cladding is 0.123 mm thick. The yield stresses are $\sigma_{Y,Cu} = 70$ MPa and $\sigma_{Y,Al} = 100$ MPa, and the Young's moduli are $E_{Cu} = 120$ GPa and $E_{Al} = 70$ GPa.

Solution

So far, we have applied the equations for axial loading only if the body has uniform material properties in a cross-section. While the properties vary over the cross-section here, this problem can be solved with using the same principles.

P_{Cu} P_{Cu}

Imagine there were two separate bodies, a tube of copper and a solid rod of aluminum, and they were subjected to tensile forces, P_{Cu} and P_{Al}, respectively. We want the stresses and strains in those separate bodies to be the same as in the respective portions of the clad wire subjected to axial force F_0.

P_{Al} P_{Al}

The copper cladding is bonded to the aluminum, so the copper tube and aluminum rod should stretch by the same amount, $\delta_{Cu} = \delta_{Al}$. The elongations are found from $\delta = PL/EA$ applied to each body.

$$\delta_{Cu} = \delta_{Al} \Rightarrow \frac{P_{Cu}L}{E_{Cu}A_{Cu}} = \frac{P_{Al}L}{E_{Al}A_{Al}} \Rightarrow P_{Cu} = P_{Al}\frac{E_{Cu}A_{Cu}}{E_{Al}A_{Al}}$$

$P_{Cu} = \sigma_{Cu}A_{Cu}$

$P_{Al} = \sigma_{Al}A_{Al}$

F_0

The total force on the two bodies is equal to the force on the wire,
$$P_{Cu} + P_{Al} = F_0.$$

Substitute the expression for P_{Cu} in terms of P_{Al} into $P_{Cu} + P_{Al} = F_0$:

$$P_{Al} = F_0\frac{E_{Al}A_{Al}}{E_{Al}A_{Al} + E_{Cu}A_{Cu}} \qquad P_{Cu} = F_0\frac{E_{Cu}A_{Cu}}{E_{Al}A_{Al} + E_{Cu}A_{Cu}}$$

$$\delta_{Cu} = \delta_{Al} = \delta \Rightarrow \frac{F_0 L}{E_{Al}A_{Al} + E_{Cu}A_{Cu}}$$

Limit the stress in each of the copper and the aluminum to 50% of its yield stress.

$$\frac{P_{Cu}}{A_{Cu}} = \frac{\sigma_{Y,Cu}}{2} = 35 \text{ MPa} \Rightarrow F_0 = 428 \text{ N} \qquad \frac{P_{Al}}{A_{Al}} = \frac{\sigma_{Y,Al}}{2} = 50 \text{ MPa} \Rightarrow F_0 = 1049 \text{ N}$$

Limit the elongation to $\delta = 0.4$ mm $\Rightarrow F_0 = 839$ N

None of these conditions are violated if F_0 is less than the minimum of the three values $\Rightarrow F_0 = 428$ N.

This problem is statically indeterminate in that the forces P_{Cu} and P_{Al} cannot be determined solely from $P_{Cu} + P_{Al} = F_0$. As we see, the material (E) and the geometry (A) must also be considered.

>>End Example Problem 3.9

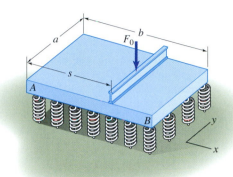

A slab lays on a foundation that is elastic. The stiffness of the foundation is described by a spring constant, which accounts for the fact that the load from the slab is distributed. The spring constant k gives the resisting force when a 1 in.^{2} area is depressed by 1 in. The stiffness of the foundation is $k = 2$ lb/in.^{3} The slab is loaded as shown, so that the loading is uniform along the y-direction. Determine the deflections of the ends A and B and the rotation of the slab. Approximate the slab as rigid. (Take $F_0 = 10^4$ lb, $a = 15$ ft, $b = 20$ ft, $s = 12$ ft.)

Solution

The foundation stiffness k is interpreted as follows: if an area A deflects by v, the foundation resists with a force $F_r = kAv$.

Uniformity of the loading in the y-direction implies that the problem is symmetric about the x-z plane, and so can be analyzed in the x-z plane. Consider the spring constant k' for the 2-D analysis. It will have units of lb/in.^{2} (in. for deflection and in. for length in the x-direction).

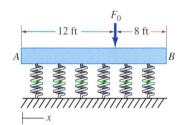

To find k', combine the springs that occupy a unit length in x (1 in.) by the entire 15 ft width of the slab in the y-direction. If all these springs are deflected by 1 in., then the force they would apply is

$$\left(2\frac{\text{lb}}{\text{in.}^3}\right)(1 \text{ in.})(15 \text{ ft})\frac{12 \text{ in.}}{1 \text{ ft}}(1 \text{ in.}) = 360 \text{ lb}$$

Therefore, the stiffness for the 2-D analysis is $k' = 360$ lb/in.^{2}

Because the slab is rigid and remains straight while it deflects and rotates, the springs provide a linearly distributed force per length $q(x)$ with values $q_A = k'v_A$ and $q_B = k'v_B$ at the ends.

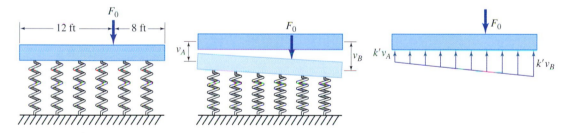

Analyze the distributed force as a uniform distribution, plus a triangular distribution:

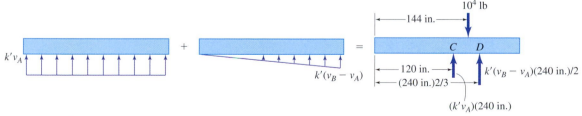

Apply equilibrium to find:

$$\sum M|_D = 0 \Rightarrow v_A = 0.0463 \text{ in.}, \quad \sum F_y = 0 \Rightarrow v_B = 0.1852 \text{ in.}$$

>>End Example Problem 3.10

Additional data on material properties needed to solve problems can be found in Appendix D or inside back cover.

A concrete column has outer dimensions of 250 mm by 250 mm. Concrete is often reinforced with steel reinforcing rods (rebar). There are nine rebars arranged parallel to the column length, and each bar has a diameter of 19.05 mm. The rebar and concrete are assumed to act as if they are bonded (equal axial strains). Because the bars are distributed symmetrically in the column, an axial force in the center results in axial compression of the concrete and rebar. Take the concrete to have an elastic modulus of 20 GPa in compression, and the steel rebar to have an elastic modulus of 200 GPa.

3.56 Determine the axial compressive force that causes the concrete to reach a safe stress level of 10 MPa.

3.57 If the column is 3 m tall and is observed to shorten by 1 mm, what must be the axial load?

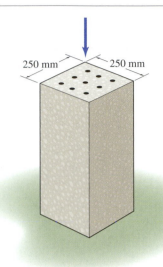

Probs. 3.56–57

3.58 The relatively rigid platform rests on three short plastic rods, which rest on a flat surface. A distributed force acts on the platform as shown. The rods have a modulus of $E = 4$ GPa and a diameter of 15 mm, but their lengths differ: $L_A = 60$ mm, $L_B = 80$ mm, and $L_C = 80$ mm. The separation between the rods is $L = 300$ mm and the distributed force $q = 40$ kN/m. Determine the deflection of the points A, B, and C where the platform rests on the rods.

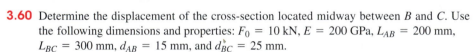

Prob. 3.58

The bar of uniform elastic modulus E has a step in diameter as shown, and is subjected to an axial force of F_0 acting at the shoulder.

3.59 Determine the axial stress in each of the segments AB and BC. Use the following dimensions and properties: $F_0 = 10$ kN, $E = 200$ GPa, $L_{AB} = 200$ mm, $L_{BC} = 300$ mm, $d_{AB} = 15$ mm, and $d_{BC} = 25$ mm.

3.60 Determine the displacement of the cross-section located midway between B and C. Use the following dimensions and properties: $F_0 = 10$ kN, $E = 200$ GPa, $L_{AB} = 200$ mm, $L_{BC} = 300$ mm, $d_{AB} = 15$ mm, and $d_{BC} = 25$ mm.

3.61 Determine the lengths of the individual segments L_{AB} and L_{BC} of the bar if the total length $L_{AB} + L_{BC} = 18$ in., and stress in the right section is to have half of the stress in the left section. Use the following dimensions: $d_{AB} = 0.75$ in. and $d_{BC} = 1$ in.

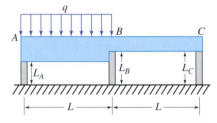

Probs. 3.59–61

The bar ACB of length $L_1 + L_2$ is subjected to an axial load F_0. The bar has a modulus of E_b and an area of A_b. In order to reduce the deflection of the end B (raise the effective stiffness), the bar is connected at C to an effectively rigid cross member, which in turn rests against plastic rods that have a modulus of E_p, an area of A_p, and a length of L_3.

3.62 Derive an expression for the deflection at point B, v_B, in terms of the variables in the problem.

3.63 Assuming the plastic rods have fixed length, area, and properties, will their stiffening effect be greater if the cross member is attached closer to $A(L_1 < L_2)$ or closer to $B(L_1 > L_2)$? Answer this question and explain for two cases of the stiffness per unit length of the bar and rod (a) $E_b A_b \ll E_p A_p$ and (b) $E_b A_b \gg E_p A_p$.

3.64 Let the cross member be connected to ACB midway along its length ($L_1 = L_2$). Derive an expression for the upper limit on the stiffness F_0/v_B that can be reached no matter how stiff the plastic rods are made.

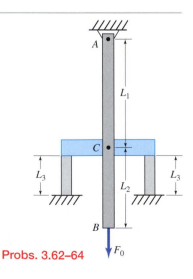

Probs. 3.62–64

3.65 The cases shown depict springs assembled in series and in parallel. The behavior of deformable mechanical elements in series or parallel is an important idea of general use in mechanics.

(a) Let v_0 denote the displacement of the plate on which the force F_0 is applied. Derive the stiffness F_0/v_0 of the assembly in terms of the individual spring constants k_1 and k_2 for the two cases of series and parallel.

(b) For the two cases of series and parallel, explain how the stiffness F_0/v_0 of the assembly depends on the stiffnesses of the two springs assuming one of them, say k_1, becomes very large compared to the other.

(c) Use the general mathematical results for the stiffnesses of two springs in series and parallel from part (a), and take the limit of k_1 large compared to k_2 to confirm your reasoning of part (b).

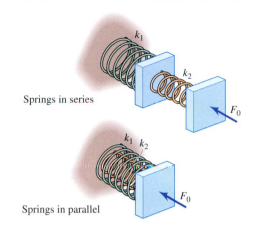

Springs in series

Springs in parallel

Prob. 3.65

3.66 The horizontal bar is pinned at F, supported by two steel cables, AB and DE, and loaded by the force $F_0 = 25$ kip. Take the cables to have an effective modulus of $E = 20 \times 10^6$ psi, and each an area of 1.25 in.^{2} Take the lengths to be $L_{AB} = 3$ ft, $L_1 = 2$ ft, $L_2 = 2$ ft, $L_3 = 4$ ft, and $\alpha = 60°$. Determine (a) the cable stresses σ_{AB} and σ_{DE} and (b) the deflection of point C. Approximate the bar $BCDF$ as rigid.

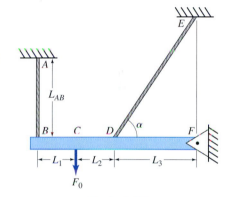

Prob. 3.66

3.67 Three springs are stretched and attached to a rigid plate to keep it in position. Consider what happens when the spring constants vary slightly from their nominally identical values. Given the gaps g_1 and g_2, determine the final forces in the three springs and the angle at which the plate tilts when the springs are attached. Take $g_1 = 30$ mm, $g_2 = 40$ mm, $a = 400$ mm, $b = 300$ mm, $k_1 = 95$ N/mm, $k_2 = 104$ N/mm, and $k_3 = 105$ N/mm.

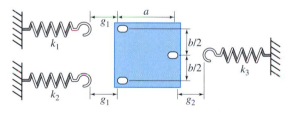

Prob. 3.67

3.68 Two springs are stretched and hooked to a plastic strip ($E = 2$ GPa). In this problem, we consider whether the deformation of the strip should be neglected in determining the resulting tension in the springs. Take $L = 98$ mm, $k = 100$ N/mm, $L_t = 400$ mm, and $L_p = 200$ mm.

Do the following calculations for each of the two cases: plastic cross-section is 7 mm $\times$ 2 mm and plastic cross-section is 25 mm $\times$ 8 mm.

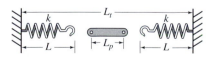

Prob. 3.68

(a) Compute the stiffness of the plastic strip, EA/L, and compare with the spring stiffness.

(b) Compute the tension in the springs by including deformation of the plastic.

(c) Compute the tension in the springs by approximating the plastic as rigid.

3.69 The system shown has two relatively rigid blocks, each of which is attached to ground by two shear pads. All pads have the same dimensions (cross-sectional area A and thickness h). The left pads have a shear modulus of G_1 and the right pads have a shear modulus of G_2. The two blocks are also connected to each other by a spring with constant k.

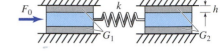

Prob. 3.69

(a) Explain how the displacement of the blocks would depend on the load in the limits of the spring stiffness being zero and infinite.
(b) Derive an expression for the displacement of block 1 in terms of the load F_0 and the parameters A, h, G_1, G_2, and k.
(c) Derive an expression for the ratio of the displacement of block 2 to that of block 1.
(d) Show that your explanation of the dependence of the block deflections in the limits of high and low spring stiffness agree with the expressions derived in parts (b) and (c) when k takes on the respective limits.

3.70 The simple spoked system shown consists of a rigid wheel and four identical pretensioned spokes connected tangentially to the edge of the hub. The tension in each spoke is 300 N. Say the hub is subjected to a torque M_0 as shown. The spokes are steel and circular in cross-section with diameter d. Determine the torsional stiffness, that is the ratio of the torque M_0 to the rotation of the hub (in radians). Assume small rotations. Take $d = 2$ mm, $R = 40$ mm, and $b = 380$ mm.

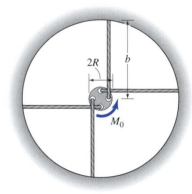

Prob. 3.70

3.71 The simple spoked system shown consists of a rigid wheel and four identical pretensioned spokes connected radially to the edge of the hub. The tension in each spoke is 2000 N. Say the hub is subjected to a torque M_0 as shown. The spokes are steel and circular in cross-section with diameter d. Determine the rotation of the hub relative to the wheel for a torque of $M_0 = 1$ N-m. Account for the reorientation of the pretensioned spoke due to the rotation of the hub, but neglect the change in the spoke length due to rotation. You will find that the torsional stiffness due to reorientation is relatively small compared to the stiffness with tangentially attached spokes. Take $d = 4$ mm, $R = 45$ mm, and $D = 620$ mm.

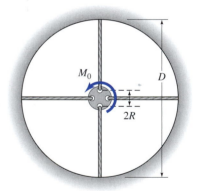

Prob. 3.71

3.72 The simple spoked system shown consists of a rigid wheel and four identical pretensioned spokes connected tangentially to the edge of the hub. The tension in the spokes is 600 N. Say the hub is subjected to a force $F_0 = 200$ N as shown. The spokes are steel and circular in cross-section with diameter d. This causes the hub to move vertically without rotating. (a) Determine the vertical deflection of the hub, accounting for the elongation of the vertically oriented spokes and the reorientation of the horizontal spokes. (b) Determine the fraction of the total resistance to the force that is due to reorientation of the horizontal spokes. Take $d = 3$ mm, $R = 42$ mm, and $b = 340$ mm.

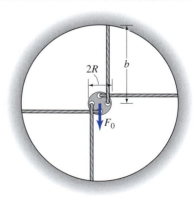

Prob. 3.72

3.73 A sheet handling system includes an elastomeric sheet, which is extended by a drive applying torque M_0. The sheet turns two rolls that resist because they are mounted on torsional springs, each with a stiffness of 0.7 N-m/rad. Say the sheet is viewed as is fixed on the right end and that the drive seeks to draw in 5 mm of the tape. Determine the torque M_0 that the drive needs to apply. The sheet has an average modulus of 4 MPa, a thickness of 1.5 mm, and a width (into the page) of 40 mm. Take the dimensions to be $L_1 = 100$ mm, $L_2 = 150$ mm, $D_1 = 25$ mm, and $D_2 = 35$ mm.

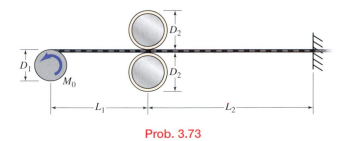

Prob. 3.73

3.74 Because the threads are right and left handed, turning a turnbuckle increases the tension in both cables. Consider here the contribution of the deformation of the turnbuckle itself to the increase in cable tension. Let the lead of each screw be 2 mm (each turn of the turnbuckle causes the screw threads to advance by 2 mm). Take the steel cable to have an effective diameter of 6 mm and an effective modulus of 165 GPa. Determine the increase in the cable tension with one turn of the turnbuckle (a) when the deformation of the turnbuckle is neglected and (b) when the deformation of the turnbuckle is accounted for. Take the dimensions to be $L_1 = 3$ m, $L_{tb} = 150$ mm, and $A_{tb} = 20$ mm^2. Neglect the deformation of the screw and the hooks.

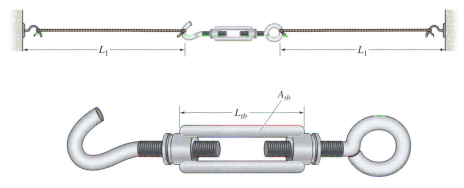

Prob. 3.74

3.75 A structure rests on a set of 30 m long vertical steel piles each having the cross-section shown. The center of mass of the 500 Mg structure, located at the point designated with the circle shown, is not in the center of the piles ($e < s/2$). (a) Determine the forces in the piles. (b) Determine the tilting of the structure associated with weight. Take the dimensions to be $s = 8$ m, $b = 9$ m, $e = 2$ m, $w = 300$ mm, and $t = 20$ mm.

3.76 A structure rests on a set of 30 m long vertical steel piles each having the cross-section shown. The 300 Mg structure has a center of mass at the center of the piles (as if $e = s/2$). The structure is then subjected to wind loading that is determined to be statically equivalent to a net horizontal force of 100 kN acting at 30 m above the foundation. The foundation piles must react to the structure weight and the overturning moment of the wind loading. Determine the maximum and minimum pile stresses.

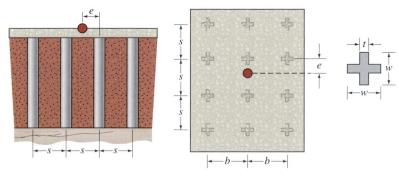

Probs. 3.75–76

3.77 Steel piles for a foundation are set on uneven ground so that the central ones are longer than the outer ones. All piles are steel pipes of 10 in. outer diameter and 0.5 in. thickness. The piles support a concrete cap and a structure (not shown) above it. The weight of the cap and structure is 400 kip with center of gravity off center from the two longer piles (the structure is still located symmetrically between two rows of piles separated by s). Determine the forces in the piles. Take the dimensions to be $b = 20$ ft, $s = 15$ ft, $c = 8$ ft, $L_1 = 50$ ft, and $L_2 = 70$ ft.

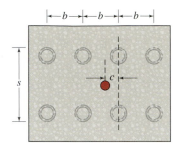
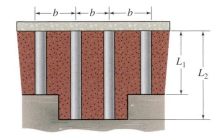

Prob. 3.77

3.78 A plate is suspended with three springs. When the springs are identical with nominal constant k_0, they are to be located around the plate center at equal distances R_0 from the center. However, only one spring had its nominal spring constant of k_0; the other two springs had different constants, k_1 and k_2. The plate must still be suspended so that it remains horizontal.

Determine altered locations of the second and third springs, R_1 and R_2, in terms of the nominal distance R_0 and the spring constants.

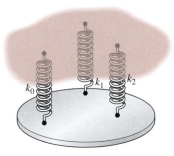
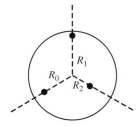

Prob. 3.78

3.79 A decorative truss-like structure is to be devised based on identical members located symmetrically about a central point at angles $\theta_1, \theta_2, \ldots \theta_i, \ldots \theta_N$ (there are $2N$ bars). The bars have modulus E, cross-sectional area A, and length L. Treat the semicircular structure to which the bars are attached as rigid. Prove that the force in the ith bar is

$$P_i = \frac{F_0 \sin\theta_i}{2\displaystyle\sum_{i=1}^{N}\sin^2\theta_i}$$

Also determine the deflection at the point where the load is applied.

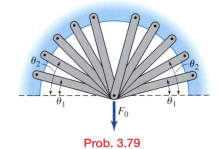

Prob. 3.79

3.80 Composite materials often consist of long fibers embedded in a matrix of another material, for example, graphite fibers embedded in plastic. When the fibers are uniformly distributed and small compared to the overall dimensions, the composite can act as a homogeneous material with uniform effective properties. Some of the effective elastic properties of such composites can be predicted to a good approximation using mechanics of materials concepts.

Prob. 3.80

The fraction of the total volume that is occupied by the fibers is named the fiber volume fraction, V_f. The volume fraction of the matrix is V_m, which satisfies the relation $V_f + V_m = 1$, since the total volume is the sum of fiber volume and matrix volume. Assume that the fibers are aligned and extend over the full length of the bar. Then, the fraction of the volume occupied by fibers (V_f) is the same as the fraction of the cross-sectional area containing fiber when the bar is cut perpendicularly to the fibers. Further, assume that the fibers are bonded to the matrix.

An overall uniaxial load P is applied to the composite. Assume the fiber and matrix are each in uniaxial tension. The composite elastic modulus E_{comp} is the proportionality between the overall stress on the composite and the overall strain. Show that E_{comp} is related to the elastic moduli of the fiber E_f and matrix E_m, and their volume fractions is given by the following so-called "rule of mixtures":

$$E_{comp} = E_f V_f + E_m V_m$$

3.81 The horizontal bar, approximated as rigid, is attached to a compliant material which provides a reaction force which opposes the deflection of the material; the force per length q is proportional to the deflection $q = kv$.

(a) A concentrated downward force is applied to the bar at the point shown. Determine the deflection of the bar at the two ends as a function of the parameters.

(b) Use the above result to determine the position a/L at which the force should be applied if the deflection at the left end is to be zero. Relate this to another result regarding distributed loads that might be familiar.

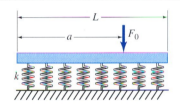

Prob. 3.81

Focused Application Problem

3.82 A small-scale wind turbine tower must pivot about its base because it is raised up from the ground by a gin pole. The tower is secured in the plane as shown by two pairs of 0.25 in. cable, each attached to an anchorage. All cables are pretensioned by equal amounts. The thrust force from the turbine's operation is balanced by increases in the tensions in the front cables and corresponding decreases in the rear cables. Determine the tension changes in the upper and lower pairs of cables due to a 250 lb thrust force. Take the effective modulus of the steel cable to be 24×10^6 psi. The distances are $w = 30$ ft, $h_0 = 45$ ft, $h_1 = 37$ ft, and $h_2 = 22$ ft.

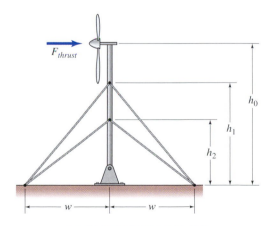

Prob. 3.82 (Appendix A6)

>>End Problems

3.5 Thermal Effects

By causing expansion and contraction, changes in temperature directly affect deformation and indirectly affect stress. Here we show how to account for this effect of temperature changes, which in some applications can be critical.

1. Most materials expand with increasing temperature approximately in proportion to the temperature increase.

No forces are applied to this bar, and it is free to expand or contract. Lines are drawn on the bar, dividing it into equal segments, and its temperature is raised by ΔT.

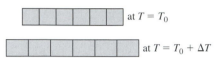

It is observed that each segment of the bar elongates by an amount proportional to its length. For example, if the whole $\Rightarrow$ bar elongates by 1 mm, half of it elongates by 0.5 mm.

Temperature increase causes a uniform thermal strain.

2. Define thermal strain and coefficient of thermal expansion.

Thermal strain, ε_{th}, is defined as the normal strain that an element, unconstrained by any neighboring elements, would undergo if the temperature is altered.

Coefficient of thermal expansion, α, captures the intrinsic tendency of a material to expand or contract due to temperature changes, and it is defined as the proportionality between the thermal strain and the change in temperature.

Thermal strain is defined as the strain that an unconstrained body undergoes when the temperature is altered. The thermal strain is approximately proportional to the temperature increase. We define the **coefficient of thermal expansion**, α, as the proportionality between thermal strain and temperature change: $\varepsilon_{th} = \alpha \Delta T$. Because ε is unitless, units for α are $(°C)^{-1}$ or $(°F)^{-1}$.

Interpret signs of equation $\varepsilon_{th} = \alpha \Delta T$.

For usual case of $\alpha > 0$:

Heating ($\Delta T > 0$) expands ($\varepsilon_{th} > 0$)

Cooling ($\Delta T < 0$) contracts ($\varepsilon_{th} < 0$)

Typical values for materials

Glass ($\sim 10^{-6} (°C)^{-1}$): $\varepsilon_{th} = 0.1\%$ needs $\Delta T = 1000°C$

Metal ($\sim 10^{-5} (°C)^{-1}$): $\varepsilon_{th} = 0.1\%$ needs $\Delta T = 100°C$

Plastic ($\sim 10^{-4} (°C)^{-1}$): $\varepsilon_{th} = 0.1\%$ needs $\Delta T = 10°C$

3. The strain due to stress is termed elastic strain, and it adds to the thermal strain.

Now, consider the effect of applying stresses while raising the temperature.

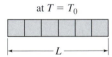
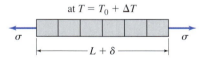

The elastic strain is defined is the strain due to stress with no change in temperature: $\varepsilon_{elastic} = \sigma/E$. One finds that the total strain due to an application of stress and a change of temperature is the sum of the elastic and thermal strains. So elastic and thermal strains can be superposed (added).

Total strain $\varepsilon = \varepsilon_{elastic} + \varepsilon_{th} = \sigma/E + \alpha \Delta T$

Definitions of stress (σ) and strain (ε) remain the same: $\sigma = P/A$ and $\varepsilon = \delta/L$

The material law, or relation between stress and strain, must be revised as follows:

No longer true: $\varepsilon = \sigma/E$ or $\delta = PL/EA$ (assumed elastic material, no temperature change)

New relations: $\varepsilon = \sigma/E + \alpha \Delta T$ or $\delta = PL/EA + \alpha \Delta T L$

Note: Properties, such as E and particularly σ_Y, can also depend on temperature. While these can be important in some applications, thermal expansion is the only effect of temperature treated here.

4. The combined effect of forces and temperature are illustrated in this situation where both vary.

This bar is initially at temperature T_0. Then, one part of the bar is heated to $T_0 + \Delta T_1$ and another part is cooled to $T_0 - \Delta T_2$. Both ΔT_1 and ΔT_2 are positive. At the same time, external forces F_0 are applied as shown to the bar. With the acting forces and with the new temperatures, what are the stresses in the bar? If the displacement at point C is zero, what are the displacements at points A, B, and D? Find these in terms of F_0, ΔT_1, ΔT_2, E, α, A, L_{AB}, L_{BC}, and L_{CD}.

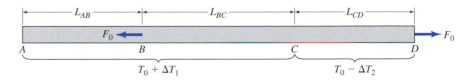

Uniform E, A, α

5. Find internal forces and from them find stresses and elastic strains.

Use FBDs of segments to find internal forces:

 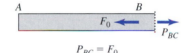 

$P_{AB} = 0$

$P_{BC} = F_0$

$P_{CD} = F_0$

$AB{:}\ \sigma = P_{AB}/A = 0 \Rightarrow \varepsilon_{elastic} = 0$

$BCD{:}\ \sigma = F_0/A \Rightarrow \varepsilon_{elastic} = F_0/EA$

6. Find thermal strains from temperature changes.

Use the temperature change in each segment to determine its thermal strain:

$$ABC{:}\ \varepsilon_{th} = \alpha\Delta T_1 \quad (>0) \qquad CD{:}\ \varepsilon_{th} = -\alpha\Delta T_2 \quad (<0)$$

7. Add the elastic strains due to stress and the thermal strains due to temperature changes to find the total strain.

In each segment, the total strain ε is the sum of elastic ($\varepsilon_{elastic}$) and thermal ($\varepsilon_{thermal}$) strains:

$$AB{:}\ \varepsilon_{AB} = 0 + \alpha\Delta T_1$$
$$BC{:}\ \varepsilon_{BC} = F_0/EA + \alpha\Delta T_1$$
$$CD{:}\ \varepsilon_{CD} = F_0/EA - \alpha\Delta T_2$$

8. Relate the elongations to the total strain.

$$\delta_{AB} = \varepsilon_{AB}L_{AB} = \alpha\Delta T_1\, L_{AB}$$
$$\delta_{BC} = \varepsilon_{BC}L_{BC} = F_0L_{BC}/EA + \alpha\Delta T_1\, L_{BC}$$
$$\delta_{CD} = \varepsilon_{CD}L_{CD} = F_0L_{CD}/EA - \alpha\Delta T_2\, L_{CD}$$

9. Relate the elongations to the displacements.

$$\delta_{AB} = u_B - u_A;\ \delta_{BC} = u_C - u_B;\ \delta_{CD} = u_D - u_C$$

We are given $u_C = 0$; therefore

$$\delta_{BC} = u_C - u_B \Rightarrow u_B = -\delta_{BC} = -F_0L_{BC}/EA - \alpha\Delta T_1 L_{BC}$$
$$\delta_{AB} = u_B - u_A \Rightarrow u_A = u_B - \delta_{AB} = -F_0L_{BC}/EA - \alpha\Delta T_1(L_{AB} + L_{BC})$$
$$\delta_{CD} = u_D - u_C \Rightarrow u_D = u_C + \delta_{CD} = F_0L_{CD}/EA - \alpha\Delta T_2 L_{CD}$$

>>End 3.5

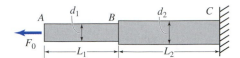

A steel bar with varying diameter is simultaneously subjected to an axial force $F_0 = 5000$ lb and is cooled by 50°F. (a) Determine the maximum stress in the bar. (b) Determine the displacements at A and B. $L_1 = 10$ in., $L_2 = 14$ in., $d_1 = 0.75$ in., $d_2 = 1$ in., $E = 30 \times 10^6$ psi, and $\alpha = 8 \times 10^{-6}$ (°F)$^{-1}$.

Solution

To help picture the effects of the temperature change and the applied force, we show their effects on deformation independently.

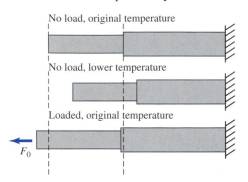

No load, original temperature

No load, lower temperature

Loaded, original temperature

We seek the displacements and stress induced by the combination of changing the temperature and applying the force.

External loads only act at the two ends. So, when F_0 is applied, there is a support reaction of F_0, regardless of whether the temperature is changed.

The internal force in both segments is F_0.

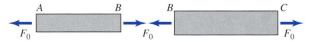

Find the stresses in each segment.

$$\sigma_1 = \frac{P_1}{A_1} = \frac{5000 \text{ lb}}{\frac{\pi}{4}(0.75 \text{ in.})^2} = 11.32 \text{ ksi} \qquad \sigma_2 = \frac{P_2}{A_2} = \frac{5000 \text{ lb}}{\frac{\pi}{4}(1 \text{ in.})^2} = 6.37 \text{ ksi}$$

The maximum stress is 11.32 ksi.

$$\varepsilon = \varepsilon_{elastic} + \varepsilon_{thermal} \qquad \delta = \delta_{elastic} + \delta_{thermal}$$

$$\delta_1 = \frac{P_1 L_1}{E A_1} + \alpha \Delta T L_1 = \frac{(5000 \text{ lb})(10 \text{ in.})}{(30 \times 10^6 \text{ in.})\frac{\pi}{4}(0.75 \text{ in.})^2} - 8 \times 10^{-6}(\text{°F})^{-1}(50\text{°F})(10 \text{ in.}) = -2.27 \times 10^{-3} \text{ in.}$$

$$\delta_2 = \frac{P_2 L_2}{E A_2} + \alpha \Delta T L_2 = \frac{(5000 \text{ lb})(14 \text{ in.})}{(30 \times 10^6 \text{ in.})\frac{\pi}{4}(1 \text{ in.})^2} - 8 \times 10^{-6}(\text{°F})^{-1}(50\text{°F})(14 \text{ in.}) = -2.63 \times 10^{-3} \text{ in.}$$

Define deflections u_A, u_B, and u_C as positive to the right.

$$\delta_1 = u_B - u_A \qquad \delta_2 = u_C - u_B$$

The end C is fixed at $\Rightarrow u_C = 0$

$$\Rightarrow u_B = -\delta_2 \Rightarrow u_B = 2.63 \times 10^{-3} \text{ in. (to the right)}$$

$$\Rightarrow u_A = u_B - \delta_1 \Rightarrow u_A = 2.86 \times 10^{-3} \text{ in. (to the right)}$$

>>End Example Problem 3.11

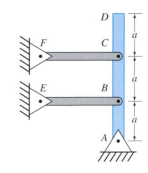

In the configuration shown, all members are at a uniform temperature, and *ABCD* is vertical. Then, the two horizontal bars are subjected to different temperatures: *BE* is warmed by ΔT_1, and *CF* by ΔT_2. Assume bar *ABCD* can pivot about point *A*, but remains rigid. Determine the displacement of the point *D*.

Solution

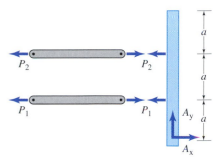

The two members, *BE* and *CF*, are two-force members (the only forces acting on them are due to pins at each end). Each member must be subjected to axial tension or compression. We draw the internal forces assuming tension and label them P_1 and P_2.

Equal and opposite forces must act on *ABCD*, since the forces on the two-force members come from the pins of *ABCD*.

From equilibrium, we find: $\quad \sum M|_A = P_2(2a) + P_1(a) = 0 \Rightarrow P_1 = -2P_2$

Notice that forces P_1 and P_2 cannot be found from equilibrium alone. The summation of forces would bring in the pin force at *A* and would not help to find P_1 and P_2.

From the material law of each bar:

$$\delta_1 = \delta_{1,elastic} + \delta_{1,thermal} = \frac{P_1 L}{EA} + \alpha \Delta T_1 L$$

$$\delta_2 = \delta_{2,elastic} + \delta_{2,thermal} = \frac{P_2 L}{EA} + \alpha \Delta T_2 L$$

This adds two more equations, but also two more unknowns (δ_1 and δ_2).

The elongations of the two bars will be related because they are attached to the pivoting rigid bar.

First relate the elongations to the displacements, noting that displacements are zero at *E* and *F*.

$$\delta_1 = u_B - u_E = u_B \qquad \delta_2 = u_C - u_F = u_C$$

Since *ABCD* rotates rigidly about *A*, the displacements at *B* and *C* must be related.

From similar triangles $\quad \dfrac{u_B}{a} = \dfrac{u_C}{2a} \Rightarrow u_C = 2u_B \Rightarrow \delta_2 = 2\delta_1$

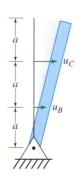

Therefore,

$$\frac{P_2 L}{EA} + \alpha \Delta T_2 L = 2\left[\frac{P_1 L}{EA} + \alpha \Delta T_1 L \right]$$

Use $P_1 = -2P_2 \Rightarrow P_2 = \dfrac{EA\alpha}{5}[2\Delta T_1 - \Delta T_2]$

Because the bar pivots about point *A*, we can again use similar triangles to find the displacement u_d.

$$\frac{u_C}{2a} = \frac{u_D}{3a} \Rightarrow u_D = \frac{3}{2}u_c = \frac{3}{2}\delta_2 = \frac{3}{2}\left[\frac{P_2 L}{EA} + \alpha \Delta T_2 L \right] = \frac{3\alpha L}{5}[\Delta T_1 + 2\Delta T_2]$$

>>End Example Problem 3.12

Additional data on material properties needed to solve problems can be found in Appendix D or inside back cover.

3.83 A steel wire of diameter 0.0625 in. is strung between a house and a utility pole. It is to remain taut during the entire year with a minimum tension of 10 lb. Say that over the year the temperature ranges from 100°F down to −20°F. Assume there is no displacement of the house and the pole. (a) At what tension should the wire be set if it is installed in the spring when the temperature is 65°F? (b) Over the course of the year, what will the maximum stress in the wire be? Take the wire to be 30 ft. long and the modulus and thermal expansion to be $E = 30(10^6)$ psi and $\alpha = 6.6(10^{-6})(°F)^{-1}$.

Prob. 3.83

3.84 The bar is composed of steel and brass segments that are connected together and then placed between two fixed supports at room temperature (25°C). The temperature is then raised to (70°C). (a) Determine the stress in each bar. (b) Determine the displacement, if any, of the midpoint of the brass portion. Take the areas and lengths to be $A_s = 100$ mm^2, $A_b = 160$ mm^2, $L_s = 200$ mm, and $L_b = 150$ mm, and material properties $E_s = 200$ GPa, $E_b = 100$ GPa, $\alpha_s = 12(10^{-6})(°C)^{-1}$, and $\alpha_b = 21(10^{-6})(°C)^{-1}$.

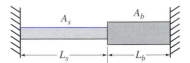

Prob. 3.84

3.85 A copper rod that is fixed at its bottom is separated from an upper surface by a gap. At room temperature (25°C) the rod has length $L = 120$ mm and the gap $g = 0.1$ mm. (a) Determine the stress in the rod if the rod is heated to 100°C. (b) To what maximum temperature can the rod be raised without developing any stress? Take the rod area, modulus, and thermal expansion to be $A = 40$ mm^2, $E = 120$ GPa, and $\alpha = 17 \times 10^{-6}(°C)^{-1}$.

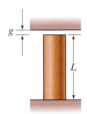

Prob. 3.85

3.86 The vertical bars, approximated as rigid, are connected by three deformable links. The links all have the same cross-sectional area A. The upper two links have the same modulus and thermal expansion, which are different from those of the lower link. Note that $\alpha_a < \alpha_b$. Let the temperature of the three be increased by ΔT.

Determine an expression for the internal force in each of the bars, and indicate if each is in tension or compression. Hint: the rigid bars can displace and rotate.

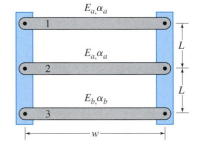

Prob. 3.86

3.87 A steel bar shown is fixed at the right end and has an out-of-plane thickness of 10 mm. Say that the temperature is increased by 20°C and a force F_0 is applied as shown midway along the bar. Use the properties $E = 200$ GPa, $\alpha = 12 \times 10^{-6}$ (°C)$^{-1}$. (a) Determine the value of the force F_0 which results in zero displacement at the left end ($x = 0$). For the value of force F_0 determined in part (a) and the temperature rise of 20°C, determine the following: (b) the stress at $x = 50$ mm, (c) the displacement at $x = 100$ mm, and (d) the stress at $x = 150$ mm.

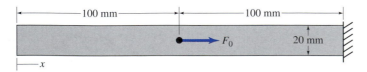

Prob. 3.87

3.88 Four elastic bars (two identical pairs), separated by the distances shown, are pinned to a single rigid member (vertical) and pin-supported at their other ends. The vertical member is pinned at its center C. Initially, the bars have no stress in them. The temperature is decreased by ΔT (ΔT is a positive number) in only the outer bars (labeled 1), which have a coefficient of thermal expansion, α. The temperature does not change in the inner bars (labeled 2).

(a) Calculate the forces in the four elastic bars, indicating whether they are tensile or compressive.
(b) Calculate the displacement of the pin at Q, indicating whether it is to the left or right.

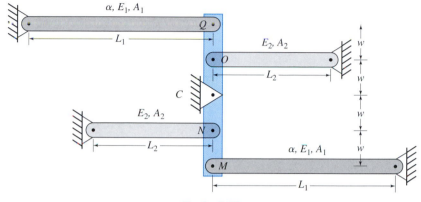

Prob. 3.88

3.89 Aluminum wire is sometimes clad with copper. Because the thermal expansions are different, stresses can develop in the wire because the bonding between the wire and the cladding requires them to strain together. Say the aluminum wire of diameter 0.125 in. has copper cladding that is 0.1 in. thick. Assume the clad wire is subjected to zero net axial force, and that the stresses are zero at 70°F. Calculate the stresses in each of the aluminum and copper wire due to a temperature increase of 200°F. Indicate the stress magnitude and whether each is tensile or compressive. Take the modulus and thermal expansion of aluminum to be $E = 10(10^6)$ psi and $\alpha = 13(10^{-6})(°F)^{-1}$, and those of copper to be $E = 17(10^6)$ psi and $\alpha = 9.5(10^{-6})(°F)^{-1}$.

3.90 The portion of a steel roadway must be allowed to expand and contract with seasonal temperature changes. The part of the roadway over the pin does not move. Temperature variations from −30°F to 120°F are to be tolerated. How much lateral movement does the rocker need to accommodate? Take the properties of steel to be $E = 30(10^6)$ psi and $\alpha = 6.6(10^{-6})(°F)^{-1}$, and the length $L = 120$ ft.

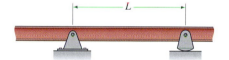

Prob. 3.90

3.91 A square Plexiglas® sheet, which is bolted to the steel frame shown, serves as a window in an outdoor structure. While the 3 mm diameter bolts fit snugly in the holes in the steel frame, oversize holes are used in the Plexiglas to avoid stresses due to thermal expansion. Say that temperatures ranging from 10°C to 50°C are to be accommodated without any stresses occurring in the Plexiglas, which has a thermal expansion coefficient of $70(10^{-6})(°C)^{-1}$. Presume that no change in the bolt separation distance of 600 mm occurs during the temperature changes. Determine the minimum required diameter of the holes in the Plexiglas, assuming they are optimally placed, for there to be zero stress in the Plexiglas over the range of temperature.

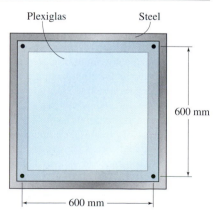

Prob. 3.91

Plexiglas® is a registered trademark of Altuglas Int'l., Arkema Inc., 2000 Market St., Philadelphia, PA 19103.

3.92 Batteries in a flashlight must be held in electrical contact even though the outer plastic housing expands differently than the battery and terminals. This is accomplished by having a compliant spring compressed as the cover is screwed on the flashlight. When the cover is screwed on at room temperature (70°F), the spring, with constant 5 lb/in., compresses by 0.25 in. The plastic casing has a thermal expansion coefficient of $85(10^{-6})(°F)^{-1}$. Let the battery and spring have a single thermal expansion coefficient like that of steel, $6.6(10^{-6})(°F)^{-1}$, and neglect the elastic deformation of the battery. If the temperature of the flashlight increases to 120°F, does the spring force increase or decrease and by how much? Take the casing wall to have an outer diameter of 1.5 in., thickness 0.125 in., length $L = 3$ in., and elastic modulus of 300 ksi.

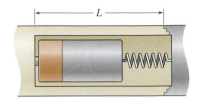

Prob. 3.92

3.93 A rubber band, which is 0.125 in. thick and 0.5 in. wide and forms a loop of diameter 0.99 in. at room temperature (70°F), is slipped around a 1 in. diameter steel rod. Take the rubber to have a modulus of 300 psi and a thermal expansion coefficient of $120(10^{-6})(°F)^{-1}$. (a) At what temperature would the rubber band just slip off the rod? (b) What would be the stress in the rubber band if the rod and rubber band were placed in ice water? Neglect the deformation of the steel due to stress, but include its thermal expansion with $\alpha = 6.6(10^{-6})(°F)^{-1}$.

3.94 The thermal expansion of a new material in the form of a 150 mm bar is measured by comparison with a 120 mm long standard copper bar with a thermal expansion coefficient of $17(10^{-6})(°C)^{-1}$. The two bars are fixed at one end to a common plate, and the change in separation of the other two ends is measured with an LVDT (linear variable displacement transducer). When the temperature is raised from 25°C to 100°C, the ends of the two bars are now separated by an *additional* 0.027 mm. Determine the thermal expansion coefficient of the new material.

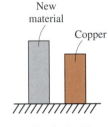

Prob. 3.94

3.95 The feeding tube for a low-temperature device shown is vacuum insulated. The feeding tube is made of two coaxial stainless-steel tubes, assembled at room temperature of 20°C, and rigidly connected only at both ends. During operation, liquid nitrogen is forced through the inner tube at −196°C, and the temperature of the outer tube is measured to be −20°C. Determine the stresses in the inner and outer tubes during operation. Consider only axial expansion of the tubes and neglect radial expansion. Other dimensions of the system are: overall length of 200 mm, 8 mm ID for the inner tube, 12 mm ID for the outer tube, and a wall thickness of 0.5 mm for each tube. Take the thermal expansion coefficient of stainless steel to be $\alpha = 17 \times 10^{-6}(°C)^{-1}$.

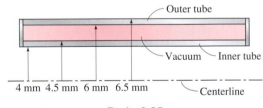

Prob. 3.95

3.96 Electrical wires are sometimes secured to a housing as shown. Because of thermal expansion mismatch between the copper wire and the surrounding steel housing and screw, such an arrangement cannot tolerate excessive changes in ambient temperature. Say that the screw has been tightened at 25°C so that the screw force is 200 N. Consider both temperature increases as well as decreases. Determine the temperature at which contact would be lost between the screw and wire. The housing is relatively bulky so approximate it as elastically rigid relative to the screw and wire (although it expands and contracts thermally like steel). Assume the wire is compressed over a length equal to the screw diameter and estimate the elastic resistance of the wire by assuming an appropriate length and area. Use the dimensions $L_s = 10$ mm, $d_s = 3$ mm, and $d_w = 2$ mm and respective properties of steel and copper: $E_s = 200$ GPa, $\alpha_s = 12(10^{-6})(°C)^{-1}$, $E_c = 120$ GPa, and $\alpha_c = 17(10^{-6})(°C)^{-1}$.

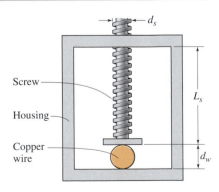

Prob. 3.96

3.97 A torch causes concentrated heating of the central portion of a bar that is initially unstressed at uniform temperature. The bar has uniform properties, E, A, α, and length $2L$. Approximate the remaining structure to which the ends are attached as rigid and immovable. Say the heating causes the bar's temperature to increase by an amount:

$$\Delta T = \Delta T_0 \exp\left[\frac{-|x|}{c}\right]$$

where x is the distance from the center, and the parameter c captures the length of the heated zone.

Determine the compressive stress in the bar as a function of the parameters.

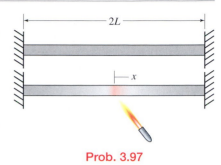

Prob. 3.97

3.98 Fiber-reinforced composites (see Problem 3.80 on the effective modulus of composites) are sometimes subjected to temperature changes. We are interested in the effective thermal expansion of the composite, along with any stresses that develop in the fiber and matrix, even if no external loads are applied.

Prob. 3.98

Assume that the fibers, which are aligned and bonded to the matrix, extend along the full length of the material. Let the volume fraction, elastic modulus, and coefficient of thermal expansion be V_f, E_f, and α_f in the fiber and V_m, E_m, and α_m in the fiber.

Derive formulas for:

(a) the stress that develops in the matrix due to a temperature rise of ΔT.
(b) the effective thermal expansion coefficient of the composite.

Show that the effective thermal expansion coefficient reduces to the rule of mixtures based on the individual thermal expansion coefficients, $\alpha_{comp} = \alpha_f V_f + \alpha_m V_m$, if the elastic moduli are equal ($E_f = E_m$).

3.99 The steel cable-stayed bridge shown is pinned at the left pylon (fixed against longitudinal motion), but free to move longitudinally at the other tower and at the end supports. If the temperature is expected to range from $-20°C$ to $40°C$, how much total longitudinal displacement do the bearings need to accommodate at (a) the other tower, (b) the left end support, and (c) the right end support. Take the dimensions to be $L = 400$ m and $s = 150$ m, and $\alpha = 12(10^{-6})(°C)^{-1}$.

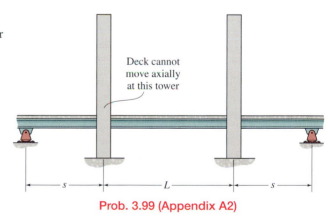

Deck cannot move axially at this tower

Prob. 3.99 (Appendix A2)

>>End Problems

3.6 Wrapped Cables, Rings, and Bands

Axial loading can also occur in long, thin bodies that are not straight. In this example, the headband is tight enough around the head to keep the flashlight in place. We consider the tension in rings and bands that wrap around other bodies and the loading that induces the tension.

---

1. Determine the strain in a circular ring from its change in arc length.

Here is a ring initially with a radius R_i. Lines have been drawn to indicate sectors. The circumference c is the length of the member. Let the ring be expanded so its radius is R_f.

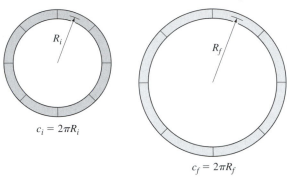

$$c_i = 2\pi R_i$$

$$c_f = 2\pi R_f$$

As the ring expands, the circumference increases. So the strain is tensile in the long direction (around the circumference), just as in a straight member in tension.

$$\text{Axial strain:} \quad \varepsilon = \frac{c_f - c_i}{c_i} = \frac{2\pi(R_f - R_i)}{2\pi(R_i)} = \frac{R_f - R_i}{R_i}$$

---

2. A tensile force that acts in the circumferential direction is in equilibrium only if there is an outward radial force on the ring.

Given the circumferential tensile strain around the ring, there should be a tensile force in the circumferential direction. Isolate an infinitesimally short segment of the ring. Equal and opposite internal tensile forces P act on this segment.

Now consider a finite segment of the ring. The internal circumferential forces at its ends are not aligned. What additional load keeps this segment in equilibrium?

In the case of the headband, the ring remains at its enlarged diameter because the head is keeping it there. We feel the pressure on our head from the headband, so the head exerts outward distributed forces on the headband.

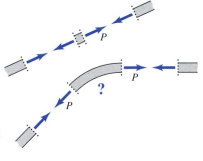

---

3. Relate the internal tensile force in the ring to the outward radial force per length q.

A uniform, outward radial force per length q keeps the circular ring in equilibrium (it produces zero net force and moment). Draw an FBD of half of the ring. To be in equilibrium, the internal forces balance q.

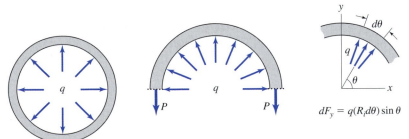

$$dF_y = q(R_i d\theta)\sin\theta$$

Integrate the distributed force q over the arc length to find the net upward (y) force and determine P from vertical equilibrium.

$$\sum F_y = -2P + \int_0^\pi q\sin\theta\, R_i\, d\theta = 0 \Rightarrow P = qR_i$$

4. Relate the radial expansion of the ring to the outward distributed force q.

$$\sigma = \frac{P}{A} = \frac{qR_i}{A} \qquad \varepsilon = \frac{\sigma}{E} = \frac{qR_i}{EA} \qquad R_f - R_i = R_i\varepsilon = \frac{qR_i^2}{EA}$$

We assume that the ring is thin: R_i and R_f are close enough that $P = qR_i$ and $P = qR_f$ would be approximately equal.

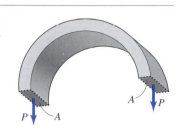

---

5. A partial circular arc will also be in tension if a radial force per length acts.

The member in tension need not form a complete circle. Whenever the direction of the tension changes along a member, there must be forces acting perpendicularly to the member to maintain equilibrium.

For example, belts that can be tightened are used to secure cargo to a truck bed, for example this set of pipes. As it is tightened, the belt goes into tension.

Where the belt wraps around a pipe, a change in the direction of tension occurs. This requires an outward distributed force q from the pipe on the curved segment of belt. The belt applies an equal and opposite inward distributed force on the pipes, which presses the pipes together and onto the truck bed.

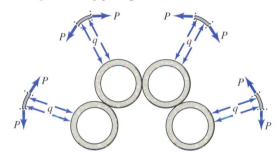

From equilibrium of the curved segment of belt, we again find that the distributed force per length $q = P/R$. By tightening the belt, we raise the tension P and, therefore, the force q on the pipes.

---

6. A body, such as a gear or a hub, can be firmly attached to a shaft if it has a circular hole with a diameter slightly smaller than the shaft diameter.

Say a body, such as a gear, with a hole originally of diameter d_i, is placed around a shaft of diameter d_s, where $d_i < d_s$. d_i and d_s would differ by only a slight percentage.

Shaft

Body with hole

To "shrink-fit" or "press-fit" a gear on this shaft, heat the gear or cool the shaft, and slip the gear onto the shaft. When the parts come to room temperature, they try to regain their initial diameters but cannot. The assembled gear and shaft press strongly against one another.

A steel insert was press-fit into this aluminum hub for an SAE Formula One car. As seen in Chapter 4, a large pressure between the hub and the insert is needed so enough torque could be transmitted via friction.

In general, we cannot compute stresses due to a shrink-fit with mechanics of materials.

---

7. Estimate the gripping force of a thin ring that is shrink-fit around an initially larger, rigid cylinder.

Say a thin ring rather than a gear is shrink-fit around a shaft. If the shaft is much stiffer than the ring (relatively rigid), only the ring deforms, and the force per length between them can be found.

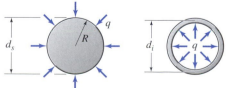

The ring expands so its inner diameter equals that of the shaft:

Circumferential strain in ring: $\varepsilon = (d_s - d_i)/d_i$

Tensile stress in ring: $\sigma = E\varepsilon$

Tensile force in ring: $P = \sigma A$

Inward force per length on shaft: $q = P/R = 2P/d_s$

>>End 3.6

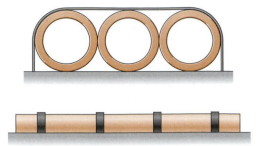

Three terracotta pipes are tied down with four rubber straps spaced along the length of the pipe. Each pipe has a 200 mm outer diameter, length of 5 m, and weighs $W = 200$ N. The straps have an unstretched length of 800 mm and a stiffness of 1 N/mm. Determine the contact force between the truck bed and each pipe and the contact force between each neighboring pair of pipes. Neglect friction between the straps and pipes and between the pipes and the truck bed.

Solution

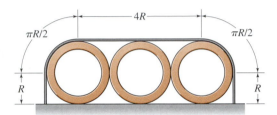

The elongated length, L_f, of the straps when they are wrapped around the pipes can be found from the geometry:

$$L_f = R + \pi R/2 + 4R + \pi R/2 + R = 6R + \pi R = 9.14R = 914 \text{ mm}$$

Use the strap stiffness to find the tension P in each strap:

$$P = k\delta = (1 \text{ N/mm})(914 \text{ mm} - 800 \text{ mm}) = 114.2 \text{ N}$$

Draw FBDs of the pipes separated from the truck bed and from each other. Keep the wrapped portion of the strap in contact with each pipe in the FBD of that pipe. Note that the force F corresponds to the total tensile force in all four straps. Hence, $F = 457$ N.

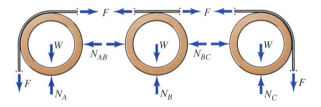

Apply equilibrium to the left pipe:

$$\sum F_x = -N_{AB} + F = 0 \Rightarrow N_{AB} = F = 457 \text{ N}$$
$$\sum F_y = -W + N_A - F = 0 \Rightarrow N_A = W + F = 657 \text{ N}$$

By symmetry or equilibrium of the right pipe, we find:

$$N_{BC} = 457 \text{ N}, N_C = 657 \text{ N}$$

Apply equilibrium to the center pipe:

$$\sum F_y = -W + N_B = 0 \Rightarrow N_B = 200 \text{ N}$$

>>End Example Problem 3.13

It is desired to slip a steel band around a concrete column. At 80°F, the steel band is 0.03 in. thick, 0.5 in. wide (into the page), and has an inner diameter of 11.997 in. The column has an outer diameter of 12 in. at 80°F.

(a) Show that if the band is warmed to 170°F, the band may be slipped over the column.

(b) Determine the stress in the band when the temperature returns to 80°F.

(c) Determine the stress in the band and the pressure (force per area) between the band and the column if the temperature rises to 120°F. Neglect the elastic deformability of column relative to the thin band.

$E_{steel} = 30 \times 10^6$ psi, $\alpha_{steel} = 9.6 \times 10^{-6}$(°F)$^{-1}$, $\alpha_{concrete} = 6.5 \times 10^{-6}$(°F)$^{-1}$,

Solution

(a) Heat up the steel from 80°F to 170°F $\Rightarrow \Delta T = 90$°F.

$\varepsilon_{th} = \alpha_{steel} \Delta T = (9.6 \times 10^{-6}$(°F)$^{-1})(90$°F$) = 8.64 \times 10^{-4}$

$\Delta d = (\varepsilon_{th})(12$ in.$) = 0.0104$ in. (Multiply ε_{th} by 11.997 in. and find nearly the same increase in diameter.) New diameter of band $= 11.997$ in. $+ 0.0104$ in. $= 12.01$ in. So the band would fit around the column.

(b) As the steel band returns to 80°F, it wants to return to its initial diameter of 11.997 in. The column is assumed to be rigid relative to the band, so it constrains the band to a diameter of 12 in. This increase in length of the band is purely elastic strain, since the thermal strain at 80°F is zero.

$\varepsilon = (12$ in. $- 11.997$ in.$)/12$ in. $= 2.50 \times 10^{-4}$. (This is also the circumferential strain since $c = \pi d$.) Since the thermal strain in the band is zero,

$\sigma = E\varepsilon = (30 \times 10^6$ psi$)(2.50 \times 10^{-4}) = 7500$ psi.

(c) Raise the temperature to 120°F: the band expands and the column expands. First, determine the diameters that the band and column would have if they could freely expand with the temperature change.

$\Delta d_{steel} = (9.6 \times 10^{-6}$(°F)$^{-1})(120$°F $- 80$°F$)(12$ in.$) = 4.608 \times 10^{-3}$ in.

$\Rightarrow d_{steel} = 11.997$ in. $+ 4.608 \times 10^{-3}$ in. $= 12.00161$ in.

$\Delta d_{concrete} = (6.5 \times 10^{-6}$(°F)$^{-1})(120$°F $- 80$°F$)(12$ in.$) = 3.120 \times 10^{-3}$ in.

$\Rightarrow d_{concrete} = 12.000$ in. $+ 3.120 \times 10^{-3}$ in. $= 12.00312$ in.

The column wants to be larger than the band, and since the column is rigid, it will constrain the band to be at the column's diameter. This diameter difference is made up by elastic strain in the band.

$\varepsilon = (12.00312$ in. $- 12.00161$ in.$)/12$ in. $= 1.258 \times 10^{-3}$

The stress in the band is related to this strain

$\Rightarrow \sigma = E\varepsilon = (30 \times 10^6$ psi$)(1.258 \times 10^{-4}) = 3775$ psi

The stress in the band at 120°F is less than the stress at 80°F, because the band expands thermally more than concrete does.

$P = \sigma A = (3775$ psi$)(0.03$ in.$)(0.5$ in.$) = 56.6$ lb

$q = P/R = (56.6$ lb$)/(6$ in.$) = 9.44$ lb/in.

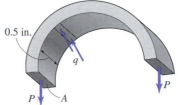

Since this q acts over the width of 0.5 in.

$\Rightarrow$ Pressure $= (9.44$ lb/in.$)/(0.5$ in.$) = 18.87$ psi

>>End Example Problem 3.14

PROBLEMS

Additional data on material properties needed to solve problems can be found in Appendix D or inside back cover.

3.100 Three cylinders, each of a diameter $D = 30$ mm, are arranged in a triangular pattern and are bound together with a rubber band. If the rubber band loop was cut and laid out flat, it would have an initial length of 70 mm, a thickness of 2 mm, and width of 6 mm (into the paper). Assume that the cylinders are nondeformable and that they make contact in pairs at one point. Determine the normal force between each pair of contacting cylinders. Take the rubber band to have a modulus of 1 MPa.

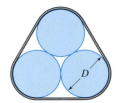

Prob. 3.100

3.101 Seven relatively rigid circular cylinders of 20 mm nominal diameter are arranged in a hexagonal pattern and are bound together with a rubber band. Let the exterior cylinders have identical diameters, which may or may not differ very slightly from the diameter of the interior cylinder. If the rubber band were laid out as a circular loop, it would have an unstretched diameter of 50 mm, a thickness of 2 mm, and a width of 8 mm (into the paper). Determine the maximum and minimum values for (a) the force between each pair of exterior cylinders and (b) the force between an exterior cylinder and the interior cylinder. Explain how the size differences between exterior and interior cylinders could affect the forces between cylinders. Take the rubber to have a modulus of 1 MPa.

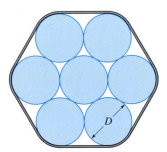

Prob. 3.101

3.102 A circular cylinder of diameter $D_2 = 75$ mm is tied to a larger cylinder of diameter $D_1 = 300$ mm. Say the tie has a strength $\sigma_f = 10$ MPa and a diameter $d = 3$ mm. Assume that the cylinders are nondeformable. Determine the maximum force with which the small cylinder can be pressed against the larger cylinder before the tie breaks. Does the force increase or decrease as D_2 is reduced relative to D_1?

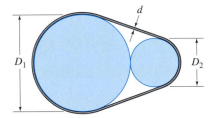

Prob. 3.102

3.103 One hundred wood members, each 8 ft long with a square cross-section of 1.5 in. by 1.5 in., are arranged in a 10×10 pattern. The members are squeezed together by 3 rubber bands spaced along the length of the members. When unstretched, the bands can each be arranged as a circle of diameter 15 in. The bands are initially 0.1875 in. thick and 1 in. wide (into the paper). Assume that the wood members are nondeformable. Determine the force that one needs to apply to extract a wood member from the center axially, if all the other wood members are prevented from moving. Take the friction coefficient between boards to be $\mu = 0.5$. Take the rubber band to have a modulus of 100 psi.

Prob. 3.103

3.104 A thin ring is composed of two materials, with distinct thermal expansion coefficients (α_1 and α_2) and elastic moduli (E_1 and E_2). When the ring is heated up and the two materials expand by different amounts (if $\alpha_1 \neq \alpha_2$), tensile or compressive stresses may develop between the rings. The materials could debond from each other, if excessive tensile stresses develop.

Approximate the two rings as very thin compared to the ring radius itself ($t_1 \ll R$ and $t_2 \ll R$). Assuming that the two rings remain bonded, show that the radial tensile stress, σ_i, that the rings exert on one another is given by the formula:

$$\sigma_i = \frac{(\alpha_2 - \alpha_2)\Delta T}{R\left[\dfrac{1}{E_1 t_1} + \dfrac{1}{E_2 t_2}\right]}$$

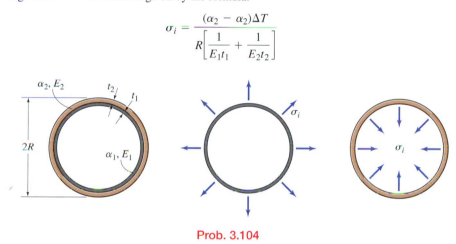

Prob. 3.104

>>End Problems

Chapter Summary

3.1 In axial loading, internal forces act parallel to the long axis of the member, through the centroid of the cross-section.

Internal normal force P is related to the **elongation** δ.

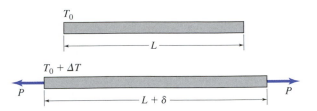

$$\delta = \frac{PL}{EA}$$

E: Young's modulus
A: Cross-sectional Area
L: Length

Displacement: Absolute motion of a single point body as a consequence of loading

u: horizontal displacement

v: vertical displacement

Elongation δ is related to the difference in displacements parallel to the bar at its ends.

$$\delta = u_2 - u_1$$

Initial: Deformed:

3.2 Multiple **external forces** (F_1, F_2, F_3) can act along the length of bar.

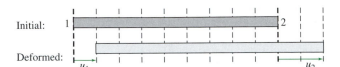

Define internal normal force P at a cross-section.

With multiple external forces, the internal force P varies along the bar.

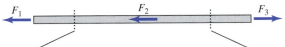

At all points between 1 and 2:
$$P = F_1 = F_3 - F_2$$

At all points between 2 and 3:
$$P = F_3 = F_1 + F_2$$

Determine stress at a particular point using P and area at that cross-section.

$$\sigma = \frac{P}{A}$$

Determine elongation in a segment between two cross-sections based on P in that segment.

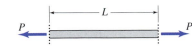

$$\delta = \frac{PL}{EA}$$

P, E, and A must be uniform in segment, which has length L.

3.3 In complex systems with multiple members in axial loading, members can rotate and displacements can act both parallel and perpendicular to members. The geometry simplifies for small rotations.

Displacement and rotation of a rigid body can dictate the displacements of deforming bodies that attach to it. The displacements of multiple bars connected to the same rigid body can be related to the displacement and rotation of that rigid body.

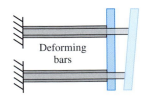

Deforming bars

When a bar elongates and rotates, displacements parallel to the bar, u_{par}, affect elongation and displacements perpendicular to the bar, u_{perp}, affect rotation.

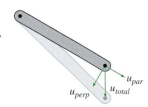

3.4 Statically Determinate Problems

In previous sections of the chapter, when external loads are prescribed, displacements can be found by applying relations we have studied in this sequence:

$$\text{External loads} \Rightarrow \text{Internal loads} \Rightarrow \text{Elongations} \Rightarrow \text{Displacements}$$

In particular, we were able to determine internal loads using equilibrium equations, that is, using statics alone (*statically determinate*).

Statically Indeterminate Problems

Sometimes there are insufficient equilibrium equations to find internal loads from external loads (*statically indeterminate*).

We use the same relations, but instead of solving sequentially, we solve simultaneously. All unknowns remain as variables until all equations are combined and then solved.

$$\text{External loads} \Leftrightarrow \text{Internal loads} \Leftrightarrow \text{Elongations} \Leftrightarrow \text{Displacements}$$

These problems can be solved, because we can always find additional relations between displacements, which compensate for the insufficient number of equilibrium equations.

3.5

Increase the temperature of an otherwise free bar of length L by ΔT. Because the elongation δ_{th} is proportional to L, define:

Thermal strain ε_{th}: $\quad \varepsilon_{th} = \dfrac{\delta_{th}}{L}$

Thermal strain is proportional to temperature increase: $\quad \varepsilon_{th} = \alpha \Delta T$

Coefficient of Thermal Expansion α depends on the material.

Simultaneously increase the temperature and apply axial forces. The total strain is equal to the **elastic strain** $\varepsilon_{elastic}$ due to stress plus the **thermal strain** ε_{th} due to ΔT.

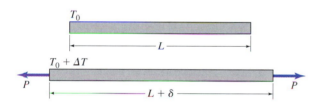

$$\varepsilon = \varepsilon_{elastic} + \varepsilon_{th} = \frac{\sigma}{E} + \alpha \Delta T$$

$$\delta = \delta_{elastic} + \delta_{th} = \frac{PL}{EA} + \alpha \Delta T$$

3.6

Radial force per length q results in a circumferential internal force P that causes a band or ring to expand.

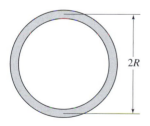 

Find the normal stress from the internal force and the ring cross-sectional area: $\sigma = P/A$.

Strain $\varepsilon = \sigma/E$ is proportional to the increase in circumference.

Torsion

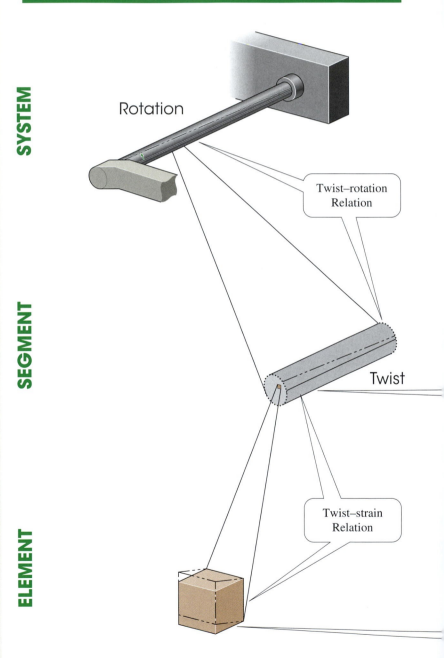

DEFORMATIONS

SYSTEM

Rotation

Twist–rotation Relation

SEGMENT

Twist

Twist–strain Relation

ELEMENT

Shear strain

FORCES

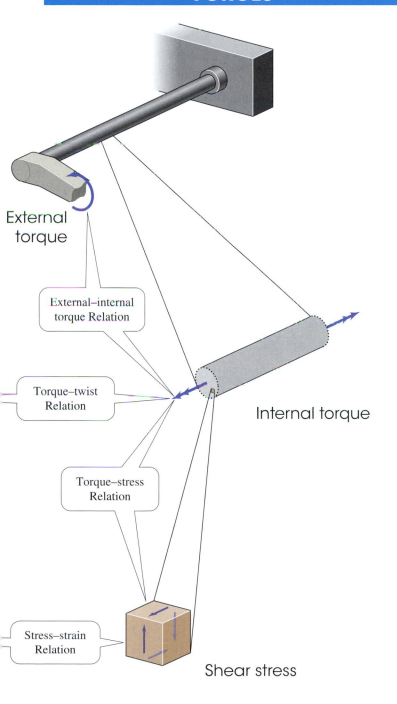

External torque

External–internal torque Relation

Torque–twist Relation

Internal torque

Torque–stress Relation

Stress–strain Relation

Shear stress

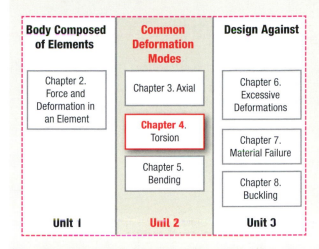

SYSTEM

SEGMENT

ELEMENT

Chapter Outline

A shaft of a torsion bar suspension deforming due to up and down displacement of a tire is an example of *twisting* or *torsional deformation*. A shaft subjected to torsion can be viewed as composed of segments or slices along the length of the member. Each cross-section rotates through some angle (**4.1**). When end faces of a segment rotate through different angles, the segment twists, giving rise to shear strains that vary in the radial direction (**4.2**). External loads on the bar result in an internal twisting moment or torque that acts oppositely on the two ends of a segment (**4.3**). The shear stresses acting on each face of a segment add up to the internal twisting moment (**4.4**). For a shaft of circular cross-section loaded only by opposite torques at its ends, the twist or relative rotation depends on the bar's length, shear modulus, and its polar moment of inertia, a property of the cross-section (**4.5–4.6**). The relation between torque and twist and maximum shear stress can also be determined for some non-circular cross-sections (**4.7–4.9**). A shaft can be subjected to several external torques along its length, with the internal torque varying in different segments (**4.10–4.14**). When a motor drives a shaft causing it to rotate at constant speed, the power, rotation speed, and torque are interrelated, but stress and deformation are related to internal torque just as for a stationary shaft (**4.15**).

137

4.1 Rotation

The pattern of deflection is a characteristic feature of a shaft in torsion. Successive cross-sections of a twisting shaft rotate by different amounts. Here, we describe this deflection pattern and how it is quantified.

1. Each cross-section of a shaft in torsion rotates about the shaft axis.

This wire was lying on the flat surface.

Now, one leg is held down and the other is pivoted. We consider the center portion, which stays straight and twists, to be the shaft.

The pivoting leg rotates by ϕ. If ϕ is small (in radians), then the end deflects by $w\phi$.

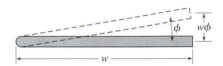

2. The rotation of one end of the shaft relative to the other captures the deformation in twisting and it is related to the torque.

Twist, $\Delta\phi$, is defined as the difference in rotation angle of one cross-section of a shaft relative to another, due to torsional deformation.

We define **twist** as the difference in rotation: the rotation angle of one end minus the other. In this case, since one leg doesn't rotate, the twist is equal to ϕ. The fingers apply forces to the legs that result in opposite torques on the central portion equal to Fw.

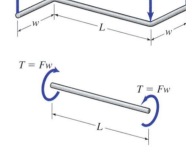

Here is a torsion bar suspension. Engineers design the torsion bar to give the right resistance to up and down motion of the tire over the road. When driving over a bump, the tire moves up relative to the rest of the vehicle, due to twisting of the torsion bar. The torsion bar is analogous to the twisting center portion of the wire. The upward motion of the tire is analogous to the end of the wire leg that lifts. If the torsion bar does not twist enough, the vehicle moves up too much (with the tire). If the twist is too great, the tire deflects too much relative to the vehicle and could contact it.

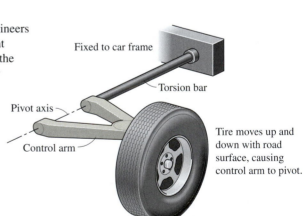

3. Successive cross-sections along the length of a twisted shaft rotate by different amounts.

A twisted shaft not only rotates at its ends. Every cross-section rotates, usually through a different angle.

The pegs on this shaft were initially aligned. The shaft is then twisted by rotating the right end through some angle.

From the new orientations of the pegs on the shaft, you can see that each cross-section rotates through a different angle.

4. For a uniform shaft twisted by torques at its ends, the rotation angle varies linearly with distance along the shaft.

A foam tube can illustrate the variation in rotation along a twisted shaft. A straight line was drawn on the side.

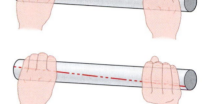

The tube was then twisted slightly, moving the line.

As viewed from the side, the line stays approximately straight. The vertical displacement of the line varies linearly along its length. So the rotations of cross-sections vary linearly.

We express the linear variation of rotation ϕ from ϕ_1 at $x = 0$ to ϕ_2 at $x = L$ as follows:

$$\phi = \phi_1 + (\phi_2 - \phi_1)\frac{x}{L}$$

where x is the distance along the shaft and L is its length. For example, halfway along the shaft (at $x = L/2$), $\phi = (\phi_1 + \phi_2)/2$, that is equal to the mean of the rotations at the ends.

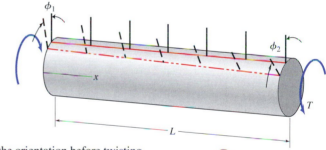

Rotation is measured relative to the orientation before twisting. A cross-section can rotate in either direction, which we distinguish with a plus or minus sign. We use the right hand rule to define positive rotation. The rotations ϕ_1 and ϕ_2 here, and at all cross-sections in between, are positive.

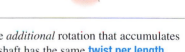

ϕ positive x positive

5. The twist per length, that is, the change in rotation angle per unit distance along the body, captures the intensity of the twisting deformation.

Every 1 mm of the shaft feels the same twisting torque. So the *additional* rotation that accumulates over every 1 mm segment is the same. Every segment of this shaft has the same **twist per length** (e.g., the same degrees of rotation per millimeter length).

$$\text{Twist per length} = \Delta\phi/L = (\phi_2 - \phi_1)/L$$

The twist per length captures the intensity of the twist.

Twist per length captures the intensity of the torsional deformation, and it is defined as the difference in rotation angle of one cross-section of a shaft relative to another, divided by the distance between the cross-sections.

>>End 4.1

4.2 Shear Strain in Circular Shafts

Rotation of successive cross-sections through different angles distorts or strains elements in a shaft. Here we look at the type of strain, its variation from point to point in the shaft, and how the overall twisting of the shaft affects the strain.

1. Rotations of successive cross-sections through different angles cause shear strain.

Rectangles were drawn on the side of this shaft.

When the shaft is twisted, the angles are no longer 90°. The rectangles have distorted or *sheared* into parallelograms.

Shear strain is due to the different rotations of successive cross-sections.

2. The shear strain, which is equal to the relative shear displacement over an element divided by its length, can be related to rotations at the ends of the shaft.

This element extends from x_1 to x_2. Follow its deformation when the shaft ends rotate.

The cross-section at x_1 rotates by $\phi(x_1)$. The cross-section at x_2 rotates by a greater angle $\phi(x_2)$.

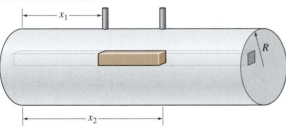

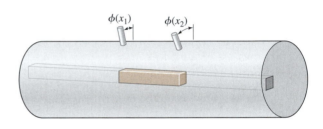

Here is the element, originally and as deformed

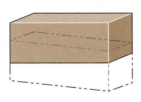

To see only the change in shape (strain), move the deformed element up by $R\phi(x_1)$.

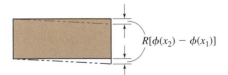

Compute the shear strain from the angle change as the rectangle distorts into a parallelogram.

Shear strain on the outer surface of shaft is

$$\gamma = \frac{R[\phi(x_2) - \phi(x_1)]}{x_2 - x_1}$$

If the element extends over the length of the shaft ($x_1 = 0$, $x_2 = L$) and has a twist $\Delta\phi$, then
$$\gamma = \frac{R\Delta\phi}{L}$$

3. Shear strain is proportional to the twist per length.

Shear strain depends on $\dfrac{\Delta\phi}{L}$, which is the twist per length. This is the intensity of twisting, or how rapidly the rotation changes over the length L of the shaft.

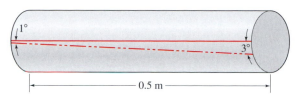

As an example, for this shaft

$$\frac{\Delta\phi}{L} = \frac{3° - 1°}{0.5} = 4\frac{\text{deg}}{\text{m}} = 6.98 \times 10^{-2}\frac{\text{rad}}{\text{m}}$$

4. Shear strain increases with radial distance from the shaft center.

So far, we have considered strain on the outside surface of the shaft. What is the strain of an element inside the shaft, located at a distance ρ from the center? Use the same reasoning as above, except now all displacements are equal to $\rho\phi$, instead of $R\phi$.

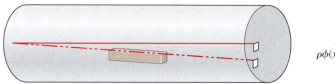

Shear strain varies with radial position ρ: $\gamma = \dfrac{\rho\Delta\phi}{L}$

5. The direction of the shear strain depends on the position of the element around the circumference of the shaft, but not along the length.

We just considered how the strain magnitude varies with radial position ρ in the shaft. The direction of the strain varies at different points around the shaft circumference.

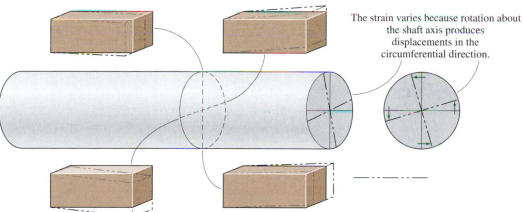

The strain varies because rotation about the shaft axis produces displacements in the circumferential direction.

Because the shear strain depends on the relative displacements of neighboring cross-sections, the direction and magnitude of the strain do not vary along the length of the shaft.

>>End 4.2

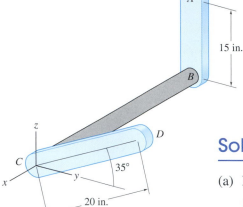

(a) The shaft *BC* twists and arm *AB* pivots rigidly about *B*. As a result, end *A* of the arm displaces by 0.1 in. perpendicularly to *AB* in the *y*-direction. Determine the axis about which the rotation occurs and the rotation in degrees.

(b) The arm *CD* is initially oriented at 35° relative to the *y*-axis. *BC* twists, causing *CD* to rotate by 0.1° about the positive *x*-axis. Determine the displacement of the point *D* along each of the coordinate axes.

Solution

(a) Here are 3-D and 2-D views of *AB* pivoting rigidly about *B*.

B stays in position, and *A* displaces by 0.1 in. Pivoting rigidly means that *AB* does not deform, but keeps its original shape. With the right hand rule, the rotation is about the negative *x*-axis.

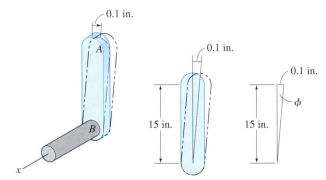

Displacement 0.1 in. $=$ 15 in. $\tan(\phi)$. Displacement 0.1 in. is small compared to 15 in. $\Rightarrow$ 0.1 in. $=$ (15 in.) ϕ, (angle ϕ in radians). Therefore, $\phi = 0.667 \times 10^{-2}$ rad $= 0.382°$.

(b) Here are 3-D and 2-D views of *CD* pivoting rigidly about *C*. The rotation is 0.1° about the positive *x*-axis. So as rotated, *BC* is 35.1° from the *y*-axis.

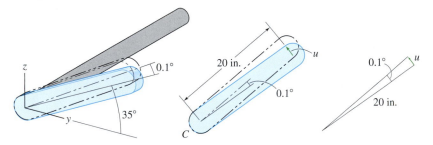

C stays in position, and *D* displaces by $u = $ (20 in.) $\tan (\phi)$.
$0.1° = 1.745 \times 10^{-3}$ radians $\ll 1 \Rightarrow$ displacement is perpendicular to the initial direction of *CD*. $u = $ (20 in.) $\phi = 3.49 \times 10^{-2}$ in.

Displacement is $-(3.49 \times 10^{-2}$ in.$)(\cos 55°) = -2.00 \times 10^{-2}$ in. along *y*.
Displacement is $(3.49 \times 10^{-2}$ in.$)(\sin 55°) = 2.86 \times 10^{-2}$ in. along *z*.

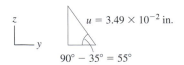

$$90° - 35° = 55°$$

>>End Example Problem 4.1

The shaft twists due to equal and opposite torques acting at its ends. End A rotates by $1.5°$ about the positive x-axis, and end B rotates by $2°$ about the negative x-axis.

Determine the shear strain of an element located 300 mm from A at a point C on the side of the cylinder. Find the magnitude of the shear strain, and draw a strained element.

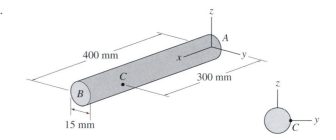

Solution

Since equal and opposite torques act at its ends, the shaft twists uniformly. The strain is uniform along the length and varies in the radial direction (from the shaft centerline).

Here are the rotation directions. The relative rotation is $\Delta\phi = 1.5° + 2° = 3.5°$.

If the rotations were in the same direction, instead we would have $\Delta\phi = 2° - 1.5° = 0.5°$.

In the formula for the shear strain $\gamma = \rho\Delta\phi/L$. $\Delta\phi$ must be in radians, so $\Delta\phi = 3.5° = 6.11 \times 10^{-2}$ rad.

On the outer surface, $\rho = R = 7.5$ mm, so
$\gamma = \rho\Delta\phi/L = (7.5 \text{ mm})(6.11 \times 10^{-2})/400 \text{ mm} = 1.145 \times 10^{-3}$.

Consider how the faces of a cubic element at point C, aligned with the coordinate axes would move.

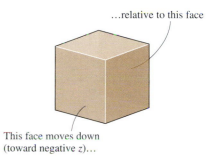

The element deforms as shown under this strain.

The direction of straining would change as one considers different elements around the circumference.

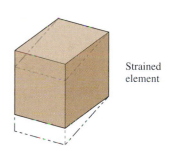

>>End Example Problem 4.2

Additional data on material properties needed to solve problems can be found in Appendix D or inside back cover.

4.1 The shaft AB twists, and BC moves rigidly. At B the shaft rotates by 1.5° about the $+x$-axis. How much does the end C displace and in what direction?

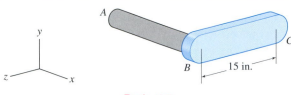

Prob. 4.1

4.2 Arm AB is initially oriented at 25° from the x-axis. The shaft BC twists, while AB moves rigidly. A displaces by 0.5 mm in the $+y$-direction. (A also displaces in the x-direction.) What is the rotation angle of the shaft at B, and about what axis is the rotation?

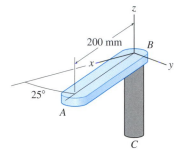

Prob. 4.2

4.3 The shaft at B and the attached disk rotate by 2° about the $-y$-axis. By how much does the point C displace? (Indicate direction and magnitude along the x- and y-axes.)

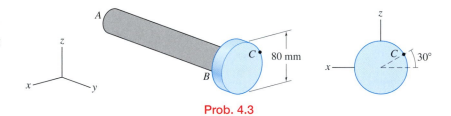

Prob. 4.3

4.4 The shaft twists due to two equal and opposite torques applied to its ends. The rotations are 0.7° about the $-z$-axis at B, and 0.1° about the $-z$-axis at D. What is the direction and magnitude of the rotation (a) at C and (b) at E?

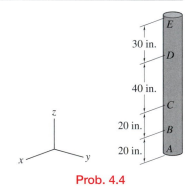

Prob. 4.4

4.5 The shaft twists due to two equal and opposite torques applied to its ends. Point C displaces 0.2 mm in the $-x$-direction. Point B displaces 0.1 mm in the $+y$-direction. Determine the direction and magnitude of the displacement of point A.

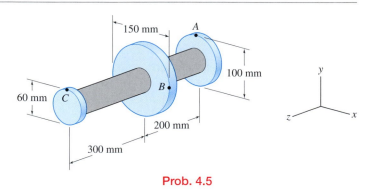

Prob. 4.5

4.6 The shaft twists due to equal and opposite torques applied to its ends. The cross-section A rotates by $0.3°$ about the $-z$-axis, and B rotates by $0.1°$ about the $+z$-axis. What is the magnitude of the shear strain at point C? Also, draw the deformed shape of an elemental cube at C.

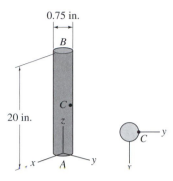

Prob. 4.6

4.7 The shaft twists due to equal and opposite torques applied to its ends. The cross-section B rotates by $1°$ about the $-x$-axis. At point C, the shear strain is 2×10^{-4} (the deformed element is shown in the diagram). What is the direction and magnitude of the rotation at A?

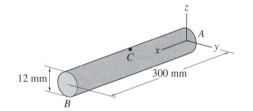
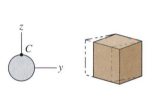

Prob. 4.7

4.8 A thin disk is bonded to each end of a shaft, which has outer and inner diameters of 1 in. and 0.75 in. Twisting is due to equal and opposite torques applied to the ends. Point A, located along the z-axis, displaces by 4×10^{-2} in. in the $-y$-direction, and C, located along the y-axis, displaces by 6×10^{-2} in. in the $+z$-direction. Determine the magnitude of the shear strain at point B, which is located at the shaft inner diameter. Also, draw the deformed shape of an elemental cube at B.

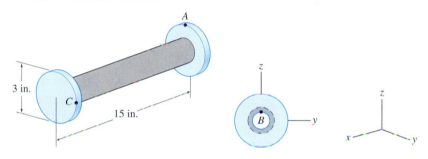

Prob. 4.8

4.9 The 16 mm diameter shaft twists due to two equal and opposite torques applied to its ends. Points A, B, and D are 40 mm, 30 mm, and 50 mm from the center of the shaft. The displacement at B is 0.3 mm in the $-x$-direction. The shear strain at C is 3×10^{-4} (deformed element is shown in the diagram). Determine the displacement of points D and A.

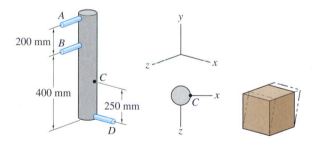

Prob. 4.9

4.10 The shaft ABC, 40 mm in diameter, twists under equal and opposite torques applied to its ends. Rotation at A is $1°$ about the $-y$-axis. Due to the rotation at C, point D of the rigid disk moves 0.2 mm in the $-z$-direction (D also displaces in the x-direction). Determine the magnitude of the shear strain at point B. Also, draw the deformed shape of an elemental cube at B.

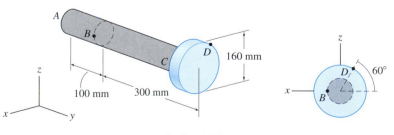

Prob. 4.10

4.11 The bottom bracket spindle undergoes torsion during pedaling. If the spindle is fixed against rotation at the chain ring and the rotation where the spindle connects to the far crank is 0.1°, by how much has the initially horizontal crank and pedal moved downwards? Take the dimensions to be $a = 70$ mm, $b = 25$ mm, $c = 9.5$ mm, $d = 80$ mm, and $L = 170$ mm.

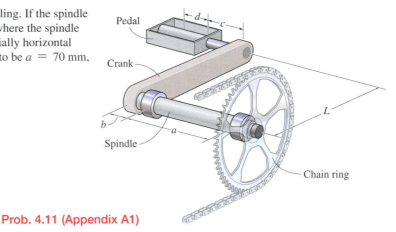

Prob. 4.11 (Appendix A1)

4.12 A large downward force on the right pedal results in a moment about the forward axis of a bike. This moment can be balanced by an opposite moment of the hands pulling up on right handlebar and pushing down on the left. Thus, two equal and opposite moments have to be absorbed by the frame. A portion of this load results in twisting of the down tube. If the tube material has allowable shear strain of 0.002, what is the maximum allowable relative angle of rotation of the head tube and the seat tube? Let the down tube be 24 in. long and have a circular cross-section with outer diameter of 1.75 in. and wall thickness of 0.125 in.

Prob. 4.12 (Appendix A1)

4.13 Consider a simplified drill string consisting of single sized drill pipe. Say the bit does not rotate, and the maximum shear strain that the pipe material can safely take is known to be 0.0012. Determine how many turns of the string are allowable at the top of the $L = 400$ m string without exceeding the maximum shear strain. Take the parameters to be $D_o = 120$ mm and $D_i = 76$ mm.

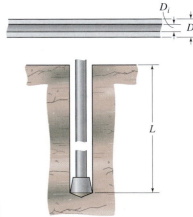

Prob. 4.13 (Appendix A3)

4.14 Say that the pivot-beam is initially horizontal and that the legs lift so that the end of the pivot-beam displaces upward by 1 in. Determine the rotation of the shaft where the pivot-beam connects to it. Take the dimensions to be $D_2 = 1$ in., $q = 6$ in., $R = 9$ in., $s = 3$ in., and $w = 17$ in.

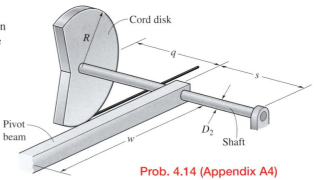

Prob. 4.14 (Appendix A4)

4.15 Say that the pivot-beam is initially horizontal and that the legs lift the padded beam by 1.5 in. If the maximum shear strain in the shaft to the cord disk is 0.001, by how much has the plate stack been lifted by the cord? Neglect any elongation in the cord. Take the dimensions to be $D_2 = 1$ in., $q = 6$ in., $R = 9$ in., $s = 3$ in., and $w = 17$ in.

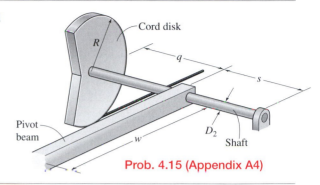

Prob. 4.15 (Appendix A4)

4.16 A torsion analysis of an external fracture fixation system under a forward load on the foot predicts that the carbon fiber rod twists about its axis by a 5° angle at the plane of the lower pin relative to the upper pin. Say that the other deflections of the rod are neglected, and the pins and bone are rigid. What is the transverse deflection of one bone fragment relative to the other at the points in the bone nearest and farthest from the rod? Approximate the bone as circular with an outer diameter of 26 mm and an inner diameter of 20 mm. Let the diameter of the solid carbon fiber bar be 10.5 mm. Take the dimensions to be $L_r = 300$ mm and $L_p = 60$ mm.

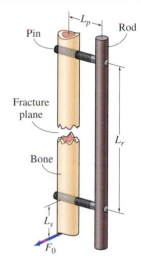

Prob. 4.16 (Appendix A5)

4.17 An intramedullary nail is fixed by screws to the bone at its two ends a distance $L = 300$ mm apart across the fracture. Approximate the bone as circular with an outer diameter of 26 mm and an inner diameter of 20 mm. The nail itself has an outer diameter of 10 mm and an inner diameter of 5.4 mm. Say the two screws rotate by a 10° angle with respect to each other due to twisting of the intermedullary nail. What is the maximum relative displacement between the two faces of the fractured bone fragments? Assume that no load is transmitted from the rod to the bone between the screws (the nail is fully within the intramedullary canal).

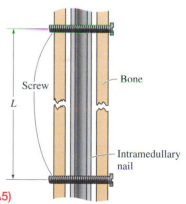

Prob. 4.17 (Appendix A5)

4.18 An intramedullary nail is fixed by screws to the bone at its two ends a distance $L = 300$ mm apart across the fracture. Approximate the bone as circular with an outer diameter of 26 mm and an inner diameter of 20 mm. The nail itself has an outer diameter of 10 mm, and an inner diameter of 5.4 mm. If the shear strain in the nail is not to exceed 0.005, what is the allowable rotation of the two bone fragments with respect to each other?

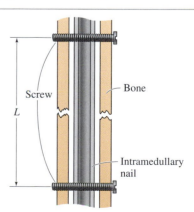

Prob. 4.18 (Appendix A5)

>>End Problems

4.3 Application and Transmission of Torque

For a shaft to twist, the applied forces must result in torques, moments, or couples acting about the long direction of the shaft (its axis). We consider several ways torque can be applied to a shaft by connected or contacting members, and we explain how torque is transmitted internally along a shaft from one end to the other.

1. Torques can be applied to the external surface of a circular shaft by circumferential friction forces.

In twisting the foam tube, we squeeze the outside and apply frictional forces to it.

Grip a golf club and drive a ball.

The aluminum hub of this SAE Formula One car was shrunk fit around a steel insert, producing enormous pressures. The insert and hub can exert torques on each other through the friction forces.

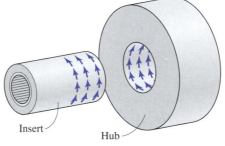

Insert

Hub

The force of the ball on the club head creates a twisting moment about the shaft. Frictional forces from the gripping hands balance that torque. A golfer must grip the club tightly enough and the friction coefficient must be high enough to provide sufficient torque.

2. Torques may be applied to the external surface of a non-circular shaft by normal forces that produce moments about the shaft axis.

It is common to exert torques by contacting non-circular surfaces of a shaft.

This box wrench contacts the hexagonal head of the bolt. Two or more forces are exerted depending on the fit.

Robertson screwdrivers are used to turn socket-headed Robertson screws. Torque is transmitted to the screw by the four corners of the screw-driver shaft.

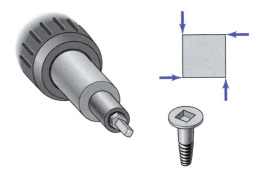

The end of the torsion bar has a hexagonal shape. The bar is inserted into a hexagonal hole in the car frame, which prevents rotation of the torsion bar.

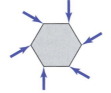

3. Torques can be applied to an axial face, e.g., the end face of a shaft, by circumferential friction forces.

The transfer of torque between a flywheel, clutch plate, and pressure plate occurs through circumferential friction forces acting on axial faces.

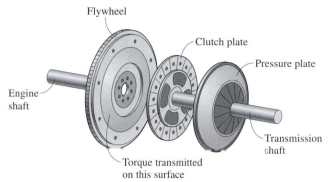

Flywheel

Clutch plate

Pressure plate

Engine shaft

Transmission shaft

Torque transmitted on this surface

When the pressure plate, clutch plate and flywheel are pressed together, the rotation of the engine shaft and flywheel causes rotation of the transmission shaft.

How exactly is the torque from the flywheel transmitted to the clutch plate and onto the transmission shaft?

Here is the flywheel rotating and the friction forces from the clutch plate that resists the rotation.

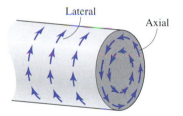

Rotation

Flywheel

Here are the equal and opposite friction forces of the flywheel that drive the clutch plate, which in turn drives the transmission shaft.

Clutch plate

To feel similar friction forces, press a flat rod against your hand and rotate the rod.

---

4. The circumferential forces that create a torque on the end face of a shaft also transmit internal torque within a shaft.

We have now seen torques transmitted by forces on both lateral and axial surfaces of a shaft.

Lateral

Axial

How is the torque transmitted across an internal cross-section of a shaft?

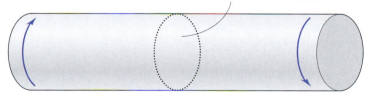

The forces that transmit torque across this internal axial cross-section act in the circumferential direction. But unlike the friction forces on the end of a shaft, these internal forces are due to shear stresses exerted by the adjacent removed material.

4.4 Shear Stress in Circular Shafts

Torque is transmitted internally within a shaft across each axial face by shear stresses that act circumferentially. Here we show how these stresses are consistent with the shear strains in a twisting shaft and are related quantitatively to the torque.

1. The directions of shear stresses are consistent with the directions of the shear strains.

We have seen that the shear strains in elements from different points around the circumference of a twisted shaft are in different directions. The shear stresses must be consistent with those shear strains. Also, as discussed in Chapter 2, equilibrium requires equal shear stresses on the four faces of an element.

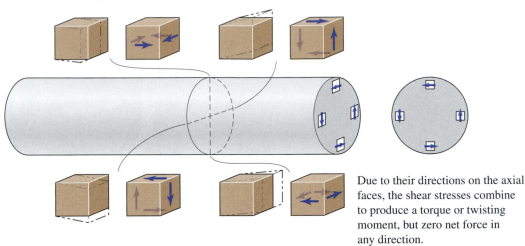

Due to their directions on the axial faces, the shear stresses combine to produce a torque or twisting moment, but zero net force in any direction.

2. Confirm the presence of shear stresses on non-axial faces of elements by imposing equilibrium on half of the shaft.

Forces on the end faces of the half-shaft create moments about the positive y-axis. If no other forces act, then $\Sigma M|_y \neq 0$.

The shear stresses that act on the longitudinal face create opposite moments, so $\Sigma M|_y = 0$.

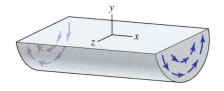

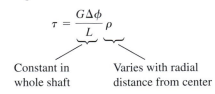

3. For an elastic material, relate the shear stress to the shear strain and hence to the relative rotation.

Say the strains are low enough for the material to be elastic. The shear stress is proportional to the shear strain: $\tau = G\gamma$.

Since the magnitude of the shear strain is given by

$$\gamma = \frac{\Delta\phi}{L}\rho$$

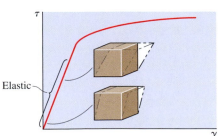

The magnitude of the shear stress is

$$\tau = \underbrace{\frac{G\Delta\phi}{L}}_{\substack{\text{Constant in} \\ \text{whole shaft}}} \underbrace{\rho}_{\substack{\text{Varies with radial} \\ \text{distance from center}}}$$

4. Find the moment about the shaft axis due to the shear stress acting on an area element of the axial face.

Find the moment due to the stress acting on a small element.

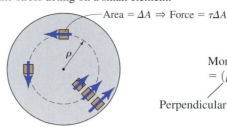

All shear stresses on the axial face act in the circumferential direction

Area $= \Delta A \Rightarrow$ Force $= \tau \Delta A$

Moment about center
$= (\rho)(\text{Force}) = \rho \, \tau \, \Delta A$

Perpendicular distance or moment arm

---

5. Integrate to add up the moments due to the stresses on the entire axial face and equate them to the torque. Define polar moment of inertia, I_p, which captures the effect of the shaft cross-sectional geometry.

Add up moments $\rho \tau \Delta A$ acting on all area elements, where $\tau = \dfrac{G \rho \Delta \phi}{L}$ varies with ρ.

$$T = \sum_{\text{all } \Delta A} \rho \tau \Delta A \qquad \Longrightarrow \qquad T = \int_A \rho \tau \, dA = \int_A \rho \frac{G \rho \Delta \phi}{L} \, dA = \frac{G \Delta \phi}{L} \int_A \rho^2 \, dA$$

Limit of many small area elements $\Delta A \Rightarrow dA$

Compare two elements of the same area dA. An area element farther from the center has greater ρ, and ρ has a double effect: stress is greater and the contribution of that stress to the moment is greater

Greater moment arm

Greater stress, hence force

Polar Moment of Inertia, I_p, captures the contribution of a circular shaft's cross-section to its resistance to twisting, and it is defined mathematically as the integral over the cross-sectional area of the square of the radial position from the center.

Polar Moment of Inertia, I_p, is defined mathematically by the integral $I_p = \int_A \rho^2 dA$. I_p has units of (length)^4. I_p, not the area, is the geometric property of the cross-section that affects torsion.

From the equation $T = \dfrac{G \Delta \phi}{L} \int_A \rho^2 \, dA$ we see the torque T depends on the twist ($\Delta\phi$), the material (G), the shaft length (L), and the shaft cross-section through I_p.

---

6. Relate the twist and stress to the torque.

$$T = \frac{G I_p}{L} \Delta \phi \quad \text{or} \quad \Delta \phi = \frac{TL}{G I_p}$$

Torque–Twist Relation
Shaft cross-section (I_p), length (L), and material (G) control twist ($\Delta\phi$) produced by a given torque (T).

Substitute $\Delta \phi$ in terms of T into

$$\tau = \frac{G \rho \, \Delta \phi}{L} \Rightarrow \tau = \frac{T \rho}{I_p}$$

Torque–Stress Relation
Torque (T) and cross-section (I_p) control the variation of shear stress (τ) with radial position (ρ).

---

7. Compute I_p for various cross-sections.

For a hollow cross-section $R_i \leq \rho \leq R$:

$$I_p = \int_A \rho^2 dA = \int_0^{2\pi} \int_{R_i}^{R} \rho^2 \rho \, d\rho \, d\theta = \int_0^{2\pi} \frac{R^4 - R_i^4}{4} d\theta = \frac{\pi}{2}\left[R^4 - R_i^4 \right]$$

General circular shaft with hole

$I_p = \frac{\pi}{2}(R^4 - R_i^4)$

Solid circular shaft (no hole, $R_i = 0$)

$I_p = \frac{\pi}{2} R^4$

Find I_p for thin ring:
$\rho \approx R$ (mean radius)
Area $\approx 2\pi R t$

$$\int_A \rho^2 dA \approx R^2(\text{Area})$$

$$I_p \approx 2\pi R^3 t$$

(For $t = 0.1 \, R$, thin-ring approximation is within 0.25% of exact I_p)

Thin circular ring ($t \ll R$)

$I_p \approx 2\pi R^3 t$

>>End 4.4

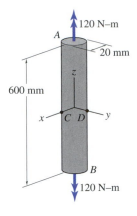

Torques equal to 120 N-m are applied to the ends of the shaft. The shaft has a shear modulus of $G = 80$ GPa. The cross-section at A rotates by $1°$ about the positive z-axis.

(a) Calculate the shear strain and stress in elements at points C and D (where x- and y-axes pierce the shaft), and draw them on elements in the appropriate directions.

(b) Find the rotation of end B of the shaft.

Solution

The key relations are:

$$\text{Stress:}\quad \tau = \frac{T\rho}{I_p} \qquad \text{Strain:}\quad \gamma = \frac{\tau}{G} = \frac{\rho \Delta \phi}{L} \qquad \text{Relative rotation:}\quad \Delta \phi = \frac{TL}{GI_p}$$

From the radius $R = 10$ mm $= 0.01$ m, we can find $I_p = \dfrac{\pi}{2} R^4 = 1.57 \times 10^{-8}\,\text{m}^4$.

(a) Both points C and D are on the outside of the shaft, so $\rho = R = 10$ mm:

$$\tau = \frac{TR}{I_p} = \frac{(120\ \text{N-m})(0.01\ \text{m})}{(1.57 \times 10^{-8}\ \text{m}^4)} = 76.4\ \text{MPa}$$

$$\gamma = \frac{\tau}{G} = \frac{(76.4\ \text{MPa})}{(80\ \text{GPa})} = 9.55 \times 10^{-4}$$

$$\Delta \phi = \frac{TL}{GI_p} = \frac{(120\ \text{N-m})(0.6\ \text{m})}{(80\ \text{GPa})(1.57 \times 10^{-8}\ \text{m}^4)} = 5.73 \times 10^{-2}\ \text{rad} = 3.28°$$

Twisting is shown in the diagram. The directions of shear stresses shown are consistent with the torque. From these, we can draw shear strains and shear stresses on elements.

(b) From $\Delta \phi = 3.28°$ and the rotation at A of $1°$, we find the rotation at B to be $= 2.28°$ about the negative z-axis.

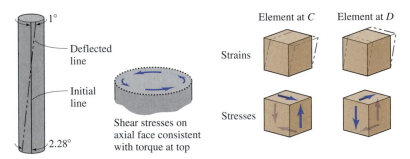

Arm wrestlers can develop their muscles, but they can only do so much to strengthen their bones. The force exerted on the hand causes torsion of the humerus (bone in upper arm). Say the cross-section of the humerus is approximated as hollow and circular, with inner and outer diameters 19 mm and 15 mm. Also, assume that the whole twisting moment is taken by the bone.

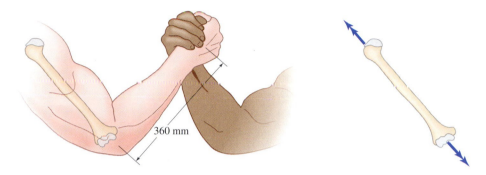

360 mm

If the humerus can withstand a shear stress $\tau_{max} = 70$ MPa, determine the maximum force the hand can apply without fracturing the humerus.

Solution

The force of the hand F_h causes twisting moment on the arm $T = F_h$ (0.36 m).

The torque-shear stress relation is $\tau = \dfrac{T\rho}{I_p}$.

Fracture occurs first where shear stress is maximum (at the outer radius), so

$$\tau_{max} = \frac{TR}{I_p}$$

The cross-sectional properties of the twisting member (humerus) can be found:

$$R = 9.5 \text{ mm} = 9.5 \times 10^{-3}\text{ m} \quad \text{and} \quad I_p = \frac{\pi}{2}\left[R^4 - R_i^4\right] = 7820 \text{ mm}^4 = 7.824 \times 10^{-9}\text{ m}^4$$

The torque T_{max}, at which the humerus fracture stress τ_{max} is reached is found to be:

$$T_{max} = \frac{\tau_{max} I_p}{R} = \frac{(70 \text{ MPa})(7.82 \times 10^{-9}\text{ m}^4)}{0.0095 \text{ m}} = 57.7 \text{ N-m}$$

From T_{max}, we can find $(F_h)_{max}$, the force the hand exerts when the humerus fractures:

$$(F_h)_{max} = T_{max}/(0.36 \text{ m}) = 160 \text{ N}$$

Note that this estimate of the load seems rather low (~36 lb). In fact, not all of the torque would be taken up by the bone; some would be taken up by the arm muscles.

>>End Example Problem 4.4

A SAE Formula One car includes an aluminum hub into which a steel insert was press-fit. There is to be no slip between the hub and the insert for any reasonably transmitted torque. If the pressure is sufficiently high, then no slip could occur. The designers determined the necessary pressure at the shrink-fit as follows. Say the hub and shaft must transmit torques up to the level at which the shaft is just at the shear yield stress ($\tau_Y = 30$ ksi). Assume that friction of at least $\mu = 0.3$ can be relied upon with aluminum against steel. The shaft diameter $d = 0.75$ in. and the length of the contact region is $w = 1.5$ in.

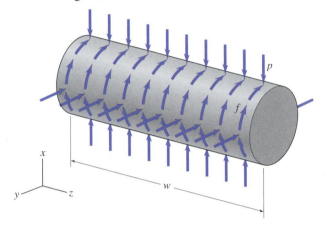

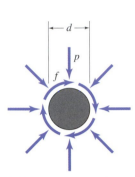

f: Friction force per unit area
p: Normal force per unit area

Find the minimum required contact pressure p.

Solution

For a shaft with diameter d, the torque to cause yielding, T_Y, would lead a maximum stress $\tau = \tau_Y$:

$$\tau_Y = \frac{T_Y(d/2)}{I_p} \Rightarrow T_Y = \frac{\tau_Y I_p}{(d/2)} = \frac{\pi \tau_Y (d/2)^3}{2} = \frac{\pi(30 \text{ ksi})(0.375 \text{ in.})^3}{2} = 2490 \text{ lb-in.}$$

The tangential friction forces must be able to transmit a torque of this magnitude to the shaft. Both the normal force and the friction forces are distributed uniformly over the contact area. We define p as the normal force per area, and f as the friction force per area. A friction force acting on an element of area ΔA causes a moment about the shaft axis, and T equals the sum of the moments:

$$T = M|_z = \sum (f)(\Delta A)\frac{d}{2} = (f)\frac{d}{2}\sum(\Delta A) = (f)\frac{d}{2}A = f\frac{\pi d^2 w}{2}$$

Therefore, to be able to transmit the maximum allowable torque we must be able to generate *at least* a friction force per area of

$$f = \frac{2T_y}{\pi d^2 w} = \frac{2(2490 \text{ lb-in.})}{\pi(0.75 \text{ in.})^2(1.5 \text{ in.})} = 1875\frac{\text{lb}}{\text{in.}^2}$$

According to the Coulomb friction law, the maximum friction force F between bodies in contact is $F_{max} = \mu N$, where μ is the coefficient of friction and N is the normal force. Since the friction force and the normal force are transmitted over the same contact area, $(F_{max}/A) = \mu (N/A)$, the forces per area are similarly related: $f_{max} = \mu p$. With $\mu = 0.3$, we must therefore have a pressure of at least

$$p = \frac{f}{\mu} = \left(\frac{1}{0.3}\right)1875\frac{\text{lb}}{\text{in.}^2} = 6250 \text{ psi}$$

>>End Example Problem 4.5

A plastic tube of inner diameter $d_i = 0.8$ in. and outer diameter $d_o = 1.0$ in. is restrained at end B. The tube has a shear modulus of $G = 100$ ksi. The tube is twisted by forces F that are applied to the ends of an embedded steel rod. Treat the steel rod as rigid relative to the tube. The ends of the rod each displace a distance of 0.02 in. in the directions of the force.

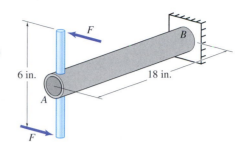

Determine the forces F applied to the rods, and determine the shear strain at the inner diameter of the tube.

Solution

The displacement 0.02 in. is small compared to the distance 3 in. from the shaft center to the tip of the steel rod.

Use the small angle approximation to determine ϕ_A:

$$0.02 \text{ in.} = 3 \text{ in.}(\phi_A) \Rightarrow \phi_A = 6.67 \times 10^{-3} \text{ rad}$$

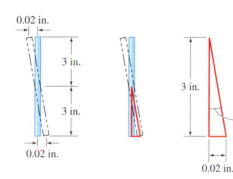

Since $\phi = 0$ at B, we find the relative rotation to be: $\Delta\phi = 6.67 \times 10^{-3}$ rad

The shaft properties are $R_i = 0.4$ in., $R = 0.5$ in., so $I_p = \dfrac{\pi}{2}\left[R^4 - R_i^4\right] = 5.80 \times 10^{-2}$ in.^{4}

Find the torque from the twist:

$$T = \frac{GI_p\Delta\phi}{L} = \frac{(100 \text{ ksi})(5.80 \times 10^{-2} \text{ in.}^4)(6.67 \times 10^{-3})}{(18 \text{ in.})} = 2.15 \text{ lb-in.}$$

The torque T is related to F by $T = F(6 \text{ in.})$, so $F = 0.358$ lb.

At the inner diameter ($\rho = 0.4$ in.), the shear strain is:

$$\gamma = \frac{\rho\Delta\phi}{L} = \frac{(0.4 \text{ in.})(6.67 \times 10^{-3})}{(18 \text{ in.})} = 1.481 \times 10^{-4}$$

>>End Example Problem 4.6

Additional data on material properties needed to solve problems can be found in Appendix D or inside back cover.

4.19 The cross-section A of the steel shaft ($G = 80$ GPa) rotates by $4°$ about the $+z$-axis, and C rotates by $1°$ about the $+z$-axis. The shaft twists due to two equal and opposite torques applied to its ends. Determine the magnitude of the shear stress at point B (located 6 mm from the shaft center in the y-direction). Also, draw the stresses acting on an elemental cube at B. ($L_1 = 600$ mm.)

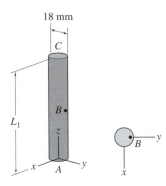

Prob. 4.19

4.20 The shaft twists due to two equal and opposite torques applied to its ends. B rotates by $1°$ about the x-axis. The shear stress in the aluminum shaft ($G = 3.8 \times 10^6$ psi) at point C is 5 ksi. The directions of the stresses acting at C are shown on an elemental cube. What is the direction and magnitude of the rotation at A? ($L_1 = 25$ in. and $d_1 = 0.5$ in.)

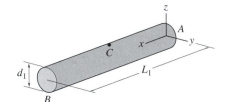
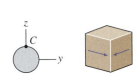

Prob. 4.20

4.21 The shaft twists due to two equal and opposite torques applied to its ends. The stress at the outer surface at point A is 75 MPa. The directions of the stresses acting at A are shown on an elemental cube. Determine the magnitude of the stress at point B, which is at the same cross-section as A, but 10 mm from the center in the $-z$-direction. Also, draw the stresses at B on an elemental cube. ($L_1 = 200$ mm and $L_2 = 350$ mm.)

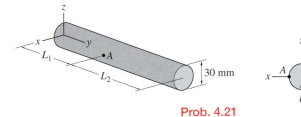

Prob. 4.21

4.22 The end A is fixed, and a torque of $T_0 = 200$ lb-in. is applied to the right end. Determine the magnitude of the stress at B. Also, draw the stresses at B on an elemental cube.

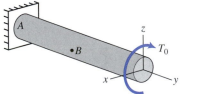
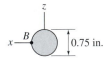

Prob. 4.22

4.23 Equal and opposite torques are applied to the shaft at the ends A and D. The point B, located at 50 mm from the shaft centerline, displaces by 1 mm in the $+y$-direction. The rotation at C is $1°$ about the $+z$-axis. Determine the magnitude and direction of the torque applied to the shaft at D ($G = 1$ GPa). ($L_1 = 300$ mm, $L_2 = 400$ mm, $L_3 = 200$ mm, $d_1 = 20$ mm, and $d_2 = 30$ mm.)

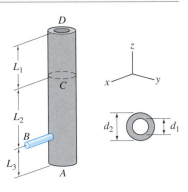

Prob. 4.23

4.24 The rubber tube ($G = 300$ psi), which is 40 in. long, is fixed at A. The shear strain at B is $\gamma = 0.02$, and a deformed elemental cube at B is shown. This straining is produced by a pair of forces (only F_1 or F_2). Determine which force (F_1 or F_2) is applied and the magnitude of the forces. ($L_1 = 8$ in., $L_2 = 40$ in., $d_1 = 1$ in., and $d_2 = 3$ in.)

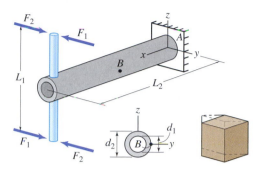

Prob. 4.24

4.25 The end A of the solid steel shaft ($G = 80$ GPa) is fixed. The tangential component of the force at D on the gear $F_t = 2500$ N causes the shaft to twist. (Bearings, not shown, apply no torque to the shaft, but they balance the transverse forces.) Determine the magnitude and direction of the displacements at points B and C, which are 80 mm and 100 mm from the shaft centerline. ($L_1 = 200$ mm, $L_2 = 150$ mm, $L_3 = 200$ mm, $d_1 = 26$ mm, and $d_2 = 120$ mm.)

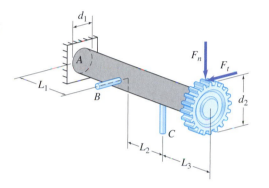

Prob. 4.25

4.26 A shrunk-fit hub applies 10000 psi of pressure over the contact region shown. The friction coefficient is 0.4. The friction forces from the hub on the shaft act in the directions shown and the torque is balanced at A. What is the maximum stress that can act at a surface element at point B of the steel shaft ($G = 11 \times 10^6$ psi)? Also, draw the shear stresses on an element at B. ($d = 1$ in. and $L = 0.3$ in.)

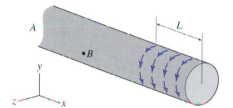

Prob. 4.26

4.27 An Allen wrench ($G = 80$ GPa) is used to turn a screw at A. Say the screw is not turning and that the fingers each apply $F = 80$ N to the wrench as shown. Determine the rotation at B due to twisting. Approximate the hexagonal cross-section as circular with 7 mm diameter. ($L_1 = 40$ mm, and $L_2 = 100$ mm.)

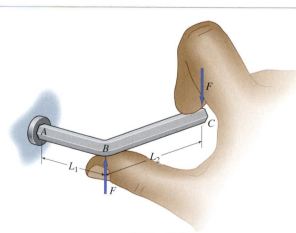

Prob. 4.27

4.28 The plastic shaft ($G = 100$ ksi, $d = 0.5$ in.) is fixed at A. The stress at C is 1500 psi with directions indicated in the diagram. The shaft is loaded by one pair of forces (only F_1 or F_2). (a) Determine which force (F_1 or F_2) is applied and the magnitude of the forces. (b) Determine the magnitude and direction of the rotation at B. ($L_1 = 8$ in., $L_2 = 12$ in., and $L_3 = 2$ in.)

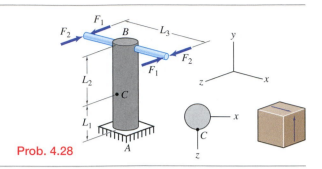

Prob. 4.28

4.29 The bent steel wire (80 GPa, $d = 2$ mm) is twisted by the four forces $F = 5$ N. Neglect bending of AB and CD due to F, and take B and C to have zero displacement. Assume the center plane E does not rotate, so A and D displace by equal amounts in the opposite directions. Determine the displacement of point D. ($L_1 = 20$ mm and $L_2 = 100$ mm.)

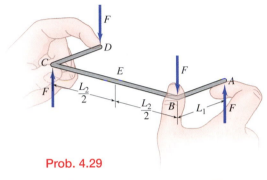

Prob. 4.29

4.30 The top of the rubber tube ($G = 100$ psi) is gripped over length L_2 where a uniform pressure is applied. B is fixed, and it is desired to rotate the upper end by 5° (measured at the bottom of the gripped zone). If the friction coefficient is 0.6, what is the minimum pressure necessary to twist the shaft the desired amount without slipping? ($L_1 = 30$ in., $L_2 = 2$ in., $d_1 = 0.5$ in., and $d_2 = 1$ in.)

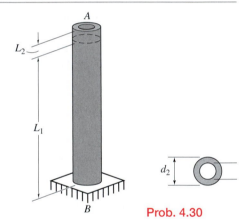

Prob. 4.30

4.31 A collar is mounted to an aluminum shaft ($G = 27$ GPa and diameter $= 20$ mm) with a setscrew that transmits the torque to the shaft. The end A is fixed. The shear strain at the point B on the surface is 5×10^{-4}. Determine the minimum normal force (in N) that must be exerted by the setscrew on the shaft if the friction coefficient between them is 0.5. The collar contacts the shaft only at the top and bottom as shown. ($L_1 = 200$ mm and $L_2 = 300$ mm.)

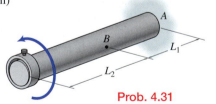

Prob. 4.31

4.32 A plastic rod of length L has been subjected to environmental degradation, which has reduced the modulus on the outside from the original nominal value of G_i to a value of G_r. Assume the modulus varies parabolically with the radial coordinate according to:

$$G = G_1 - r^2 G_2$$

(Determine G_1 and G_2 in terms of G_i and G_r.)

(a) Demonstrate that the torque T is related to the relative rotation according to:

$$T = \frac{G_i I_p}{L} \left[1 - \frac{2}{3} \frac{G_i - G_r}{G_i} \right] \phi$$

(b) Say the outer modulus is reduced by 20% below the nominal value. By how much does the normalized stiffness $(TL)/(G_i I_p \phi)$ decrease and by much does the average modulus decrease? Explain the difference between these two decreases.

4.33 When two bodies are joined by many rivets, one cannot determine the forces carried by each rivet exactly. Instead, the following approximation is used. The applied load is expressed as a single force acting at the center of the rivet pattern, plus a couple. Here we consider the rivet forces when the applied load is only a couple.

The approximation, appropriate for a symmetric arrangement of rivets, is inspired by the pattern of stress and strain in torsion. We picture the couple as attempting to rotate the riveted body about its center. Each rivet is assumed to react to this load by applying a force that acts perpendicular to the line segment from the pattern center to that rivet, with a magnitude proportional to the line segment length. The proportionality between force and length is the same for all rivets. This proportionality is determined from setting the total moment due to the rivets equal to the applied moment.

Using this approach, determine the forces in each of the rivets under the loading shown.

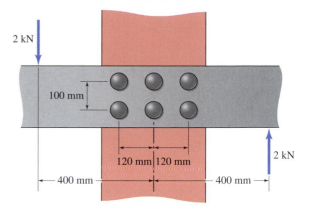

Prob. 4.33

Focused Application Problems

4.34 In some bikes, the crank, chain ring, and bottom bracket spindle are integral, and the spindle is joined to the second crank via a spline. Say the spline has 6 ridges, and that the mean diameter of the spline faces is $D_s = 0.95$ in. Let the width of the splined crank be $w = 1.25$ in. If the torque to be transmitted is 3000 lb-in., find the force per length on each groove in the crank.

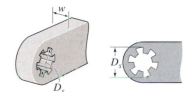

Prob. 4.34 (Appendix A1)

4.35 The bottom bracket spindle undergoes torsion during pedaling. The spindle is hollow with outer diameter 27 mm and inner diameter 21 mm, and has a shear modulus of 80 GPa. Determine the maximum shear stress in the spindle if it is fixed against rotation at the chain ring, and the pedal is moved down by 0.4 mm. Take the dimensions to be $a = 70$ mm, $b = 25$ mm, $c = 9.5$ mm, $d = 80$ mm, and $L = 170$ mm.

4.36 The bottom bracket assembly has a hollow spindle with outer diameter 27 mm and inner diameter 21 mm. If the maximum shear stress due to torsion alone is to remain below 200 MPa, what is the maximum vertical force that can be applied to the pedal? Take the dimensions to be $a = 70$ mm, $b = 25$ mm, $c = 9.5$ mm, $d = 80$ mm, and $L = 170$ mm.

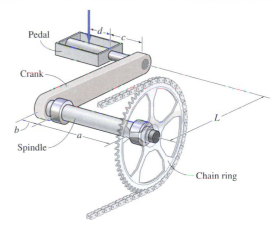

Probs. 4.35–36 (Appendix A1)

4.37 When biking on a rough road, the arms, if held stiff, can apply significant forces to the drop bars. Depending on the orientation of the hands and their placement on the drop bars, this could result in significant twisting of the handlebar. Let each hand apply 300 N to the drop bars. If the hands were oriented to produce maximum torsional stress in the handlebar, what would be the maximum torsional shear stress? Take the handlebar to be tubular, with outer diameter of $D = 24$ mm and a wall thickness of 1.8 mm. Approximate the drop bars as circular in shape with radius R. Take the dimensions to be $a = 16.8$ mm, $b = 200$ mm, $d = 29$ mm, and $R = 100$ mm.

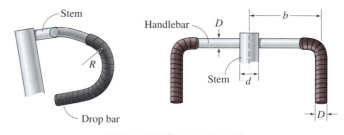

Prob. 4.37 (Appendix A1)

4.38 A large downward force on the right pedal results in a moment about the forward axis of a bike. This moment can be balanced by an opposite moment of the hands pulling up on the right handlebar and pushing down on the left. Thus, two equal and opposite moments have to be absorbed by the frame. Since it is the most direct and stiffest pathway, the down tube, rather than the top tube, takes most of this moment. Some of the moment is directed along the tube producing torsion, and some of it will produce bending. Let the total moment from the feet and the hands each equal 1500 lb-in. Resolve the moment to determine the twisting moment, and then determine the maximum shear stress in the tube and the rotation of the head tube relative to the seat tube. Take the down tube to be 24 in. long aluminum, oriented at angle $\theta = 48°$, with circular cross-section having outer diameter of 1.75 in. and wall thickness of 0.125 in.

Prob. 4.38 (Appendix A1)

4.39 Daily traffic patterns into and out of a city accessed by a bridge can result in many more vehicles on one side of the deck compared to the other side. Each lane of the deck is $w = 7.5$ m. As an extreme case, say there are bumper to bumper 10 Mg trucks each 10 m long on both lanes on one side of the deck. This arrangement of vehicles results in a moment per length on the deck about its centerline, which tends to twist the deck. Calculate the moment per length (in N-m per meter).

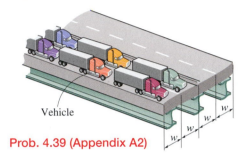

Prob. 4.39 (Appendix A2)

4.40 A drill string is sometimes stuck because the mud surrounding it hardens. The string consists of steel collars, with outer and inner diameters 7 in. and 2.81 in., having a total length of 200 ft, above which is 1200 ft of drill pipe with outer and inner diameters 4.5 in. and 2.75 in. Say the hardened mud has a shear yield strength of 0.5 psi. Determine the torque that must be applied to the top of the drill pipe to break the string away from the hardened mud.

Prob. 4.40 (Appendix A3)

4.41 This type of intramedullary nail fits tightly inside the intramedullary canal of the fractured bone. Some distance from the fracture plane both bone and nail share the torque, with the bone carrying a torque of 1 N-m. No torque is carried by the bone at the fracture plane, and torque is transmitted over some length via friction from the nail to bone. The inner and outer diameters of the bone are 20 mm and 26 mm; the titanium nail has an outer diameter of 20 mm and a wall thickness of 1 mm. Given that the compressive radial stress is 0.15 MPa and the friction coefficient is 0.2, determine the length over which the nail transmits to the bone the 1 N-m torque that it carries away from the fracture plane.

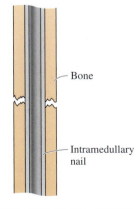

Prob. 4.41 (Appendix A5)

4.42 A torsion analysis of an external fracture fixation system under a forward load on the foot predicts that the carbon fiber rod twists about its axis. The lower pin rotates by an angle of 2° relative to the upper pin. Consider only the twisting deformation in the rod produced by this loading. Determine the maximum shear stress in the solid carbon fiber rod, if it is approximated as isotropic with $G = 50$ GPa and has a diameter of 10.5 mm. Take the dimensions to be $L_r = 300$ mm and $L_p = 60$ mm.

4.43 A torsion analysis of an external fracture fixation system is to be conducted under conditions of a forward force of magnitude $F_0 = 100$ N. Determine (a) the rotation of the near pin relative to the far pin and (b) the maximum shear stress in the carbon fiber rod. Consider only the torsional deformation of the carbon fiber rod, which is approximated as isotropic with $G = 90$ GPa and has a diameter of 10.5 mm. Take the dimensions to be $L_r = 300$ mm and $L_p = 60$ mm.

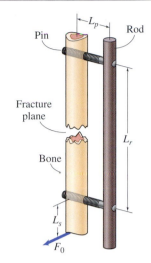

Probs. 4.42–43 (Appendix A5)

4.44 This type of intramedullary nail is connected to the fractured bone by screws. Say an 8 N-m torque is applied to the bone far from fracture. The torque is picked up by the nail via the screw at one end, carries the torque across the fracture plane, and then transmits the torque via the other screw back to the bone. So the bone carries no torque between the screws. Approximate the bone as circular with an outer diameter of 26 mm and an inner diameter of 20 mm. The nail itself has an outer diameter of 10 mm and an inner diameter of 5.4 mm. Take the shear moduli of the nail and bone to be 43 GPa and 3.3 GPa, respectively. If the maximum relative displacement of the outer diameter of the bone fragments across the fracture plane is to be 1 mm, determine the maximum separation distance between the screws.

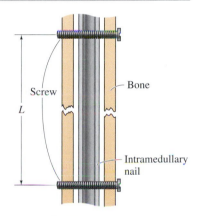

Prob. 4.44 (Appendix A5)

4.45 This type of intramedullary nail fits tightly inside the intramedullary canal of the fractured bone. Some distance from the fracture plane both bone and nail rotate through the same angle and share the 10 N-m applied torque, with the nail and the bone carrying torques of 7.071 N-m and 2.929 N-m respectively. Approximate the bone as circular with an outer diameter of 26 mm and an inner diameter of 20 mm. The nail itself has an outer diameter of 20 mm and a wall thickness of 1 mm. Take the shear moduli of the nail and bone to be 43 GPa and 3.3 GPa, respectively, and take the pressure between the nail and bone to be 0.3 MPa and the friction coefficient to be 0.2. At the fracture plane, the bone carries zero torque. Under these conditions, torque is transmitted via friction to the bone over a slip zone, which extends 78 mm on both sides of the fracture plane. In the slip zone the rotation of the bone and nail differ and vary continuously. By integrating the twist per length in the nail and in the bone, determine the angle of rotation of the bone surfaces.

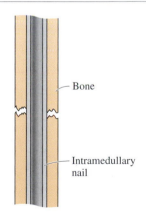

Prob. 4.45 (Appendix A5)

>>End Problems

4.5 Strength and Stiffness

Strength and stiffness are important concepts that are carefully distinguished in mechanics. Here we define strength and stiffness generally, and we give examples of computing them in different circumstances.

1. In general, stiffness is the proportionality between load and deflection assuming linear elastic behavior, and strength is the load at which the system either fails or ceases to act elastically.

Strength is a general term that captures the maximum load that a body can safely carry, but it is defined in each circumstance with regard for the type of loading and how the load is defined.

Stiffness is a general term that captures the proportionality between the load and the deformation when a body deforms elastically, but it is defined in each circumstance with regard for the type of loading and how the load and deformation are defined.

Consider a body that deflects under the action of a load. For example, hold down the wire shown and apply a force to lift one end. For small forces, the deflection is proportional to the load, and returns to zero when the load is removed. When the force is so large that the clip becomes permanently bent, the strength has been reached.

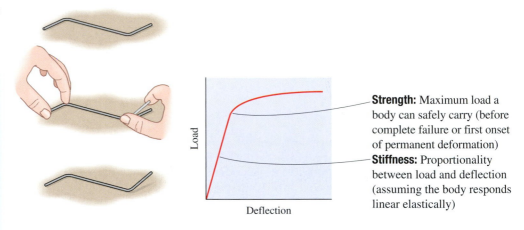

Strength: Maximum load a body can safely carry (before complete failure or first onset of permanent deformation)

Stiffness: Proportionality between load and deflection (assuming the body responds linear elastically)

In any particular situation, the *load* and *deflection* must be defined based on the needs of the situation. Then, strength and stiffness can be computed given those definitions.

---

2. In this example, we determine the strength and stiffness when the load is defined to be the torque on a shaft and the deflection is defined to be the relative rotation of the ends of the shaft.

Find the torsional strength and stiffness of a steel golf shaft that is 40 in. long, where the yield stress and shear modulus are:

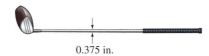

0.375 in.

$$\tau_Y = 30000 \text{ psi} \quad G = 11 \times 10^6 \text{ psi}$$

Say we define the load as the torque T and the deflection as the twist $\Delta\phi$.

Use the stress–torque relation: $\tau = \dfrac{T\rho}{I_p}$

Use the torque–twist relation: $\Delta\varphi = \dfrac{TL}{GI_p}$

For solid shaft: $I_p = \dfrac{\pi R^4}{2}$

Strength: At what torque is yield reached, that is, at what torque is $\tau = \tau_Y$ somewhere in the shaft?

If the material is still elastic up to the load at which τ_Y first reached, $\tau = \dfrac{T\rho}{I_p}$

τ is greatest at $\rho = R \Rightarrow$ Strength is torque T_Y at which $\tau = \tau_Y$ at $\rho = R \Rightarrow \tau_Y = \dfrac{T_Y R}{I_p} = \dfrac{2T_Y}{\pi R^3}$

$$\Rightarrow T_Y = \frac{\pi \tau_Y R^3}{2} = \frac{\pi (30000 \text{ psi})(0.1875 \text{ in.})^3}{2} = 311 \text{ lb-in.}$$

Stiffness: If the shaft is elastic, the torque is proportional to the twist. The stiffness is defined as the ratio of load/deflection or $T/\Delta\phi$.

Torque–twist relation:

$$T = \frac{GI_p}{L}\Delta\phi \Rightarrow \frac{T}{\Delta\phi} = \frac{GI_p}{L} = \frac{(11 \times 10^6 \text{ psi})\pi(0.1875 \text{ in.})^4}{2(40 \text{ in.})} = 534 \frac{\text{lb-in.}}{\text{rad}} = 9.32 \frac{\text{lb-in.}}{\text{deg}}$$

3. In this example, a twisting shaft is the only deforming member in a larger system, but the stiffness and strength are defined based on the load on, and the deflection of, the larger system.

Sometimes, a twisting bar is the deforming part of a larger system. For that system, it may not be useful to define strength as the maximum allowable torque or stiffness as the torque per twist. For example, in this torsion bar suspension, the load and deflection of interest to the engineer are the force on the wheel and the vertical displacement of the wheel relative to the chassis.

Because the deforming member is the twisting shaft, we relate the quantities of interest to the torque and the twist.

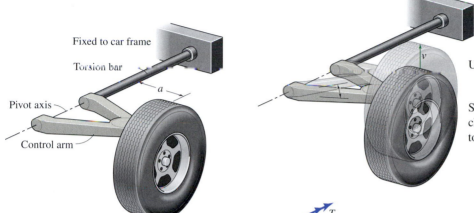

Upward deflection of wheel: v

$$v = \phi a \ (\phi \ll 1 \text{ in radians})$$

Since the other end of the torsion bar is fixed to the chassis, the rotation at the control arm, ϕ, is equal to the twist $\Delta\phi$ of the torsion bar.

The moment due to the wheel force F about the shaft axis is equal to the torque T on the torsion bar

$$T = Fa$$

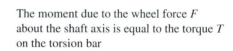

4. Define stiffness as the load of interest necessary to cause a unit deflection of interest.

For this application, the load and deflection of interest are the wheel force F and the wheel displacement v, so we define stiffness as

$$\text{Stiffness} = \frac{\text{Wheel force}}{\text{Wheel displacement}} = \frac{F}{v} = \frac{T/a}{\phi a} = \frac{T}{(\Delta\phi)a^2} = \frac{GI_p}{La^2}$$

Take the dimensions and properties to be $L = 1300$ mm, $a = 400$ m, $R = 16$ mm, and $G = 80$ GPa.

$$\text{Stiffness} = \frac{GI_p}{La^2} = \frac{(80 \text{ GPa})}{(1.3 \text{ m})(0.4 \text{ m})^2} \frac{\pi(0.016 \text{ m})^4}{2} = 39600 \text{ N/m} = 39.6 \text{ kN/m}$$

5. Define strength as the load of interest necessary to cause failure or yielding of the system.

Since in this application the load of interest is the wheel force F, we define strength as

$$\text{Strength} = \text{Force to Yield} = F_Y = \frac{T_Y}{a} = \frac{\tau_Y I_p}{aR}$$

If $\tau_Y = 500$ MPa $\Rightarrow$ Strength $= \dfrac{(500 \text{ MPa})\dfrac{\pi}{2}(0.016 \text{ m})^3}{0.4 \text{ m}} = 8040$ N

>>End 4.5

4.6 Dependence of Stiffness and Strength on Shaft Properties

The torque per twist (stiffness) and the torque to reach a critical stress τ_Y (strength) depend on several aspects of the shaft, including the length, the material, and the cross-section. We highlight how these aspects affect stiffness and strength to develop insight into the design of mechanical systems with twisting members.

1. The stiffness and strength of a shaft under torsion can depend on the shaft length, material, and cross-section.

	Formula	Length	Material	Cross-Section
Stiffness	$\dfrac{T}{\Delta\phi} = \dfrac{GI_p}{L}$	Decreases with L	Increases with shear modulus G	Increases with I_p
Strength	$T_Y = \dfrac{\tau_Y I_p}{R}$	Independent of L	Increases with yield strength τ_Y	Increases with I_p/R

Length: For a longer shaft, there is more twist for a given torque (less stiff). If other aspects of a design restrict shaft length, one cannot adjust length to change stiffness.

Material: If the material itself is stiffer (higher G) or stronger (higher τ_Y), then the shaft will be stiffer and stronger. For the full range of available materials, G can vary easily by a factor 10^5; the maximum stress also varies significantly. If there is freedom to choose from a wide range of materials, the engineer can radically alter the stiffness and strength. However, cost, manufacturability or operating conditions often influence material choice.

Cross-Section: The shape of the cross-section strongly affects stiffness and strength in torsion. For a circular shaft, I_p captures the important effect of cross-section. Later we will see how the cross-section affects torsion for shafts of shapes other than circular.

---

2. Gain insight into the polar moment of inertia I_p by comparing hollow and solid cylinders of the same area.

I_p is very different from area. These solid and hollow shafts have nearly the same area.

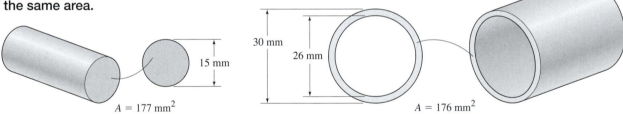

$A = 177 \text{ mm}^2$

30 mm · 26 mm · 15 mm

$A = 176 \text{ mm}^2$

The hollow shaft is stiffer than the solid one by a factor of $\dfrac{(I_p)_{hollow}}{(I_p)_{solid}} = \dfrac{30^4 - 26^4}{15^4} = 6.97$.

If the wall were even thinner and the radius larger (fixed area), the tube would be even stiffer.

---

3. Recognize that a hollow shaft with a very thin wall, while very stiff in torsion given its cross-sectional area, is prone to failing in other ways.

While a larger radius and thinner wall can make a shaft stiffer in torsion, a shaft with an extremely thin wall could be readily dented or could buckle when twisted.

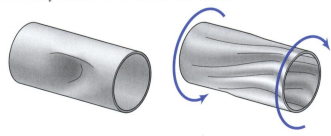

4. The internal torque depends on I_p, rather than on the area, because the stress is nonuniformly distributed, and the same shear stress contributes differently to the torque depending on its radial distance from the center.

To see why the torque is not proportional to area, compare how stress relates to net load on the cross-section for axial and torsional loads.

Axial load

normal stress
Stress σ is uniform at all points of cross-section
Each area ΔA contributes same force $\sigma \Delta A$

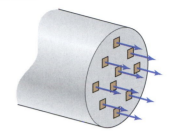

Forces perpendicular to plane of cross-section

Torsional load

shear stress
Stress τ increases with ρ: $\tau = \dfrac{G\Delta\phi}{L}\rho$
Each area ΔA contributes different moment $\rho\tau\Delta A$

Forces in plane of cross-section

To summarize, the stress in torsion is not uniform but increases with radial position. Also, for a given stress, each area element's contribution to the torque depends on the radial position of the element.

5. Understand the dependence on I_p by comparing the contributions from different parts of a solid shaft to the total torque.

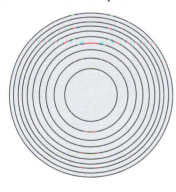

This shaft of radius 1 is divided into 10 rings of equal area $\pi(1)^2/10$.

Compare the individual contributions of individual rings to I_p (or to the total torque).

	1	2	3	4	5	6	7	8	9	10
R_{inner}	0	0.32	0.45	0.55	0.63	0.71	0.78	0.84	0.89	0.95
R_{outer}	0.32	0.45	0.55	0.63	0.71	0.78	0.84	0.89	0.95	1
% I_p	1	3	5	7	9	11	13	15	17	19
Accum. I_p	1	4	9	16	25	36	49	64	81	100

The outer ring with 10% of the area produces 19 times as much torque as the inner 10%. One gets more stiffness (more torque for given twist) using outer portions rather than inner portions.

Alternatively, if one removes the inner 50% of the shaft area (up to $R_{inner} = 0.71$) the area (and weight) are reduced by 50%. But only 25% of the stiffness is lost and 75% remains.

Why is the polar moment of inertia, I_p, due to an outer ring so much greater than that of an inner ring of the same area?

Because I_p is equal to ρ^2 integrated over the area

$$I_p = \iint \rho^2 \, dA$$

A portion of cross-section located farther out (ρ greater) has greater stress and contribution to moment than an inner portion

with the same area

>>End 4.6

4.7 General Guidelines for Torsional Stiffness of Non-Circular Cross-Sections ⸻

Often, torsion occurs in members that have non-circular cross-sections. Their stiffness (and strength) also depends on shaft length, material, and cross-section. Here we extend the reasoning applied to shafts of circular cross-section to rationalize the stiffness of general cross-sections.

1. Length, shear modulus, and shear strength affect shafts of non-circular cross-section in the same ways as they affect shafts of circular shafts.

Stiffness is proportional to shear modulus G and inversely proportional to length L.

Strength is proportional to the maximum allowable stress τ_Y.

2. Assuming the area is constant, the following guidelines on cross-section are useful: hollow is stiffer than solid, compact is stiffer than spread out, and closed is stiffer than non-closed.

In the previous section, we learned why circular shafts depend on I_p instead of on area. The reasoning involved the variation of the shear stress in the cross-section and the differing contributions to the moment. We apply this reasoning to devise rules-of-thumb for the stiffness of different cross-sections of the *same* area.

- Moving material far from the center also stiffens shafts of non-circular cross-sections.

 is much stiffer than

- Having all outer material equally distant from the center stiffens the shaft compared to letting some material be farther and some closer.

 is much stiffer than

- A closed thin-walled cross-section is much stiffer than an open thin-walled section. (This will be rationalized later.)

 is much stiffer than

The relative stiffness of various cross-sections is shown next. Methods for calculating the stiffness and strength for some non-circular cross-sections are given in the following sections.

3. Quantify the relative stiffness of various cross-sections of the same area.

We take a circular tube with 8/7 ratio of outer to inner diameter as a baseline (I). All cross-sections to be compared here have the same area as the baseline. The thin-walled cross-sections (II and III shown below) all have the same wall thickness as the baseline.

The baseline thin-walled circular cross-section is the most efficient one for torsion. It has maximum stiffness for its area. We denote the stiffness of this cross-section as 100%. All other cross-sections are listed relative to this baseline.

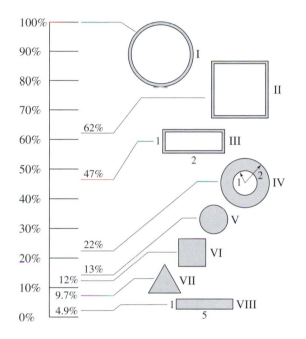

Stiffness of Cross-Sections of Equal Area

Percentages are based on $OD/ID = 8/7$ for baseline (I)

Stiffness of thin-walled sections (I, II, and III) relative to each other will be as shown if areas and wall thicknesses are equal.

Stiffness of non-thin-walled sections (IV, V, VI, VII, and VIII) relative to each other will be as shown if areas are equal.

Stiffness of thin-walled sections relative to non-thin-walled sections increases as wall thickness decreases for same area. But, if walls are too thin, shafts can easily dent or buckle.

4. Torsional stiffness can be increased through complex combinations of material and geometry.

We have only considered bodies with uniform material properties that are equal in all directions. Very sophisticated design goes into snowboards, for example, to give them high torsional stiffness, while keeping them relatively flexible in bending. This design involves more complex materials arranged in various ways.

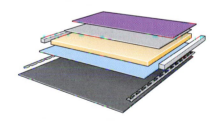

>>End 4.7

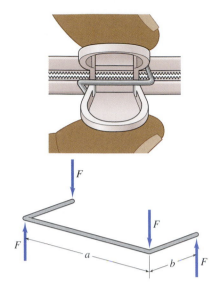

This "chip clip" for keeping bags of snacks closed and fresh is based on a torsional spring. The torsional resistance of this spring is due in part to the twisting of the central portion of the wire. In the simple model below, the forces cause the right arm to pivot up, and the left to pivot down by the same angle. (The forces F would be related to the finger forces on the clip.)

To simplify the analysis, consider only the twisting of the central portion of length a; assume the arms move rigidly. The wire is 2 mm in diameter; $a = 30$ mm, $b = 15$ mm. Take $G = 80$ GPa.

We define the stiffness here as the force F needed to produce one degree of rotation of each arm (in N/deg). The strength is defined as the rotation (of one side) at which the shear yield stress of $\tau_Y = 500$ MPa is first reached.

Determine the stiffness and the strength.

Solution

The torque applied to each end of the central portion is given by $T = Fb$.

The pivot angle of each arm ϕ_{arm} equals the rotation at that end of the central portion.

Each arm rotates by ϕ_{arm}, so the relative rotation of central portion $\Delta\phi = 2\phi_{arm}$.

Use the torque–twist relation $\Delta\phi = \dfrac{TL}{GI_p}$ with $I_p = \dfrac{\pi R^4}{2}$ to write the stiffness as

$$\text{Stiffness} = \frac{F}{\phi_{arm}} = \frac{T/b}{\Delta\phi/2} = \frac{2GI_p}{ab} = \frac{\pi G R^4}{ab} = \frac{\pi(80\text{ GPa})(1\text{ mm})^4}{(30\text{ mm})(15\text{ mm})}$$

$$= 559\frac{\text{N-m}}{\text{rad}} = 9.75\frac{\text{N-m}}{\text{deg}}$$

The shear strength is reached when the maximum τ (at the outer surface) first reaches τ_Y.

From the torque–shear stress relation,

$$\tau = \frac{T\rho}{I_p} \Rightarrow \tau_Y = \frac{T_Y R}{I_p} \Rightarrow T_Y = \frac{\tau_Y I_p}{R} = 0.785\text{ N-m}$$

From the force to cause yield is $F_Y = \dfrac{T_Y}{b} = 52.4$ N, we can find the arm rotation at yield

$$(\phi_{arm})_Y = \frac{F_Y}{\text{stiffness}} = \frac{52.4\text{ N}}{9.75\text{ N/deg}} = 5.37°$$

>>End Example Problem 4.7

A structure to be lifted into space was originally designed based on using 2024 aluminum rod 20 mm in diameter. To save weight, tubing of the same material is being contemplated. The tubing must be at least as stiff in torsion as the original rod. Consider tubing with a 3 mm wall thickness that is available in outer diameters of 12 mm, 16 mm, 20 mm, and so forth increasing in increments of 4 mm.

Find the minimum size tube that is at least as stiff in torsion as the original rod.

Determine the tube strength relative to the rod.

Solution

The stiffness, or torque per twist, of a circular shaft is $\dfrac{T}{\Delta\phi} = \dfrac{GI_p}{L}$.

Since we are comparing members with the same shear modulus G and length L, we need to compare only the moments of inertia, I_p.

For solid rod, diameter = 20 mm, $I_p = \dfrac{\pi R^4}{2} = \dfrac{\pi (10\text{ mm})^4}{2} = 1.571 \times 10^4 \text{ mm}^4$

For tubing, $I_p = \dfrac{\pi}{2}\left[R^4 - R_i^4\right]$, we use a spreadsheet, or do the calculations by hand and find:

OD (mm)	ID (mm)	I_p (mm⁴)
12	6	1908.518
16	10	5452.234
20	14	11936.48
24	18	22266.04
28	22	37345.68

The tube with the smallest OD with I_p that exceeds $1.571 \times 10^4 \text{ mm}^4$ is 24 mm. It is $22266/15710 = 1.42$ times as stiff as the solid rod.

The maximum shear stress in a circular shaft is $\tau_{\max} = \dfrac{TR}{I_p}$.

To compare strengths, we compare I_p/R.

    Rod: $I_p/R = 1.571 \times 10^4/10 = 1571 \text{ mm}^3$.

    Tube: $I_p/R = 2.23 \times 10^4/12 = 1856 \text{ mm}^3$.

I_p of the tube is $1856/1571 = 1.18$ times as large as for the solid rod. $\tau_{\max}$ will therefore be smaller for the tube, and so the tube is 1.18 times stronger than the solid rod.

Since density is the same, we can compare the weights per length by comparing the areas. Rod area = 314 mm²; tube area = 197 mm². So the tube does weigh less than the rod.

Note that since the rod costs about $20/kg vs. the tubing that costs approximately $60/kg, the tubing is more expensive. However, since the cost of transport into space is very high, conversion to tubing may very well make sense.

>>End Example Problem 4.8

An Allen wrench has a hexagonal cross-section that is approximated here as circular with 6 mm diameter. The steel of the wrench yields when $\tau_Y = 250$ MPa. The dimensions are $a = 70$ mm and $b = 45$ mm. Define the strength of the wrench as the largest finger forces F that can be applied before first yielding the wrench. Simplify the problem by ignoring the bending.

Determine the strength of the wrench.

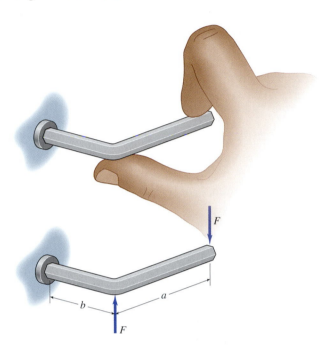

Solution

The portion of the wrench that twists is the part of length b. The torque applied to each end: $T = (F)(a)$.

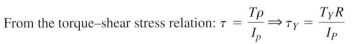

The strength is reached when the maximum shear $\tau_{max} = \tau_Y$.

From the torque–shear stress relation: $\tau = \dfrac{T\rho}{I_p} \Rightarrow \tau_Y = \dfrac{T_Y R}{I_P}$

We can find the torque at yield

$$T_Y = \frac{\tau_Y I_p}{R} = \frac{\pi \tau_Y (R)^3}{2} = \frac{\pi (250 \text{ MPa})(3 \text{ mm})^3}{2} = 10.6 \text{ N-m}$$

The strength, or force to yield, F_Y, is found from $F_Y = \dfrac{T_Y}{a} = \dfrac{10.6 \text{ N-m}}{70 \text{ mm}} = 151.4 \text{ N}.$

Note: Allen wrenches of larger cross-sections do tend to have larger arm lengths, consistent with the fact that they can withstand larger torques without yielding.

>>End Example Problem 4.9

A hollow circular steel tube 6 ft long with an outer diameter of 2.0 in. and 0.125 in. wall thickness serves as a post and supports twisting moments equal to 500 lb-ft. The twist of the tube is of concern. For design purposes, several other cross-sections, all of the same cross-sectional area (hence, weight per length), are examined as replacements for the hollow circular post. Take $G = 11 \times 10^6$ psi for all posts.

(a) Determine the rotation in degrees for the original hollow circular post.

Using the chart in Section 4.7.3, determine the additional rotation in degrees associated with substituting the following cross-sections: (b) hollow square, (c) solid circular, and (d) solid square.

Solution

(a) The twist of the circular tube can be found from

$$\Delta\phi = \frac{TL}{GI_p} = \frac{(500 \text{ lb-ft})(6 \text{ ft})(12 \text{ in./ft})^2}{(11 \times 10^6 \text{ psi})\frac{\pi}{2}[(1 \text{ in.})^4 - (0.875 \text{ in.})^4]} = 0.0604 \text{ rad} = 3.46°$$

The tube in question has an inner diameter/outer diameter ratio of 0.875 in., and so the values from the chart in Section 4.7.3, which compare various cross-sections, can be used directly.

(b) From the chart, a hollow square cross-section is 62% as stiff as a hollow circular
  $\Rightarrow (\Delta\phi)_{\text{hollow square}} = 3.46°/0.62 = 5.6°$, or an increase of 2.1°.

(c) From the chart, a solid circular cross-section is 13% as stiff as a hollow circular
  $\Rightarrow (\Delta\phi)_{\text{solid circular}} = 3.46°/0.13 = 27°$, or an increase of 23°.

(d) From the chart, a solid square cross-section is 12% as stiff as a hollow circular
  $\Rightarrow (\Delta\phi)_{\text{solid square}} = 3.46°/0.12 = 29°$, or an increase of 25°.

>>End Example Problem 4.10

Additional data on material properties needed to solve problems can be found in Appendix D or inside back cover.

4.46 One end of a solid aluminum shaft ($G = 27$ GPa, length $= 700$ mm) is fixed, and a torque is applied to the other end. Let the stiffness of the shaft be defined as the torque relative to the rotation. If the stiffness of the shaft is to be 5 N-m/deg, what should be the diameter?

4.47 The shaft is loaded with two forces as shown and is fixed at A. The strength is defined as the maximum allowable force F that does not produce shear stress above the material's maximum allowable value of 4 ksi. Determine the strength. By what factor would the strength increase if the shaft diameter were doubled? ($L_1 = 12$ in., $L_2 = 6$ in., and $d = 0.5$ in.)

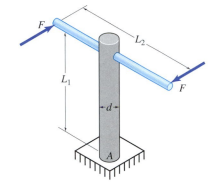

Prob. 4.47

4.48 One design includes a solid shaft of diameter 1 in. consisting of aluminum with an allowable shear stress of 15 ksi. There is space for a larger shaft, up to 2 in. in diameter. By switching to a hollow shaft, more torque can be carried without increasing the weight. Hollow shafts of the same material with outer diameter 2 in. are available with wall thicknesses in 0.0625 in. increments. If the hollow shaft must weigh no more than the original shaft, how many additional lb-in. of torque can an acceptable hollow shaft support?

4.49 To save expense, a solid shaft is to replace a hollow shaft of the same material. The hollow shaft has inner and outer diameters of 32 mm and 40 mm. The solid shaft is available in diameters of 1 mm, 2 mm, 3 mm, and so forth. What is the smallest diameter solid shaft that has at least the same stiffness as the original hollow one? By what percentage does the weight of the solid shaft exceed the hollow shaft?

4.50 The system composed of a single bent piece of wire deforms primarily by twisting of the center portion BC (neglect bending). The wire has a circular cross-section with diameter d. Define the stiffness as the ratio of the force F applied at each of the four points to the deflection at point A or D. Take the center cross-section E not to rotate. Derive a formula for the stiffness.

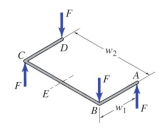

Prob. 4.50

4.51 The torsion spring system consists of a plastic shaft ($G = 200$ ksi). The material has a maximum allowable shear stress of 3 ksi. Take the arm BC to rotate rigidly. The stiffness of the system, defined as the force at C relative to the downward deflection there, should be 2 lb/in. What diameter of solid shaft gives this stiffness? What is the maximum allowable deflection at C if the allowable shear stress in the shaft is not to be exceeded? ($L_1 = 4$ in. and $L_2 = 3$ in.)

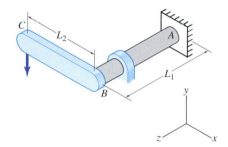

Prob. 4.51

4.52 To save weight in a torsion spring system, a solid shaft 20 mm in diameter is to be replaced by a hollow shaft of the same outer diameter and a wall thickness of 6 mm. The stiffness of the hollow shaft is what fraction of the solid shaft?

4.53 The hollow shaft with shear modulus G, inner diameter d_i, and outer diameter d_o, is fixed at one end and rotated by a rigid rod at the other end. A force F is applied to each end of the rod, and the displacement at each end is u. Determine the stiffness of the system, F/u. Assume the rotation angles are small.

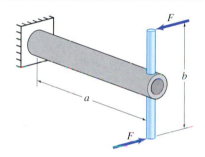

Prob. 4.53

4.54 The device shown monitors pressure on the surface A by measuring the angle of rotation of the upper end of the shaft ($G = 1$ GPa). The proportionality of pressure to rotation is to be 100 kPa/deg. Determine the shaft diameter. Take the pressure to act uniformly over the surface, and consider only the shaft to deform. ($L_1 = 500$ mm, $L_2 = 50$ mm, $L_3 = 40$ mm, and $L_4 = 30$ mm.)

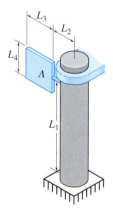

Prob. 4.54

4.55 Wrenches are sized so that the bolt is not readily torqued by hand to yield. Say a wrench with jaws separated by $s = 16$ mm has length of $L = 140$ mm. The bolt head that fits the opening has a shank of mean diameter of $d = 9$ mm. If the shear yield strength of the bolt is 250 MPa, how large a force could be applied to the end of the wrench without yielding in the bolt?

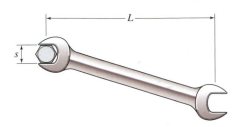

Prob. 4.55

4.56 The dimensions w_1 and w_2 of an Allen wrench increase with the hexagonal cross-section. Say each finger is capable of applying a force of 10 lb. What fraction of the shear yield stress of 35 ksi can the fingers produce in twisting the portion AB of the wrench? Approximate the hexagonal cross-section as responding in torsion identically to a circle of diameter 0.3125 in. ($w_1 = 4$ in. and $w_2 = 1.5$ in.)

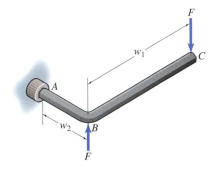

Prob. 4.56

4.57 A hollow shaft of outer diameter of 36 mm, inner diameter of 30 mm, and length of 700 mm twists by 1°. Consider a new shaft subjected to the same twist, with the same outer diameter, but the inner diameter reduced to 26 mm. Consider the maximum strain, maximum stress, and torque: will each of these increase, decrease, or remain unchanged? If there is an increase or decrease, give the fractional change = [(new-old)/old].

4.58 The original design for a sign post involves a hollow circular tube with outer diameter of 200 mm and inner diameter of 175 mm. When a 50 mph wind blows, the post twists by 2°. Consider an alternative post based on a hollow square cross-section of the same area and wall thickness. How much would this alternative post twist under the same wind load. (The results in Section 4.7.3 apply since all the dimensions have been equally scaled up.)

4.59 A solid circular steel shaft ($G = 11.6 \times 10^6$ psi) with diameter 1.8 in. is to be replaced by a solid shaft of square cross-section (same area and material). In either case, the shaft is 50 in. long and is subjected to a torque of 4000 lb-in. By approximately how many more degrees will the square shaft rotate? (Use table in Section 4.7.3.)

4.60 Stiffness and strength can be critical in high performance bicycles. Bottom bracket spindles were originally all solid and are now hollow in many designs. Consider a solid spindle of 20 mm diameter and a hollow spindle of 27 mm outer diameter and 21 mm inner diameter, both of identical materials. Determine these ratios of hollow to solid spindles: (a) weight, (b) torsional stiffness, and (c) torsional strength.

4.61 The down tube needs to have sufficient torsional stiffness. The down tube in one bike has an outer diameter of 1.75 in. and wall thickness of 0.125 in. There is room to enlarge the *OD* to 2 in. If the torsional stiffness is to be maintained, to what thickness can the wall thickness be reduced?

4.62 Vibration of a drill string can cause fatigue failure. As part of the analysis of drill string vibrations, the stiffness of a steel string needs to be determined. Consider a pipe of length 30 ft with outer and inner diameters of 3.5 in. and 2.992 in., respectively. Determine the torsional stiffness of the steel pipe.

4.63 In order to dampen oscillations in a wind turbine associated with changes in load every time a blade passes the tower, a polymeric high speed shaft (connected to the generator) is proposed. The shaft should have a torsional stiffness of $k_{torsion} = 2800$ N-m/rad. If the solid shaft has length $L = 2$ m and a shear modulus $G = 900$ MPa, what must be the shaft diameter?

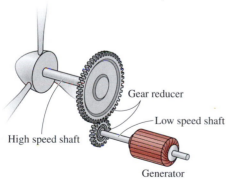

High speed shaft
Gear reducer
Low speed shaft
Generator

Prob. 4.63 (Appendix A6)

4.8 Torsion of Shafts with Rectangular Cross-Sections

Since the torsional stiffness and strength of shafts with non-circular cross-sections are also needed occasionally, these have been tabulated in engineering handbooks. Here we explain general differences in the twisting of circular and non-circular shafts. Also, we demonstrate how tables in engineering handbooks can be used to quantify the response of non-circular shafts, in particular shafts with rectangular cross-sections.

1. Non-circular and circular shafts both feature relative rotation of successive cross-sections and non-uniform distributions of shear stresses and strains, but there are other important differences.

In some ways twisting shafts with non-circular cross-sections behave similarly to shafts of circular cross-sections:

- Successive cross-sections rotate relative to each other about the shaft axis.
- The twist per length dictates the intensity of deformation.

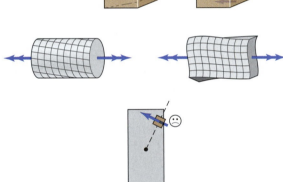

- Shear stresses and strains are produced, and while they vary in the cross-section, they do not vary along the shaft length.

Important differences between circular and non-circular cross-sections are:

- Cross-sections do not stay flat; they warp or deform out of plane.
- Shear stress in general does not act perpendicularly to lines emanating radially from the shaft center.

2. Because the lateral faces of the shaft are free of forces, the shear stress must be tangential to the edge.

The lateral surfaces of a shaft subjected to torsion are free of forces. Recall that shear stresses always act on four faces of an element. To ensure that the stresses on the axial face do not lead to shear stress on a lateral surface, the direction of the shear stresses at the edges of an axial face must be *tangential* to the edge, regardless of the cross-section.

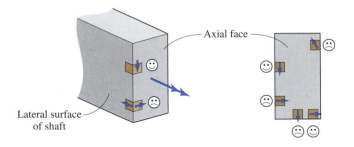

Axial face

Lateral surface of shaft

3. More complex methods are needed to analyze shafts of general cross-section, but the results of interest to strength and stiffness can be put in the same form as for circular shafts.

Because of the warping and the unknown directions of the shear stress inside the cross-section, non-circular cross-sections generally require more sophisticated methods of analysis (elasticity theory). However, the most commonly needed results are tabulated in handbooks for many cross-sections.

For non-circular cross-sections, we still want τ_{max} and $\Delta\phi$, but we must find quantities to replace R and I_p in the circular shaft torsion equations $\tau_{max} = \dfrac{TR}{I_p}$ and $\Delta\phi = \dfrac{TL}{GI_p}$.

4. Key results for non-circular cross-sections typically are given in handbooks by presenting formulas with variables and tables of numerical coefficients.

Handbooks typically display the cross-section and its dimensions as variables, together with formulas for τ_{max} and $\Delta\phi$ that include dimension variables and tabulated coefficients

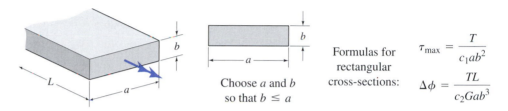

Choose a and b so that $b \le a$

Formulas for rectangular cross-sections:
$$\tau_{max} = \frac{T}{c_1 ab^2}$$
$$\Delta\phi = \frac{TL}{c_2 Gab^3}$$

Variation in coefficients c_1 and c_2 with a/b in formulas for τ_{max} and $\Delta\phi$ for rectangular cross-sections:

a/b	1.00	1.50	2.00	3.00	5.00	10.0	∞
c_1	0.208	0.231	0.246	0.267	0.291	0.312	0.333
c_2	0.141	0.196	0.229	0.263	0.291	0.312	0.333

5. Follow this example of applying the formulas and table of coefficients above.

A steel shaft 40 in. long, with cross-section 0.250 in. by 0.150 in. and $G = 11.5 \times 10^6$ psi, is twisted by a torque $T = 200$ lb-in. Find τ_{max} and $\Delta\phi$.

General Step	Step for Particular Example
Define a and b for the cross-section; choose b to be smaller than a	$a = 0.150$ in., $b = 0.250$ in.
Compute a/b	$a/b = (0.150)/(0.250) = 3$
Get c_1, c_2 from table	$c_1 = 0.267$, $c_2 = 0.263$
Evaluate τ_{max} and $\Delta\phi$	$\tau_{max} = T/c_1 ab^2 = 16000$ psi $\Delta\phi = TL/c_2 Gab^3 = 0.226$ rad $= 12.9°$

6. Apply explicit, simplified formulas for maximum stress and twist for rectangular cross-sections that are very thin relative to their width ($a/b \gg 1$).

Notice from the table that the coefficients c_1 and c_2 approach finite values for ($a/b \Rightarrow \infty$). Therefore, we can write down special formulas for rectangular cross-sections that are thin and wide:

$$\tau_{max} = \frac{3T}{ab^2} \qquad \Delta\phi = \frac{3TL}{Gab^3}$$

From these formulas, notice:

- Thickness b has a stronger effect than a: double b and the strength increases by factor of 4 (τ_{max} decreases) and stiffness increases by factor of 8 ($\Delta\phi$ decreases).
- Even if a/b is as low as 5, the approximation for a thin, wide cross-section is still accurate to within $(0.333 - 0.291)/0.291 = 14\%$.

>>End 4.8

4.9 Torsion of Shafts with Thin-Walled Cross-Sections

We saw earlier that moving material outward from the shaft center increases its stiffness and strength. Since such shafts are also lightweight and easily fabricated, it is common to use members with non-circular cross-sections that have thin walls, such as this lacrosse stick. Here we present formulas for the stiffness and strength of such shafts, and we demonstrate their use.

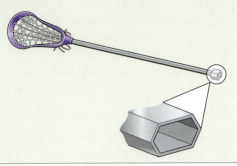

1. The shear stress acts tangentially to the perimeter and is uniform across the thickness.

The formulas given here apply to any cross-section formed by a single closed curve with thickness t (open cross-sections are addressed very approximately in Section 4.9.4).

Here is a shaft with a general thin-walled cross-section. A shaft is thin-walled if the thickness t is much less than other dimensions that describe the cross-section.

To keep the lateral face of the shaft free of force, τ must be tangential to the boundary. Since the thickness is small, we approximate τ to be uniform across thickness.

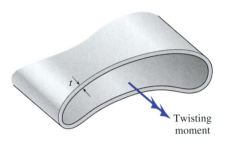

2. For a shaft with a uniform wall thickness, the shear stress is uniform around the perimeter.

Like the entire tube, the partial tube must also be in equilibrium.

$$\sum F_x = -\tau_1 Lt + \tau_2 Lt = 0 \Rightarrow \tau_1 = \tau_2$$

Since any two longitudinal faces could have been chosen for the partial tube, $\tau_1 = \tau_2$ holds true for any two points on the perimeter.

Therefore, τ is uniform over the entire cross-section.

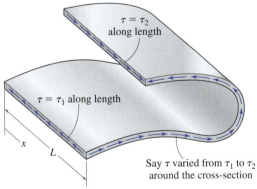

In the unusual event that the thickness t varies around the perimeter, then the product $(\tau)(t)$, referred to as the shear flow, is constant around the perimeter.

3. **The shear stress and the twist can be related to the torque and the geometry of the cross-section.**

Since τ is uniform and tangential, one can calculate $\tau_{max}(=\tau)$ and $\Delta\phi$ in terms of the torque T, shaft length L, shear modulus G, and the cross-section. We only present the final equations here (see Appendix I-1).

Besides the thickness t, formulas depend on two geometric measures of the cross-section.

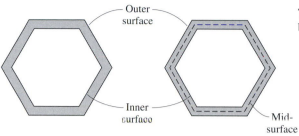

The mid-surface is the line halfway between the outer and inner surfaces.

s_m is the length (perimeter) of the mid-surface.

A_m is the area enclosed by the mid-surface.

The formulas for maximum shear stress and for twist of closed thin-walled shafts are:

$$\tau_{max} = \frac{T}{2A_m t} \qquad \Delta\varphi = \frac{TLs_m}{4GA_m^2 t}$$

---

4. **Base approximate formulas for shear stress and twist of non-closed thin-walled shafts on those for a thin rectangular cross-section.**

Some shafts have cross-sections like these that are not closed. They have relatively low torsional stiffness.

For a non-closed cross-section, the shear stress has opposite directions across the thickness, since the stress has to be tangential to all edges.

A tube with mid-section path length s_m and thickness t responds in torsion roughly like a rectangular bar of width s_m and thickness t. Since the rectangle is long and thin, we can use the approximation from Section 4.8.6 for a rectangular cross-section with a/b that is large compared to 1.

s_m: length of the mid-section path

Therefore, very approximate formulas for maximum shear stress and for twist of non-closed thin-walled shafts are:

$$\tau_{max} = \frac{3T}{s_m t^2} \qquad \Delta\phi = \frac{3TL}{Gs_m t^3}$$

---

5. **Experience the remarkably lower stiffness in a non-closed cross-section compared to a closed cross-section.**

One can readily experience the effect of replacing a closed cross-section with an open one. Try to twist a cardboard tube from an empty roll of paper towels. Thin-walled, closed cross-sections of even flimsy materials can be quite stiff.

Now, cut the tube along the entire length, and try to twist again. It may be hard to grip, but try it this way. It is much less stiff than the uncut (closed) tube. Unlike the closed circular tube, the cut tube warps significantly, that is, there are axial displacements that vary over the cross-section.

>>End 4.9

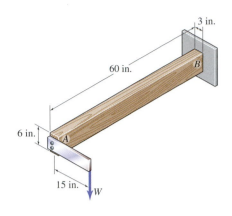

3 in.

60 in.

B

6 in.

A

15 in.

W

A 3 in. × 6 in. wood member AB is rigidly supported at B. A 15 in. bracket is attached at the other end and the load W is applied. We define the stiffness here as the end force to produce a unit downward deflection of the bracket at the load. Take G = 500 ksi (G for wood is not usually tabulated). Neglect the deformation of the bracket. Do not include the bending of the wood member, only its twisting.

Determine the stiffness and the maximum allowable end deflection if the shear stress in the board is not to exceed 900 psi.

Solution

The load W causes a torque $T = W(15 \text{ in.})$.

The downward deflection v of the bracket where the load is applied corresponds to a rotation of AB at end A: $\phi_A = v/15$ in. The end B is fixed, so $\phi_B = 0$.

Key relations for the twisting of this member of rectangular cross-section are:

$$\tau_{max} = \frac{T}{c_1 ab^2} \qquad \Delta\phi = \frac{TL}{c_2 Gab^3}$$

Use the table in Section 4.8.4. The dimensions a and b for this cross-section ($a > b$) are:

$$a = 6 \text{ in.}, b = 3 \text{ in.}, a/b = 2$$

From the table, $c_1 = 0.246$ and $c_2 = 0.229$.

Replace T and $\Delta\phi$ in terms of W and $v \Rightarrow$

$$\frac{v}{15} = \frac{W(15 \text{ in.})L}{c_2 Gab^3} = \frac{W(15 \text{ in.})(60 \text{ in.})}{0.229(500 \text{ ksi})(6 \text{ in.})(3 \text{ in.})^3}$$

From this relation, we find the stiffness: $\dfrac{W}{v} = 1374$ lb/in.

Since $\tau_{max} = 900$ psi, $T_{max} = c_1 ab^2 \tau_{max} = 1.196 \times 10^4 \text{ lb} \Rightarrow W_{max} = 797$ lb.

Using the stiffness, $v_{max} = (797 \text{ lb})/(1374 \text{ lb/in.}) = 0.580$ in.

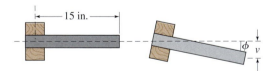

15 in.

ϕ v

An automobile frame consists of aluminum tubular members with the dimensions shown. The wall thickness is 8 mm. Such members are subjected to twisting.

Determine the torsional stiffness, defined as $T/\Delta\phi$ (in N-m/deg), of a 1 m length of such tubing. Take $G = 27$ GPa.

Say a 1 m length must carry a twist of $1°$ without yielding in shear. Determine the maximum shear stress that the material must carry.

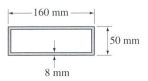

Solution

For hollow thin-walled members, formulas for maximum shear stress and for twist are:

$$\tau_{max} = \frac{T}{2A_m t} \qquad \Delta\phi = \frac{TLs_m}{4GA_m^2 t}$$

Parameters s_m and A_m are based on the arc length and area enclosed by the path midway between the inside and outside walls. The dimensions of the mid-path are as shown.

$$s_m = 2(152 \text{ mm} + 42 \text{ mm}) = 388 \text{ mm} \qquad A_m = (152 \text{ mm})(42 \text{ mm}) = 6384 \text{ mm}^2$$

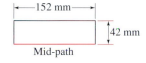

Mid-path

$$\Rightarrow \frac{T}{\Delta\phi} = \frac{4GA_m^2 t}{Ls_m} = \frac{4(27 \text{ GPa})(6384 \text{ mm}^2)^2(8 \text{ mm})}{(1 \text{ m})(388 \text{ mm})} = 1584 \frac{\text{N-m}}{\text{deg}}$$

Use the stiffness to find the torque when $\Delta\phi = 1°$:

$$T = \left(1584 \frac{\text{N-m}}{\text{deg}}\right)(1 \text{ deg}) = 1584 \text{ N-m}$$

From the torque–shear stress relations, we find

$$\tau_{max} = \frac{T}{2A_m t} = \frac{(1584 \text{ N-m})}{2(6384 \text{ mm}^2)(8 \text{ mm})} = 15.51 \text{ MPa}$$

So the material must carry a maximum shear stress of at least 15.51 MPa, if it is not to yield when twisted by $1°$.

>>End Example Problem 4.12

Additional data on material properties needed to solve problems can be found in Appendix D or inside back cover.

4.64 A steel shaft ($G = 80$ GPa) has cross-section 20 mm by 10 mm and is 600 mm long. A torque of 40 N-m is applied. Determine the maximum shear stress and twist of the shaft.

4.65 A wood beam ($G = 10^6$ psi), which is 1.5 in. by 4.5 in., is fixed at A. A steel plate is bolted to the far end of the beam, and a weight $W = 60$ lb is hung from C. By how much will end C of the steel plate displace relative to B due to the twisting of AB? (Account only for the torsion of AB, and neglect the bending of AB and BC.) What is the maximum shear stress in the wood beam? ($L_1 = 90$ in. and $L_2 = 40$ in.)

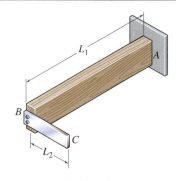

Prob. 4.65

4.66 Cables are attached to the steel post (80 GPa) as shown. The post is fixed at its base and has cross-sectional dimensions 20 mm by 100 mm. When the cables are tightened with turnbuckles, the post twists. Assume the cable tensions are each equal to F. Calculate the stiffness, which is defined as the ratio of tension in one cable relative to the twist of the post (N/rad). Note that only the component of the cable tensile force perpendicular to the post axis produces a twisting moment. ($L = 2$ m and $\theta = 30°$.)

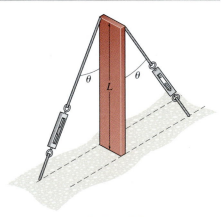

Prob. 4.66

4.67 The stress in the solid square aluminum shaft ($G = 3.8 \times 10^6$ psi), with cross-section 0.5 in. by 0.5 in., is to be kept at or below 5000 psi. End A will rotate by 1° about the $-x$-axis. What is the allowable range of rotations of end B about the $+x$- and $-x$-axes, and what is the maximum torque that can be applied? ($L = 60$ in.)

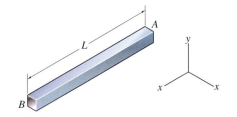

Prob. 4.67

4.68 A bar of mild steel (allowable shear stress 130 MPa) with dimensions 60 mm by 6 mm by 1 m long is subjected to torsion. A higher strength cold-rolled steel (allowable shear stress of 250 MPa) is contemplated as a replacement. How much additional torque (in N-m) can the higher strength steel bar withstand?

4.69 A thin plate 3 in. wide and 60 in. long must be able to withstand twists of up to 70° without exceeding the allowable shear stress of 18 ksi. What is the maximum permissible plate thickness that can be tolerated?

4.70 The torsional stiffness of a snowboard needs to be carefully tuned. Let the snowboard be 20 mm thick and composed of a hard plastic ($G = 2$ GPa). (In practice, complex combinations of materials are used to give snowboards suitable mechanical properties.) Each foot applies a torque to the board through the bindings separated by L. Of interest is the stiffness defined as the torque applied by each foot divided by the angle of rotation of one binding relative to the other. Make two estimates of the stiffness: one corresponding to a uniform width w_1 and one to a uniform width w_2. ($w_1 = 100$ mm, $w_2 = 200$ mm, and $L = 1$ m.)

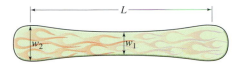

Prob. 4.70

Focused Application Problem

4.71 The downward force on the right pedal results in a complicated loading of the crank. Consider only the tendency to twist the crank, and the resulting shear stresses at the cross-section a distance x from the pedal spindle. The crank is approximated as rectangular in cross-section. If the maximum shear stress due to torsion is not to exceed 10000 psi, what is the maximum force that the foot can apply? Assume the force is applied in the middle of the pedal of width a. Take the dimensions to be $a = 3.625$ in., $b = 0.375$ in., $c = 1.5$ in., $w = 0.5$ in., and $x = 6$ in.

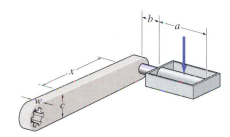

Prob. 4.71 (Appendix A1)

4.72 An aluminum tube ($G = 3.8 \times 10^6$ psi and length $= 3$ ft) with a thin-walled triangular cross-section 0.0625 in. thick is subjected to a torque of 100 lb-in. Determine the maximum shear stress in the tube, and the rotation of one end of the tube relative to the other. The dimensions of the mid-path are $w_1 = 3$ in. and $w_2 = 2$ in.

Prob. 4.72

4.73 A steel shaft ($G = 80$ GPa and length $= 2$ m) is to be twisted by 2°. The shaft is hollow with outer dimensions shown ($w = 25$ mm and $c = 15$ mm) and wall thickness 2 mm. Determine the shear stress in the tube.

Prob. 4.73

4.74 An aluminum ($G = 3.8 \times 10^6$ psi) shaft 24 in. long must have a torsional stiffness of 2000 lb-in./deg. One manufacturer has hollow square tubes with wall thickness of 0.125 in. and a wide range of outer dimensions in increments of 0.5 in. From among the sizes available, find the tube of minimum dimensions with torsional stiffness exceeding that required.

4.75 For aesthetic purposes, a steel post ($G = 80$ GPa) with a hollow hexagonal cross-section is proposed. Tubing is found that has wall thickness of 4 mm, with mid-path dimension as shown. The material has a maximum allowable shear stress of 70 MPa. What torque can this post sustain without exceeding the allowable stress? If the post must twist by up to 2° without exceeding the allowable stress, what is the shortest length of tubing that should be used? ($w = 100$ mm.)

Prob. 4.75

4.76 An aluminum lacrosse stick ($G = 27$ GPa) has a cross-section with mid-path dimensions shown and wall thickness of 2 mm. Such a stick can be subjected to torsion. If a shear stress of up to 35 MPa can be tolerated, what would be the maximum allowable twist over a length $L = 1.75$ m? ($w_1 = 20$ mm, $w_2 = 8$ mm, and $h = 22$ mm.)

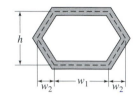

Prob. 4.76

Focused Application Problems

4.77 The geometry shown is proposed for a bridge girder. One important property of a girder is its stiffness against twisting $J = TL/(G\Delta\phi)$. Estimate J by treating the girder as a thin-walled tube. Take the dimensions to be $w_1 = 2.5$ m, $w_2 = 2$ m, $t = 40$ mm, and $h = 500$ mm.

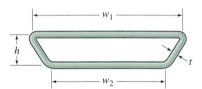

Prob. 4.77 (Appendix A2)

4.78 In using this pectoral fly machine, the hand grips and applies a force of $F_0 = 50$ lb to the handle as shown. The steel pivoting arm is approximated here as horizontal. Consider only the torsion due to this loading, and use the approximation for a thin-walled shaft. (a) Estimate the shear stress in the pivoting arm. (b) Assuming the pivoting arm is fixed at the far end, estimate the rotation of the pivoting arm where it attaches to the swinging arm. Take the dimensions to be $w = 6$ in., $L = 24$ in., $a = 1.5$ in., $b = 2$ in., $t = 0.25$ in., and $c = 20$ in.

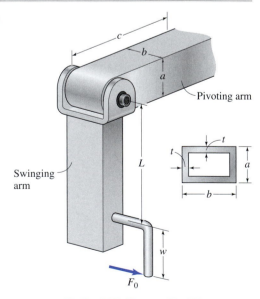

Prob. 4.78 (Appendix A4)

4.78 In using this pectoral fly machine, the hand grips and applies a force of $F_0 = 50$ lb to the handle as shown. The steel pivoting arm is approximated here as horizontal. Consider only the torsion due to this loading, and use the approximation for a thin-walled shaft. (a) Estimate the shear stress in the pivoting arm. (b) Assuming the pivoting arm is fixed at the far end, estimate the rotation of the pivoting arm where it attaches to the swinging arm. Take the dimensions to be $w = 6$ in., $L = 24$ in., $a = 1.5$ in., $b = 2$ in., $t = 0.25$ in., and $c = 20$ in.

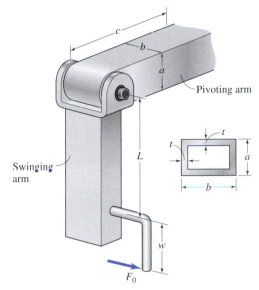

Prob. 4.78 (Appendix A4)

4.79 During operation of the leg curl machine, the pivot beam can twist. Say that the twisting rotation of the steel pivot beam where it connects to the shaft is neglected and that the pivot beam rotates by $0.3°$ where it connects to the connector links. Determine the maximum shear stress due to torsion in the pivot beam. Take the dimensions to be $h = 2$ in., $a = 3$ in., $r = 3$ in., $t_b = 0.25$ in., and $w = 17$ in.

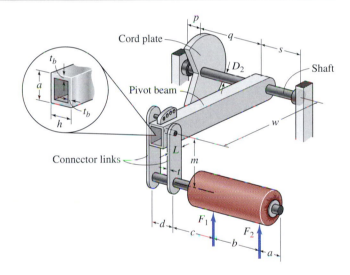

Prob. 4.79 (Appendix A4)

4.80 Intramedullary nails have been made of slotted cross-sections. Consider a titanium nail that has an outer diameter of 20 mm and a wall thickness of 1 mm. Determine the ratio of the twisting stiffness of a slotted nail with that of a complete tubular nail with the same inner and outer diameter. Estimate the twisting stiffness of the slotted cross-section using the general approximation for thin-walled open cross-sections given in Section 4.9.4.

Prob. 4.80 (Appendix A5)

4.10 Shafts with Non-Uniform Twisting Along Their Lengths

So far we have considered only shafts subjected to equal and opposite torques at their ends. But in many instances, there are more than two torques applied to a shaft. Fortunately, everything we have studied so far is still applicable. However, we need to consider discrete segments of a shaft separately, and to track their torques and twist. Here we explain a process for analyzing such shafts.

1. Bodies with complex loadings are often twisted by multiple torques along their length.

In this hand-cranked ice cream maker, you can find many of the loadings that we study in *Mechanics of Materials*.

Here is a subsystem consisting of the handle, shaft, and paddles.

Here we identify the points at which other bodies contact the subsystem.

Handle

Shaft

Paddles

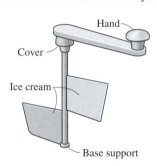

Hand

Cover

Ice cream

Base support

2. When forces are moved to a common axis, they exert both forces and couples or torques.

Here is a free body diagram of the handle, shaft, and paddles. The forces due to the hand, the cover, the ice cream, and the base support are modeled as point forces.

Each force not acting through the shaft can be replaced with a statically equivalent force acting through the shaft and a couple.

The forces acting through the shaft cause it to bend, which we study in Chapter 5. Here we show only the twisting moments. We only draw the shaft since the handle and paddles now play no role.

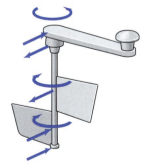

In this process, we have modeled the loads due to the contacting parts, and then identified those loads that cause twisting of the shaft. The twisting can now be quantified by analyzing a shaft with several torques applied along its length.

3. To track rotations, imagine a straight line drawn along the shaft before the torques are applied.

With this tube we will illustrate the effect of applying multiple torques to a shaft. To track their twisting effect, we draw a straight line on the shaft surface parallel to the axis. In this first diagram, the hands shown are just gripping the tube, but not yet applying torque.

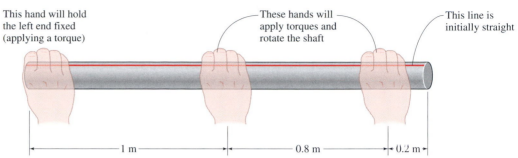

This hand will hold the left end fixed (applying a torque)

These hands will apply torques and rotate the shaft

This line is initially straight

1 m   0.8 m   0.2 m

4. Observe the displacement of the initially straight line, and notice that rotation occurs all along the shaft, not just where the torques are applied.

The hands are now applying torques to twist the tube. The torques are labeled. The tube is in equilibrium, since the total moment acting on it is zero.

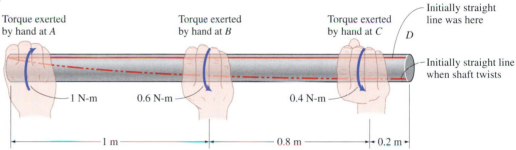

Torque exerted by hand at A

Torque exerted by hand at B

Torque exerted by hand at C

D

Initially straight line was here

Initially straight line when shaft twists

1 N-m   0.6 N-m   0.4 N-m

1 m   0.8 m   0.2 m

Torques are applied at *B* and *C*, and the tube also rotates at those cross-sections. At *A*, there is a torque applied (by the hand that resists), but no rotation. At most other cross-sections, there is rotation but no torque applied externally (by hands).

Given the applied torques, we typically want to find the rotation at any cross-section and the stress throughout the shaft, in terms of the applied torques and the geometry and material of the shaft.

5. There is no simple relation between the external torque at a cross-section and the rotation there.

Do not think that the rotation at a cross-section, say *B*, is directly related to the torque applied there. It is not! Students often make this mistake. In order to find rotations and stress, we first introduce the concept of *internal torque*, which is related to the *twist* of a given segment of the shaft.

>>End 4.10

4.11 Internal Torque and the Relation to Twist and Stress

We continue to consider the shaft with three torques applied to it. In analyzing the twisting of such a shaft, the concept of internal torque is extremely useful. Here we explain internal torque and show how it is used in analyzing torsion.

1. Follow displacements of the initially straight line to gage the intensity of twisting in different segments of the shaft.

From the motion of the surface line, the segments from A to B and from B to C each appear like a shaft with two equal and opposite torques at its ends.

The intensity of twist in AB is greater than in BC because the initially straight line is more tilted in AB. Therefore AB somehow feels a greater torque than BC. We must learn to identify the torque felt *within* a shaft, which causes its deformation and stress.

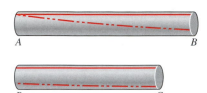

2. Distinguish external torques from internal torques.

External Torque: Torque applied by an external body, often to the surface of the shaft (here by hands).

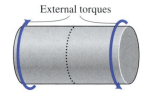

External torques

External torques

Shear stresses acting on the internal surface

Internal Torque: Torque applied via shear stresses by one portion of the shaft to the neighboring portion across an internal axial cross-section.

We can find the internal torques from the external torques by imposing equilibrium on FBD's of segments of the shaft.

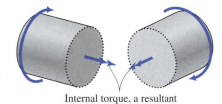

Internal torque, a resultant of the shear stresses

3. Draw FBDs of portions of the shaft to find the internal torques that must be present to maintain equilibrium.

Consider the tube with hands applying torques to it as composed of several parts. Each part must be in equilibrium.

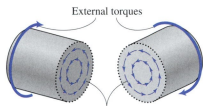

1.0 N-m   0.6 N-m   0.4 N-m

For these segments to be in equilibrium, there must be additional torques: These are the internal torques acting on internal faces!

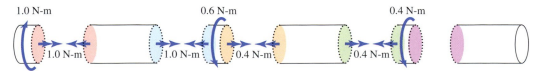

1.0 N-m   0.6 N-m   0.4 N-m

1.0 N-m   1.0 N-m   0.4 N-m   0.4 N-m

These internal torques keep every segment in equilibrium. They also satisfy Newton's 3rd Law that the forces (torques) between contacting bodies are equal and opposite.

4. Any segment with only equal and opposite torques at its ends has a uniform internal torque.

We designate the three long segments between the applied torques as *AB*, *BC*, and *CD*. Each has just equal and opposite internal torques applied to its ends. If we considered only the left half of *AB*, the internal torque at the new right end is also 1 N-m. So, the internal torque is uniform within each segment.

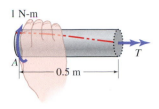

5. The stress in, and twist of, a segment with uniform internal torque are the same as in a shaft with equal and opposite torques at its ends.

Stresses

In segments *AB*, *BC*, and *CD*, we find stress with $\tau = TR/I_p$, just like a simple shaft with opposite torques at its ends. But what *T* do we use? Since the internal torque is the result of the shear stresses, we use the *internal* torque for *T*.

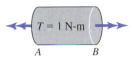

Rotation

Above, the short segments along the shaft on which the external torques act were taken to be infinitesimal. They do not twist, but each has a rotation.

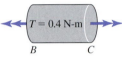

So the rotations ϕ_A, ϕ_B, ϕ_C, and ϕ_D are the rotations at the ends of the long, twisting segments *AB*, *BC*, and *CD*. Use the formula for the relative rotation for the simply twisted shaft. Here we use $GI_p = 10$ N-m^2, typical of a foam rubber tube 100 mm in diameter.

Segment *AB*: $\Delta\phi = \phi_B - \phi_A = \dfrac{T_{AB}L_{AB}}{GI_p} = \dfrac{(1 \text{ N-m})(1 \text{ m})}{10 \text{ N-m}^2} = 0.1$ rad

Segment *BC*: $\Delta\phi = \phi_C - \phi_B = \dfrac{T_{BC}L_{BC}}{GI_p} = \dfrac{(0.4 \text{ N-m})(0.8 \text{ m})}{10 \text{ N-m}^2} = 0.032$ rad

Segment *CD*: $\Delta\phi = \phi_D - \phi_C = \dfrac{T_{CD}L_{CD}}{GI_p} = \dfrac{(0 \text{ N-m})(0.2 \text{ m})}{10 \text{ N-m}^2} = 0$ rad

6. Find the absolute rotations at various points using the relative rotations found from the internal torques and the known rotation at one point.

From the internal torques we get the rotations of *B* relative to *A*, *C* to *B*, and *D* to *C*. Depending on ϕ_A, the initial horizontal line could have many positions.

Different values for ϕ_A

But we know the hand at *A* prevents rotation there, so $\phi_A = 0$. Therefore, $\phi_B = 0.1 + \phi_A = 0.1$ rad. Then, $\phi_C = 0.032 + \phi_B = 0.132$ rad, and $\phi_D = \phi_C = 0.132$ rad.

7. In summary, carefully apply the formula $\Delta\phi = TL/GI_p$ to each segment, recognizing that the internal torque, *T*, in each segment relates the relative rotation $\Delta\phi$ of that segment's ends.

- Break the shaft into segments in which internal torque is constant.
- Use the internal torque in that segment as *T* and the length of the segment as *L*.
- Treat the relative rotation of the ends of the segment as $\Delta\phi$ in the relation $\Delta\phi = TL/GI_p$.

4.12 Relation Between Senses and Signs of Internal Torque, Twist, and Stress

Sometimes torques may cause different segments of a shaft to twist in different directions. Here we show how to use the signs of the internal torque and the rotation to keep track of direction.

1. Account for the sense of the rotation and twisting.

The two segments of this shaft twist in opposite directions. We will use a sign convention (+ or −) to distinguish:

• Direction of rotation.
• Direction of twist and torque.

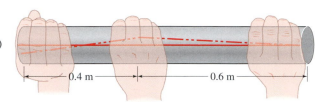

2. Use the sign of ϕ to describe the sense of the rotation and the sign of T to describe the senses of the twisting moments.

Think of a shaft as lying horizontally (like the one above). If necessary, reorient the shaft to make it horizontal. Draw the x-axis as pointing to the right. Follow these conventions for the meaning of the signs of ϕ and T.

Rotation:

• ϕ can be < 0, $= 0$, or > 0.

$\phi > 0$ if the rotation is positive according to the right hand rule about the x-axis.

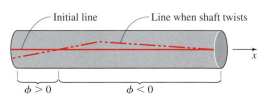

Internal Torque:

• T can be < 0, $= 0$, or > 0.

Regardless of whether $T > 0$ or $T < 0$, torques at the ends of a segment must be opposite. Rather, the sign of the internal torque signals the combination of torque senses at the ends.

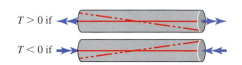

For this shaft with three external torques, $T > 0$ in the left portion and $T < 0$ in the right portion.

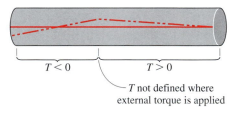

T not defined where external torque is applied

3. If the signs of T and ϕ follow the convention just described, then the signs of T and $\Delta\phi$ in the torque–twist relation will be consistent.

By assigning senses to signs of T and ϕ as above, the signs of the two sides of the torque–twist relation $\Delta\phi = TL/GI_P$ will be consistent no matter how the shaft twists. Here are all possibilities:

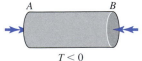

$$T > 0$$

$$\Delta\phi = \phi_B - \phi_A = \frac{TL}{GI_P} > 0$$

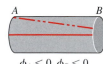

$$\phi_A < 0, \phi_B < 0$$
$$|\phi_A| > |\phi_B|$$
$$\Rightarrow \phi_B - \phi_A > 0$$

$$\phi_A < 0, \phi_B > 0$$
$$\Rightarrow \phi_B - \phi_A > 0$$

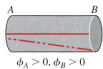

$$\phi_A > 0, \phi_B > 0$$
$$|\phi_B| > |\phi_A|$$
$$\Rightarrow \phi_B - \phi_A > 0$$

$$T < 0$$

$$\Delta\phi = \phi_B - \phi_A = \frac{TL}{GI_P} < 0$$

$$\phi_A < 0, \phi_B < 0$$
$$|\phi_B| > |\phi_A|$$
$$\Rightarrow \phi_B - \phi_A < 0$$

$$\phi_A > 0, \phi_B < 0$$
$$\Rightarrow \phi_B - \phi_A < 0$$

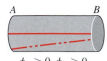

$$\phi_A > 0, \phi_B > 0$$
$$|\phi_A| > |\phi_B|$$
$$\Rightarrow \phi_B - \phi_A < 0$$

4. When determining the internal torques by applying equilibrium to portions of the shaft, make sure to follow the sign convention above for *T*.

Break the shaft into segments in which the internal torque is constant and short (infinitesimal) segments on which external torques are applied.

Look carefully at directions and signs of internal torques

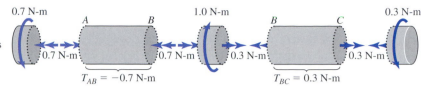

$T_{AB} = -0.7$ N-m $\qquad$ $T_{BC} = 0.3$ N-m

---

5. Apply the torque–twist relation, $\Delta\phi = TL/GI_p$, letting the sign of $\Delta\phi$ equal to the sign of *T*.

To make the calculations clear, we use these specific properties and dimensions:

$$L_1 = 0.4 \text{ m}, \ L_2 = 0.6 \text{ m}, \ GI_p = 10 \text{ N m}^2 \quad (100 \text{ mm diameter foam rubber tube})$$

Apply the torque–twist relation

$$\phi_B - \phi_A = \frac{T_{AB}L}{GI_p} = \frac{(-0.7 \text{ N-m})(0.4 \text{ m})}{10 \text{ N-m}^2} = -0.028 \text{ rad} \qquad \phi_C - \phi_B = \frac{T_{BC}L}{GI_p} = \frac{(+0.3 \text{ N-m})(0.6 \text{ m})}{10 \text{ N-m}^2} = 0.018 \text{ rad}$$

Notice that the twists in the two segments have different signs, which match respective internal torques.

---

6. Determine the rotation at all points from the twists and from the rotation at one point.

Find rotations at *A* and *B*, given that *C* is fixed

$$\phi_C = 0 \Rightarrow \phi_B = \phi_C - 0.018 = -0.018 \qquad \phi_A = \phi_B + 0.028 = 0.01$$

Interpret the positive and negative signs of ϕ_A and ϕ_B, consistent with the general rule for senses of rotations.

Here is the displaced line on the side of twisted tube:

---

7. Calculate the maximum shear stress in each segment using the magnitude of the internal torque.

Calculate the shear stress in a segment from $\tau = TR/I_p$. Use internal torque *T* from that segment. Use only the magnitude of *T*, from which only the magnitude of τ will be found.

The direction of the shear stress depends on the direction of twisting (related to the sign of *T*) and on the position around the circumference. The directions of the shear stresses were discussed in Section 4.4. Here we consider elements from the front faces of the two shaft segments:

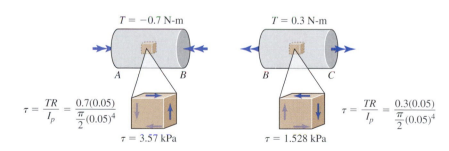

$$\tau = \frac{TR}{I_p} = \frac{0.7(0.05)}{\frac{\pi}{2}(0.05)^4} \qquad \qquad \tau = \frac{TR}{I_p} = \frac{0.3(0.05)}{\frac{\pi}{2}(0.05)^4}$$

$$\tau = 3.57 \text{ kPa} \qquad\qquad\qquad \tau = 1.528 \text{ kPa}$$

>>End 4.12

4.13 Shafts with Varying Cross-Sections

Sometimes the cross-section of a twisting member varies along the length. For example, engineers often design shafts with step changes in diameter to help in locating gears, sprockets, and bearings. Or, a member could simply have a gradual change in the cross-section, such as in bones. Here we show how the methods of calculating rotations and stresses learned so far are applied to such members.

1. Even if the shaft cross-section changes gradually, the general pattern of twisting, strain, and stress is the same as for a shaft of uniform cross-section.

The simple deformation pattern, with stress and strain proportional to radial distance ρ from the center, that we learned in Sections 4.2 and 4.4, is also a good approximation for gradually varying cross-sections.

The shaft to the right has regions of uniform and gradually varying cross-section, where the simple pattern is correct.

Simple deformation holds in:

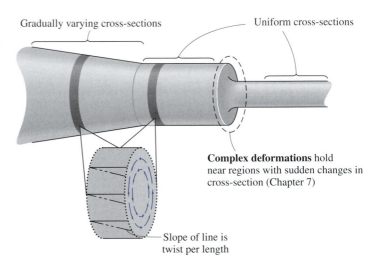

Gradually varying cross-sections   Uniform cross-sections

Where the shaft has a sudden change in cross-section, the stress and twist are not simply related to the torque.

Even with a gradually varying cross-section, each segment still twists with the rotations of neighboring cross-sections varying along the shaft length.

Complex deformations hold near regions with sudden changes in cross-section (Chapter 7)

Slope of line is twist per length

The shear stress still varies linearly with radial position and depends in the same way on the internal torque and the cross-sectional geometry.

2. Relate the shear stress to the internal torque with the same formula, except use the properties (e.g., I_p) of the cross-section of interest.

The basic distribution of shear stress in a shaft with gradually varying cross-section is the same as in a shaft of uniform cross-section.

$\tau = \dfrac{T\rho}{I_p}$ is still correct, except we use T, ρ, and I_p that pertain to the cross-section of interest.

3. Since the relative rotation of two neighboring cross-sections depends on I_p, but I_p varies along the shaft, the twist per length is not constant even if the internal torque is constant.

Instead of applying $\Delta\phi = \dfrac{TL}{GI_p}$ to a segment of finite length, we apply it to a short, potentially infinitesimal, segment Δx:

$$\Delta\phi = \left(\dfrac{T}{GI_p}\right)\Delta x \text{ where } \dfrac{T}{GI_p} \text{ is constant over length } \Delta x$$

Therefore, $\dfrac{\Delta\phi}{\Delta x} = \dfrac{T}{GI_p}$ is the twist per unit length. It corresponds to the local slope of a straight line

that was drawn parallel to the shaft axis before twisting. So, even if T is constant, the twist per length varies along the shaft because I_p varies.

4. For a shaft with segments of uniform cross-section and sudden changes between segments, apply the torsion formulas over one straight segment at a time.

For equal and opposite torques at the ends, the internal torque is uniform. Assume G is also uniform:

In AB: $(I_p)_1 = \pi(d_1)^2/32$

In BC: $(I_p)_2 = \pi(d_2)^2/32$

$\phi_B - \phi_A = TL_1/G(I_p)_1$

$\phi_C - \phi_B = TL_2/G(I_p)_2$

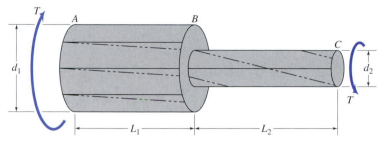

The slope of lines change, consistent with the varying twist per length

For this shaft $I_{p_1} > I_{p_2}$, so the twist per length $\Delta\phi/\Delta x$ is greater in BC because $T/G(I_p)_2 > T/G(I_p)_1$.

Compute the twist of the whole shaft: $\phi_C - \phi_A = (\phi_B - \phi_A) + (\phi_C - \phi_B) = TL_1/G(I_p)_1 + TL_2/G(I_p)_2$

Note that the simple formula $\Delta\phi = TL/GI_p$ can be used on a segment of length L only if T, G, and I_p are all uniform within the segment.

5. Use integration if the diameter, internal torque, or modulus varies continuously over the length.

The radius or shear modulus of a shaft may vary continuously with x. Also, when external torques are applied in a distributed manner on a shaft, the internal torque varies continuously with x: $T(x)$.

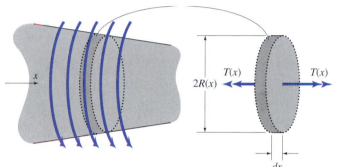

The same relations between torque, stress, and relative rotation still apply to an infinitesimal slice of length dx:

$$\tau_{max}(x) = \frac{T(x)R(x)}{I_p(x)} \qquad d\phi = \frac{T(x)dx}{G(x)I_p(x)}$$

The rotation of one cross-section at x_B, relative to another at x_A, is found from integration:

$$\Delta\phi = \phi_B - \phi_A = \int_{x_A}^{x_B} \frac{T(x)}{G(x)I_p(x)}dx$$

6. Twisting of the tibia is an example of a shaft of gradually varying cross-section under torsion.

If the lower end of the tibia was rotated and the upper end held fixed, then the internal torque would be equal at all cross-sections of the tibia.

At each cross-section, $\tau_{max} = TR/I_p$. So, over the entire bone, the maximum shear stress is where R/I_p is maximum. This occurs where R is minimum.

Observation: fracture does occur over the minimum cross-section, where τ_{max} is maximum.

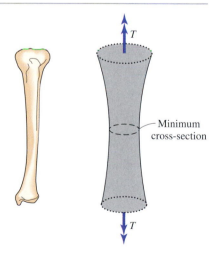

Minimum cross-section

>>End 4.13

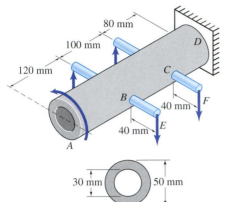

An actuator turns a rubber tube ($G = 2$ MPa) by applying a torque of 2 N-m at A. The tube is fixed at D, and is partially restrained against twisting by springs attached to the ends of rigid rods through the tube. The forces at each end of the rod that terminates at E are 5 N and the forces at each end of the rod that terminates at F are 8 N. The shaft dimensions are shown.

Determine the rotation of the end A.

Solution

Forces at E and F produce torques applied to the shaft at B and C.

The shaft is drawn with the torques acting. An FBD is drawn to find, for example, internal torque T_{BC}.

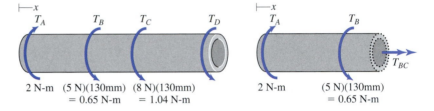

Equilibrium of entire shaft implies:

$$\Rightarrow \sum M = -(2 \text{ N-m}) + 0.65 \text{ N-m} + 1.04 \text{ N-m} + T_D = 0 \Rightarrow T_D = 0.31 \text{ N-m}$$

Find internal torques from FBDs of portions of the shaft (such as one at right above):

$$T_{AB} = 2.0 \text{ N-m}, T_{BC} = 2 - 0.65 = 1.35 \text{ N-m}$$
$$T_{CD} = 2 - 0.65 - 1.04 = 0.31 \text{ N-m}$$

From the tube cross-section,

$$I_p = \frac{\pi}{2}\left[R^4 - R_i^4\right] = \frac{\pi}{2}\left[\left(25 \text{ mm}^4\right) - \left(15 \text{ mm}^4\right)\right] = 5.34 \times 10^{-7} \text{ m}^4$$

Apply the torque–twist relation for segments where the internal torque is constant:

$$\phi_B - \phi_A = \frac{(2 \text{ N-m})(0.12 \text{ m})}{(2 \text{ MPa})(5.34 \times 10^{-7} \text{ m}^4)} = 0.225 \text{ rad}$$

$$\phi_C - \phi_B = \frac{(1.35 \text{ N-m})(0.10 \text{ m})}{(2 \text{ MPa})(5.34 \times 10^{-7} \text{ m}^4)} = 0.1264 \text{ rad}$$

$$\phi_D - \phi_C = \frac{(0.31 \text{ N-m})(0.08 \text{ m})}{(2 \text{ MPa})(5.34 \times 10^{-7} \text{ m}^4)} = 0.0232 \text{ rad}$$

Since the end D is fixed, $\phi_D = 0$.

So, $\phi_C = -0.0232$ rad, $\phi_B = \phi_C - 0.1264 = -0.1496$ rad, and
$\phi_A = \phi_B - 0.225 = -0.375$ rad

>>End Example Problem 4.13

The shaft ($G = 11 \times 10^6$ psi) is fixed at the lower end and subjected to two torques as shown. (a) Determine the maximum shear stress and specify where this maximum is reached. (b) Is there a cross-section besides the lower end where the shaft does not rotate? If so, determine that cross-section.

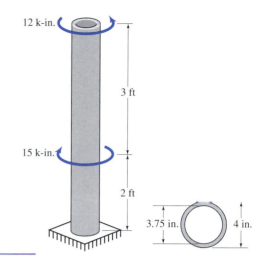

Solution

The shaft is redrawn horizontally with torques acting. Torque at the support C is unknown.

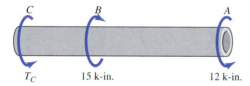

From the equilibrium of entire shaft, we find $T_C = 3$ k-in.

Find internal torques (note signs): $T_{AB} = 12$ k-in., $T_{BC} = -3$ k-in.

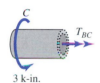 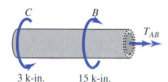

From the tube cross-section,

$$I_p = \frac{\pi}{2}\left[R^4 - R_i^4\right] = \frac{\pi}{2}\left[(2 \text{ in.})^4 - (1.875 \text{ in.})^4\right] = 5.72 \text{ in.}^4$$

Cross-section is uniform, so shear stress is maximum where internal torque is maximum:

$$\tau_{max} = \frac{TR}{I_p} = \frac{(12 \text{ k-in.})(2 \text{ in.})}{5.72 \text{ in.}^4} = 4160 \text{ psi at outer surface between } A \text{ and } B.$$

Apply the torque–twist relations in segments where the internal torque is constant:

$$\phi_B - \phi_C = \frac{(-3 \text{ k-in.})(24 \text{ in.})}{(11 \times 10^6 \text{ psi})(5.72 \text{ in.}^4)} = -0.001145 \text{ rad} \qquad \phi_A - \phi_B = \frac{(12 \text{ k-in.})(36 \text{ in.})}{(11 \times 10^6 \text{ psi})(5.72 \text{ in.}^4)} = 0.00687 \text{ rad}$$

Since end C is fixed, $\phi_B = -1.145 \times 10^{-3}$ rad $\Rightarrow \phi_A = 5.72 \times 10^{-3}$ rad

See the variation in rotation with x. Angle ϕ varies linearly from ϕ_B to ϕ_A through zero. Let Δx be distance from B toward A.

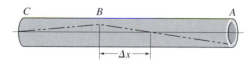

Between B and A: $\phi = \phi_B + \dfrac{\phi_A - \phi_B}{(36 \text{ in.})}\Delta x \Rightarrow \phi = 0$, where

$$\Delta x = (36 \text{ in.})(1.145 \times 10^{-3})/(5.72 \times 10^{-3} + 1.145 \times 10^{-3}) = 6.00 \text{ in.}$$

So $\phi = 0$ at the cross-section located 6 in. to the right of B.

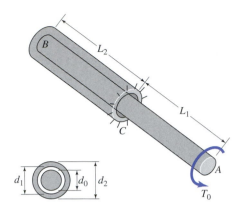

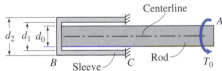

A compact design for a torsion spring is shown. It consists of a rod AB and a sleeve BC, which is closed at the end B. The rod AB is bonded to the closed end of the sleeve at B. There is a gap between the rod and the sleeve ($d_0 < d_1$), so the rod and sleeve only touch at B. The sleeve is prevented from rotating at C, and a torque T_0 is applied to the rod at A. The rod and sleeve have shear moduli G_r and G_s.

Determine the torsional stiffness of the spring, which we define as T_0/ϕ_0, where ϕ_0 is the rotation where the torque is applied at A.

Solution

The rod AB only has torques acting on it at A and B. The sleeve only has torques at B and at C (from the support restraining the rotation of the sleeve at C). Here are the rod and sleeve separated, showing torques and rotations. The rod and sleeve are connected at B, so:

- The torques where they are connected are equal and opposite (T_B).
- Rotations are equal at B where they are connected (ϕ_B).

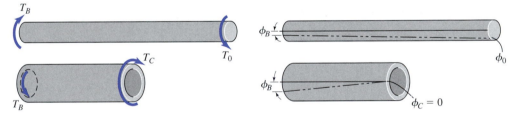

From equilibrium of the rod, $T_B = T_0$.

From equilibrium of the sleeve, $T_C = T_B = T_0$.

Internal torques in the rod and sleeve are: $T_{AB} = T_0$ and $T_{BC} = -T_0$

From the rod and sleeve cross-sections: $(I_p)_{rod} = \dfrac{\pi}{32}d_0^4$ $\qquad (I_p)_{sleeve} = \dfrac{\pi}{32}\left[d_2^4 - d_1^4\right]$

Apply the torque–twist relations:

$$\phi_A - \phi_B = \frac{T_0(L_1 + L_2)}{\dfrac{\pi}{32}Gd_0^4} \qquad \phi_C - \phi_B = \frac{-T_0L_2}{\dfrac{\pi}{32}G\left[d_2^4 - d_1^4\right]}$$

Because the sleeve is fixed at C, $\phi_C = 0$. Find ϕ_B and then ϕ_A and set $\phi_A = \phi_0$ to obtain:

$$\frac{T_0}{\phi_0} = \frac{G\pi}{32\left[\dfrac{L_1 + L_2}{d_0^4} + \dfrac{L_2}{\left[d_2^4 - d_1^4\right]}\right]}$$

>>End Example Problem 4.15

Several torques are applied to the stepped shaft ($G = 80$ GPa) shown. Neglect any stress concentrations associated with the sudden change in cross-section.

(a) Determine the maximum shear stress in the shaft.
(b) Determine the rotation of the end E relative to the end A.

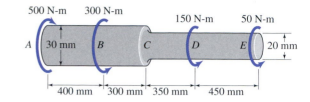

Solution

Notice that the shaft is in equilibrium, given the external torques.

Find the internal torques in each segment AB, BC, etc., by isolating portions of the shaft:

$$T_{AB} = 500 \text{ N-m}, \ T_{BC} = 200 \text{ N-m}, \ T_{CD} = 200 \text{ N-m}, \ T_{DE} = 50 \text{ N-m}$$

Apply the torque–shear stress relation for each segment $\tau_{max} = TR/I_p$, $I_p = \dfrac{\pi}{2}R^4$.

In each segment, find the stress using the internal torque T and radius R of that segment:

In AB and BC, $R = 15$ mm $\Rightarrow$ $(\tau_{max})_{AB} = 94.3$ MPa, $(\tau_{max})_{BC} = 37.7$ MPa

In CD and DE, $R = 10$ mm $\Rightarrow$ $(\tau_{max})_{CD} = 127$ MPa, $(\tau_{max})_{DC} = 31.8$ MPa

Note: even though the maximum internal torque is T_{AB}, the shear stress is maximum in CD. The smaller cross-section of CD (smaller I_p/R) compensates for the smaller internal torque in CD.

From the torque–twist relation $\Delta\phi = \dfrac{TL}{GI_p}$ and internal torque T and I_p in each segment, we find:

$\phi_B - \phi_A = 0.0314$ rad $\qquad \phi_C - \phi_B = 9.43 \times 10^{-3}$ rad

$\phi_D - \phi_C = 5.57 \times 10^{-2}$ rad $\qquad \phi_E - \phi_D = 1.790 \times 10^{-2}$ rad

Find

$$\phi_E - \phi_A = (\phi_E - \phi_D) + (\phi_D - \phi_C) + (\phi_C - \phi_B) + (\phi_B - \phi_A) = 0.1145 \text{ rad} = 6.56°.$$

>>End Example Problem 4.16

Additional data on material properties needed to solve problems can be found in Appendix D or inside back cover.

4.81 The steel shaft ($G = 11.5 \times 10^6$ psi and diameter 0.5 in.) is fixed at A, and is subjected to torques applied at cross-sections B, C, and D ($T_1 = 100$ lb-in., $T_2 = 80$ lb-in., and $T_3 = 90$ lb-in.). Determine the shear stresses at (a) point E, which is midway between A and B and (b) point F which is midway between B and C. For each point, draw the shear stresses on a cube aligned with the x-y-z axes.

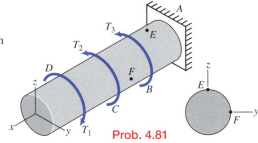

Prob. 4.81

4.82 The aluminum shaft ($G = 27$ GPa and diameter of 20 mm) is fixed at A and is subjected to torques applied at the cross-sections B, C, and D ($T_1 = 200$ N-m, $T_2 = 90$ N-m, and $T_3 = 40$ N-m). Determine the magnitude of shear strains (a) at point E, which is midway between B and C and (b) at point F which is midway between A and B. For each point, also draw a deformed element that displays the directions of the shear strains of a cube originally aligned with the x-y-z axes.

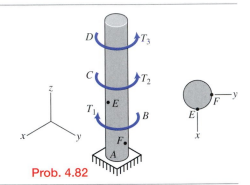

Prob. 4.82

4.83 The steel shaft ($G = 11.5 \times 10^6$ psi and diameter 0.75 in.) is supported at A, and is subjected to torques $T_1 = 600$ lb-in., $T_2 = 300$ lb-in., and $T_3 = 100$ lb-in. applied at the cross-sections B, C, and D. Determine the rotations of (a) the cross-section at B and (b) the cross-section midway between C and D. In both cases indicate the angle of rotation and whether the rotation is about the $+x$- or $-x$-axis. ($L_1 = 15$ in., $L_2 = 20$ in., and $L_3 = 30$ in.)

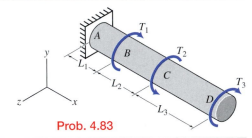

Prob. 4.83

4.84 A steel rod ($G = 80$ GPa and diameter $= 30$ mm) is used to open and close a valve that cannot be directly accessed. The valve requires a torque of 300 N-m to turn. The two supports that hold up the shaft also resist the turning of the rod, each with a torque of 40 N-m. If the valve is to be turned by 90° what must be the rotation at the actuator? Also, determine the maximum shear stress in the shaft. ($L_1 = 200$ mm, $L_2 = 300$ mm, and $L_3 = 400$ mm.)

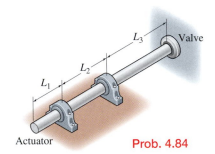

Prob. 4.84

4.85 A steel post ($G = 11.5 \times 10^6$ psi) with a sign on each side twists when wind blows. (Disregard the bending of the post, and approximate the signs as rigid.) Say a uniform pressure of 20 psi acts on the signs in the $-z$-direction. (a) Determine the maximum shear stress in the post. (b) Determine the deflections of the far edges of the signs in the x-direction due to twisting of the post and rigid rotation of the signs. ($w_1 = 40$ in., $w_2 = 30$ in., $s = 18$ in., $h = 20$ in., $L_1 = 48$ in., $L_2 = 120$ in., $d_1 = 9$ in., and $d_2 = 10$ in.)

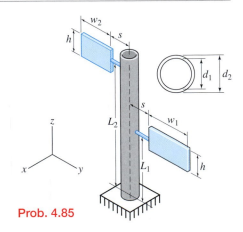

Prob. 4.85

4.86 A motor drives a plastic shaft ($G = 1$ GPa and diameter = 30 mm) at steady speed, which in turn drives a pair of gears. The resisting torques on the two gears at B and C are $T_1 = 30$ N-m and $T_2 = 20$ N-m. Since timing is important, the relative rotation between gears and motor is needed. Determine the rotation of (a) gear C relative to the motor and (b) gear B relative to C. ($L_1 = 60$ mm and $L_2 = 70$ mm.)

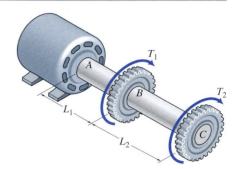

Prob. 4.86

4.87 The aluminum shaft ($G = 3.7 \times 10^6$ psi and diameter = 0.5 in.) is fixed at C, and torque $T_A = 100$ lb-in. is applied. If the rotation of the shaft at all cross-sections is not to exceed $10°$, determine the range of positive values for the torque T_B. ($L_1 = 20$ in. and $L_2 = 30$ in.)

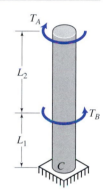

Prob. 4.87

4.88 The shaft ($G = 80$ GPa) is fixed at C, and a torque $T_B = 400$ N-m is applied. Torque T_A can be in the direction shown or opposite; let $T_A > 0$ denote the direction shown. If the shear stress in the shaft is not to exceed 50 MPa, determine the range of values, positive and negative for the torque T_A. ($L_1 = 400$ mm, $L_2 = 300$ mm, $d_1 = 34$ mm, and $d_2 = 40$ mm.)

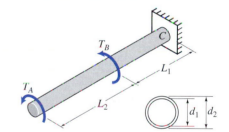

Prob. 4.88

4.89 The shaft ($G = 3.7 \times 10^6$ psi) is fixed at one end, and two torques T_A and T_B are applied as shown. Find values for T_A and T_B which produce the same maximum shear stress of 10 ksi along the entire length (excluding any stress concentration at the change in cross-section). ($L_1 = 15$ in., $L_2 = 10$ in., $d_1 = 1$ in., and $d_2 = 1.5$ in.)

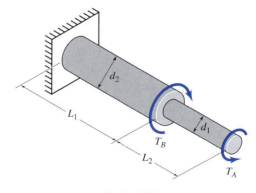

Prob. 4.89

4.90 The stepped shaft ($G = 1$ GPa) is fixed at end A, and several torques $T_1 = 8$ N-m, $T_2 = 5$ N-m, and $T_3 = 15$ N-m are applied as shown. Determine the maximum shear stress in the shaft and the rotation of the end E. ($L_1 = 120$ mm, $L_2 = 100$ mm, $L_3 = 160$ mm, $L_4 = 130$ mm, $d_1 = 15$ mm, and $d_2 = 20$ mm.)

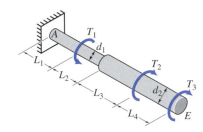

Prob. 4.90

4.91 A sheet is drawn in by two rollers driven by torques T_A and T_B. The sheet has a uniform tension with net force of 400 lb. Assume that both rollers exert the same uniform frictional force per length on the sheet. (a) Determine the torque per length applied by the sheet on each roller and the torques T_A and T_B. (b) Determine the maximum torsional shear stress in the upper roller at the cross-section $x = 6$ in. ($L = 8$ in., $d_1 = 0.5$ in., and $d_2 = 0.4$ in.)

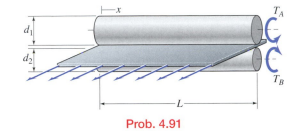

Prob. 4.91

4.92 A cylinder turns in a viscous liquid, which exerts a shear stress $\tau_1 = 200$ kPa on the side surface of cylinder. (a) Determine the torque T necessary to drive the cylinder. (b) If the drive shaft is to have a shear stress no greater than 250 MPa, what must be its minimum diameter? ($L = 2$ m and $d = 200$ mm.)

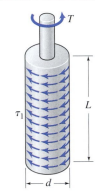

Prob. 4.92

4.93 A transmission system consists of three belts wrapped around a cylinder ($G = 11.5 \times 10^6$ psi) with diameter that increases linearly along its length. The torques exerted by the belts are $T_A = 200$ lb-in., $T_B = 100$ lb-in., $T_C = 300$ lb-in. What is the maximum shear stress in the cylinder, and at what cross-section is that maximum reached? ($L_1 = 10$ in., $L_2 = 15$ in., $d_1 = 1$ in., and $d_2 = 1.5$ in.)

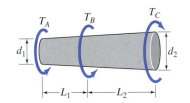

Prob. 4.93

4.94 A solid shaft of modulus G and length L is formed which has a radius that increases linearly along the length from R_1 to R_2. Say the shaft is subjected to equal and opposite torques T on its ends.

Show that the twist is related to the torque T according to:

$$\phi = \frac{2TL}{3\pi GR_1^4} \frac{R_1}{(R_2 - R_1)} \left[1 - \left(\frac{R_1}{R_2} \right)^3 \right]$$

Compare the twist of a shaft of linearly varying radius with the twist of a uniform shaft with radius equal to that of the non-uniform shaft midway along its length. Take the torque and G to be equal in the two shafts and assume the radius of the non-uniform shaft to be 20% greater at one end compared to the other.

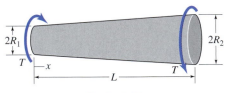

Prob. 4.94

4.95 When the angle of attack of wind is not parallel to the deck surface, it produces an applied moment per unit length, M, that tends to twist the deck

$$M = 0.5\rho U^2 B^2 C_M$$

where ρ is the density of air, U is the wind velocity, B is the width of the deck, and C_M is the aerodynamic moment coefficient (which depends on the direction of the wind). Say the relevant density of air is 1.2 kg/m^3. At a 10° angle of attack, $C_M = 1.1$. Determine rotation at the center of a span of length $L = 1400$ m when a 30 m/s wind blows. Assume that the resistance to twisting is due only to the steel girder which is 30 m wide. The twisting stiffness of the girder is $J = TL/(G\Delta\phi) = 8.89$ m^4. Assume that the towers take up the twisting moment equally and that the rotation at the towers is zero. Do not include the tendency for cables which are attached to the sides of the deck to resist twist (which they can).

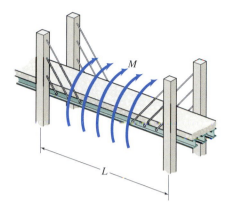

Prob. 4.95 (Appendix A2)

4.96 The wind turbine tower measures $h = 70$ m from its base to the blade shaft, and consists of an annular cylinder with an outer diameter that varies from 2.134 m at the base to 1.219 m at the top. The wall thickness is 25.4 mm. The plane of the blades is $s = 5$ m from the centerline of the tower. One failure mode that must be designed for is the loss of a blade. Let the loss of a blade correspond to horizontal force $F_{blade} = 60$ kN in the plane of the blades perpendicular to the blade shaft. Consider only the twisting of the tower due to such a load, and determine the maximum shear stress in the wall due to torsion. Where in the tower is this maximum shear stress reached?

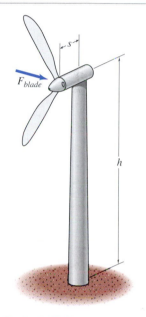

Prob. 4.96 (Appendix A6)

4.97 As part of a vibration analysis of a wind turbine, the angular position of the blades at one end of the low speed shaft relative to the far end of the high speed shaft where it enters the generator must be determined. The reduction in speed is accomplished by a gear box, which is simplified as a single large gear driving a small gear with a 45:1 reduction. Thus, the gears have a diameter ratio of 45:1. (In practice, there might be three-stages of reduction.) Say that the end of the high speed shaft at the generator is viewed as fixed, and that a torque $T_0 = 10$ kN-m is transmitted from the blades to the low speed shaft. The low speed shaft is steel of length $L_{ls} = 2$ m and diameter $d_{ls} = 100$ mm, and the high speed shaft is polymeric ($G = 1$ GPa) of length $L_{hs} = 1.2$ m and diameter $d_{hs} = 50$ mm. Determine the rotation of the blades relative to the generator. Include the relative rotation of the two meshing gears.

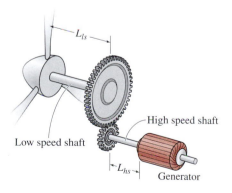

Prob. 4.97 (Appendix A6)

4.14 Statically Indeterminate Structures Subjected to Torsion

Sometimes twisting members are prevented from rotating at more than one cross-section or support. Since the unknown torque reactions cannot be determined by the single equilibrium equation, these systems are called statically indeterminate. We now learn to analyze such members by applying the same principles involving deformation, but solving the equations in a different order.

1. A body held at its two ends and twisted in the middle is statically indeterminate because the end reactions cannot be solved with the single available equilibrium equation.

The skis on this elliptical machine are designed to flex during use. Because the force exerted by the foot is off center, a twisting moment also acts on the ski. A bolt and a bearing prevent the ends from rotating and exert supporting torques. We need more than the single equilibrium condition to find the two end torques.

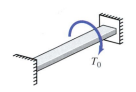

Model ski as a shaft. Consider torque, T_0, due to foot force as given.

Unknown torque exerted by bolt. Shaft rotation here is zero.

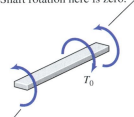

Ski

Unknown torque exerted by bearing. Shaft rotation here is zero.

2. Use the same principles and equations involving equilibrium and deformation to solve statically indeterminate problems, but combine the equations rather than solve them sequentially.

For any mode of deformation, axial (3.4), torsion, or bending (5.19), we must relate four types of **quantities** using three types of **relations**, as depicted in this schematic.

Here are those quantities and relations for the case of torsion.

For the *statically determinate* problems that we have studied so far, we applied these relations from left to right: first finding internal torques from external loads, then twists, and then rotations.

For *statically indeterminate* problems, there are not enough equilibrium equations to solve for all the torques. But we can still write down all the same equations symbolically and solve them simultaneously.

There are enough total equations to solve *statically indeterminate* problems: extra relations between the rotations compensate for the extra torques that cannot be found from equilibrium.

3. Compare the solutions of statically determinate and statically indeterminate problems and note that they involve the same variables and relations, with different variables as unknowns.

Statically determinate	**Statically indeterminate**
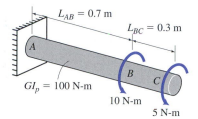	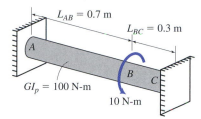

$L_{AB} = 0.7$ m, $L_{BC} = 0.3$ m, $GI_p = 100$ N-m, 10 N-m, 5 N-m

$L_{AB} = 0.7$ m, $L_{BC} = 0.3$ m, $GI_p = 100$ N-m, 10 N-m

Known: $\phi_A = 0$, $T_B = 10$ N-m, $T_C = 5$ N-m
Unknown: T_A, ϕ_B, ϕ_C

Known: $\phi_A = 0$, $T_B = 10$ N-m, $\phi_C = 0$
Unknown: T_A, ϕ_B, T_C

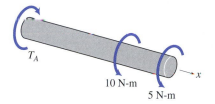

T_A, 10 N-m, 5 N-m, x

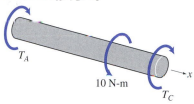

T_A, 10 N-m, T_C, x

Whichever way an unknown *external* torque (T_A or T_C) is drawn, that is the direction the torque acts if that variable turns to be positive. We need not draw unknown external torques in any particular direction.

Equilibrium of whole shaft

$\Sigma M|_x = -T_A + 10 + 5 = 0$
so can determine T_A

$\Sigma M|_x = -T_A + 10 - T_C = 0$
so cannot determine T_A and T_C

Equilibrium of segments to find internal torques T_{AB} and T_{BC} (draw in positive twisting directions).

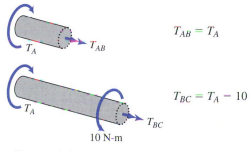

T_A, T_{AB}

$T_{AB} = T_A$

T_A, T_{BC}, 10 N-m

$T_{BC} = T_A - 10$

T_A, T_{AB}

$T_{AB} = T_A$

T_A, T_{BC}, 10 N-m

$T_{BC} = T_A - 10$

Since T_A is known, internal torques can be found.

Since T_A is unknown, internal torques are unknown.

Relate internal torques to changes of ϕ along segments where T, G, and I_p are constant.

$\phi_B - \phi_A = T_{AB}L_{AB}/GI_p$
$\phi_C - \phi_B = T_{BC}L_{BC}/GI_p$

$\phi_B - \phi_A = T_{AB}L_{AB}/GI_p$
$\phi_C - \phi_B = T_{BC}L_{BC}/GI_p$

Note where rotation ϕ is known

$\phi_A = 0$

$\phi_A = \phi_C = 0$

Solve equations

Sequentially find: T_A, T_{AB}, T_{BC}, $\phi_B - \phi_A$, $\phi_C - \phi_B$, ϕ_B, $\phi_C \Rightarrow$
$\phi_B = 0.105$ rad, $\phi_C = 0.120$ rad

In relations for $\phi_B - \phi_A$ and $\phi_C - \phi_B$, substitute for T_{AB} and T_{BC} in terms of T_A; set $\phi_A = \phi_C = 0$; equate expressions for ϕ_B giving one equation for T_A:

$$\frac{T_A L_{AB}}{GI_p} = \frac{-(T_A - 10)L_{BC}}{GI_p}$$

Calculate quantities:

$$\Rightarrow T_A = 3 \text{ N-m}, T_C = 7 \text{ N-m}, \phi_B = 0.021 \text{ rad}$$

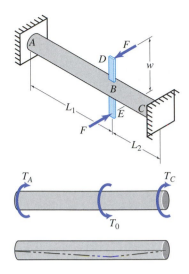

Snap-fits are commonly used to lock plastic parts together. A design for a torsional snap-fit is shown. The rod ABC is fixed at both ends and a tab attached to the rod pivots under the action of forces shown. The rod diameter is 4 mm, and $G = 1$ GPa. Treat the tabs as rigid. ($L_1 = 60$ mm, $L_2 = 40$ mm, $w = 30$ mm.) (a) Determine the rotation of the cross-section B and the displacement of end D if the force $F = 5$ N. (b) Determine the maximum shear stress in the rod.

Solution

Forces F produce an applied torque $T_0 = F(w) = 0.15$ N-m.

The torques on the rod and its rotations are shown.

For equilibrium of the rod: $T_A + T_C = T_0$. This problem is statically indeterminate, since the reaction torques cannot be found from equilibrium.

Find the internal torques from equilibrium of portions of shaft.

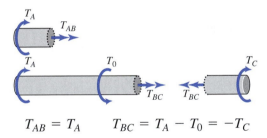

$$T_{AB} = T_A \qquad T_{BC} = T_A - T_0 = -T_C$$

Apply the torque–twist relations:

$$\phi_B - \phi_A = \frac{T_A(0.06 \text{ m})}{GI_p} \qquad \phi_C - \phi_B = \frac{(T_A - 0.15)(0.04 \text{ m})}{GI_p}$$

Since both ends are fixed, $\phi_A = 0$, $\phi_C = 0$, so

$$\phi_B = \frac{T_A(0.06 \text{ m})}{GI_p} = \frac{-(T_A - 0.15)(0.04 \text{ m})}{GI_p}$$

Solve to find $T_A = 0.06$ N-m. Then, $T_C = 0.09$ N-m, and $\phi_B = 0.1432$ rad $= 8.21°$.

Find displacement at point D: $u = (w/2)(\phi_B) = (15 \text{ mm}) \, 0.1432 \text{ rad} = 2.15$ mm.

Maximum stress is in BC where internal torque is greatest: $T_{BC} = T_C = 0.09$ N-m

$$\tau_{max} = \frac{TR}{I_p} = \frac{2T}{\pi R^3} = \frac{2(0.09 \text{ N-m})}{\pi(2 \text{ mm})^3} = 7.16 \text{ MPa}$$

>>End Example Problem 4.17

A tube is fixed at one end C and is rotated by an angle ϕ_0 at end A. A rigid rod that is embedded in the tube resists rotation because it is attached to a pair of springs. The tube has a polar moment of inertia I_p and shear modulus G. The springs each have a spring constant k.

Determine the torque applied at A.

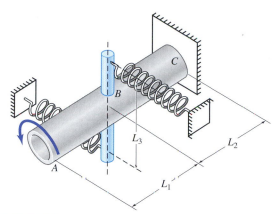

Solution

The shaft is redrawn with torques acting and rotations. No torques are known; $\phi_C = 0$ and $\phi_A = -\phi_0$ is given. Because of the sense of the rotation at A, the springs will stretch.

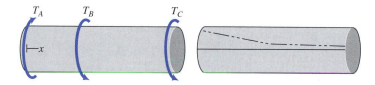

Relate the torque T_B due to the rigid rod at B to the spring force (F_s) and the displacement u at each end of the rod: $T_B = F_s L_3 = ku L_3$. (Define $u > 0$ as corresponding to the spring stretching.)

The displacement u is also related to the shaft rotation at B: $u = -(L_3/2)\phi_B$. Note the negative sign: u is positive (spring stretches) when ϕ_B is negative. Rotations are all negative, because positive rotation corresponds, as usual, to the right hand rule sense about $+x$-axis.

Internal torques (drawn assuming positive) can be related to the applied torques:

$$T_{AB} = T_A, \quad T_{BC} = T_A - T_B$$

Apply the torque–twist relation in each segment:

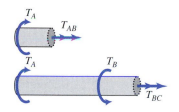

$$\phi_B - \phi_A = \frac{T_{AB}L_1}{GI_p} = \frac{T_A L_1}{GI_p} \qquad \phi_C - \phi_B = \frac{T_{BC}L_2}{GI_p} = \frac{(T_A - T_B)L_2}{GI_p}$$

Combine the equations with the goal of finding one equation and one unknown. Substitute for T_B in terms of k and ϕ_B; set $\phi_C = 0$ and $\phi_A = \phi_0$. Equate two expressions for ϕ_B, to find one equation with ϕ_0 and T_A.

$$\phi_B = -\phi_0 + \frac{T_A L_1}{GI_p} \qquad \phi_B = \frac{-T_A L_2/GI_P}{\left[1 + \dfrac{KL_3^2 L_2}{2GI_P}\right]}$$

Solve for T_A in terms of ϕ_0.

$$T_A = \frac{GI_p \phi_0}{\left[L_1 + \dfrac{L_2}{1 + kL_3^2 L_2/(2GI_p)}\right]}$$

Note special limits:
$k \Rightarrow 0$: (no spring) so tube has length $L_1 + L_2$
$k \Rightarrow \infty$: (springs hold rod rigidly in place) so tube has length L_1

Additional data on material properties needed to solve problems can be found in Appendix D or inside back cover.

4.98 The steel shaft ($G = 80$ GPa and diameter $= 20$ mm) is fixed at both ends, and a torque $T_1 = 100$ N-m is applied to it. Determine the maximum shear stress in the shaft and the rotation at the cross-section B. ($L_1 = 300$ mm and $L_2 = 200$ mm.)

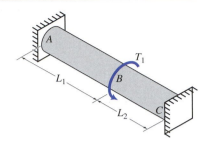

Prob. 4.98

4.99 The aluminum shaft ($G = 3.8 \times 10^6$ psi and diameter $= 0.5$ in.) is fixed at both ends. A torque T_1 is applied to it that produces a rotation of $0.5°$. Determine the torque T_1 and the maximum shear stresses in the two segments AB and BC. ($L_1 = 14$ in. and $L_2 = 18$ in.)

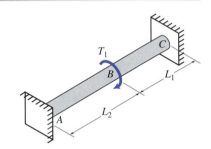

Prob. 4.99

4.100 The rubber tube ($G = 1$ MPa) is fixed at both ends and is turned by a rod. Define the stiffness as the force F divided by the displacement at one of the ends. Calculate the stiffness (neglect bending of the rod). ($L_1 = 500$ mm, $L_2 = 300$ mm, $L_3 = 200$ mm, $d_1 = 30$ mm, and $d_2 = 40$ mm.)

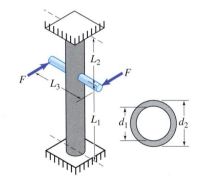

Prob. 4.100

4.101 The ends of the ski from an elliptical exercise machine are fixed against rotation. Because the weight is applied at a distance from the centerline of the ski, the ski can twist. Determine the maximum distance s at which the weight $P = 1400$ N can be offset without exceeding a rotation of $2°$. ($L_1 = 650$ mm, $L_2 = 450$ mm, $w = 70$ mm, and $h = 7$ mm.)

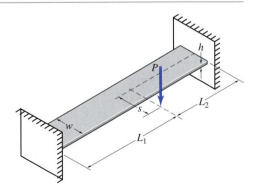

Prob. 4.101

4.102 The shaft of diameter d and shear modulus G is subjected to equal and opposite torques T_0 applied as shown. Derive a formula for (a) the rotation at B and (b) the maximum shear stress between B and C.

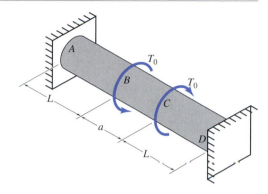

Prob. 4.102

4.103 The plastic tube ($G = 1$ GPa) is being twisted with the wrench. Say the ends A and C are fixed, and the tube is supported against bending. A force $F = 40$ N is applied perpendicularly to the length of the wrench. Determine (a) the displacement of the end D of the wrench and (b) the maximum shear stress in the tube. Treat the wrench as rigid. ($L_1 = 100$ mm, $L_2 = 150$ mm, $L_3 = 250$ mm, $d_1 = 24$ mm, and $d_2 = 30$ mm.)

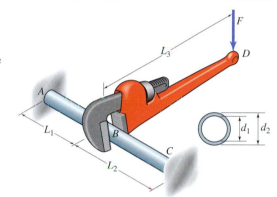

Prob. 4.103

4.104 The plastic shaft ($G = 2$ GPa and diameter $= 10$ mm) is fixed at A, and a torque $T_1 = 1$ N-m is applied at B. Rotation at end C is resisted by two identical springs ($k = 2$ N/mm) connected to a rigid rod through the shaft. Determine the deflection of one end of the rigid rod. ($L_1 = 80$ mm, $L_2 = 60$ mm, and $L_3 = 70$ mm.)

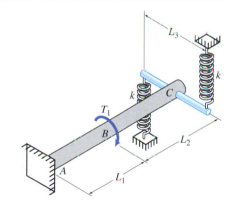

Prob. 4.104

4.105 Two aluminum tubes, BC and BD ($G = 3.8 \times 10^6$ psi), with inner and outer diameters d_1 and d_2, are welded to a plate at B. Then both ends (C and D) are fixed. A solid steel shaft ($G = 11.5 \times 10^6$ psi and a diameter of 0.75 in.) is welded to the plate at B. A torque $T_1 = 600$ lb-in. is applied at A. (a) Determine the rotation of the shaft at A. (b) Determine the maximum shear stress in the two tubes. ($L_1 = 6$ in., $L_2 = 4$ in., $L_3 = 2$ in., $d_1 = 0.875$ in., and $d_2 = 1$ in.)

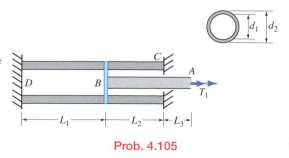

Prob. 4.105

4.106 A shaft has four holes symmetrically placed as shown running along its entire length. Estimate the shaft's torsional stiffness using the following approximate approach. Treat the shaft as consisting of three portions: *A* (inner solid), *B* (middle with holes), and *C* (outer solid). To capture the effect of the holes roughly, reduce the modulus of the middle portion (*B*) fractionally by the fraction of area removed to form the holes. Let the shear modulus of the material be G_0. Determine the proportionality between torque *T* and twist per length $\Delta\phi/L$.

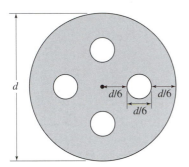
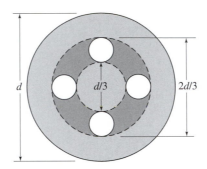

Prob. 4.106

4.107 The duct is held in position with four tension wires. Treat the duct as thin-walled with mean radius *R* and wall thickness *c*. When the system is at a uniform temperature, there is no stress. Two of the wires on each side (as shown) are cooled by ΔT_0.

Derive a formula for the stress in the duct.

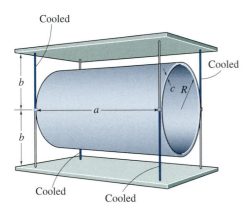
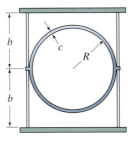

Prob. 4.107

4.108 This type of intramedullary nail fits tightly inside the intramedullary canal of the bone. Assuming there is sufficient friction, the intermedullary nail and bone twist together as a single composite shaft. Approximate the bone as circular, with an outer diameter of 26 mm and an inner diameter of 20 mm. The titanium nail has an outer diameter of 20 mm and a wall thickness of 1 mm. Take the shear moduli of bone and the titanium nail to be 3.3 GPa and 43 GPa, respectively. If a total torque of 8 N-m is applied to the combination of bone and nail, determine the maximum stresses in the bone and in the nail.

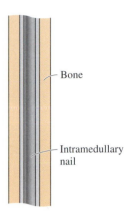

Bone

Intramedullary nail

Prob. 4.108 (Appendix A5)

4.109 This type of intramedullary nail fits tightly inside the intramedullary canal of the fractured bone. Far from the fracture plane both bone and nail rotate together as a single composite shaft with a total torque of 8 N-m. At the fracture plane, the bone carries zero torque, and so the nail carries the full 8 N-m torque. Because of friction between the nail and bone, torque is transferred from the nail to the bone, until each carries the torque it has when the two twist as a composite. Assume that the pressure between the nail and the bone is 0.2 MPa. The inner and outer radii of the reamed out bone is 20 mm and 26 mm; the titanium nail has an outer diameter of 20 mm and a wall thickness of 1 mm. Take the shear moduli of bone and the titanium nail to be 3.3 GPa and 43 GPa, respectively. If the bone rotates through a different angle than the nail (they slip) over a distance of 120 mm on both sides of the fracture, determine the coefficient of friction between the bone and the nail.

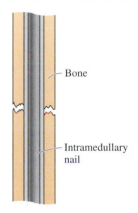

Bone

Intramedullary nail

Prob. 4.109 (Appendix A5)

>>End 4.14

4.15 Power-Torque-Speed Relations for Rotating Shafts

Everything we have learned so far about torsion is also relevant to a shaft that is driven at constant speed by a motor. Here we discuss how the torque in the shaft is related to other important parameters: the motor speed of rotation and the power it delivers.

1. Both the motor that turns a shaft and the system driven by the shaft exert twisting moments on the shaft.

Here is a motor connected to a drum by a shaft. As the shaft turns the drum, it draws in a cable. The cable passes over a pulley at the end of a boom and raises the load.

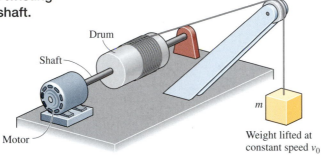

Drum

Shaft

Motor

Weight lifted at constant speed v_0

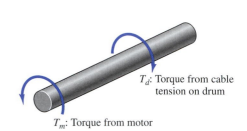

T_d: Torque from cable tension on drum

T_m: Torque from motor

2. The power, or work per time, transmitted by a motor to a shaft is proportional to the torque and rotational speed.

Work done by motor turning shaft by angle $d\theta$: $dW_m = T_m d\theta$
Angle change in time dt: $d\theta = \omega dt$ (rotation speed ω, in rad/sec)
Power (work per time) output of motor: $P_{motor} = dW_m/dt = T_m \omega$

General relation between power, torque, and speed: $P = T\omega$

Work done in time dt raising weight: $dW_{raise\_weight} = (\text{Force})(\text{distance}) = (mg)(v_0 dt)$
Power (work per time) raising weight: $P_{raise\_weight} = (dW_{raise\_weight})/(dt) = mg\, v_0$

Recall torque–work relation:

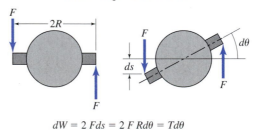

$dW = 2\,Fds = 2\,F\,Rd\theta = Td\theta$

3. The power output of the motor equals the power to lift the weight, assuming no losses.

With the following assumptions of constant speed and no frictional resistance, we can show that the power output by the motor equals the power needed to raise the weight.

Weight raised at constant speed:

- Cable tension $= mg$ (because weight in equilibrium)
- Velocity at drum surface $= \omega c$ (c is drum radius)

Pulley is frictionless:

- Tension $= mg$ in entire cable $\Rightarrow$ moment on drum $T_d = mgc$

Bearings do not resist rotation and a shaft rotating at constant rate is in equilibrium:

- $\sum M = 0 \Rightarrow T_m = T_d = mgc$

  $\Rightarrow P_{motor} = T_m \omega = mgc\omega = mg\, v_0 = P_{raise\_weight}$

4. Since energy flows steadily from the motor through the shaft to the driven body, no energy remains in the shaft, and so positive and negative work done on the shaft are equal and opposite.

For any body, such as the shaft or cable, which is moving at constant speed, there is no change in its energy. Zero net work must be done on it. Consider the work done by torques and forces:

Motor torque on shaft acts in same direction as rotation ⇒ positive work

Torque of cable on drum acts in direction opposite to rotation ⇒ negative work

Net work on shaft = 0

Direction of torques on shaft/drum

Direction of rotation of shaft/drum

Torque due to cable on drum

Motor torque

Tension in cable due to drum acts in same direction as cable motion ⇒ positive work

Tension in cable due to weight acts in direction opposite cable motion ⇒ negative work

Net work on cable = 0

Direction of forces on cable

Direction of motion of cable

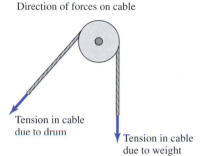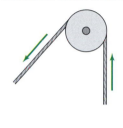

Tension in cable due to drum

Tension in cable due to weight

5. Apply the same equations to relate torque, stress, and twist of spinning shafts and non-spinning shafts.

Aside from $P = T\omega$, all the same relations between external and internal torques, stresses, and relative rotations still hold for spinning shafts, since they are still in equilibrium.

With the shaft spinning, the absolute rotation, ϕ, is meaningless (it keeps rotating). But due to the torque, two cross-sections separated by a distance L have a constant difference in rotation angle, $\Delta\phi$, which is still given by $\Delta\phi = TL/GI_p$.

Line drawn on unloaded shaft

Same line on loaded shaft (time: t)

Same line on loaded shaft (time: $t + dt$)

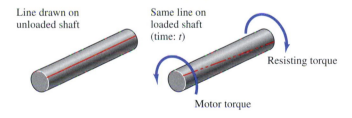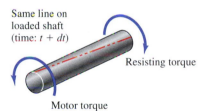

Resisting torque

Motor torque

Resisting torque

Motor torque

>>End 4.15

An electric version of an ice cream maker has a shaft with paddles, which is driven by a motor from the bottom, rather than a handle from the top. The torque of ice cream on each paddle is 2.5 N-m. The solid shaft is plastic with allowable shear strength of 10 MPa.

(a) Determine the minimum acceptable diameter for shaft.
(b) If the unit is to turn at 60 rotations per minute, what power is required of the motor?

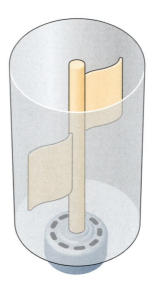

Solution

See shaft redrawn with torques acting. The resisting torques on the paddles act opposite to the motor torque T_M.

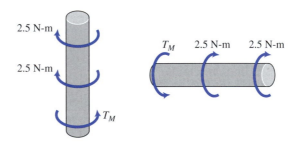

From equilibrium of entire shaft, find the motor torque $T_M = 2.5 + 2.5 = 5$ N-m.

The internal torques in the shaft are 5 N-m and 2.5 N-m. The maximum shear stress is in the segment where $T = 5$ N-m (between the motor and the nearest paddle).

$$\tau_{max} = \frac{TR}{I_p} = \frac{2T}{\pi R^3} \Rightarrow R = \left[\frac{2T}{\pi \tau_{max}} \right]^{1/3} = \left[\frac{2(5) \text{ N-m}}{\pi (10 \text{ MPa})} \right]^{1/3} = 6.83 \text{ mm}$$

So the diameter must be at least 13.66 mm.

The rotation speed of 60 rpm must be converted to units of radians/second.

$$\omega = 60 \frac{\text{rot}}{\text{min}} \left(\frac{2\pi \text{ rad}}{1 \text{ rot}} \right) \left(\frac{1 \text{ min}}{60 \text{ sec}} \right) = 6.28 \frac{\text{rad}}{\text{sec}}$$

From the power–torque–speed relation, $P = T\omega$, find the motor power required.

$P = T_M \omega = (5 \text{ N-m})(6.28 \text{ rad/sec}) = 31.4 \text{ W}$ (very minimal power requirement).

>>End Example Problem 4.19

Additional data on material properties needed to solve problems can be found in Appendix D or inside back cover.

4.110 A steel shaft ($G = 80$ GPa) is driven by a motor that supplies 3 kW of power at 300 rpm. If the steel has a maximum allowable shear stress of 60 MPa, what is the allowable range of diameters?

4.111 A steel shaft ($G = 11.5 \times 10^6$ psi, diameter $= 0.75$ in.) must deliver 2 hp. If the stress is to remain below 7 ksi, what is the allowable speed range?

4.112 A steel shaft ($G = 80$ GPa and $d = 20$ mm) drives two timing belts in the same direction. The rotations of the belts must be closely synchronized for a range of loads. The motor delivers 1200 W at 60 rpm to the shaft. 40% of the power is delivered to the belt at A and 60% at B. Determine the rotation of A relative to B and B relative to the motor. ($L_1 = 80$ mm and $L_2 = 100$ mm.)

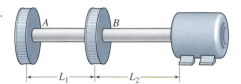

Prob. 4.112

4.113 An object to be polished is pressed against the abrasive plate with diameter $d = 8$ in. near its perimeter with a force $F = 20$ lb. The sander runs at 400 rpm, and the effective friction coefficient for polishing is 0.5. Let the allowable shear stress of the steel composing the shaft be 8 ksi. Determine (a) the minimum allowable diameter of the shaft and (b) the necessary power of the motor. ($G = 11.5 \times 10^6$ psi.)

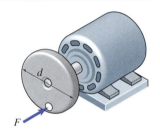

Prob. 4.113

4.114 The cylinder turns in a viscous liquid, which at 60 rpm, exerts a shear stress $\tau_1 = 30$ psi on the side surface of the cylinder. Say the drive shaft is steel that has an allowable shear stress of 25 ksi. Determine (a) the minimum diameter of the drive shaft and (b) the required motor power. ($L = 6$ ft and $d = 10$ in.)

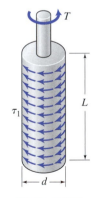

Prob. 4.114

4.115 The motor drives a drum of diameter 700 mm on which a cable is wound. The shaft is 80 mm in diameter. The shear stress in the shaft is not to exceed 40 MPa. (a) Determine the largest mass that can be lifted. (b) If the motor power is 10 kW, what is the highest velocity at which the mass can be lifted?

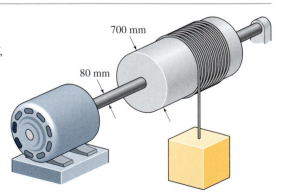

Prob. 4.115

4.116 Some pedal designs allow the leg to apply constant force to the pedal. Say the pedals are turning at 100 rpm, and that the steady force applied to the pedal is 200 N. Determine the power input by the bicyclist, and the maximum shear stress in the spindle. The bottom bracket assembly has a hollow spindle with an outer diameter of 27 mm and an inner diameter of 21 mm. Take the dimensions to be $a = 70$ mm, $b = 25$ mm, $c = 9.5$ mm, $d = 80$ mm, and $L = 170$ mm.

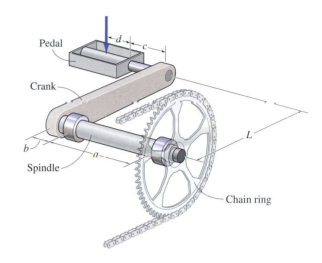

Prob. 4.116 (Appendix A1)

4.117 A wind turbine rotates at 40 rotations per minute and it produces 150 kW of power. Determine the average torque applied by the wind to the blades and, hence, the shaft. Determine the minimum diameter of the solid steel shaft which could carry this torque with the maximum shear stress remaining below 100 MPa.

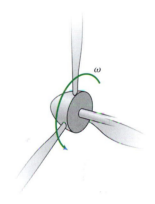

Prob. 4.117 (Appendix A6)

4.118 By using a polymeric high speed shaft (from the gear box to the generator) oscillations in a wind turbine can be reduced. As part of a vibration analysis, the twist of the shaft must be determined. Say the shaft is 1600 mm long with a diameter of 70 mm and shear modulus of 950 MPa. If the shaft delivers 100 kW to the generator at 1800 rpm, what must be the twist in this high speed shaft?

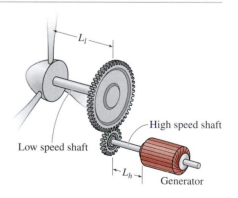

Prob. 4.118 (Appendix A6)

>>End Problems

Chapter Summary

4.1 Due to twisting deformation, each cross-section of a shaft rotates by a different angle ϕ.

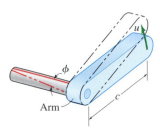 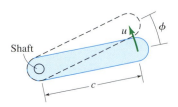

The displacement of a point on a body (e.g., gear or arm) mounted on a shaft, and rotating with it, is $u = c\phi$ and is directed perpendicularly to the arm (assuming $\phi \ll 1$).

4.2 **Shear strain** γ increases linearly with radial distance ρ from the center of the shaft and is proportional to the twist per length $(\phi_2 - \phi_1)/L$: $\quad \gamma = \dfrac{\rho\Delta\phi}{L} = \dfrac{\rho(\phi_2 - \phi_1)}{L}$

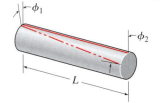

The direction of the shear strain depends on the circumferential position of the element.

4.3 **External torques** are applied by other bodies that are attached to the shaft, often on the shaft's outer surface.

Internal torques act between two parts of the shaft that are separated at an internal cross-section.

External torques

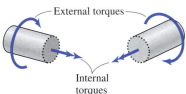

Internal torques give rise to stress and deformation.

Internal torques

4.4 **Shear stress** τ at each point has a magnitude equal to the shear modulus times the shear strain.

$$\tau = G\gamma = \frac{G\rho\Delta\phi}{L}$$

The element faces on which the shear stress act and the stress directions are consistent with the shear strain, and so depend on the circumferential position of the element.

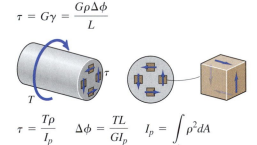

When combined, the stresses over the cross-section must be statically equivalent to the internal torque T. This implies relations between stress, twist, and torque and the dependence on cross-section through the polar moment of inertia I_p.

$$\tau = \frac{T\rho}{I_p} \qquad \Delta\phi = \frac{TL}{GI_p} \qquad I_p = \int_A \rho^2 dA$$

4.5 We distinguish carefully between strength and stiffness.

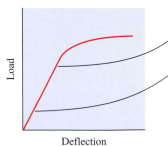

Strength refers to the maximum load (e.g., maximum force, maximum torque) before failure.

Stiffness refers to the proportionality between load and deflection (e.g., force and displacement, torque and rotation) when the body behaves elastically.

We relate strength and stiffness to loads and deflections in different ways, depending on the context.

4.6–4.7 The shape of the cross-section affects the response to torsion. As a rough rule of thumb, the stiffness increases as the material is placed farther from the center.

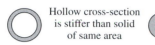
Hollow cross-section is stiffer than solid of same area

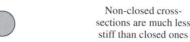

Non-closed cross-sections are much less stiff than closed ones

4.8 Formulas for calculating the twist and maximum shear stress for non-circular shafts are tabulated in handbooks. Results for shafts of length L with rectangular cross-sections are:

$$\tau_{max} = \frac{T}{c_1 a b^2} \qquad \Delta\phi = \frac{TL}{c_2 G a b^3}$$

Tabulated values for c_1, c_2 depend on b/a.

4.9 Approximate analysis of a shaft of length L with a thin-walled, closed cross-section leads to these formulas for maximum shear stress and twist:

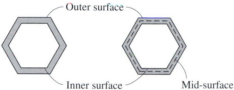

Outer surface

Inner surface — Mid-surface

s_m: mid-surface length

A_m: area within mid-surface

$$\tau_{max} = \frac{T}{2A_m t} \qquad \Delta\phi = \frac{TLs_m}{4GA_m^2 t}$$

4.10–4.13 When multiple external torques act on a shaft, the internal torques vary along the length. Use the internal torque to determine the shear stress at a cross-section and the twist of a segment of the shaft.

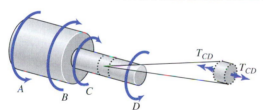

$$\tau = \frac{T\rho}{I_p}$$

Use T and I_p at the cross-section of interest.

$$\Delta\phi = \frac{TL}{GI_p}$$

T, G, I_p must be constant in segment of length L.

4.14 When there are too many supports (two shown here), one cannot find the internal torques in terms of the given external torque T_0 from equilibrium alone: the problem is statically indeterminate. One needs to solve for the support reactions, internal torques, twists, and rotations simultaneously rather than sequentially. There are always additional conditions on rotations that compensate for the extra support reactions.

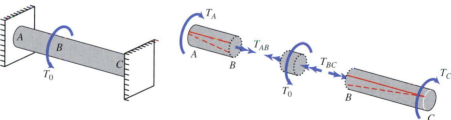

4.15 A shaft that delivers power from a motor is under torsion and spins. If the rate of spin is steady, then the internal torque is related to the shear stress, τ, and twist, $\Delta\phi$, exactly as in a non-spinning shaft.

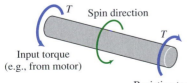

Input torque (e.g., from motor)

Spin direction

Resisting torque

T: Torque
ω: Rotation speed
P: Power

$$P = T\omega$$

Chapter 5

Bending

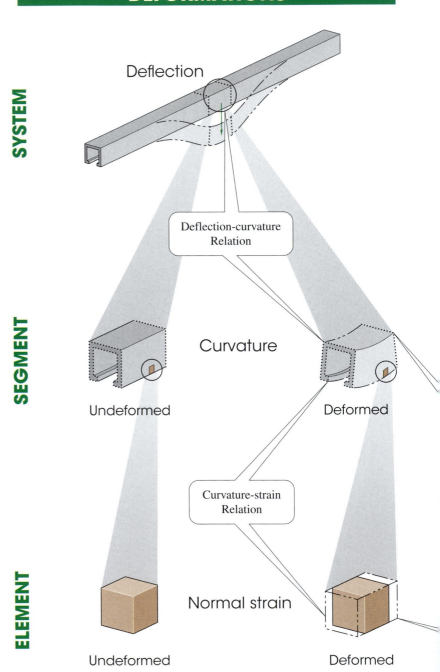

DEFORMATIONS

SYSTEM

Deflection

Deflection-curvature Relation

SEGMENT

Curvature

Undeformed

Deformed

Curvature-strain Relation

ELEMENT

Normal strain

Undeformed

Deformed

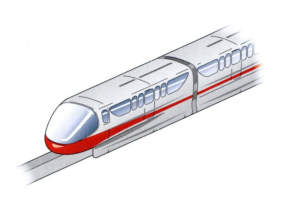

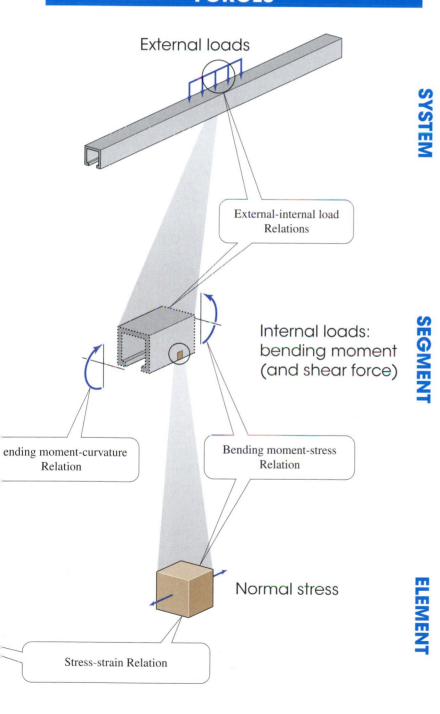

External loads

External-internal load Relations

SYSTEM

Internal loads: bending moment (and shear force)

SEGMENT

ending moment-curvature Relation

Bending moment-stress Relation

Normal stress

Stress-strain Relation

ELEMENT

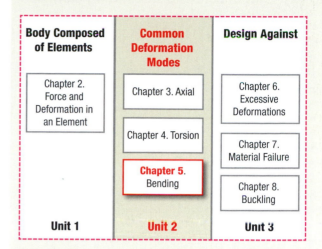

Body Composed of Elements	Common Deformation Modes	Design Against
Chapter 2. Force and Deformation in an Element	Chapter 3. Axial	Chapter 6. Excessive Deformations
	Chapter 4. Torsion	Chapter 7. Material Failure
	Chapter 5. Bending	Chapter 8. Buckling
Unit 1	Unit 2	Unit 3

Chapter Outline

The track deforming due to the weight of the passenger car is an example of *bending deformation*. Deflection of the track and weight of the passenger car are deformation and load at the **system level**. Stresses at a point in the track are at the **element level** and are needed to determine failure. To relate the track deflection and stresses at a point to the weight of the passenger car, we must consider deformation and loads at an intermediate level: the track as composed of thin **segments** cuts transversely to the track length, with each segment composed of elements. In this chapter, we describe **loads** and **deformations** at these three levels and find relations between them. External loads at the system level, such as the passenger car weight, are related to the internal loads (bending moment and shear force) at the segment level (**5.1–5.5**). Stresses due to bending moments are found by interrelating deformations and loads at the segment and element levels (**5.6–5.13**). Stresses due to shear forces are found from the stresses due to bending moments and the effect of the shear force on equilibrium of a segment (**5.14–5.15**). Deflections are then related to external forces by combining relations at the system, segment, and element levels, and efficient means of determining deflections from loads are studied (**5.16–5.20**).

5.1 Deformation in Bending

A part undergoes bending if it is long and slender, and it deflects perpendicularly to the long dimension. Usually, the part is initially straight. Here we show how to quantify bending deformation.

1. Here is a typical example of bending.

This wood ruler is held flat against the table at the left, and fingers are poised to press against it. When the fingers apply forces, the ruler deflects, primarily up or down. Whenever a part deforms in this way, we say that it acts like a "beam."

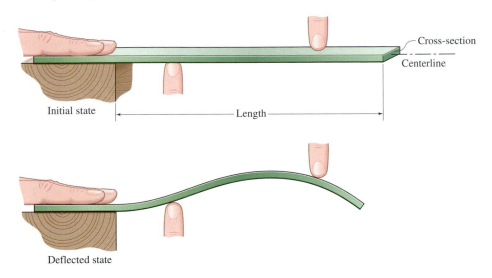

Initial state — Length

Cross-section
Centerline

Deflected state

In this chapter, we learn to determine the stresses and deflections produced by the forces and how they depend on the beam cross-section, length, and material properties.

2. Quantify the movement of each point of the line running through the center of a straight beam.

Here are two ways of quantifying the movement of the beam due to forces:

Deflection: Distance, v, that a point along the centerline moves perpendicularly to the centerline.

Slope: Angle, θ, through which the centerline rotates or tilts at a point relative to its initial orientation.

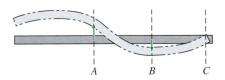

Point A: deflects up and slopes clockwise (CW)
Point B: deflects down and has zero slope
Point C: has zero deflection and slopes counter-clockwise (CCW)

3. Take a thin ruler and move and/or bend it. For what variations in deflection along the length must you apply forces (other than those needed to support the ruler's weight)?

Deflection uniform: no forces needed

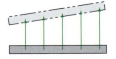

Deflection linear, slope uniform: no forces needed

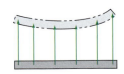

Deflection non-linear, beam curved: forces needed

In the deflected ruler above, different portions curve upward, downward, or remain straight.

4. Here is one deflected shape for which the bending deformation is constant.

Radius of curvature, ρ, captures the intensity with which a segment of a beam has deformed in bending, and ρ is defined as the radius of the circle into which a longitudinal line of the initially straight beam has deformed. The reciprocal, $\kappa = 1/\rho$, is referred to as the curvature.

Start with a straight beam, and draw straight lines and rectangles on its side as shown. Wrap this beam around a circular cylinder. The top and bottom of the beam become circular arcs. The lines initially perpendicular to the long dimension (red) are still straight, but they emanate out from the center of the circular cylinder. We take ρ to be the **radius of curvature** measured approximately to the center of the beam.

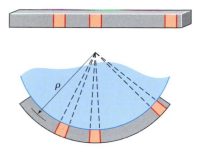

5. The bending deformation is constant along a beam deformed into a circular arc.

Each shaded rectangle on the undeformed beam is one face of a segment: a short length of beam between two neighboring cross-sections. The segment deforms into a small sector of a circle. Trace and cut out the three deformed shaded regions and reorient them; they have the same shape. Since each segment deforms into the same shape, the deformation is uniform along the length of the beam.

6. Test how the radius of curvature controls the intensity of the deformation.

Take two flat sheets of paper (each is a beam!) and roll them up. Make one very loose (large ρ) and one very tight (small ρ). When released, the loose one will be flat again, while the tight one stays rolled up. Bending into such a small radius produces strains that are high enough to deform the paper permanently.

Intensity of bending deformation depends on the radius of curvature ρ or the curvature $\kappa = 1/\rho$.

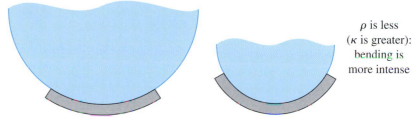

ρ is greater (κ is less): bending is less intense

ρ is less (κ is greater): bending is more intense

7. One can readily apply loads that produce a uniform radius of curvature ρ.

Take a flexible strip, such as a thin ruler, and apply equal forces with your fingers as shown. Each hand applies a couple or moment (equal and opposite forces a distance apart). The couples of the two hands must be equal and opposite. Between the thumbs, the strip has deformed into a circular arc.

Later we will define an internal bending moment at each cross-section along the beam. For the loading shown here, just as the deformation is uniform, so the internal bending moment is uniform, equal to the moment applied by each hand.

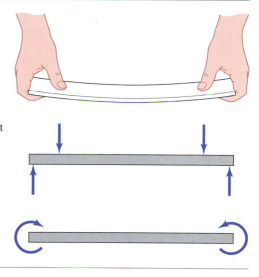

8. The moment and the curvature depend on one another.

If one increases the moment, for example, by increasing the forces applied by the fingers while keeping them the same distance apart, the radius of curvature decreases.

In the second part of this chapter, we will study the quantitative relation between the bending moment and the curvature and how it depends on the material and on the beam cross-section. We first consider how beams are loaded and how those applied loads determine the bending moment.

>>End 5.1

5.2 Beams, Loads, and Supports

We will derive equations that relate curvature, bending moment, and stress. These equations are useful approximations for real bodies. Here we discuss when and how a real body can be modeled as a beam.

1. A body is modeled as a beam, if it is shaped as a beam and loaded as a beam.

Shape: A beam is long in one direction compared to the other directions. Beam analysis gives accurate results when the cross-section is constant or varies slowly in the long direction.

Loads: A body is modeled as a beam if the external loads cause it to deflect normal to its long dimension.

2. Here are two examples of bodies that can be treated as beams and why.

Shape: The corkscrew handles are long compared to their height. Beam analysis is accurate where the cross-section varies gradually. More complex analysis is needed elsewhere (near gear teeth).

Loads: Fingers act perpendicularly to the length, tending to bend it.

Shape: The tower is long compared to the radius. Beam prediction is accurate since the cross-section varies gradually.

Loads: The weight force compresses the tower axially along the length, and does not cause bending. But forces due to the wind tend to bend the tower.

3. Distinguish two types of influences on beams: constraints (movement) and loads (forces).

The vise sets the motion (to be zero); it provides forces as necessary to maintain zero motion. (*Constraint*)

A weight sets the force; the amount of deflection depends on the resulting bending. (*Load*)

A finger can set the force or the deflection; it depends on what the finger's owner wants to do. (*Constraint or Load*)

4. Here are the symbols for the constraints and loads and whether the displacement (v), slope (v'), external force (F_e), and external moment (M_e) are given or to be determined (TBD).

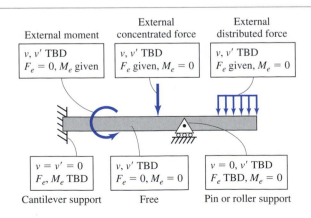

External moment
v, v' TBD
$F_e = 0, M_e$ given

External concentrated force
v, v' TBD
F_e given, $M_e = 0$

External distributed force
v, v' TBD
F_e given, $M_e = 0$

$v = v' = 0$
F_e, M_e TBD
Cantilever support

v, v' TBD
$F_e = 0, M_e = 0$
Free

$v = 0, v'$ TBD
F_e TBD, $M_e = 0$
Pin or roller support

5. Consider these examples of modeling loads and supports.

Diving board

A board bolt attaches the board to the stand and constrains horizontal and vertical motion—model it as a pin support.

The fulcrum constrains only vertical motion—model it as a roller support.

The weight of the diver is actually distributed over the length of his feet, which is small compared with the board—model it as a concentrated force.

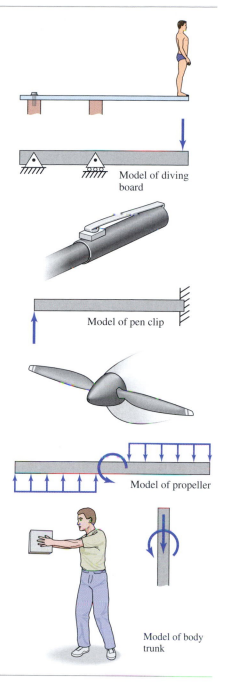

Model of diving board

Clip holding pen in pocket

The clip inserts firmly into the top of the pen—model the clip as cantilevered at the right where it meets the pen top.

Model the clipped object (it could be a shirt pocket) as imposing a force or a displacement (they are related by the clip deformation).

Model of pen clip

Airplane propeller

The shaft is small compared to the propeller—model it as applying a concentrated moment at the propeller center.

The air pressure results in drag forces on the propeller—model it as a distributed force. The distributed force is drawn here as uniform, but, like the air velocity, it is likely to vary with distance from the shaft.

Model of propeller

Worker holding a weight off to his side

If we model the trunk of the body as a beam, then we can model the weight off to the side with a statically equivalent force and couple applied at the shoulders. The couple would cause bending; the force would cause axial loading.

Model of body trunk

---

6. Recognize common support arrangements, and when appropriate draw only reactions relevant to transverse loading.

Beams can be supported in many ways, but two forms of support are so common as to be named. Here we show a cantilevered beam and a simply supported beam. Such beams can remain in equilibrium for any loading in the x-y plane. Often, however, beams are subjected only to transverse forces (F_y) or moments (M_z). In those cases, the x-force reaction, either at the cantilever support or at the pin at A, would be zero. To simplify free body diagrams, and eliminate the zero non-transverse reaction force, we will often represent only the reactions that are potentially non-zero.

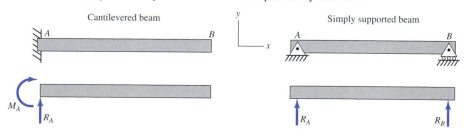

Cantilevered beam

Simply supported beam

>>End 5.2

5.3 Internal Loads in Beams ───────────

From the deformation of a beam we can recognize the presence of internal forces. We define the internal loads—shear force and bending moment—that are relevant to bending, and we agree on how their signs are related to their senses.

1. Variation in curvature implies a varying bending moment.

The ruler shown at the start of the chapter curves upward in some portions and downward in others. The bending moment, which is directly related to the curvature, has opposite directions in those portions and clearly varies along the ruler.

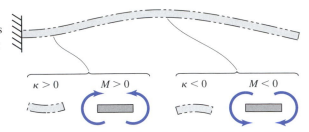

$\kappa > 0 \quad M > 0 \qquad \kappa < 0 \quad M < 0$

2. To see that more than bending moments are present, compare the motion of the cross-section for a single solid beam and for a stack of thin beams.

Draw vertical lines on the side of a soft, plastic sheet, grip the ends, and bend it. The lines rotate with the beam and stay perpendicular to the local beam centerline.

Now draw vertical lines on a deck of cards, and try to bend the deck. The lines do not rotate, but remain vertical.

The cards slip by each other, as can be seen from the close-up drawing of a few of the cards.

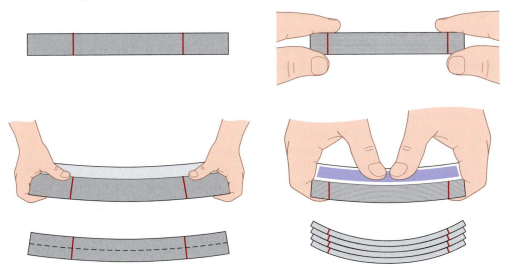

3. Explain the cards slipping as a lack of resistance to shear.

The cards slipping past one another when bending are trying, with little success, to exert shear stresses on each other.

When a beam is one solid material, shear stresses act to prevent successive bonded layers of material from sliding past one another. The shear stresses, which resist slippage, enable the body to respond as a single beam rather than as a stack of layers.

We track the shear force, which is the integrated effect of the shear stresses over the cross-section. Later we will study how the shear stresses are related to the shear force.

Sometimes we engineer assemblies of parts that together should act like a single beam (e.g., tall buildings, cranes). The beam must withstand the shearing, otherwise it will be soft like a deck of cards.

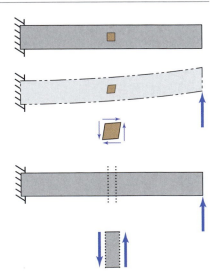

4. Recognize the presence of shear forces and bending moments by considering equilibrium of a portion of a beam.

This strip sags around the fingers that support it. The weight is modeled as a distributed force, and the fingers are modeled as point forces.

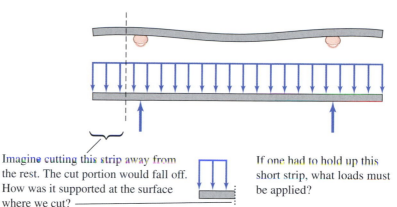

Imagine cutting this strip away from the rest. The cut portion would fall off. How was it supported at the surface where we cut?

If one had to hold up this short strip, what loads must be applied?

View the beam as composed of two adjacent portions that meet at an imaginary plane or line passing transversely through the beam. The **shear force**, *V*, is defined as the internal force that the two portions exert on each other in the direction transverse to the beam. The **bending moment**, *M*, is defined as the internal moment or couple that the two portions exert on each other about the axis perpendicular to the plane of bending.

Both a **shear force**, V, and a **bending moment**, M, must act at the "cut" to keep the strip in equilibrium. These internal loads are exerted by the rest of the beam, so an equal and opposite force and moment are exerted by the strip on the rest of the beam.

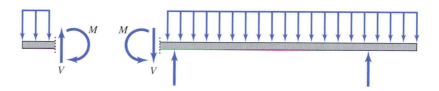

5. Think about a short segment of the beam. *V* and *M* act on both cut faces.

Draw internal loads on both cut faces of an infinitesimally short segment of the beam. V and M do not vary over such a short length. Material just to the left and right exert the forces and moments that are drawn on the segment.

The directions of V and M vary along the beam. A segment taken from elsewhere in the beam likely has different internal force and moment. We will develop formal methods for finding the directions and magnitudes of V and M.

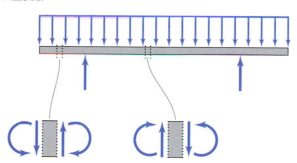

6. Agree to use the sign of *V* and *M* to signal their directions.

We use this convention (agreement) to relate the sign of V on a segment and its direction.

If $V > 0$, we mean   If $V < 0$, we mean

Do not describe the shear force as up or down. There is always an equal and opposite pair.

We use this convention (agreement) to relate the sign of M on a segment and its direction.

If $M > 0$, we mean   If $M < 0$, we mean

Do not describe the bending moment as CW or CCW.
There is always an equal and opposite pair.

5.4 Internal Loads by Isolating Segments

As will be seen, the stresses and deflections depend on the shear force *V* and bending moment *M* along the beam. Here we show how to obtain formulas for *V(x)* and *M(x)*, where *x* is the location of a cross-section from the left end.

1. Find *V* and *M* at each cross-section *x* by applying equilibrium to a portion of the beam from *x* to the left or right end.

1. Find the support loads from statics and then draw the entire beam with all loads. Define *x* from the left end.

2. Draw a beam portion that extends from either end to the cross-section *x* where *V* and *M* are to be found. For this problem, cut the beam at a cross-section to the left of the center load (*a* and *b*) and to the right of the center load (*c* and *d*).

Use either portion *a* or *b* to find *V* and *M* when *x* is *anywhere* in $0 < x < L/2$. Since the forces acting on the FBD change as *x* passes $L/2$, use either *c* or *d* when *x* is *anywhere* in $L/2 < x < L$.

3. Draw the FBDs, showing all loads on each beam portion. Draw *V* and *M* at each cut, assuming positive senses.

4. Write down $\Sigma F_y = 0$ and $\Sigma M = 0$ for each portion and solve for *V* and *M*. It is convenient, but not necessary, to take $\Sigma M \mid_{cut} = 0$, that is moments about the cut. *V* produces no moment since it acts through the cut.

(a) $\Sigma F_y = -P - V = 0 \Rightarrow V = -P$

   $\Sigma M \mid_{cut} = Px + M = 0 \Rightarrow M = -Px$

(b) Results for *V* and *M* same as (a)

(c) Results for *V* and *M* same as (d)

(d) $\Sigma F_y = V - P = 0 \Rightarrow V = P$

   $\Sigma M \mid_{cut} = -M - P(L - x) = 0 \Rightarrow M = -P(L - x)$

5. From the formulas for *V(x)* and *M(x)*, plot *V* and *M* vs. *x* along the beam. These are called the shear force and bending moment diagrams.

From their signs and the sign convention, one could determine the actual directions of *V* and *M*.

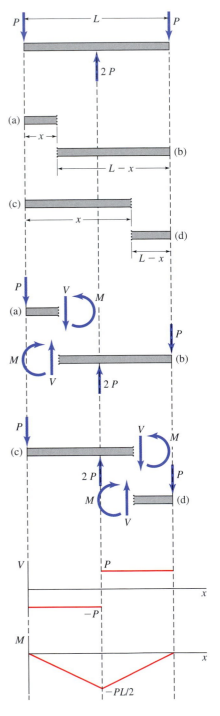

2. Draw a distributed force on the FBD, and only replace with a statically equivalent force at the correct stage in the analysis.

Consider this beam. We replace the supports with the unknown reactions, but we keep the distributed load in the drawing.

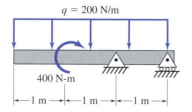

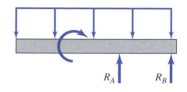

$q = 200$ N/m

400 N-m

|— 1 m —|— 1 m —|— 1 m —|

R_A R_B

---

3. Once the FBD is completed, replace the distributed force with its statically equivalent load.

With the FBD of the whole beam complete, we impose the equilibrium conditions: zero net force and moment. We can replace the distribution with its statically equivalent force of $(200 \text{ N/m})(3 \text{ m}) = 600$ N at the center, which has the same net force and moment as the distribution. Then, we impose equilibrium on the beam.

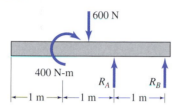

600 N

400 N-m

R_A R_B

|— 1 m —|— 1 m —|— 1 m —|

$$\sum M|_B = -400 + 600(1.5)$$
$$- R_A(1) = 0 \Rightarrow R_A = 500 \text{ N}$$

$$\sum F_y = -600 + R_A + R_B = 0$$
$$\Rightarrow R_B = 100 \text{ N}$$

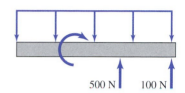

500 N   100 N

---

4. Draw new FBDs for portions to find V and M, and draw the distributed force acting on each portion.

We isolate different portions of the beam to find V and M in $0 < x < 1$, in $1 < x < 2$, and in $2 < x < 3$. We choose to isolate the left or right portion depending on which has fewer loads. This choice is obvious for $0 < x < 1$ and $2 < x < 3$, and it probably doesn't matter for $1 < x < 2$.

In each case, since we are considering a new FBD, draw the distributed force on the portion.

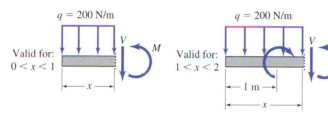

Valid for: $0 < x < 1$

Valid for: $1 < x < 2$

Valid for: $2 < x < 3$

$q = 200$ N/m

$3 \text{ m} - x$ 100 N

Not valid for $x > 1$ m (FBD excludes 400 N-m moment)

Not valid for $x < 1$ m (FBD includes moment), nor for $x > 2$ (FBD excludes R_A)

Not valid for $x < 2$ m (FBD excludes R_A and moment)

---

5. Once FBDs for portions are complete, replace the distributed force with its statically equivalent load and then use equilibrium to find V and M.

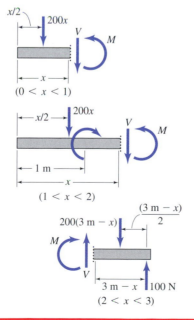

$x/2$ 200x

V M

x

$(0 < x < 1)$

$x/2$ 200x

V M

1 m

x

$(1 < x < 2)$

$(3 \text{ m} - x)/2$

200(3 m − x)

M V

$3 \text{ m} - x$ 100 N

$(2 < x < 3)$

$$\sum F_y = -(200x) - V = 0$$
$$\sum M|_{cut} = (200x)(x/2) + M = 0$$
$$V = (-200x) \text{ N}$$
$$M = \left(-200x^2/2\right) \text{ N-m}$$

$$\sum F_y = -(200x) - V = 0$$
$$\sum M|_{cut} = (200x)(x/2) - 400 + M = 0$$
$$V = (-200x) \text{ N}$$
$$M = \left(-200x^2/2 + 400\right) \text{ N-m}$$

$$\sum F_y = -200(3 - x) + V + 100 = 0$$
$$\sum M|_{cut} = -200(3 - x)(3 - x)/2 + 100(3 - x) - M = 0$$
$$V = (200(3 - x) - 100) \text{ N}$$
$$M = \left(-200(3 - x)^2/2 + 100(3 - x)\right) \text{ N-m}$$

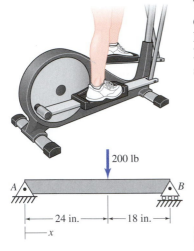

The ski in this elliptical machine can be modeled as simply supported – a pin at one end and a roller at the other. A force of 200 lb due to a person's weight acts as shown. Determine the shear force and bending moment in the ski by isolating segments. Draw the shear force and bending moment in correct directions on an infinitesimal segment at $x = 30$ in.

200 lb

A ⎯⎯ B

24 in. — 18 in.

x

Solution

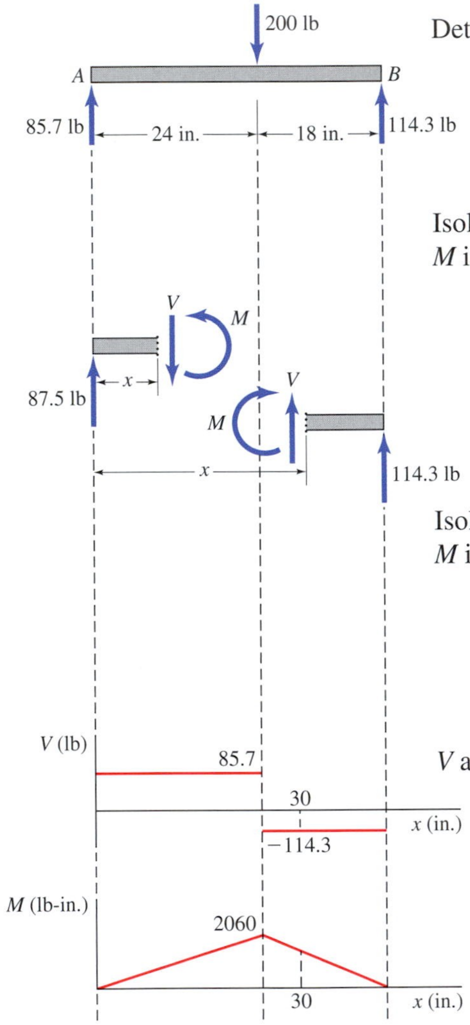

Determine the support loads from statics.

$$\sum F_y = R_A + R_B - 200\text{ lb} = 0$$
$$\sum M|_A = R_B(42\text{ in.}) - (200\text{ lb})(24\text{ in.}) = 0$$
$$\Rightarrow R_A = 85.7\text{ lb},\ R_B = 114.3\text{ lb}$$

Isolate a segment from A to x, where x is between A and the 200 lb load. Draw V and M in positive senses.

$$\sum F_y = 85.7\text{ lb} - V = 0$$
$$\sum M|_{cut} = -(85.7\text{ lb})x + M = 0$$
$$\Rightarrow V = 85.7\text{ lb},\ M = 85.7x\text{ lb-in.}$$
$$(\text{for } 0 < x < 24\text{ in.})$$

Isolate a segment from x to B, where x is between the 200 lb load and B. Draw V and M in positive senses.

$$\sum F_y = 114.3\text{ lb} + V = 0$$
$$\sum M|_{cut} = (114.3\text{ lb})(42\text{ in.} - x) - M = 0$$
$$\Rightarrow V = -114.3\text{ lb},\ M = (114.3)(42 - x)\text{ lb-in.}$$
$$(\text{for } 24 < x < 42\text{ in.})$$

V and M diagrams have been drawn.

At $x = 30$ in., $V = -114.3$ lb,
$$M = (114.3\text{ lb})(42\text{ in.} - 30) = 1371\text{ lb-in.}$$

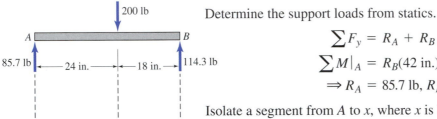

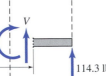

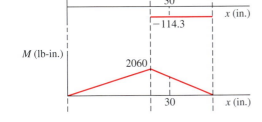

Since these are directions if V and M are positive

These are the actual directions of V and M with the magnitudes labeled

114.3 lb   1371 lb-in.

Notice, you could not find the maximum M by setting $dM/dx = 0$, since M is not a single smooth function. You must find $M_{max} = (85.7\text{ lb})(24\text{ in.}) = 2060$ lb-in. by inspecting the diagram.

>>End Example Problem 5.1

An aluminum ruler 1 m in length is clamped over 200 mm at its left end and is free to deflect under its own weight. The mass density is 2750 kg/m³. Determine the shear force and bending moments for the 800 mm unclamped portion. Determine the magnitudes of the maximum shear force and bending moment, and draw them on an infinitesimal segment in the correct directions.

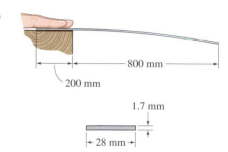

Solution

The force per length due to the weight of the ruler is

$$q_0 = \frac{\rho g V}{\text{Length}} = \frac{(2750 \text{ kg/m}^3)(9.81 \text{ m/s}^2)(0.028 \text{ m})(0.0017 \text{ m})(1 \text{ m})}{1 \text{ m}} = 1.284 \text{ N/m}$$

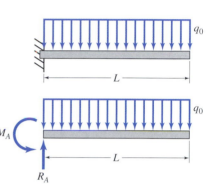

Over the clamped region, the deflection and rotation of the ruler is zero. We can view the clamped portion as exerting cantilever support reactions (force and moment) on the left end of the unclamped portion ($L = 0.8$ m).

While the reactions R_A and M_A could be found, V and M can be found without them, if we isolate from x to the right end. We draw the distributed force that acts on the isolated portion.

When we are about to sum F_y and M, we can replace the distributed force with the statically equivalent load $q_0(L - x)$ acting in the center of the segment of length $(L - x)$.

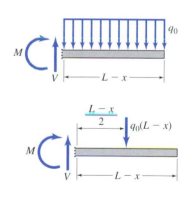

Then, we impose equilibrium equations

$$\sum F_y = V - q_0(L - x) = 0$$

$$\sum M|_{cut} = -M - q_0(L - x)^2/2 = 0$$

$$\Rightarrow V = q_0(L - x), \quad M = -q_0(L - x)^2/2$$

$$(\text{for } 0 < x < L)$$

From the expressions for $V(x)$ and $M(x)$, V and M diagrams have been drawn.

Inspect V and M diagrams: they both have maximum magnitudes at $x = 0$.

$$V(0) = q_0 L = (1.284 \text{ N/m})(0.8 \text{ m}) = 1.027 \text{ N}$$

$$M(0) = -q_0 L^2/2 = -(1.284 \text{ N/m})(0.8 \text{ m})^2/2 = -0.411 \text{ N-m}$$

Given the signs of V and M at $x = 0$, their actual directions are

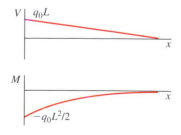

1.027 N

0.411 N-m

>>End Example Problem 5.2

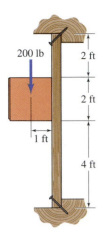

200 lb

2 ft

2 ft

1 ft

4 ft

Cabinets are hung from studs in the wall. The weight of the cabinet supported by the stud shown is 200 lb. Model the effect of the cabinet weight as a statically equivalent couple acting midway between the cabinet's top and bottom attachment points. Even though the stud is securely nailed to other members at the top and the bottom, treat it as simply supported. Determine and draw the shear force and bending moment diagrams for the stud. Determine the magnitudes of the shear force and bending moment on both sides of the applied moment.

Solution

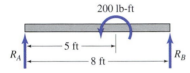

200 lb-ft

5 ft

8 ft

R_A

R_B

Find the support reactions.

$$\sum F_y = R_A + R_B = 0$$

$$\sum M|_A = R_B(8 \text{ ft}) + (200 \text{ lb-ft}) = 0 \Rightarrow R_A = 25 \text{ lb}, R_B = -25 \text{ lb}$$

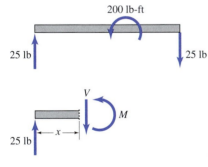

200 lb-ft

25 lb

25 lb

V

M

25 lb

x

Isolate the segment from A to x, where x is located between A and the 200 lb-ft moment.

$$\sum F_y = 25 \text{ lb} - V = 0$$

$$\sum M|_{cut} = -(25 \text{ lb})x + M = 0$$

$$\Rightarrow V = 25 \text{ lb}, M = (25x) \text{ lb-ft}$$

$$(\text{for } 0 < x < 5 \text{ ft})$$

Isolate the segment from A to x, where x is located between the 200 lb-ft moment and B.

$$\sum F_y = 25 \text{ lb} - V = 0$$

$$\sum M|_{cut} = -(25 \text{ lb})x + 200 \text{ lb-ft} + M = 0$$

$$\Rightarrow V = 25 \text{ lb}, M = (25x - 200) \text{ lb-ft}$$

$$(\text{for } 5 \text{ ft} < x < 8 \text{ ft})$$

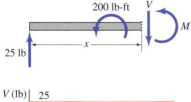

200 lb-ft

V

M

25 lb

x

V and M diagrams have been drawn.

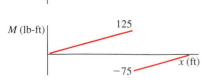

V (lb)   25

x (ft)

M (lb-ft)

125

x (ft)

−75

Notice that M jumps in value across the applied moment. Again, one cannot find the maximum M by setting $dM/dx = 0$.

To the left of $x = 5$ ft,
$V = 25$ lb, $M = 125$ lb-ft

To the right of $x = 5$ ft,
$V = 25$ lb, $M = -75$ lb-ft

25 lb   125 lb-ft

25 lb   75 lb-ft

>>End Example Problem 5.3

A wooden dam of 4 m width holds back stagnant water. The hydrostatic pressure exerted by the water, p, is proportional to depth, $p = \gamma z$, where $\gamma = 9810$ N/m^3 is the weight per volume of water. Model the dam as a simply supported beam and determine the shear force and bending diagrams. Determine the maximum bending moment.

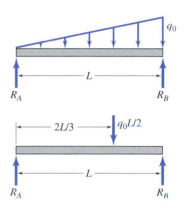

Solution

Let q_0 be the maximum force per length at the bottom of the dam:

$$q_0 = (p)(\text{width}) = (9810 \text{ N/m}^3)(3 \text{ m})(4 \text{ m})$$

The force per length is linearly distributed $q(x) = q_0(x/L)$.

For a linear (triangular) force distribution:

Net force = area under distribution = $(q_0)(L)/2$
Net force acts at $2L/3$ (that is, 2/3 of the way toward max $q(x)$)

Find the support reactions

$$\sum F_y = R_A - q_0 L/2 + R_B = 0$$
$$\sum M|_A = R_B(L) - (q_0 L/2)(2L/3) = 0 \Rightarrow$$
$$R_A = q_0 L/6, \; R_B = q_0 L/3$$

Find V and M at distance a from A (a can be any value in $0 < a < L$).

Isolate segment from A to a. This is a new free body diagram: we must draw on the distributed force again.

$$q(x) = q_0 x/L, \text{ in } 0 < x < a; \text{ max value of } q(x) \text{ is } q_0 \, a/L \text{ at } x = a$$

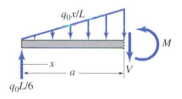

We are about to sum $\sum F_y$ and $\sum M|_{cut}$ so we can replace the distribution with the net force of $(q_0 a/L)(a)/2$ at point $2a/3$ from left

$$\sum F_y = q_0 L/6 - q_0 a^2/2L - V = 0$$
$$\sum M|_{cut} = -(q_0 L/6)a + (q_0 a^2/2L)(a/3) + M = 0$$

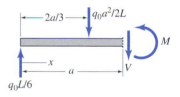

Since a referred to an arbitrary cross-section, we can replace a with x

$$V(x) = (q_0 L/6) - q_0 x^2/2L, \; M(x) = (q_0 L x/6) - q_0 x^3/6L$$

Since M is a single smooth function of x with maximum not at the ends, find $M_{\max}$ by differentiation:

$$\frac{dM}{dx} = \frac{q_0 L}{6} - \frac{q_0 x^2}{2L} = 0 \Rightarrow x = \frac{L}{\sqrt{3}} \Rightarrow M_{\max} = q_0 \left[\frac{L^2}{6\sqrt{3}} - \left(\frac{L}{\sqrt{3}} \right)^3 \frac{1}{6L} \right] = \frac{q_0 L^2}{9\sqrt{3}}$$

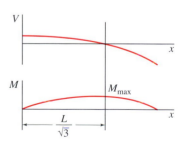

$$\Rightarrow M_{\max} = \frac{(9810 \text{ N/m}^3)(3 \text{ m})(4 \text{ m})(3 \text{ m})^2}{9\sqrt{3}} = 6.80 \times 10^4 \text{ N-m}$$

>>End Example Problem 5.4

5.5 Variation of Internal Loads with Applied Loads

We often seek the maximum values of *V* and *M* along a beam. It can be tedious to find the maximum from the equations we just derived. Instead, we want to draw *V* and *M* diagrams directly with few equations and identify the maximum from the diagram.

1. To draw *V* and *M* directly, we follow this overall procedure: find *V* and *M* at the left end, then determine how *V*, and later *M*, vary along the beam.

To draw the *V* and *M* diagrams directly, the procedure we follow consists of the following steps:

1. Find support reactions (there may be reactions at both ends, but only R_A and M_A are non-zero in this problem).
2. Draw *V*, going from the left end to the right.
3. Draw *M*, going from the left end to the right.

Below we relate *V* and *M* to the support reactions and find how they vary gradually or jump in value at certain points.

In the derivations, unknown internal loads are always drawn in positive directions. We refer to numbered regions in the diagram at the right.

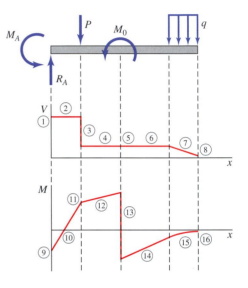

2. Find values *V*(0) and *M*(0) at the left end.

Isolate an infinitesimal segment at the left end.

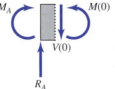

The segment is so thin that R_A and $V(0)$ create zero moment.

$$\sum F_y = R_A - V(0) = 0 \Rightarrow V(0) = R_A$$
V(0) is proportional to R_A. (See *V* at ①.)

$$\sum M = -M_A + M(0) = 0 \Rightarrow M(0) = M_A$$
M(0) is proportional to M_A. (See *M* at ⑨.)

3. Find changes in *V* and *M* at a concentrated force *P* or concentrated moment M_0.

Isolate an infinitesimal segment surrounding a concentrated load. *V* and *M* may have different values (V_L and V_R) and (M_L and M_R) on the two sides of the load. Again, the forces create no moment since the segment is so thin.

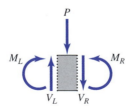

$$\sum F_y = V_L - P - V_R = 0 \Rightarrow V_R = V_L - P \Rightarrow$$
V jumps by *P*. (See *V* at ③.)

$$\sum M = -M_L + M_R = 0 \Rightarrow M_R = M_L \Rightarrow$$
No jump in *M* at *P*. (See *M* at ⑪.)

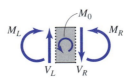

$$\sum F_y = V_L - V_R = 0 \Rightarrow V_R = V_L \Rightarrow$$
No jump in *V* at M_0. (See *V* at ⑤.)

$$\sum M = -M_L + M_0 + M_R = 0 \Rightarrow M_R = M_L - M_0 \Rightarrow$$
M jumps by M_0. (See *M* at ⑬.)

4. Find variation of V and M over a portion of the beam with no external loads.

Isolate a portion that has no external loads from x_L to $x_L + \Delta x$. Find values at the right end, V_R and M_R, in terms of values at the left end, V_L, M_L, and Δx.

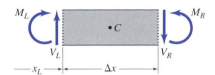

$\sum F_y = V_L - V_R = 0 \Rightarrow V_R = V_L \Rightarrow$
No change in V. (See V in ②, ④, ⑥.)

$\sum M|_C = -M_L - (V_L + V_R)(\Delta x/2) + M_R = 0$
Since $V_R = V_L \Rightarrow M_R = M_L + V_L(\Delta x) \Rightarrow$
M varies linearly with slope $V = V_R = V_L$.
M increases by area under curve of V vs. x.
(See M in ⑩, ⑫, ⑭.)

---

5. Find variation of V and M for distributed force.

Isolate a portion that has distributed force q_0 from x_L to $x_L + \Delta x$. Find values at the right end, V_R and M_R, in terms of values at the left end, V_L and M_L, q, and Δx.

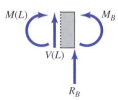

$\sum F_y = V_L - q_0(\Delta x) - V_R = 0 \Rightarrow V_R = V_L - q_0(\Delta x) \Rightarrow$
V has slope $-q_0$. ($dV/dx = -q_0$)
V is linear if q_0 is constant over a finite segment. (See V in ⑦.)

$\sum M|_C = -M_L - (V_R + V_L)(\Delta x)/2 + M_R = 0$

$M_R - M_L = (V_R + V_L)(\Delta x)/2 \Rightarrow$
M increases by the area under curve of V vs. x.
Consider limit when Δx is small.
M has slope V. ($dM/dx = V$)
Since $M_R = M_L + V_L(\Delta x) - q_0(\Delta x)^2/2$,
M is parabolic if q_0 is constant over a finite segment.
(See M in ⑮.)

---

6. V and M vary together in these different ways. For simplicity, M is chosen to start at the same value at the left end of each variation. In summary, the slope of M at each point equals V, and the area under V equals the change in M. Positive area under V equals positive change (increase) in M.

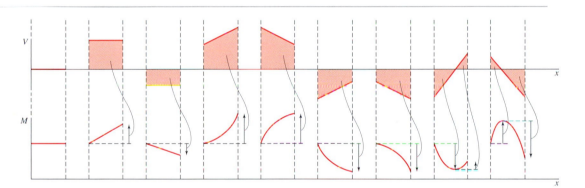

---

7. Check if values reached at the right end of beam, $V(L)$ and $M(L)$, agree with external loads there.

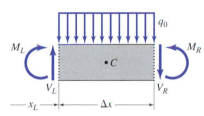

$\sum F_y = V(L) + R_B = 0 \Rightarrow V(L) = -R_B$
$V(L)$ is proportional to R_B. (See V at ⑧.)

$\sum M = -M(L) + M_B = 0 \Rightarrow M(L) = M_B$
$M(L)$ is proportional to M_B. (See M at ⑯.)

If $V(L)$ and $M(L)$ are not related correctly to R_B and M_B, then an error was made somewhere.

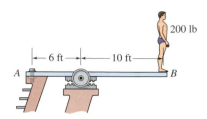

A 200 lb diver stands on the end of the diving board. Determine the shear force and bending moment diagrams by using the relations between external loads and the changes in the internal loads.

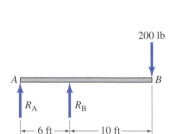

Solution

Determine support loads from statics: $R_A = -333$ lb, $R_B = 533$ lb.

Determine $V(0)$ and $M(0)$ from left support.

$$R_A = -333 \text{ lb}, M_A = 0$$
$$V(0) = -333 \text{ lb}, M(0) = 0$$

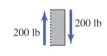

Determine V along beam.

Start from $V(0) = -333$ lb.

$0 < x < 6$ ft, no load $\Rightarrow V$ constant, $V = -333$ lb

Just before 533 lb force, at $x = 6^-$, $V(6^-) = -333$ lb

$x = 6$ ft, concentrated force, $V(6^-) = -333$ lb

$\sum F_y = -333 \text{ lb} + 533 \text{ lb} - V(6^+) = 0 \Rightarrow V(6^+) = 200 \text{ lb}$

$6 < x < 16$ ft, no load $\Rightarrow V$ constant $\Rightarrow V = 200$ lb

$V(16)$ is consistent with external force at B

Determine M along beam.

Start from $M(0) = 0$

$0 < x < 6$ ft, $V = -333$ lb (constant) $\Rightarrow$

M is linear with slope -333 lb
ΔM from $x = 0$ to $x = 6$ ft is area under V vs. x
$\Delta M = (-333 \text{ lb})(6 \text{ ft}) \Rightarrow M(6^-) = -2000 \text{ lb-ft}$

$x = 6$ ft, concentrated force $\Rightarrow$ no change in $M \Rightarrow$
$M(6^+) = M(6^-) = -2000$ lb-ft

$6 < x < 16$ ft, $V = 200$ lb (constant) $\Rightarrow$

M is linear with slope 200 lb
ΔM from $x = 6$ to $x = 16$ ft is area under V vs. x
$\Delta M = (200 \text{ lb})(10 \text{ ft}) \Rightarrow M(16) = -2000 \text{ lb-ft} + 2000 \text{ lb-ft} = 0.$
It is consistent with zero moment reaction at B.

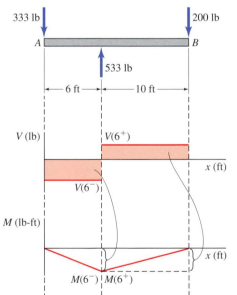

>>End Example Problem 5.5

Determine the shear force and bending moment diagrams by using the relations between external loads and the changes in the internal loads.

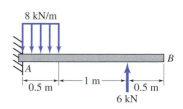

Solution

Determine support loads from statics.

$$R_A = (8 \text{ kN/m})(0.5 \text{ m}) - (6 \text{ kN}) = -2 \text{ kN}$$

$$M_A = -(8 \text{ kN/m})(0.5 \text{ m})(0.25 \text{ m}) + (6 \text{ kN})(1.5 \text{ m}) = 8 \text{ kN-m}$$

Determine $V(0)$ and $M(0)$ from support at A.

$$V(0) = -2 \text{ kN}, \, M(0) = 8 \text{ kN-m}$$

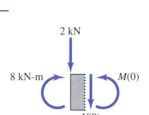

Determine V along beam.

Start from $V(0) = -2$ kN.

$0 < x < 0.5$ m, q constant ($q = 8$ kN/m down) $\Rightarrow$
V has slope $= -8$ kN/m

$$V(0.5) = -2 - (8 \text{ kN/m})(0.5 \text{ m}) = -6 \text{ kN}$$

0.5 m $< x <$ 1.5 m, no load $\Rightarrow V$ constant $\Rightarrow V = -6$ kN

$x = 1.5$ m, concentrated force $\Rightarrow$
$V(1.5^+) = -6$ kN $+ 6$ kN $= 0$

1.5 m $< x <$ 2 m, no load $\Rightarrow V$ constant, $V = 0$

$V(2) = 0$ is consistent with zero external force at the right end.

Determine M along beam.

Start from $M(0) = 8$ kN-m

$0 < x < 0.5$ m, $V < 0$ and decreases $\Rightarrow M$ decreases and slope becomes more negative. M is parabola that is concave down.
$\Delta M =$ area under $V = (0.5)(-2 \text{ kN} - 6 \text{ kN})(0.5 \text{ m}) = -2$ kN-m
$M(0.5) = 8$ kN-m $- 2$ kN-m $= 6$ kN-m

0.5 m $< x <$ 1.5 m, $V = -6$ kN $\Rightarrow M$ has constant slope
$\Delta M =$ area under $V = (-6 \text{ kN})(1 \text{ m})$
$M(1.5) = 6$ kN-m $- 6$ kN-m $= 0$

1.5 m $< x <$ 2 m, $V = 0$, M is constant ($M = 0$)
$M(2) = 0$ is consistent with zero external moment at right end.

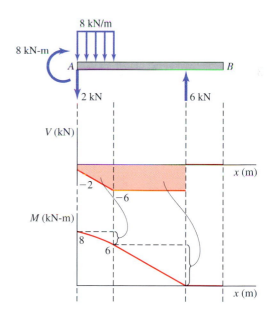

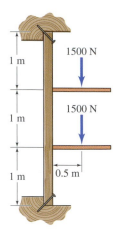

1 m

1500 N

1 m

1500 N

1 m

0.5 m

Weights are carried on a pair of shelves. The loads experienced by one stud are shown. Model each force as resulting in a concentrated moment on the beam. Model the stud as a simply supported beam. Determine the shear force and bending moment diagrams by using the relations between external loads and the changes in the internal loads.

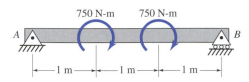

750 N-m   750 N-m

A   B

1 m   1 m   1 m

Solution

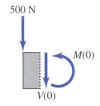

500 N

$M(0)$

$V(0)$

Determine support loads from statics: $R_A = -500$ N, $R_B = 500$ N.

Determine $V(0)$ and $M(0)$ from left support.

$$R_A = -500 \text{ N}, M_A = 0$$
$$V(0) = -500 \text{ N}, M(0) = 0$$

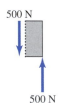

500 N

500 N

Determine V along beam.

Start from $V(0) = -500$ N

$0 < x < 3$ m, no external force $\Rightarrow V$ constant, $V = -500$ N

$V(3)$ is consistent with support force at B.

Determine M along beam.

Start from $M(0) = 0$

$0 < x < 1$ m, $V = -500$ N (constant) $\Rightarrow$
M is linear with slope -500 N
ΔM from $x = 0$ to $x = 1$ m is area under V vs. x
$\Delta M = (-500 \text{ N})(1 \text{ m}) \Rightarrow M(1^-) = -500$ N-m

$x = 1$ m, concentrated moment $\Rightarrow$ jump in $M \Rightarrow$
$M(1^+) = -500$ N-m $+ 750$ N-m $= 250$ N-m

$1 < x < 2$ m, M has slope of -500 N $\Rightarrow$
$M(2^-) = 250$ N-m $- 500$ N(1 m) $= -250$ N-m

$x = 2$ m, concentrated moment $\Rightarrow$ jump in $M \Rightarrow$
$M(2^+) = -250$ N-m $+ 750$ N-m $= 500$ N-m

$2 < x < 3$ m, M has slope of -500 N $\Rightarrow$
$M(3) = 500$ N-m $- 500$ N(1 m) $= 0$;
agrees with zero external moment at B.

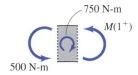

750 N-m

$M(1^+)$

500 N-m

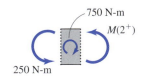

750 N-m

$M(2^+)$

250 N-m

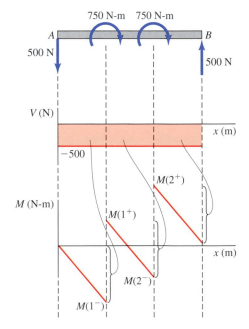

750 N-m   750 N-m

A   B

500 N   500 N

V (N)

x (m)

-500

M (N-m)

$M(2^+)$

$M(1^+)$

x (m)

$M(2^-)$

$M(1^-)$

>>End Example Problem 5.7

A book shelf is constructed from a 12 ft board placed on bricks at 2 ft from each end. Books, each 1 in. wide and weighing 2 lb, are placed side by side along the entire length of the board. Consider the load due to the books only. Determine the shear force and bending moment diagrams by using the relations between external loads and changes in the internal loads.

Solution

Board is 12 ft = 144 in. ⟹ 144 books
Total weight = 288 lb, q = 24 lb/ft

Support reactions from bricks must each be (24 lb/ft)(12 ft)/2 = 144 lb

Determine $V(0)$ and $M(0)$ from external loads at A.

$$V(0) = 0, M(0) = 0$$

Determine V along beam.

Start from $V(0) = 0$

$0 < x < 2$ ft, q constant, slope of $V = -24$ lb/ft
$\Rightarrow V(2) = 0 - (24 \text{ lb/ft})(2 \text{ ft}) = -48$ lb

$x = 2$ ft, concentrated force ⟹
$V(2^+) = -48$ lb + 144 lb = 96 lb

2 ft $< x < 10$ ft, q constant, slope of $V = -24$ lb/ft
$\Rightarrow V(10^-) = 96 - (24 \text{ lb/ft})(8 \text{ ft}) = -96$ lb

$x = 10$ ft, concentrated force ⟹
$V(10^+) = -96$ lb + 144 lb = 48 lb

10 ft $< x < 12$ ft, q constant, slope of $V = -24$ lb/ft
$\Rightarrow V(12) = 48 - (24 \text{ lb/ft})(2 \text{ ft}) = 0$

$V(12) = 0$ is consistent with zero external force at B.

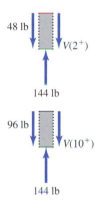

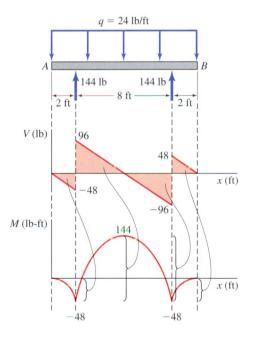

Determine M along beam.

Start from $M(0) = 0$

$0 < x < 2$ ft, $V < 0$ and decreases ⟹ M has zero slope at $x = 0$ and becomes increasingly negative. ΔM = area under V = $(-48 \text{ lb})(2 \text{ ft})/2 \Rightarrow$
$M(2) = 0 - 48$ lb-ft = -48 lb-ft

2 ft $< x < 10$ ft, V varies from positive to negative. M increases where $V > 0$ and has a maximum at $V = 0$ at $x = 6$ ft. ΔM = area under V = $(96 \text{ lb})(4 \text{ ft})/2$. $M(6) = -48$ lb-ft + 192 lb-ft = 144 lb-ft. M decreases where $V < 0$, M = area under V = $(-96 \text{ lb})(4 \text{ ft})/2$. $M(10)$ = 144 lb-ft − 192 lb-ft = -48 lb-ft.

10 ft $< x < 12$ ft, $V > 0$ and decreases ⟹ M increases and has zero slope at $x = 12$ ft. ΔM = area under V = $(48 \text{ lb})(2 \text{ ft})/2$, $M(12) = -48$ lb-ft + 48 lb-ft = 0.

$M(12)$ is consistent with zero external moment at B.

>>End Example Problem 5.8

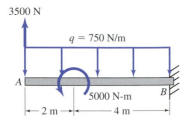

The beam is loaded as shown. Determine the shear force and bending moment diagrams by using the relations between external loads and the changes in the internal loads.

Solution

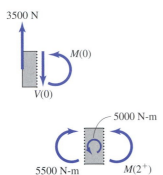

Support reactions are $R_B = 1000$ N, $M_B = 2500$ N-m

Determine $V(0)$ and $M(0)$ from external loads at A.

$$V(0) = 3500 \text{ N}, M(0) = 0$$

Determine V along beam.

Start from $V(0) = 3500$ N

> $0 < x < 6$ m, $q = $ constant, slope of $V = -750$ N/m $\Rightarrow$
> $V(6) = 3500$ N-m $- (750 \text{ N/m})(6 \text{ m}) = -1000$ N

> $V(6) = -1000$ N is consistent with R_B.

Determine M along beam.

Start from $M(0) = 0$

> $0 < x < 2$ m, $V > 0$ and decreases $\Rightarrow M$ increases but at a decreasing rate.
> At $x = 2$, $V = 3500$ N-m $- (750 \text{ N})(2 \text{ m}) = 2000$ N-m
> $\Delta M = $ area under $V = 0.5(3500 \text{ N} + 2000 \text{ N})(2 \text{ m})$
> $M(2^-) = 5500$ N-m

> $x = 2$ m, concentrated moment $\Rightarrow$ jump in M
> $\Rightarrow M(2^+) = 5500$ N-m $- 5000$ N-m $= 500$ N-m

> 2 m $< x < 6$ m, V varies from positive to negative. M has a maximum at x_m, where $V = 3500$ N $- (750 \text{ N/m})x_m = 0 \Rightarrow x_m = 4.67$ m.
> $\Delta M = $ area under V from $x = 2$ m to $x_m \Rightarrow$
> $M(4.67) = 500$ N-m $+ (2000 \text{ N})(4.67 \text{ m} - 2 \text{ m})/2 = 3167$ N-m

> 4.67 m $< x < 6$ m, V is negative, M decreases
> $\Delta M = $ area under V from $x = 4.67$ m to 6 m $\Rightarrow$
> $M(6) = 3167$ N-m $+ (-1000)(6 \text{ m} - 4.67 \text{ m})/2 = 2500$ N-m

> $M(6) = 2500$ N-m is consistent with the support reaction M_B.

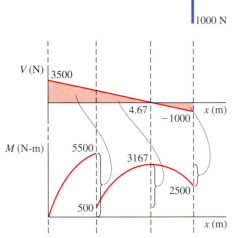

>>End Example Problem 5.9

Plastic tabs deflect and grip a *CD* when placed in a case. To release the *CD*, one presses on the top to deflect the tab. Consider this simplified model of the tab based on three beams. Determine the shear force and bending diagrams for the segments *AB*, *BC*, and *CD*.

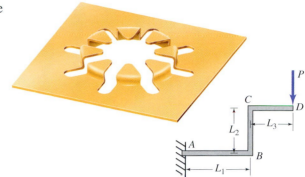

Solution

Break the tab into its straight segments, find the force and moment acting between each segment, and then analyze each segment as a single beam.

Here are the three segments separated, with the loads that act between them.

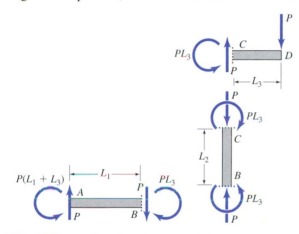

Notice that force P for BC is axial and so does not affect shear force and bending diagrams.

Redraw all beams as horizontal.

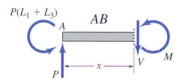

$$\sum F_y = P - V = 0$$
$$\sum M \big|_{cut} = P(L_1 + L_3) - Px + M = 0$$
$$\Rightarrow V = P, M = -P(L_1 + L_3) + Px$$

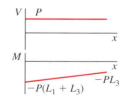

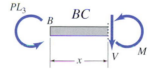

$$\sum F_y = -V = 0$$
$$\sum M \big|_{cut} = PL_3 + M = 0$$
$$\Rightarrow V = 0, M = -PL_3$$

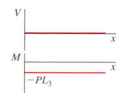

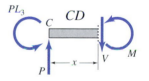

$$\sum F_y = P - V = 0$$
$$\sum M \big|_{cut} = PL_3 - Px + M = 0$$
$$\Rightarrow V = P, M = -PL_3 + Px$$

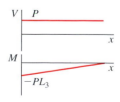

>>End Example Problem 5.10

Additional data on material properties needed to solve problems can be found in Appendix D or inside back cover.

5.1 Given $q = 200$ N/m, consider the internal loads (shear force and bending moment) at the cross-section located 2 m from the left end. Determine the internal loads by (a) finding the support reaction and then isolating the segment from $x = 0$ to $x = 2$ m and (b) isolating the segment from $x = 2$ m to the right end. (Take the lengths to be $L_1 = 1.5$ m and $L_2 = 3$ m.)

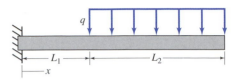

Prob. 5.1

5.2 Given $M_0 = 80$ kip-ft, find the support reactions and consider the internal loads (shear force and bending moment) at the cross-section located 3 ft from the left end. Determine the internal loads by (a) isolating the segment from $x = 0$ to $x = 3$ ft and (b) isolating the segment from $x = 3$ ft to the right end. (Take the lengths to be $L_1 = 6$ ft and $L_2 = 4$ ft.)

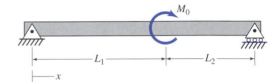

Prob. 5.2

5.3 Given $P = 20$ kN, consider the internal loads (shear force and bending moment) at the cross-section located 6 m from the left end. Determine the internal loads by (a) finding the support reactions and isolating the segment from $x = 0$ to $x = 6$ m and (b) isolating the segment from $x = 6$ m to the right end. (Take the lengths to be $L_1 = 5$ m and $L_2 = 3$ m.)

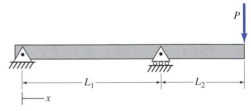

Prob. 5.3

5.4 Given $q = 10$ kN/m, determine the shear force and bending moment at the cross-section located at $x = 4$ m from the left end. (Take the lengths to be $L_1 = 1$ m, $L_2 = 2$ m, and $L_3 = 4$ m.)

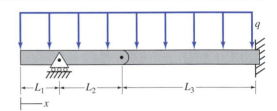

Prob. 5.4

5.5 Given $q = 300$ lb/in. and $P = 8000$ lb, determine the shear force and bending moment at the cross-section located at $x = 50$ in. from the left end. (Take the lengths to be $L_1 = 60$ in., $L_2 = 30$ in., $L_3 = 15$ in., and $L_4 = 50$ in.)

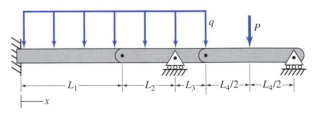

Prob. 5.5

5.6 A portion of length $L = 4$ m of a pipe is embedded in the ground. Where the pipe emerges from the ground, it is loaded by a moment M_0 and a transverse force $F_0 = 30$ kN. Assume the ground reacts with a linearly varying force per length of maximum magnitude q as shown. Determine the moment M_0, as well as the shear force and bending moment at a distance of $x = 2$ m from the bottom of the pipe.

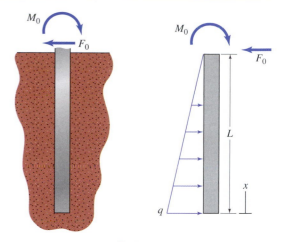

Prob. 5.6

5.7 A skier's weight $W = 180$ lb rests on the point $c = 44$ in. from the left end of the 6 ft long ski. The distributed force of the snow on the ski is approximated as varying linearly. (a) Determine the force per length at the ends q_1 and q_2. (b) Determine the shear force and bending moment just to the left of the point at which the skier's weight acts.

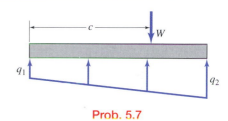

Prob. 5.7

5.8 Take $M_0 = 400$ N-m and $q = 100$ N/m. Draw the shear force and bending moment diagrams for the beam shown. Specify the maximum positive and negative shear force and bending moment. (Take the lengths to be $L_1 = 2$ m, $L_2 = 1.5$ m, and $L_3 = 3$ m.)

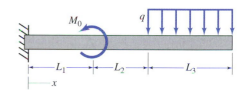

Prob. 5.8

5.9 Take $P_1 = 700$ lb and $P_2 = 700$ lb. Draw the shear force and bending moment diagrams for the beam shown. Specify the maximum positive and negative shear force and bending moment. (Take the lengths to be $L_1 = 8$ ft, $L_2 = 4$ ft, and $L_3 = 4$ ft.)

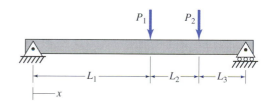

Prob. 5.9

5.10 Take $P = 200$ lb and $q = 10$ lb/in. Draw the shear force and bending moment diagrams for the beam shown. Specify the maximum positive and negative shear force and bending moment. (Take the lengths to be $L_1 = 8$ in. and $L_2 = 14$ in.)

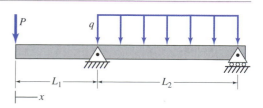

Prob. 5.10

5.11 A post is buried in the ground and subjected to a transverse force $P = 500$ N at the top. A simple model for the reaction of the ground consists of two uniform distributions of force per length q_1 and q_2. (a) Determine the values for the forces per length q_1 and q_2. (b) Draw the beam on its side (with the applied force P acting upward at the left end $x = 0$) and draw shear force and bending moment diagrams. Specify the maximum positive and negative shear force and bending moment. (Take the lengths to be $L_1 = 200$ mm, $L_2 = 200$ mm, and $L_3 = 4$ m.)

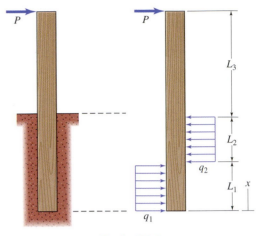

Prob. 5.11

5.12 Take $q = 100$ N/m. Draw the shear force and bending moment diagrams for the beam shown. Specify the maximum positive and negative shear force and bending moment. (Take the lengths to be $L_1 = 300$ mm and $L_2 = 300$ mm.)

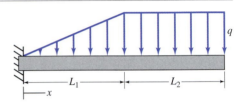

Prob. 5.12

5.13 Take $P = 200$ lb, $M_0 = 1500$ lb-ft, and $q = 100$ lb/ft. Draw the shear force and bending moment diagrams for the beam shown. Specify the maximum positive and negative shear force and bending moment. (Take the lengths to be $L_1 = 3$ ft, $L_2 = 2$ ft, and $L_3 = 5$ ft.)

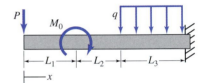

Prob. 5.13

5.14 Take $q = 100$ N/m, $P = 600$ N, and $M_0 = 1000$ N-m. Draw the shear force and bending moment diagrams for the beam shown. Specify the maximum positive and negative shear force and bending moment. (Take the lengths to be $L_1 = 5$ m, $L_2 = 2$ m, and $L_3 = 2$ m.)

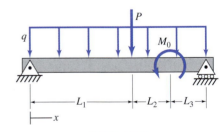

Prob. 5.14

5.15 The bending moment magnitude must not exceed 1000 N-m. Determine the maximum allowable force per length (assuming $q > 0$), and draw the shear force and bending moment diagrams for the beam shown. Specify the maximum positive and negative shear force and bending moment. (Take the lengths to be $L_1 = 0.75$ m and $L_2 = 0.75$ m.)

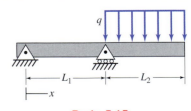

Prob. 5.15

5.16 The bending moment magnitude must not exceed 300 lb-ft. Determine the maximum allowable force per length q, and draw the shear force and bending moment diagrams for the beam shown. Specify the maximum positive and negative shear force and bending moment. (Take the lengths to be $L_1 = 3$ ft and $L_2 = 5$ ft.)

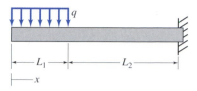

Prob. 5.16

5.17 The bending moment magnitude must not exceed 3 N-m. Determine the maximum allowable force P, and draw the shear force and bending moment diagrams for the beam shown. Specify the maximum positive and negative shear force and bending moment. (Take the lengths to be $L_1 = 80$ mm and $L_2 = 30$ mm.)

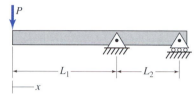

Prob. 5.17

5.18 The bending moment magnitude must not exceed 800 lb-in. Determine the maximum allowable moment M_0, and draw the shear force and bending moment diagrams for the beam shown. Specify the maximum positive and negative shear force and bending moment. (Take the lengths to be $L_1 = 15$ in. and $L_2 = 9$ in.)

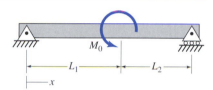

Prob. 5.18

5.19 Given $q = 1000$ N/m, determine the allowable range of M_0 if the magnitude of the bending moment is everywhere to remain less than 500 N-m. (Consider both positive and negative bending moments.) M_0 acts in the direction shown. (Take the lengths to be $L_1 = 1200$ mm, $L_2 = 800$ mm, and $L_3 = 500$ mm.)

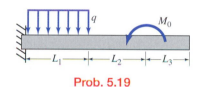

Prob. 5.19

5.20 Given $M_0 = 800$ lb-ft, determine the allowable range of P if the magnitude of the bending moment is everywhere to remain less than 600 lb-ft. (Consider both positive and negative bending moments.) P acts in the direction shown. (Take the lengths to be $L_1 = 4$ ft, $L_2 = 3$ ft, and $L_3 = 3$ ft.)

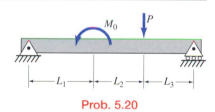

Prob. 5.20

5.21 Given $q = 1600$ N/m, determine the allowable range of M_0 if the magnitude of the bending moment is everywhere to remain less than 2500 N-m. (Consider both positive and negative bending moments.) M_0 acts in the direction shown. (Take the lengths to be $L_1 = 2$ m and $L_2 = 1$ m.)

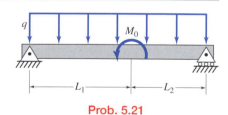

Prob. 5.21

5.22 The moment along the beam, which is subjected to a uniform force per length q, may be positive and/or negative, depending on the support separation a. It is useful to place the support to make the maximum magnitude of the moment as small as possible. Determine a relative to L at which the maximum bending moment is minimized.

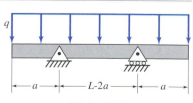

Prob. 5.22

5.23 Analyze the frame shown to determine the loads acting between various parts. Then, considering only the transverse loading (not axial), draw the shear force and bending moment diagrams for member *EDC*. (Reorient *EDC* to be horizontal, with *E* at the left.) Specify the maximum positive and negative shear force and bending moment.

5.24 Analyze the frame shown to determine the loads acting between various parts. Then, considering only the transverse loading (not axial), draw the shear force and bending moment diagrams for member *ABC*. (Reorient *ABC* to be horizontal, with *A* at the left.) Specify the maximum positive and negative shear force and bending moment.

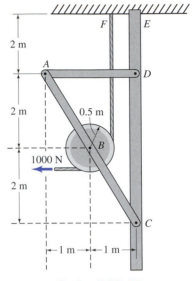

Probs. 5.23–24

5.25 Analyze the frame shown to determine the loads acting between various parts. Then, considering only the transverse loading (not axial), draw the shear force and bending moment diagrams for member *FDCB*. Specify the maximum positive and negative shear force and bending moment.

5.26 Analyze the frame shown to determine the loads acting between various parts. Then, considering only the transverse loading (not axial), draw the shear force and bending moment diagrams for member *EDA*. (Reorient *EDA* to be horizontal, with *E* at the left.) Specify the maximum positive and negative shear force and bending moment.

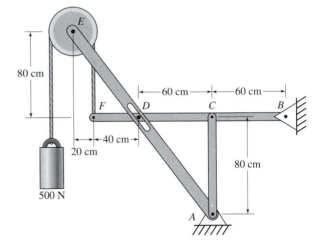

Probs. 5.25–26

5.27 A beam is supported as shown. Determine the loads that must act if the shear force and bending moment diagrams are to be as shown.

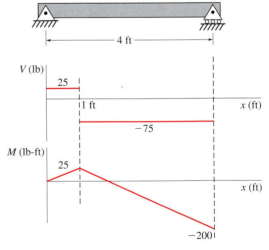

Prob. 5.27

5.28 A beam is supported as shown. Determine the loads that must act if the shear force and bending moment diagrams are to be as shown.

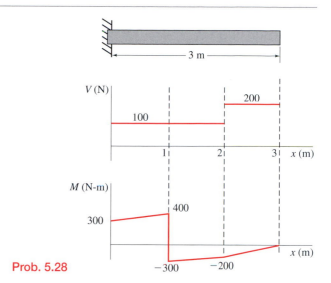

Prob. 5.28

5.29 A beam is supported as shown. Determine the loads that must act if the shear force and bending moment diagrams are to be as shown.

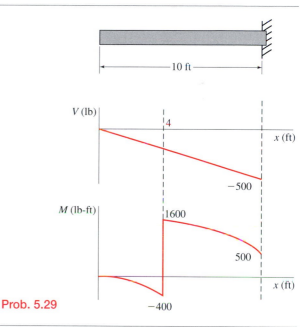

Prob. 5.29

5.30 A beam is supported as shown. Determine the loads that must act if the shear force and bending moment diagrams are to be as shown.

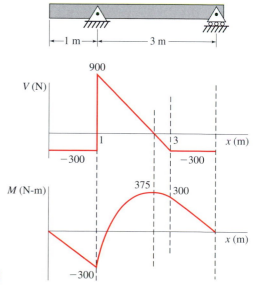

Prob. 5.30

5.31 A beam is supported as shown. Determine the loads that must act if the shear force and bending moment diagrams are to be as shown.

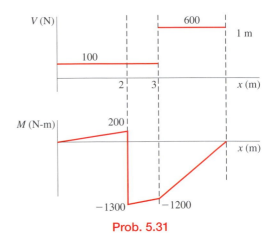

Prob. 5.31

5.32 Under the conditions depicted, the force on the rack that drives this corkscrew is 400 N. Dimensions shown are in millimeters. The guidepost keeps the mechanism aligned by sliding in a hole. It can exert a moment, but no force. Assume the screw applies only a vertical force. Draw the shear force and bending moment diagrams for the outlined upper horizontal member.

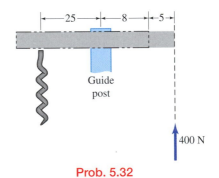

Prob. 5.32

5.33 The portion of the telephone pole above the road surface is 40 ft long, weighs 1000 lb, and leans at 5° due to the long term action of a wire that exerts a 100 lb tension to the left at a point 30 ft from the base. To counteract the leaning, a stabilizing cable is attached to a point 5 ft from the top. The cable makes a 45° angle with the horizontal road surface and carries a tension of 130 lb. Draw the shear force and bending moment diagrams for the transverse loadings on the pole. Be careful to resolve forces perpendicular and parallel to the pole.

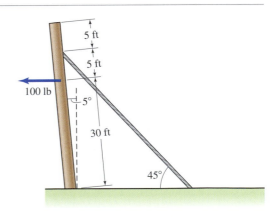

Prob. 5.33

5.34 The set of 18 large books are displayed in three groups symmetrically on the shelving unit. Each book is 50 mm wide and has a mass of 3 kg. Supports are located at points 1/3 of the length from each end. Draw the shear force and bending moment diagrams for the shelf.

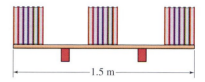

Prob. 5.34

5.35 The simply supported arch is subjected to a vertical force at the angle θ_0. Derive formulas for the shear force $V(\theta)$, the normal force $N(\theta)$, and the bending moment $M(\theta)$. There will be two different formulas for each depending on whether $\theta < \theta_0$ or $\theta > \theta_0$.

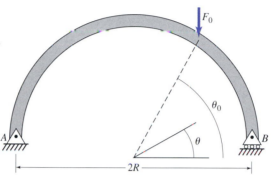

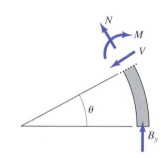

Prob. 5.35

Focused Application Problems

5.36 Pedaling produces bending and twisting of the spindle. Consider here only the transverse loading (bending), and ignore the torsion. Draw the shear force and bending moment diagrams for the spindle from the bearing near the crank (labeled $x = 0$) to the bearing by the chain ring. There is no force applied to the other pedal, and the chain tensions act in the same plane as the bearing, so they do not contribute to the bending. Let the pedal force be 1000 N. Take the dimensions to be $a = 70$ mm, $b = 25$ mm, $c = 9.5$ mm, $d = 80$ mm, and $L = 170$ mm.

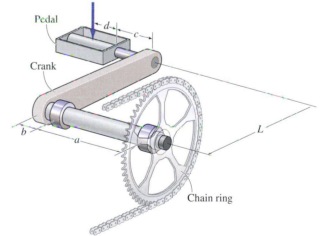

Prob. 5.36 (Appendix A1)

5.37 This cable-stayed footbridge has cables attached at the points on the pylons as shown. Let the tension be 100 kN in the cables supporting the deck. Determine the tension in the rear stay so that the moment at the base of the pylon is zero (so a narrow support can be used). Then, determine the shear force and bending moment diagrams for the pylon. Take the dimensions to be $w_1 = 10$ m, $w_2 = 50$ m, $w_3 = 30$ m, $w_4 = 90$ m, $h_1 = 40$ m, $h_2 = 30$ m, and $h_3 = 20$ m.

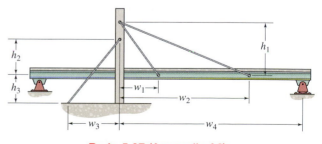

Prob. 5.37 (Appendix A2)

5.38 This reinforced concrete, cable-stayed bridge has only a single cable supporting the deck at its center. Take the loading of the bridge to be uniform 5 Mg/m. Determine the cable tension that minimizes the maximum bending moment along the deck and the value of the maximum bending moment. The deck does not rest on the pylon, but is only supported at its ends and by the cable. Take the dimensions to be $w_1 = 50$ m, $w_2 = 20$ m, $h_1 = 40$ m, and $h_2 = 30$ m.

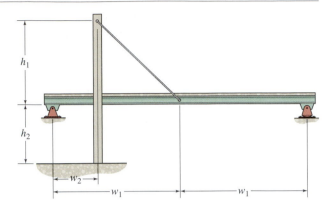

Prob. 5.38 (Appendix A2)

5.39 In using this pectoral fly machine, the hand grips and applies a force to the handle as shown. Assume that the force of the hand is 50 lb and is approximated as uniformly distributed over the gripped section of length w. Draw the shear force and bending moment diagrams of the handle and give the maximum shear force and bending moment. Let the upper end of the vertical part of the handle correspond to $x = 0$ on the diagrams. Take the dimensions to be $w = 4$ in. and $h = 3$ in.

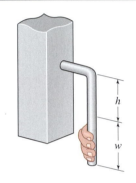

Prob. 5.39 (Appendix A4)

5.40 Say that each leg exerts an upward force of $F_1 = F_2 = 50$ lb. Model the connector links as simple supports (exerting only vertical forces). Draw the beam with loads on it, and then draw the shear force and bending moment diagrams. Take the dimensions to be $a = 4$ in., $b = 16$ in., $c = 4$ in., and $d = 3.5$ in.

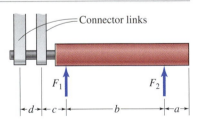

Prob. 5.40 (Appendix A4)

5.41 Say that a single leg exerts on upward force of $F_2 = 75$ lb ($F_1 = 0$). Consider bending in the vertical plane in the shaft that connects the pivot beam and the cord plate. Ignore the twisting of the shaft and the bending in the horizontal plane due to the cord tension. The cord that is attached to the cord plate and balances the twisting of the shaft does not contribute to bending in the vertical plane, because it is oriented parallel to the pivot beam in the position shown. Draw this shaft as a beam with loads on it and then draw the shear force and bending moment diagrams. Take the dimensions to be $a = 4$ in., $b = 16$ in., $c = 4$ in., $d = 3.5$ in., $D_1 = 1.5$ in., $L = 1.75$ in., $p = 2$ in., $q = 6$ in., $m = 5$ in., $s = 3$ in., and $t = 0.25$ in.

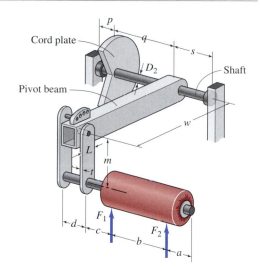

Prob. 5.41 (Appendix A4)

5.42 The blades of a wind turbine are subjected to the load due to the weight of the blade itself. Because the blade rotates, this loading changes with time and hence results in fatigue loading. The weight loading is maximum when the blade is horizontal. Let the cross-section be approximated as a uniform ellipse with outer dimensions $c_1 = 1.3$ m and $c_2 = 0.3$ m, and a uniform wall thickness of 15 mm. The blade is fiberglass with a mass density of 1700 kg/m^3. Let the blade be 20 m long. (a) Determine the force per length due to the weight. (b) Draw the shear force and bending moment diagrams, and determine the maximum shear force and bending moments.

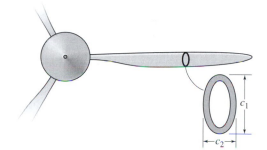

Prob. 5.42 (Appendix A6)

5.43 This 45 ft wind turbine tower pivots about its base and is secured by two pairs of cables, which are initially under equal tensions. Due to the thrust force of 300 lb on the turbine, the tension in the upper cables each changes by 166 lb (the front cable increases and the rear cable decreases). Determine the changes in tension in the lower pair of cables, and then draw the shear force and bending moment diagrams for the tower. The distances are $w = 30$ ft, $h_0 = 45$ ft, $h_1 = 37$ ft, and $h_2 = 22$ ft.

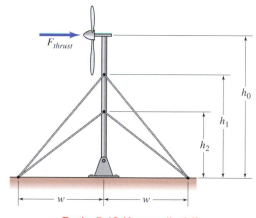

Prob. 5.43 (Appendix A6)

>>End Problems

5.6 Strain Distribution in Bending

We now quantify the strains within an initially straight beam that is bent to a known radius of curvature, ρ. This will lead us to the stress and ultimately to the bending moment. For simplicity, we show the deformation for a beam of rectangular cross-section, although this picture of deformation holds generally. The effect of cross-section is addressed later.

1. Strains in bending are described with respect to a coordinate system aligned with the beam.

The strains vary from point to point in a bending beam. This variation is readily described when x-y-z coordinates are defined relative to the beam in its straight unbent shape. The x-axis is through the center of the cross-section in the long direction. The y- and z-axes are perpendicular to the other faces. When the beam bends, the centerline lies in the x-y plane. We say the beam bends in the x-y plane.

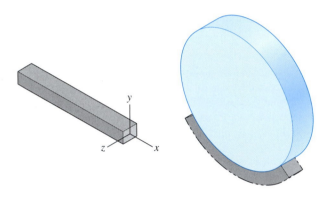

2. Draw squares on the surface of the beam and follow their changes in shape with bending.

Imagine different line elements throughout the beam, all initially parallel to the x-axis. Their lengths change, producing normal strain along x. The strains differ for elements at different y-positions, but not for elements at different x- or z-positions.

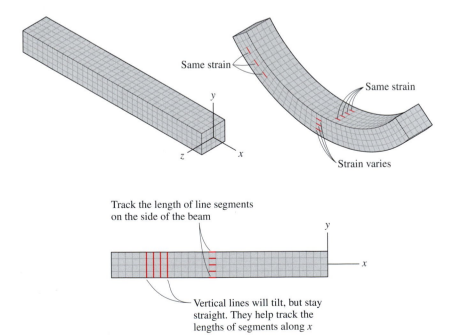

Same strain

Same strain

Strain varies

Track the length of line segments on the side of the beam

Vertical lines will tilt, but stay straight. They help track the lengths of segments along x

3. Do horizontal line elements at different positions y, initially all equal in length, get longer or shorter?

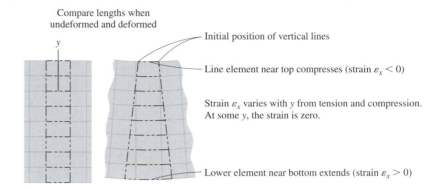

Compare lengths when undeformed and deformed

Initial position of vertical lines

Line element near top compresses (strain $\varepsilon_x < 0$)

Strain ε_x varies with y from tension and compression. At some y, the strain is zero.

Lower element near bottom extends (strain $\varepsilon_x > 0$)

4. Where is the strain zero?

The **neutral plane** is defined as the plane that experiences no strain when a beam is bent. The **neutral axis** (here x) is the intersection between the plane of bending (here x-y) and the neutral plane (here x-z).

A beam can have any cross-section (this one is trapezoidal). Here we consider cross-sections that are symmetric about the x-y plane. Regardless of the (symmetric) cross-section, there is one plane through the beam at which the strain is zero while bending. This is called the **neutral plane**. In Section 5.8, we will determine the location of the neutral plane. For now, we assume it is known.

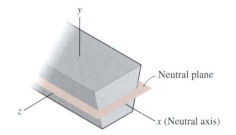

Neutral plane

x (Neutral axis)

We locate the x-axis to coincide with the **neutral axis**, where the neutral plane intersects the symmetry plane. Since $y = 0$ on the x-axis (on the neutral plane), $\varepsilon_x = 0$ at $y = 0$.

5. Determine ε_x to vary linearly with y.

Consider one beam segment that is initially of length Δx. When the beam is bent, the sides of the segment tilt, deforming the segment into a sector of angle $\Delta\theta$. Since the red arc was the neutral axis, $\varepsilon_x = 0$ there. Let the radius ρ be measured from the center of the imaginary cylinder around which the beam is wrapped to the neutral plane. Use geometry of circular arcs to find the strain of the green line at a general position y.

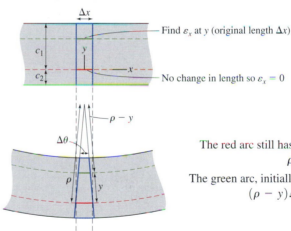

Δx

Find ε_x at y (original length Δx)

No change in length so $\varepsilon_x = 0$

$\rho - y$

$\Delta\theta$

The red arc still has length Δx, so the arc length
$$\rho\Delta\theta = \Delta x$$

The green arc, initially of length Δx, now has length
$$(\rho - y)\Delta\theta = \rho\Delta\theta - y\Delta\theta$$

Strain of green arc: $\varepsilon_x = \dfrac{\text{Change in length}}{\text{Initial length}} = \dfrac{(\rho\Delta\theta - y\Delta\theta) - \rho\Delta\theta}{\rho\Delta\theta} = \dfrac{-y}{\rho} \Rightarrow \varepsilon_x = \dfrac{-y}{\rho}$ Strain variation through thickness

6. Maximum and minimum strains are at the top and bottom of the beam, at distances c_1 and c_2 from $y = 0$.

The changes in length in the diagrams above are exaggerated. Usually, ρ is large compared to the height ($c_1 + c_2$) of the beam, so c_1/ρ and c_2/ρ will both be very small ($\ll 1$). The distance ρ to the neutral plane would only be slightly larger than the radius of the wrapped cylinder.

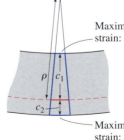

Maximum compressive strain: $\varepsilon_x = -c_1/\rho$

Location of the neutral plane, and then c_1 and c_2, will be determined in Section 5.8 for a given cross-section.

ρ c_1

c_2

Maximum tensile strain: $\varepsilon_x = c_2/\rho$

>>End 5.6

5.7 Stresses in Bending

From the strains, we now find the stresses, which also vary with *y*. We also show how this stress distribution is qualitatively consistent with loads on the body.

1. Stresses on cubic elements from the top and bottom of the beam are compressive and tensile.

No forces act on the sides of the beam. So the only normal stress is in the *x*-direction (σ_x), and it is related to strain ε_x.

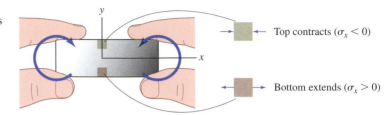

Top contracts ($\sigma_x < 0$)

Bottom extends ($\sigma_x > 0$)

2. Visualize the compressive and tensile stresses at the top and bottom of the beam by thinking of the behavior of a small crack at the edge.

Bend a beam with a small crack on one edge. If the crack is on the bottom, it opens, due to tension. If it is on the top, the crack closes, due to compression.

Bottom   Top

3. Compressive and tensile stresses at top and bottom also make sense because they form a couple.

Separate the beam at a cross-section, and apply the stresses exerted by the adjacent removed elements.

The stresses produce opposite forces on the upper and lower parts of the cross-section, that is, a net couple or bending moment.

4. Relate the stress to the strain assuming the material is elastic.

From the elastic relation $\sigma_x = E\varepsilon_x$, and $\varepsilon_x = \dfrac{-y}{\rho}$, we find $\sigma_x = \dfrac{-Ey}{\rho}$

Stress variation through the thickness

If the material is not elastic, that is the stress is not a linear function of strain, then while the strain would still vary linearly with *y* over the cross-section, the stress would not.

5. Consider different possible positions for the neutral plane, where the stresses change from tensile to compressive, and see how the net load due to the stresses would change.

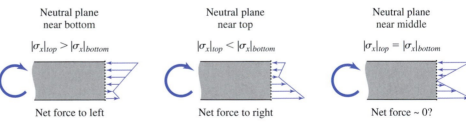

Neutral plane near bottom

$|\sigma_x|_{top} > |\sigma_x|_{bottom}$

Net force to left

Neutral plane near top

$|\sigma_x|_{top} < |\sigma_x|_{bottom}$

Net force to right

Neutral plane near middle

$|\sigma_x|_{top} = |\sigma_x|_{bottom}$

Net force ~ 0?

Stresses should produce a moment and zero net axial force. This condition sets the location of the neutral plane.

But if the neutral plane is in the middle, it is still not quite right for this cross-section.

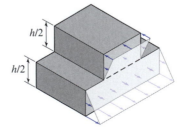 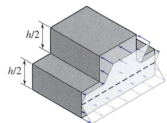

If the neutral plane is in the middle, tension acts on a larger area and produces a net tensile force.

If the neutral plane is below the middle, the forces due to tension and compression could balance.

6. Notice that it is much easier to bend the same beam in one plane vs. another, and explain the difference in terms of the stress variation with *y*.

Bend a rectangular beam (e.g., a ruler), starting from two different orientations, about the same cylinder.

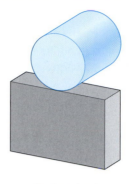 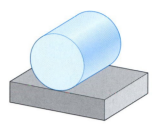

Harder to bend   Easier to bend

We can explain why the moment to bend a beam into a given radius is different in the two orientations. Consider a beam with a 4 mm × 40 mm cross-section. (The neutral plane is in the middle for a rectangular cross-section.)

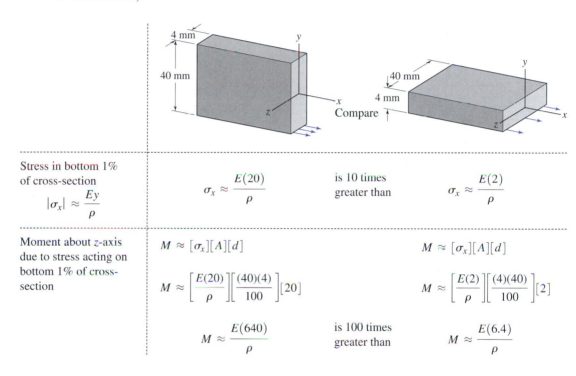

Stress in bottom 1% of cross-section $\lvert\sigma_x\rvert \approx \dfrac{Ey}{\rho}$	$\sigma_x \approx \dfrac{E(20)}{\rho}$	is 10 times greater than	$\sigma_x \approx \dfrac{E(2)}{\rho}$
Moment about *z*-axis due to stress acting on bottom 1% of cross-section	$M \approx [\sigma_x][A][d]$ $M \approx \left[\dfrac{E(20)}{\rho}\right]\left[\dfrac{(40)(4)}{100}\right][20]$ $M \approx \dfrac{E(640)}{\rho}$	is 100 times greater than	$M \approx [\sigma_x][A][d]$ $M \approx \left[\dfrac{E(2)}{\rho}\right]\left[\dfrac{(4)(40)}{100}\right][2]$ $M \approx \dfrac{E(6.4)}{\rho}$

7. Summarize how the contribution of the stress to the moment depends on the element location in the cross-section.

For a fixed material (E) and radius of curvature (ρ), the moment produced is greater for an element farther from the neutral plane because:

1. Stresses are greater $\quad \sigma = \dfrac{-Ey}{\rho}$ *y* is greater if material is farther from the neutral plane

2. Moment arm is greater $\quad M = [\sigma][A][d]$ *d* is greater if material is farther from the neutral plane

We will see that we make a beam stiff (hard to bend), by having more material farther from the neutral axis.

>>End 5.7

Points A, C, and E are located at the edges of a beam of rectangular cross-section, and points B and D are midway along the sides. The bar is bent with uniform curvature as shown.

What is the axis along which the elements experience the primary bending strain, and what is the axis along which the strain varies?

For each of the elements A, B, C, D, and E, is the primary bending strain tensile, compressive, or zero?

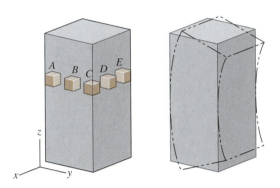

Solution

The primary bending strain is along the z-axis, and it varies along the y direction

A: Tension
B: Zero
C: Compression
D: Compression
E: Compression

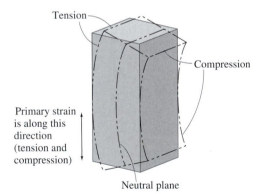

>>End Example Problem 5.11

Points A, C, and E are located at the edges of a beam of rectangular cross-section, and points B and D are midway along the sides. The beam shown bends in the x-z plane with the concave side pointed down (toward $-z$).

What is the axis along which the elements experience the primary bending strain, and what is the axis along which the strain varies?

For each of the elements A, B, C, D, and E, is the primary bending strain tensile, compressive, or zero?

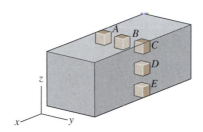

Solution

The primary bending strain is along the x-axis, and it varies along the z direction

A: Tension
B: Tension
C: Tension
D: Zero
E: Compression

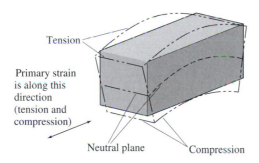

>>End Example Problem 5.12

The *x-y-z* origin is at the lower rear corner of the beam of rectangular cross-section. The beam bends into a radius of curvature of 4 m due to the application of forces by the fingers as shown.

Determine the strain along the *y*-direction at the points $(x, y, z) = (0, 50, 30)$; $(15, 50, 0)$; $(10, 50, 25)$; and $(20, 50, 5)$.

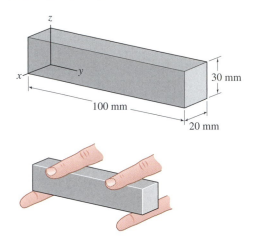

Solution

The bending strain varies along the *z*-axis, with tension on the bottom, and compression on the top (because of the positions of the fingers).

Bending strain will depend on distance in the *z*-direction from the *x-y* plane through the bar center, since the neutral plane is in the middle for rectangular cross-section.

The origin of the (x, y, z) axes is at the back corner.

$(x, y, z) = (0, 50, 30)$
$\varepsilon_y = -(15 \text{ mm})/(4000 \text{ mm}) = -3.75 \times 10^{-3}$

$(x, y, z) = (15, 50, 0)$
$\varepsilon_y = (15 \text{ mm})/(4000 \text{ mm}) = 3.75 \times 10^{-3}$

$(x, y, z) = (10, 50, 25)$
$\varepsilon_y = -(10 \text{ mm})/(4000 \text{ mm}) = -2.50 \times 10^{-3}$

$(x, y, z) = (20, 50, 5)$
$\varepsilon_y = (10 \text{ mm})/(4000 \text{ mm}) = 2.50 \times 10^{-3}$

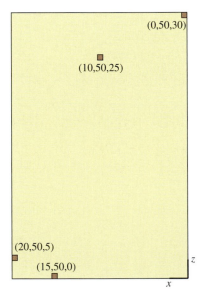

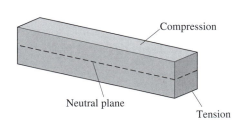

>>End Example Problem 5.13

The *x-y-z* origin is at the lower rear corner of the beam of rectangular cross-section. The beam shown bends in the *x-y* plane, with a radius of curvature of 10 m and the concave side pointed toward $+y$.

Determine the strain along the *x*-direction at the points $(x, y, z) = (200, 0, 0)$; $(250, 50, 40)$; $(200, 25, 80)$; $(250, 40, 20)$; and $(200, 5, 70)$.

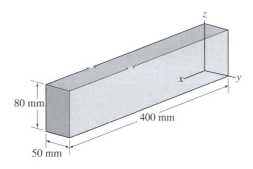

80 mm

400 mm

50 mm

Solution

The bending varies along *y* direction. Tension is where $y = 0$, and compression is where $y = 50$.

Bending strain will depend on distance in the *y*-direction from *x-z* plane through bar center, since the neutral plane is in the middle for a rectangular cross-section.

Origin of (x, y, z) axes is at the back corner.

$(x, y, z) = (200, 0, 0)$
$\varepsilon_y = (25 \text{ mm})/(10000 \text{ mm}) = 2.50 \times 10^{-3}$

$(x, y, z) = (250, 50, 40)$
$\varepsilon_y = (25 \text{ mm})/(4000 \text{ mm}) = -2.50 \times 10^{-3}$

$(x, y, z) = (200, 25, 80)$
$\varepsilon_y = (0)/(4000 \text{ mm}) = 0$

$(x, y, z) = (250, 40, 20)$
$\varepsilon_y = -(15 \text{ mm})/(4000 \text{ mm}) = -1.50 \times 10^{-3}$

$(x, y, z) = (200, 5, 70)$
$\varepsilon_y = (20 \text{ mm})/(4000 \text{ mm}) = 2.00 \times 10^{-3}$

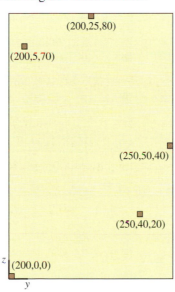

(200,25,80)

(200,5,70)

(250,50,40)

(250,40,20)

z (200,0,0)

y

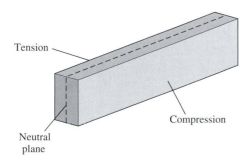

Tension

Compression

Neutral plane

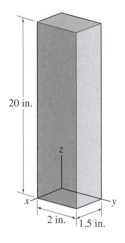

The x-y-z origin is at the lower rear corner of the beam of rectangular cross-section. The elastic modulus is 30×10^6 psi. The beam shown bends in the x-z plane with the concave side pointed toward $-x$ and a radius of curvature of 10^3 in.

For the elements at the points $(x, y, z) = (0, 2, 10)$; $(0.75, 0, 12)$; $(1.2, 1, 8)$; $(0.1, 0.1, 10)$; and $(1.5, 1.8, 10)$:

Indicate direction of stresses (a, b, c, d, e, f, or none).

Determine the magnitude of the stress and indicate if it is tensile or compressive.

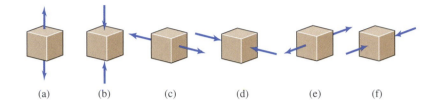

(a)   (b)   (c)   (d)   (e)   (f)

Solution

The primary bending strain is along z-axis and varies along the x direction, from compression where $x = 0$ to tension where $x = 1.5$ in.

$(x, y, z) = (0, 2, 10)$: compression; stress direction (b)
$|\sigma| = (30 \times 10^6 \text{ psi})(0.75 \text{ in.})/(1000 \text{ in.}) = 22500 \text{ psi}$

$(x, y, z) = (0.75, 0, 12)$: on neutral plane – zero stress

$(x, y, z) = (1.2, 1, 8)$: tension; stress direction (a)
$|\sigma| = (30 \times 10^6 \text{ psi})(1.2 \text{ in.} - 0.75 \text{ in.})/(1000 \text{ in.}) = 13500 \text{ psi}$

$(x, y, z) = (0.1, 0.1, 10)$: compression; stress direction (b)
$|\sigma| = (30 \times 10^6 \text{ psi})(0.75 \text{ in.} - 0.1 \text{ in.})/(1000 \text{ in.}) = 19500 \text{ psi}$

$(x, y, z) = (1.5, 1.8, 10)$: tension; stress direction (a)
$|\sigma| = (30 \times 10^6 \text{ psi})(1.5 \text{ in.} - 0.75 \text{ in.})/(1000 \text{ in.}) = 22500 \text{ psi}$

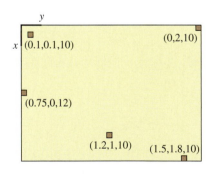

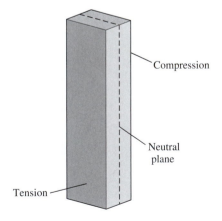

The *x-y-z* origin is at the lower rear corner of the beam of rectangular cross-section. The elastic modulus is 70 GPa. The beam shown bends due to the application of fingers as shown. The radius of curvature of 5 m.

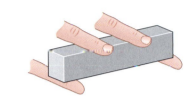

For the elements at the points (x, y, z) = (40, 0, 0); (40, 10, 15); (40, 5, 5); (40, 15, 10); and (40, 10, 20):

Indicate the direction of stresses (a, b, c, d, e, f, or none).

Determine the magnitude of the stress and indicate if it is tensile or compressive.

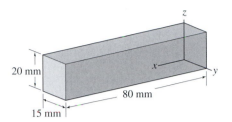

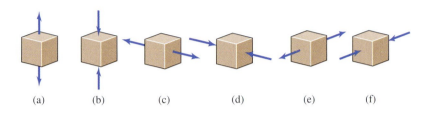

(a)   (b)   (c)   (d)   (e)   (f)

Solution

The primary bending strain is along the *x*-axis and varies along the *z* direction, from tension where $z = 0$ to compression where $z = 20$ mm.

(x, y, z) = (40, 0, 0): tension; stress direction (e)
$|\sigma| = (70 \times 10^9 \text{ Pa})(10 \text{ mm})/(5 \text{ m}) = 140 \times 10^6 \text{ Pa} = 140 \text{ MPa}$

(x, y, z) = (40, 10, 15): compression; stress direction (f)
$|\sigma| = (70 \times 10^9 \text{ Pa})(15 \text{ mm} - 10 \text{ mm})/(5 \text{ m}) = 70 \text{ MPa}$

(x, y, z) = (40, 5, 5): tension; stress direction (e)
$|\sigma| = (70 \times 10^9 \text{ Pa})(10 \text{ mm} - 5 \text{ mm})/(5 \text{ m}) = 70 \text{ MPa}$

(x, y, z) = (40, 15, 10): on neutral plane – zero stress

(x, y, z) = (40, 10, 20): compression; stress direction (f)
$|\sigma| = (70 \times 10^9 \text{ Pa})(20 \text{ mm} - 10 \text{ mm})/(5 \text{ m}) = 140 \text{ MPa}$

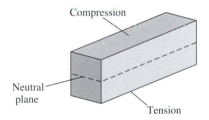

Compression

Neutral plane

Tension

>>End Example Problem 5.16

Additional data on material properties needed to solve problems can be found in Appendix D or inside back cover.

5.44–49 The bars are deformed as shown, each bent only in either the x-y, x-z, or y-z plane. Points A, C, and E are located at the edges, and points B and D are located midway along the sides. For each of the following problems, answer these questions. What is the axis (x, y, or z) along which this body has its main bending strain? Indicate if the bending strain is tensile, compressive, or zero at points A, B, C, D, and E.

5.44 Diagram of Deformation 1
5.45 Diagram of Deformation 2
5.46 Diagram of Deformation 3
5.47 Diagram of Deformation 4
5.48 Diagram of Deformation 5
5.49 Diagram of Deformation 6

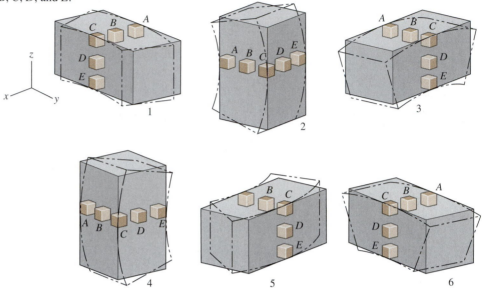

Probs. 5.44–49

5.50–55 Fingers exert forces as shown on the bars causing them to bend. Points A, C, and E are located at the edges, and points B and D are located midway along the sides. For each of the following problems, answer these questions. What is the axis (x, y, or z) along which this body has its main bending strain? Indicate if the bending strain is tensile, compressive, or zero at points A, B, C, D, and E.

5.50 Diagram of Loading 1
5.51 Diagram of Loading 2
5.52 Diagram of Loading 3
5.53 Diagram of Loading 4
5.54 Diagram of Loading 5
5.55 Diagram of Loading 6

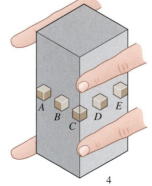
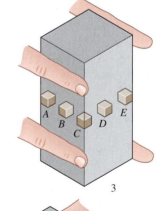
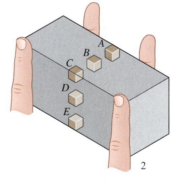
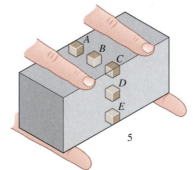
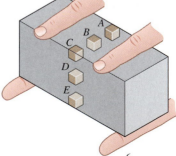

Probs. 5.50–55

5.56–61 The bars shown are 20 mm × 30 mm × 80 mm. The deformed shapes of the bars are indicated, the curvature is $\kappa = 10^{-2}\ \text{m}^{-1}$, and the elastic modulus is 200 GPa. Points A, C, and E are located at the edges, and points B and D are located midway along the sides.

For each of the following problems, compute the bending stress at points A, B, C, D, and E using the notation of positive as tensile and negative as compressive. Also, for each point where the stress is not zero, draw the stresses on an element in one of these ways.

5.56 Diagram of Deformation 1
5.57 Diagram of Deformation 2
5.58 Diagram of Deformation 3
5.59 Diagram of Deformation 4
5.60 Diagram of Deformation 5
5.61 Diagram of Deformation 6

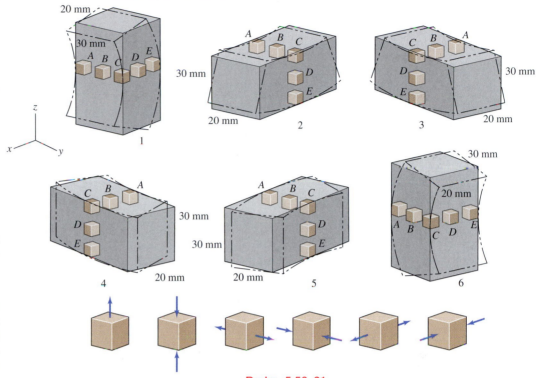

Probs. 5.56–61

5.62–67 The bars shown are 20 mm × 30 mm × 80 mm. Fingers exert forces as shown on the bars causing them to bend. The curvature is $\kappa = 10^{-2}\ \text{m}^{-1}$, and the elastic modulus is 200 GPa. Points A, C, and E are located at the edges, and points B and D are located midway along the sides.

For each of the following problems, compute the bending stress at points A, B, C, D, and E using the notation of positive as tensile and negative as compressive. Also, for each point where the stress is not zero, draw the stresses on an element in one of these ways.

5.62 Diagram of Loading 1
5.63 Diagram of Loading 2
5.64 Diagram of Loading 3
5.65 Diagram of Loading 4
5.66 Diagram of Loading 5
5.67 Diagram of Loading 6

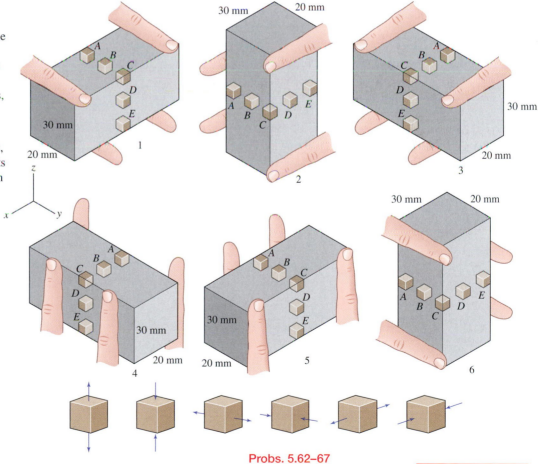

Probs. 5.62–67

>>End Problems

5.8 Bending Equations

We derive the major equations for bending by making the stress distribution just found consistent with zero net force and moment M on the beam.

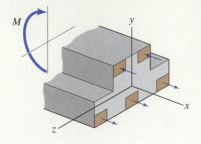 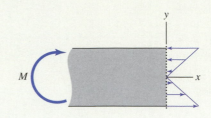

---

1. The stresses must sum to zero net force: this sets the position of the neutral axis at the centroid.

$$\text{Net Force} = \sum F_x = 0 \implies \int_A \sigma_x \, dA = \int_A \frac{-Ey}{\rho} dA = 0 \implies \int_A y \, dA = 0$$

Limit of many small area elements E and ρ are constant over area

Recall $\bar{y} = \dfrac{1}{A} \displaystyle\int_A y \, dA$ is the position of the centroid from the bottom of the cross-section.

$\implies$ To satisfy $\displaystyle\int_A y \, dA = 0$, define $y = 0$ (neutral axis) at the centroid!

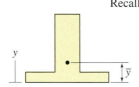

---

2. The moments due to the stresses must sum to moment M.

Moment of inertia, I, captures the contribution of a beam's cross-section to its resistance to bending, and it is defined mathematically as the integral over the cross-sectional area of the square of the distance away from the neutral plane.

$M = $ Sum of moments due to force created by stress (σ_x) on each area element (dA), times moment arm (y)

Limit of many small area elements

$$M = \int_A -y\sigma_x \, dA = \frac{E}{\rho}\int_A y^2 \, dA$$

Tension ($\sigma_x > 0$) acting on $y > 0$ produces $M < 0$ about z-axis

The geometry of the cross-section affects bending through:

$$I = \int_A y^2 \, dA$$

I is defined as the **Second Moment of Inertia** about the neutral plane.

I quantifies how distant material is on average from the neutral plane.

---

3. Recognize the significance of the terms in the relation between moment M (internal load or segment level force) and curvature ρ (segment level deformation).

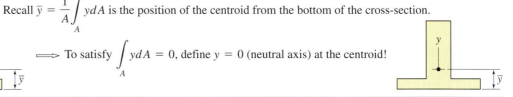

Stiffness of material   Geometry of cross-section

$$M = \frac{EI}{\rho}$$

Moment–curvature relation

---

4. Relate moment M (segment level force) and stress σ (element level force).

Combine $\sigma_x = \dfrac{-Ey}{\rho}$ and $M = \dfrac{EI}{\rho} \implies \sigma_x = \dfrac{-My}{I}$ Moment–curvature relation

The magnitude of σ_x increases with bending moment M and distance y from the neutral plane; it decreases with I.

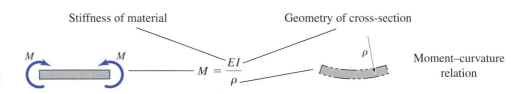

Rationalize the sign of σ_x

$M > 0$ $\sigma_x = \dfrac{-My}{I} \implies$ compression ($\sigma_x < 0$) on top of beam ($y > 0$)
tension ($\sigma_x > 0$) on bottom of beam ($y < 0$)

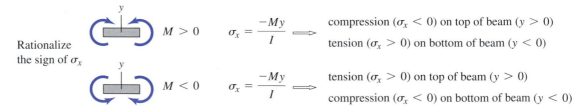

$M < 0$ $\sigma_x = \dfrac{-My}{I} \implies$ tension ($\sigma_x > 0$) on top of beam ($y > 0$)
compression ($\sigma_x < 0$) on bottom of beam ($y < 0$)

5. Calculate I for a rectangle, a very important and common cross-section.

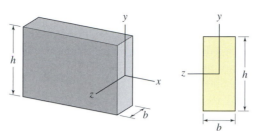

The centroid is at the center of $b \times h$ cross-section.

$$I = \int_A y^2 dA = \int_{y=-h/2}^{y=h/2} \int_{z=-b/2}^{z=b/2} y^2 dz dy =$$

$$\int_{y=-h/2}^{y=h/2} \left[y^2 z \big|_{-b/2}^{b/2} \right] dy = \frac{by^3}{3} \bigg|_{-h/2}^{h/2} = \frac{bh^3}{12}$$

In $I = \frac{1}{12}bh^3$, h is in the plane of bending, and b is perpendicular to the plane of bending.

6. Make sense of the difference between bending the ruler one way vs. the other.

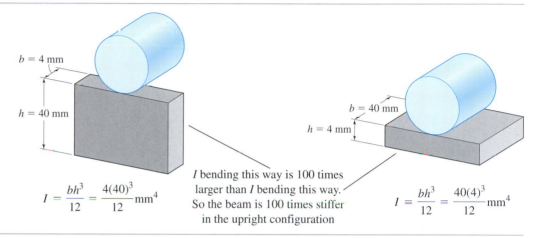

$b = 4$ mm

$h = 40$ mm

$$I = \frac{bh^3}{12} = \frac{4(40)^3}{12} \text{mm}^4$$

I bending this way is 100 times larger than I bending this way. So the beam is 100 times stiffer in the upright configuration

$b = 40$ mm

$h = 4$ mm

$$I = \frac{bh^3}{12} = \frac{40(4)^3}{12} \text{mm}^4$$

7. Calculate I for a circle, another very important and common cross-section.

$2c$

The centroid is in the center.
Use polar coordinates to find I.

Use trig identity:

$$(\sin \theta)^2 = \frac{1}{2}[1 - \cos(2\theta)]$$

$$I = \int_A y^2 dA = \int_{\theta=0}^{y=2\pi} \int_{r=0}^{x=c} (r \sin \theta)^2 r dr d\theta = \int_{\theta=0}^{\theta=2\pi} \left[\frac{r^4}{4} \bigg|_0^c \right] \frac{1}{2} (1 - \cos 2\theta) d\theta = \frac{\pi c^4}{4}$$

8. Find $\bar{y}$ and I for other common cross-sections from tables (see Appendix C).

For example, here are the area properties for the case of an isosceles triangle. Note that tables with properties of areas, such as Appendix C, commonly use x-y coordinates, rather than the y-z coordinates that form our beam cross-section.

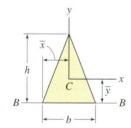

$$A = \frac{bh}{2} \qquad \bar{x} = \frac{b}{2} \qquad \bar{y} = \frac{h}{3}$$

$$I_x = \frac{bh^3}{36} \qquad I_y = \frac{hb^3}{48}$$

$$I_{xy} = 0 \qquad I_{BB} = \frac{bh^3}{12}$$

9. This analysis of bending applies only to symmetric cross-sections.

  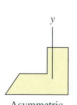 

Symmetric   Asymmetric

Because $\sigma_x = \dfrac{-Ey}{\rho}$, there would be a bending moment about the y-axis in an asymmetric cross-section

because $M_y = \int_A -\sigma z dA \neq 0$. We study asymmetric cross-sections in Section 5.13.

>>End 5.8

The steel bar of rectangular cross-section is subjected to a bending moment in one of the two directions shown. A strain gage is attached to the side of the bar as shown and measures the strain along the length of the bar. Determine the strain (a) if $M_1 = 300$ N-m and $M_2 = 0$ and (b) if $M_1 = 0$ and $M_2 = 300$ N-m. Take $E = 200$ GPa.

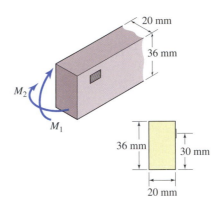

Solution

Properties of the cross-section that are needed are the centroid $\bar{y}$ and the moment of inertia I about the centroid.

The centroid is at the center of the rectangular cross-section.

I about the centroid is $I = \dfrac{bh^3}{12}$

The dimensions b and h must be chosen carefully, depending on the direction of the bending moment.

(a) For a moment in direction of M_1, the stress and strain are compressive where the strain gage is attached.

Definition of b and h for bending in direction of M_1

$(b = 20 \text{ mm}, h = 36 \text{ mm})$

$$\varepsilon_x = \frac{\sigma_x}{E} = \frac{-M_1 y}{EI} = \frac{-(300 \text{ N-m})(30 \text{ mm} - 18 \text{ mm})}{(200 \text{ GPa})\dfrac{(20 \text{ mm})(36 \text{ mm})^3}{12}} = -2.31 \times 10^{-4}$$

(b) For a moment in direction of M_2, the stress and strain are tensile where the strain gage is attached.

Definition of b and h for bending in direction of M_2

$(b = 36 \text{ mm}, h = 20 \text{ mm})$

$$\varepsilon_x = \frac{\sigma_x}{E} = \frac{-M_2 y}{EI} = \frac{-(300 \text{ N-m})(-10 \text{ mm})}{(200 \text{ GPa})\dfrac{(36 \text{ mm})(20 \text{ mm})^3}{12}} = 6.25 \times 10^{-4}$$

>>End Example Problem 5.17

The neck of a guitar has a semi-circular shape. Due to various loadings it could sustain bending moments in the two directions shown. If the compressive stress must be no larger than 1000 psi, determine the maximum allowable bending moment (a) in the direction M_1 and (b) in the direction M_2.

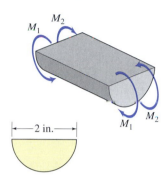

Solution

Properties of the cross-section that are needed are the centroid $\bar{y}$ and I about the centroid.

Appendix C contains geometric properties for various cross-sections. Use properties for the semi-circle.

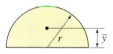

$$\bar{y} = \frac{4r}{3\pi} \qquad I = 0.1098\, r^4$$

Evaluate $\bar{y}$: $\quad \bar{y} = \frac{4r}{3\pi} = \frac{4(1\text{ in.})}{3\pi} = 0.424$ in.

Note, the centroid given is measured from the flat surface.

Evaluate I about the centroid: $\quad I = 0.1098(1\text{ in.})^4 = 0.1098$ in.^{4}

Maximum stresses will be at the top and the bottom. Determine distances of the top and the bottom from centroid (they are not equal).

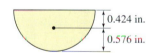

(a) For a moment in direction of M_1, the stress is compressive on the bottom ($c_1 = 0.576$ in.)

$$\sigma_x = \frac{M_1 c_1}{I} \Rightarrow M_1 = \frac{\sigma_x I}{c_1} = \frac{(1000\text{ psi})(0.1098\text{ in.}^4)}{0.576\text{ in.}} = 190.8 \text{ lb-in.}$$

(b) For a moment in direction of M_2, the stress is compressive on the top ($c_2 = 0.424$ in.)

$$\sigma_x = \frac{M_2 c_2}{I} \Rightarrow M_2 = \frac{\sigma_x I}{c_2} = \frac{(1000\text{ psi})(0.1098\text{ in.}^4)}{0.424\text{ in.}} = 259 \text{ lb-in.}$$

>>End Example Problem 5.18

Additional data on material properties needed to solve problems can be found in Appendix D or inside back cover.

5.68 A bending moment $M_1 = 150$ N-m acts on the bar with circular cross-section. Determine the maximum tensile stress if the diameter is $d = 24$ mm. At what point (A, B, C, or D) is this maximum tensile stress reached?

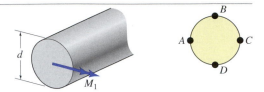

Prob. 5.68

5.69 A bending moment $M_1 = 600$ lb-in. acts on the bar with rectangular cross-section. Determine the maximum compressive stress if $a = 2.5$ in. and $w = 0.5$ in. At what point (A, B, C, or D) is this maximum compressive stress reached?

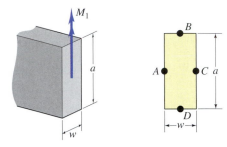

Prob. 5.69

5.70 A bending moment $M_1 = 30$ N-m acts on the bar with a thin semi-circular annular cross-section. Determine the maximum tensile stress if $R = 18$ mm and $c = 2$ mm. At what point (A or B) is this maximum tensile stress reached?

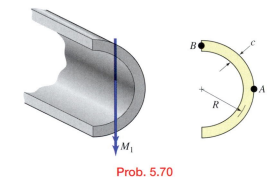

Prob. 5.70

5.71 A bending moment $M_1 = 75$ lb-in. acts on the bar with a semi-circular cross-section. Determine the maximum compressive stress if $d = 0.75$ in. At what point (A or B) is this maximum compressive stress reached?

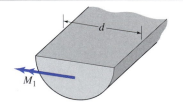
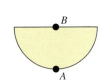

Prob. 5.71

5.72 A strain gage measuring strain along the length of the rectangular steel bar is attached to the point shown. Take the dimensions to be $h = 40$ mm, $a = 50$ mm, and $w = 14$ mm. Determine the strain that would be measured at this point for a moment of 200 N-m that acts (a) in the direction 1 and (b) in the direction 2.

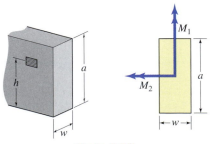

Prob. 5.72

5.73 A strain gage measuring strain perpendicular to the length of the rectangular aluminum bar is attached to the point shown (at the edge). Take the dimensions to be $h = 1.5$ in. and $a = 0.125$ in. Determine the transverse strain that would be measured by the strain gage at this point for a moment of 80 lb-in. that acts (a) in the direction 1 and (b) in the direction 2.

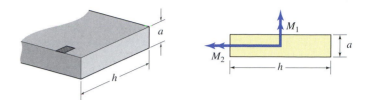

Prob. 5.73

5.74 The beam of triangular cross-section has a maximum allowable compressive stress of 175 MPa. Take the dimensions to be $c = 40$ mm and $a = 30$ mm. Determine the maximum allowable bending moment if the moment acts (a) in the direction of M_1 and (b) in the direction of M_2.

Prob. 5.74

5.75 The beam with a thin semi-circular annular cross-section has a maximum allowable tensile stress of 7000 psi. Take the dimensions to be $c = 0.125$ in. and $R = 1.2$ in. Determine the maximum allowable bending moment if the moment acts (a) in the direction of M_1 and (b) in the direction of M_2.

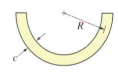

Prob. 5.75

5.76 The extruded beam with trapezoidal cross-section has a maximum allowable tensile stress of 10 MPa. Take the dimensions to be $a = 30$ mm, $b = 10$ mm, and $c = 15$ mm. Determine the maximum allowable bending moment if the moment acts (a) in the direction of M_1 and (b) in the direction of M_2.

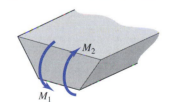

Prob. 5.76

5.77 The straight beam of semicircular cross-section with a diameter of $d = 30$ mm is bent into a circular arc of radius $c = 2.5$ m concave down, under a moment $M = 6$ N-m. Determine the elastic modulus E and the bending strain at the top A of the beam.

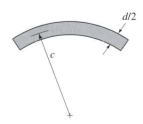

Prob. 5.77

5.78 The straight beam of rectangular cross-section with $a = 0.5$ in. and $b = 1.5$ in. is bent into a circular arc of radius $c = 30$ ft concave down, under a moment $M = 10^4$ lb-in. Determine the elastic modulus E and the bending strain at point A of the beam.

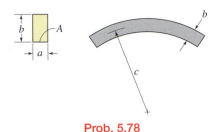

Prob. 5.78

5.79 The straight beam with a cross-section in the form of thin semi-circular arc with $c = 0.125$ in. and $R = 3$ in. is bent into a circular arc of radius $b = 25$ ft concave upward, under a moment $M = 100$ lb-ft. Determine the elastic modulus E and the bending strain at the point A of the beam.

Prob. 5.79

5.80 A standard channel cross-section C10 $\times$ 30 can be subjected to both directions of bending moments (see Appendix E). The beam height and the location of the neutral plane are shown; the relevant $I = 3.94$ in^4. Say the allowable stress in tension is 15 ksi and the allowable stress in compression is 18 ksi. What is the bending moment of the lowest magnitude at which the stress will exceed either allowable? Which bending moment (M_1 or M_2) is critical, which stress (tensile or compressive) is reached, and at what points?

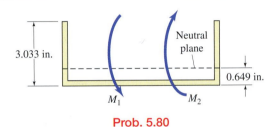

Prob. 5.80

5.81 A strain gage to measure the bending strain is attached at the bottom of the flange of standard steel wide-flange cross-section gage W12 $\times$ 87 (see Appendix E). Some dimensions of the beam are given, and the relevant moment of inertia for bending in the vertical plane is 740 in^4. When the loading of the beam shifts, the strain gage reading changes from 450×10^{-6} to 330×10^{-6}. Determine the change in bending moment that caused this change in bending strain.

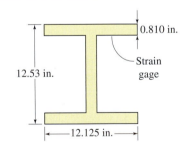

Prob. 5.81

5.82 The steel wide-flange cross-section W16 $\times$ 100 (see Appendix E) can be subjected to bending moments in either direction as shown. (a) If only moment $M_1 = 3.4(10^5)$ lb-in. acts, determine the maximum tensile stress. (b) If only moment $M_2 = 1.5(10^6)$ lb-in. acts, determine the maximum tensile stress.

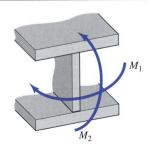

Prob. 5.82

5.83 Beams are often subjected to bending moments that change senses along the length of the beam. A standard channel cross-section C15 × 50 beam has a moment of inertia of 11.0 in^4 (see Appendix E). Some dimensions of the beam cross-section are shown, including the location of the neutral plane. (a) For a bending moment of $M_A = 2000$ lb-ft, determine the maximum tensile and compressive stresses and where in the cross-section each occurs. (b) For a bending moment of $M_B = 2800$ lb-ft, determine the maximum tensile and compressive stresses and where in the cross-section each occurs.

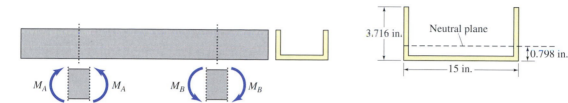

Prob. 5.83

5.84 The beam with a T-shaped cross-section built up from wood members is loaded so that the bending moment varies. Some dimensions of the beam cross-section are shown, including the location of the neutral plane. The moment of inertia is $2(10^8)$ mm^4. (a) For a bending moment of $M_A = 30$ kN-m, determine the maximum tensile and compressive stress and where in the cross-section each occurs. (b) For a bending moment of $M_B = 42$ kN-m, determine the maximum tensile and compressive stress and where in the cross-section each occurs.

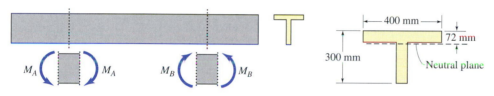

Prob. 5.84

5.85 A moment $M = 5$ N-m is applied to the plastic extrusion of the complex cross-section shown. The plastic has a Young's modulus of 2.6 GPa. Strains at A and B are measured and found to have values $\varepsilon_A = 0.0022$ and $\varepsilon_B = -0.0017$. Determine the location of the centroid of the cross-section relative to the bottom and the moment of inertia I.

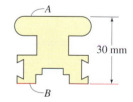

Prob. 5.85

5.86 A moment $M = 30$ lb-in. is applied to the plastic extrusion of the complex cross-section shown. The plastic has a Young's modulus of 300 ksi. Strains at A and B are measured and found to have values $\varepsilon_A = -0.0012$, and $\varepsilon_B = 0.0019$. Determine the location of the centroid of the cross-section relative to the bottom and the moment of inertia I.

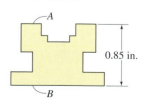

Prob. 5.86

5.87 Nanowires of various cross-sections can be produced. A nanowire of equilateral triangular cross-section 400 nm across is subjected to bending. From experiments, it is determined that the modulus of this material is 80 GPa, and that the stress to fracture the wire is 1 GPa. (a) Determine the smallest radius into which the wire can be bent before fracture. (b) Determine the bending moment to fracture the wire.

Prob. 5.87

5.88 A stiletto heel has a semi-circular cross-section with radius of 0.25 in. Due to walking over irregular surfaces and shifts in body weight, moments of up to 30 lb-in. can act where the heel attaches to the shoe. Determine the maximum tensile or compressive stress in the heel.

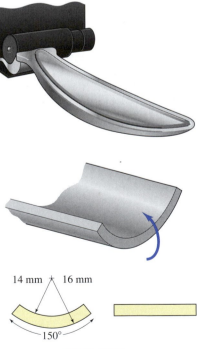

Prob. 5.88

5.89 Cross-sections of tool handles are chosen not only to be ergonomically and esthetically pleasing but also to have higher resistance to bending. Compare the curved cross-section shown with a flat rectangular cross-section that has the same thickness and area. With a bending moment acting in the direction shown, determine (a) the ratio of the maximum tensile stress in the curved handle to that in the flat handle and (b) the ratio of the curved handle's curvature due to bending to that in the flat handle.

14 mm   16 mm

150°

Prob. 5.89

5.90 Consider an external fracture fixation system under the condition that the foot is caught while the leg is trying to move up. This results in a downward 100 N force on the foot. Translate this into a bending moment on the solid carbon fiber rod that has a 10.5 mm diameter. Determine the bending stress in the rod. Take the dimensions to be $L_r = 300$ mm and $L_p = 60$ mm.

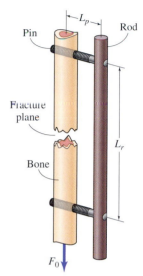

Prob. 5.90 (Appendix A5)

5.91 The cross-sectional properties of the bridge girder shown have been determined. The overall dimensions are $h = 3.5$ m and $w = 30$ m. The moment of inertia I associated with bending in the vertical plane is 4 m^4, and the neutral plane is located at $c = 1.5$ m from the top. If the maximum bending stress in tension and compression is 460 MPa, determine the maximum allowable bending moment if the factor of safety is to be 2.

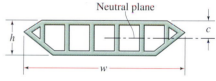

Prob. 5.91 (Appendix A2)

5.92 The wind turbine tower measures 70 m from base to the blade axis and consists of an annular cylinder with outer diameter that varies from 2.134 m at the base to 1.219 m at the top. The wall thickness is 25.4 mm. Let a steady torque of 5 MN-m be applied by the wind as it turns the blades. This torque applies a bending moment to the tower. At what cross-section of the tower, from top to bottom, does this moment create the greatest bending stress, and what is that maximum bending stress?

Prob. 5.92 (Appendix A6)

>>End Problems

5.9 Bending of Composite Cross-Sections

We often fabricate members loaded in bending to have complex cross-sections, such as this aluminum extrusion. In general, the properties $\bar{y}$ and I for complex cross-sections are not tabulated, and integration may be difficult or tedious. But if the cross-section is the sum or difference of simple shapes with tabulated properties, $\bar{y}$ and I can be easily determined.

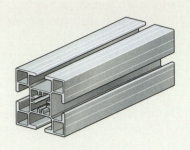

---

1. As a first step in finding its centroid, try to add and subtract simple shapes to make a complex cross-section.

---

2. What is $\bar{y}$ for a composite cross-section in general?

We must integrate ydA over the area A to find $\bar{y}$. We can integrate ydA over each individual shape and add or subtract the results. But, integrating ydA over each shape gives $\bar{y}$ for that shape, times its area:

$$\bar{y} = \frac{1}{A}\left[\int_{A_1+A_2-A_3} ydA\right] = \frac{1}{A}\left[\int_{A_1} ydA + \int_{A_2} ydA - \int_{A_3} ydA\right] = \frac{[A_1\bar{y}_1 + A_2\bar{y}_2 - A_3\bar{y}_3]}{A} = \frac{[A_1\bar{y}_1 + A_2\bar{y}_2 - A_3\bar{y}_3]}{[A_1 + A_2 - A_3]}$$

---

3. Here we compute $\bar{y}$ for the particular composite cross-section shown above.

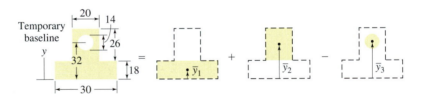

Find centroid for each shape	$\bar{y}_1 = 9$ mm	$\bar{y}_2 = 18 + 13 = 31$ mm	$\bar{y}_3 = 32$ mm
Find area for each shape	$A_1 = 540$ mm^2	$A_2 = 520$ mm^2	$A_3 = \pi(14)^2/4 = 153.9$ mm^2

Find area of entire cross-section $\quad A = A_1 + A_2 - A_3 = 906$ mm^2

Find $\bar{y}$ of entire cross-section $\quad \bar{y} = \dfrac{A_1\bar{y}_1 + A_2\bar{y}_2 - A_3\bar{y}_3}{A} = 17.72$ mm

---

4. What is I for a composite cross-section in general?

Recall $\sigma_x = \dfrac{-Ey}{\rho}$ where $y = 0$ at the centroid. So we want $I = \displaystyle\int_A y^2 dA$

calculated with $y = 0$ at the centroid $\bar{y}$ of the *entire* cross-section.

For a cross-section that is a composite of simple shapes, the integral is a sum or difference of integrals.

$$I = \int_{A_1+A_2-A_3} y^2 dA = \int_{A_1} y^2 dA + \int_{A_2} y^2 dA - \int_{A_3} y^2 dA = I_1 + I_2 - I_3$$

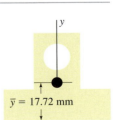

$\bar{y} = 17.72$ mm

5. Find I_1, I_2, I_3 recalling that each is the moment of inertia about the centroid of the *entire* cross-section.

The I tabulated for a simple shape (e.g., Appendix C) is usually about the centroid of *that* shape.

Say we want $I_1 = \int_{A_1} y^2 dA$, but with $y = 0$ at $\bar{y}$.

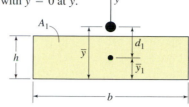

Let I_{c1} be the tabulated I of A_1 about its centroid $\bar{y}_1$.

Use *Parallel Axis Theorem* (see Appendix B):
$I_1 = I_{c1} + A_1 d_1^2$
where $d_1 = |\bar{y} - \bar{y}_1|$

Follow the example of A_1, which is a rectangle: $I_{c1} = \dfrac{bh^3}{12}$ is tabulated I about $\bar{y}_1$.

We must add a term $A_1 d_1^2$ to I_{c1}
$A_1 = bh$
$d_1 = |17.72 \text{ mm} - 18/2 \text{ mm}|$

To find I for each individual area, we add its I_c and its Ad^2. To form the total I for the whole area, the moments of inertia for individual areas are added or subtracted, depending on whether that area is present or absent from the composite cross-section.

6. Here we compute I for the particular composite cross-section shown above.

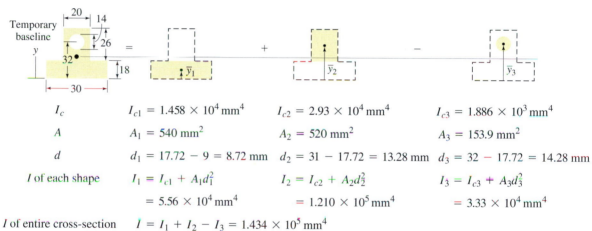

I_c	$I_{c1} = 1.458 \times 10^4 \text{ mm}^4$	$I_{c2} = 2.93 \times 10^4 \text{ mm}^4$	$I_{c3} = 1.886 \times 10^3 \text{ mm}^4$
A	$A_1 = 540 \text{ mm}^2$	$A_2 = 520 \text{ mm}^2$	$A_3 = 153.9 \text{ mm}^2$
d	$d_1 = 17.72 - 9 = 8.72 \text{ mm}$	$d_2 = 31 - 17.72 = 13.28 \text{ mm}$	$d_3 = 32 - 17.72 = 14.28 \text{ mm}$
I of each shape	$I_1 = I_{c1} + A_1 d_1^2$	$I_2 = I_{c2} + A_2 d_2^2$	$I_3 = I_{c3} + A_3 d_3^2$
	$= 5.56 \times 10^4 \text{ mm}^4$	$= 1.210 \times 10^5 \text{ mm}^4$	$= 3.33 \times 10^4 \text{ mm}^4$

I of entire cross-section $\quad I = I_1 + I_2 - I_3 = 1.434 \times 10^5 \text{ mm}^4$

>>End 5.9

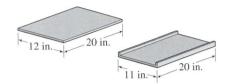

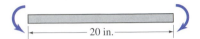

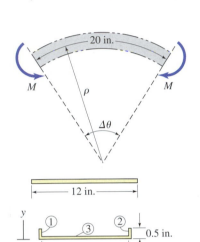

Compare the flexibility of two aluminum baking sheets, each 0.0625 in. thick by determining the angle of bending $\Delta\theta$ due to 50 in.-lb bending moments applied at their far edges. Compare (a) the flat 12 in. × 20 in. sheet and (b) the same size sheet, but with 0.5 in. folded up along the two 20 in. edges.

Solution

Bending produces an arc of radius ρ. The arc angle $\Delta\theta = (20 \text{ in.})/\rho$.

Use the moment–curvature relation: $M = \dfrac{EI}{\rho}$

(a) For the rectangular cross-section (unfolded):

$$I = \frac{bh^3}{12} = \frac{(12 \text{ in.})(0.0625 \text{ in.})^3}{12} = 2.44 \times 10^{-4} \text{ in.}^4$$

$$\Delta\theta = \frac{20 \text{ in.}}{\rho} = \frac{(20 \text{ in.})M}{EI} = \frac{(20 \text{ in.})(50 \text{ lb-in.})}{(10^7 \text{ psi})(2.44 \times 10^{-4} \text{ in.})} = 0.410 \text{ rad} = 23.5°$$

(b) For the cross-section with folded edges:

Find centroid of the composite cross-section.

$\bar{y}_1 = \bar{y}_2 = 0.25$ in. and $A_1 = A_2 = (0.5 \text{ in.})(0.0625 \text{ in.})$
$\bar{y}_3 = (0.0625 \text{ in.})/2$ and $A_3 = (11 \text{ in.})(0.0625 \text{ in.})$

$$\bar{y} = \frac{2(0.0625 \text{ in.})(0.5 \text{ in.})(0.25 \text{ in.}) + (0.0625 \text{ in.})(11 \text{ in.})(0.03125 \text{ in.})}{2(0.0625 \text{ in.})(0.5 \text{ in.}) + (0.0625 \text{ in.})(11 \text{ in.})}$$

$$= 0.0496 \text{ in.}$$

Use the parallel axis theorem to find I for each region and combine:

$$I = 2I_1 + I_3 = 2[(0.0625 \text{ in.})(0.5 \text{ in.})^3/12 + (0.0625 \text{ in.})(0.5 \text{ in.})(0.25 \text{ in.} - 0.0495 \text{ in.})^2]$$
$$+ [(11 \text{ in.})(0.0625 \text{ in.})^3/12 + (0.0625 \text{ in.})(11 \text{ in.})(0.03125 \text{ in.} - 0.0495 \text{ in.})^2] = 4.27 \times 10^{-3} \text{ in.}^4$$

$$\Delta\theta = \frac{20 \text{ in.}}{\rho} = \frac{(20 \text{ in.})M}{EI} = \frac{(20 \text{ in.})(50 \text{ lb-in.})}{(10^7 \text{ psi})(4.27 \times 10^{-3} \text{ in.})} = 0.0234 \text{ rad} = 1.342°$$

Notice that the folded sheet is far stiffer than the flat sheet ($I = 4.27 \times 10^{-3}$ in.^{4} vs. $I = 2.44 \times 10^{-4}$ in.^{4}). In the folded sheet, there is material from all regions 1, 2, and 3 that is relatively far from the neutral plane. By comparison, material in the flat sheet is no more than 0.03125 in. from the neutral plane.

>>End Example Problem 5.19

Concrete can withstand very little bending because it is weak in tension. The block shown has three equally spaced holes, each 30 mm in diameter. If the tensile stress is limited to 2 MPa, determine the maximum bending moment that can be applied in directions (a) and (b).

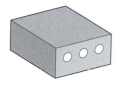
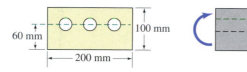
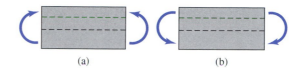

Solution

To determine stress for a given bending moment, cross-sectional properties, $\bar{y}$ and I, are needed.

Note that the three holes are identical and are located at same y-position.

$$\bar{y} = \frac{(200\text{ mm})(100\text{ mm})(50\text{ mm}) - 3\pi(15\text{ mm})^2(60\text{ mm})}{(200\text{ mm})(100\text{ mm}) - 3\pi(15\text{ mm})^2} = 48.81\text{ mm}$$

The removed material is above the center; so as expected, the centroid (neutral axis) is below center.

Use the parallel axis theorem to find I for each region and combine.

$$I = I_1 - 3I_2 = (200\text{ mm})(100\text{ mm})^3/12 + (200\text{ mm})(100\text{ mm})(50\text{ mm} - 48.81\text{ mm})^2$$
$$- 3[\pi(15\text{ mm})^4/4 + \pi(15\text{ mm})^2(60\text{ mm} - 48.81\text{ mm})^2] = 1.631 \times 10^7\text{ mm}^4$$

Use the relation between bending moment and stress

$$\sigma_x = \frac{-My}{I} \Rightarrow M = \frac{-\sigma_x I}{y}$$

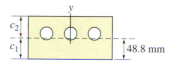

(a) Moment causes tension on the bottom ($y = -c_1$).

$$M = \frac{-\sigma_x I}{y} = \frac{-(2 \times 10^6\text{ N/m}^2)(1.631 \times 10^{-5}\text{ m}^4)}{(-0.0448\text{ m})} = 728\text{ N-m}$$

(b) Moment causes tension on the top ($y = c_2$).

$$M = \frac{-\sigma_x I}{y} = \frac{-(2 \times 10^6\text{ N/m}^2)(1.631 \times 10^{-5}\text{ m}^4)}{(0.0552\text{ m})} = -591\text{ N-m}$$

When the side with tension is closer to the neutral axis (case (a)), it requires a larger moment to produce the same tension. The difference between the two moments would be larger if the removed material (holes) shifted the centroid and neutral plane $\bar{y}$ more from the center.

>>End Example Problem 5.20

Additional data on material properties needed to solve problems can be found in Appendix D or inside back cover.

5.93 The steel beam with cross-section shown is bent to have a radius of curvature of $c = 45$ m. Determine (a) the strain at point A and (b) the bending moment that must be applied.

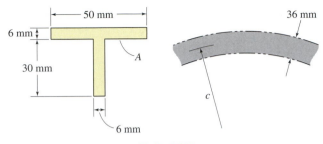

Prob. 5.93

5.94 The aluminum beam with cross-section shown is bent to have a radius of curvature of $b = 32$ ft. Determine (a) the strain at point A and (b) the bending moment that must be applied.

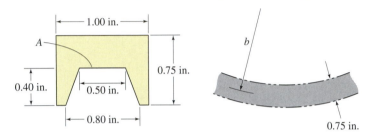

Prob. 5.94

5.95 The molded plastic beam ($E = 3.2$ GPa) has a cross-section in the form of a triangle with a hole of 8 mm in diameter tangent to the lower surface. The beam is bent to have a curvature of $c = 35$ m. Determine (a) the maximum tensile and compressive stresses and (b) the bending moment that must be applied.

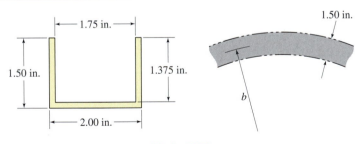

Prob. 5.95

5.96 A steel sheet 0.125 in. thick is folded to form a beam with the cross-section shown. The beam is bent to have a curvature of $b = 120$ ft. Determine (a) the maximum tensile and compressive stresses and (b) the bending moment that must be applied.

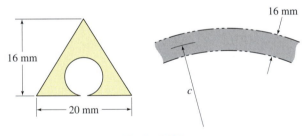

Prob. 5.96

5.97 A cast concrete form with the cross-section shown has holes that are 80 mm in diameter. A bending moment $M_0 = 10^3$ N-m is applied to the form. Determine the maximum tensile and compressive stresses in the concrete.

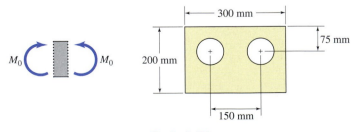

Prob. 5.97

5.98 A wood beam is built up from two boards nailed together. The beam is subjected to the bending moment shown with $M_0 = 7000$ lb-in. Determine the maximum tensile and compressive stresses in the wood beam.

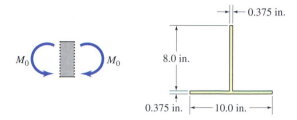

Prob. 5.98

5.99 The steel beam with the cross-section shown is subjected to bending. If the tensile stress is not to exceed 150 MPa, determine the maximum allowable bending moment assuming both senses are possible. Which is the sense of the bending moment (M_A or M_B) at which the maximum tensile stress is reached, and where in the cross-section is it reached?

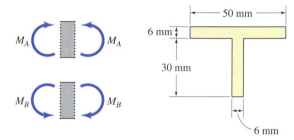

Prob. 5.99

5.100 The aluminum beam with the cross-section shown is subjected to bending. If the compressive stress is not to exceed 20 ksi, determine the maximum allowable bending moment assuming both senses are possible. Which is the sense of the bending moment (M_A or M_B) at which the maximum compressive stress is reached, and where in the cross-section is it reached?

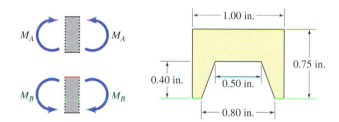

Prob. 5.100

5.101 The molded plastic beam ($E = 3.2$ GPa) has a cross-section in the form of a triangle with a hole of 8 mm in diameter tangent to the lower surface. The beam is loaded so that the bending moment varies. (a) For a bending moment of $M_A = 1.5$ N-m in the sense shown, determine the maximum tensile and compressive stress and where each occurs. (b) For a bending moment of $M_B = 2.2$ N-m in the sense shown, determine the maximum tensile and compressive stress and where each occurs.

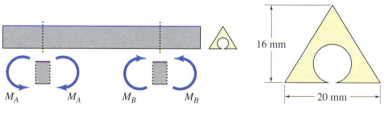

Prob. 5.101

5.102 A steel sheet 0.125 in. thick is folded to form a beam with the cross-section shown. The compressive stress during bending is not to exceed 10 ksi. (a) If the bending is in the direction of M_A, determine the maximum allowable bending moment. (b) If the bending is in the direction of M_B, determine the maximum allowable bending moment.

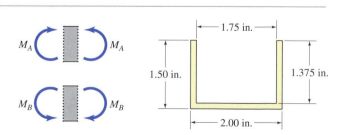

Prob. 5.102

5.103 A cast concrete form ($E = 4$ GPa) with cross-section shown has holes that are 80 mm in diameter. A beam with this cross-section form is seen to sag and develop a curvature consistent with the bending moment M_0 shown. Say the tensile stress in the concrete is not to exceed 3 MPa. (a) Determine the maximum curvature in the sag. (b) Determine the maximum bending moment M_0.

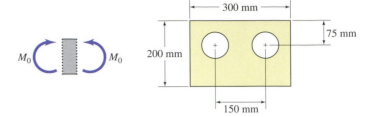

Prob. 5.103

5.104 One design of a tennis racquet, an application in which stiffness is critical, consists of titanium ($E = 17(10)^6$ ksi) having the hollow cross-section shown. Approximate the triangle as equilateral, with an outer dimension of 1 in. on each side, and a wall thickness of 0.0625 in. Under a bending moment of 100 lb-in., determine the change in angle over a 20 in. length.

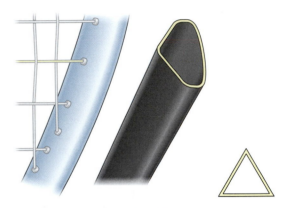

Prob. 5.104

5.105 Extruded plastic snow poles are formed with the cross-section shown. The outer diameter is 1.3 in., the outer wall thickness is 0.09 in., and cross-members have a thickness of 0.06 in. Let the plastic have a modulus of 430 ksi. The pole is bent so that over 6 ft, its angle changes by 30°. Determine (a) the bending moment in the pole and (b) the maximum stress.

Prob. 5.105

5.106 A 1 m long inflated cylindrical rubber balloon has an outer diameter of $d_0 = 40$ mm and a thickness of 0.5 mm. A bending moment of 0.5 N-m is required to bend the balloon into a quarter circle. Determine the effective modulus of the rubber in this inflated state.

5.107 Cardboard paper towel rolls are an example of using a highly efficient cross-section to create an acceptably stiff product with small amounts of a relative low stiffness material. A roll 12 in. long has a 1.5 in. outer diameter and a thickness of 0.0625 in. Compare this to an unrolled sheet 12 in. long that would come from slitting the roll along its length and unfurling it. Say both are subjected to a bending moment of 5 lb-in. Take the elastic modulus to be 1.5×10^6 psi, typical of paper and wood products. Determine the angle change due to bending over the 12 in. length for (a) the roll and (b) the flat sheet.

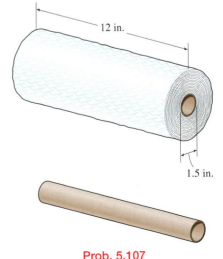

Prob. 5.107

5.108 Each of the fork blades on a bicycle has a cross-section that is modeled with an elliptical outer boundary and a uniform wall thickness. Forks are loaded primarily in the plane of the tire by loads transmitted from the tire to the axle. Stresses in the fork blades are associated predominantly with bending. If we seek to limit the bending stress in this carbon fiber composite fork blade to 85 ksi, what is the maximum allowable bending moment on the blade? Take the dimensions to be $a = 0.75$ in., $b = 1.25$ in., and $t = 0.1$ in.

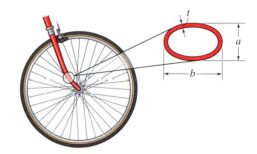

Prob. 5.108 (Appendix A1)

5.109 The girder supporting the bridge consists of a box section. The outer height and width are $h = 4$ m and $w = 28$ m. The upper plate and lower plates are each 20 mm thick, the 5 webs are each 15 mm thick, and the holes are of equal size. (a) Determine the moment of inertia I for this cross-section associated with bending of the girder in the vertical plane (due to its weight and cable tensions). (b) Compare I with the estimate based on having only the upper and lower plates at their current positions.

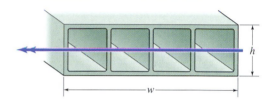

Prob. 5.109 (Appendix A2)

5.110 Exercise equipment must also be designed for unexpected usages, such as hanging on the equipment with one's full weight. Say someone hanging on the swinging arm in this pectoral fly machine produces a bending moment of 1500 lb-in. on the arm. If the cross-section is as shown, determine the bending stress at the points A and B indicated. Take the dimensions to be $a = 1.5$ in., $b = 2$ in., and $t = 0.25$ in.

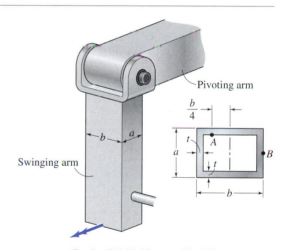

Prob. 5.110 (Appendix A4)

>>End Problems

5.10 Bending Stresses Under a Non-Uniform Bending Moment

We now bring together the analysis of shear force and bending moment from the first part of this chapter with the analysis of stress in this second part.

1. Use the bending moment at a cross-section to calculate stresses and curvature at that cross-section.

Strain and stress in bending were studied for the case of a uniform bending moment, which occurs when opposite bending moments are applied at the ends of a beam. We have seen, however, that the shear force and bending moment, $V(x)$ and $M(x)$, for a general transverse loading, vary with x along the beam. How are strains and stresses in bending affected?

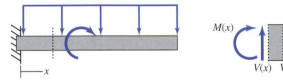

By comparing predictions to experimental results for long slender members, one finds that the bending stress, σ_x, and the curvature, κ, are still given, to a very good approximation, by

$$\sigma_x = \frac{-M(x)y}{I} \qquad \kappa = \frac{M(x)}{EI}$$

Except, now we use the bending moment $M(x)$ that pertains to the cross-section of interest. So κ can vary with x along the beam. σ_x depends on x because of $M(x)$, and on the distance y from the neutral plane.

The shear force, $V(x)$, negligibly affects σ_x and κ. The shear stresses due to $V(x)$ will be studied in Sections 5.14 and 5.15.

---

2. Consider the variation of the bending moment along the beam to determine the location of the maximum bending stress.

These two cases provide a contrast as to the location of the maximum stresses relative to the applied load.

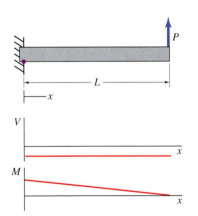
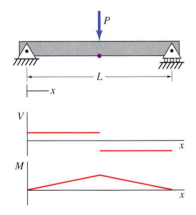

Bending moment, and hence stress, is maximum at the support, not near the applied load. Maximum tensile stress due to bending is at •.

Bending moment, and hence stress, is maximum at the center, near the applied load. Maximum tensile stress due to bending is at •.

There is no general rule about whether stresses are maximum near or far from an applied load. Instead, always consider the internal load, which is the bending moment in the case of bending.

3. When the distances from the neutral plane to the top and bottom of the beam differ, carefully consider both positive and negative bending moments.

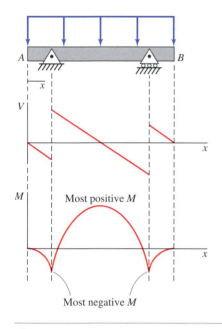

Some loadings, such as that at the left, produce both positive and negative $M(x)$. We pay particular attention to this when the distances from the neutral plane to the top and bottom of the beam differ. Then, positive and negative moments of the same magnitude will produce different maximum tensile and compressive stresses.

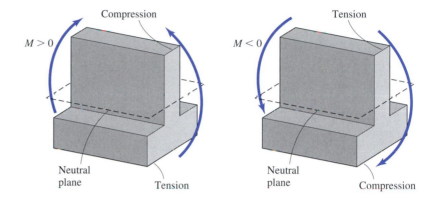

If we are interested in only the maximum tensile or compressive stress in a beam, we might not find it at the cross-section where $|M|$ is maximum. A lesser moment of the opposite sign might produce a greater tensile or compressive stress, because distance from the neutral plane also affects bending stress.

4. Analyze bending in the *x-z* plane similarly to bending in the *x-y* plane, but recognize that the neutral plane and the moment of inertia differ.

We have focused primarily on bending in the *x-y* plane. Although we designated the moment as simply M, it is a bending moment about the *z*-axis: M_z. The same ideas apply when the bending moment acts instead about the *y*-axis: M_y. In that case, the neutral plane is the *x-y* plane, and the stress varies with the coordinate z (away from the neutral plane). Tension occurs on the side of the beam corresponding to $z > 0$, provided $M_y > 0$. A different moment of inertia I_y, in contrast to I_z, applies to bending in the *x-z* plane.

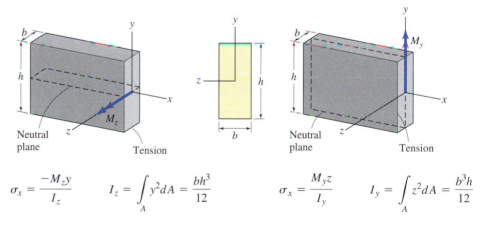

$$\sigma_x = \frac{-M_z y}{I_z} \qquad I_z = \int_A y^2 dA = \frac{bh^3}{12} \qquad \sigma_x = \frac{M_y z}{I_y} \qquad I_y = \int_A z^2 dA = \frac{b^3 h}{12}$$

When moments M_y and M_z act simultaneously, the normal stress σ_x is the sum of the normal stresses due to the individual moments. As we will see in Chapter 6, we can add normal stresses due to two loadings if those normal stresses act on the same plane. We will study bending of a non-symmetric cross-section in Section 5.13. Bending in the *x-y* and *x-z* planes will then be coupled.

>>End 5.10

To compute bending stresses, the ski in this elliptical machine is modeled as a beam pinned at its two ends. The person's weight of 280 lb acts at the point shown, and the reactions at the supports are given. The ski is 3 in. wide and 0.375 in. thick. (a) Determine the strain at the top of the ski at a distance 18 in. from the left end. (b) Determine the maximum tensile stress due to bending and its location. The elastic modulus is 30×10^6 psi.

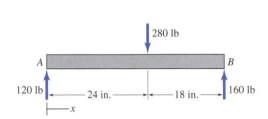

Solution

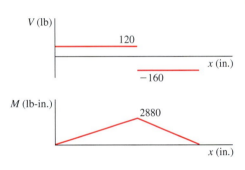

Shear force and bending moment diagrams have been found for this type of problem before.

To evaluate the bending stress at any point (x, y), we determine M at the cross-section (x) and then substitute into $\sigma = \dfrac{-My}{I}$

Strain can be found from $\varepsilon = \dfrac{\sigma}{E}$

For this rectangular cross-section, we find

$$I = \frac{bh^3}{12} = \frac{(3 \text{ in.})(0.375 \text{ in.})^3}{12} = 1.318 \times 10^{-2} \text{ in.}^4$$

At $x = 18$ in., $M = (120 \text{ lb})(18 \text{ in.}) = 2160$ lb-in.

The top of the beam is at $y = 0.5\,(0.375 \text{ in.}) = 0.1875$ in.

$$\sigma = \frac{-My}{I} = \frac{-(2160 \text{ lb-in.})(0.1875 \text{ in.})}{1.318 \times 10^{-2} \text{ in.}^4} = -3.07 \times 10^4 \text{ psi} \Rightarrow \varepsilon = \frac{\sigma}{E} = -1.024 \times 10^{-3}$$

$M > 0$ everywhere, so the top is in compression and the bottom in tension.

The maximum tensile stress is at the bottom of the beam at the cross-section where M is *maximum*. By inspection of the diagram, $M_{\max} = (2880 \text{ lb-in.})$ at $x = 24$.

$$\sigma = \frac{-My}{I} = \frac{-(2880 \text{ lb-in.})(-0.1875 \text{ in.})}{1.318 \times 10^{-2} \text{ in.}^4} = 4.10 \times 10^4 \text{ psi}$$

Notice that one cannot find the maximum M, and hence the maximum stress, by setting $dM/dx = 0$. The maximum moment is often at a point where the derivative is not defined (the slope is discontinuous). The maximum M must be found from the bending moment diagram.

>>End Example Problem 5.21

A concrete form with holes of 3 in. diameter overhangs a support. Concrete weighs 145 lb/ft^3. Determine the longest overhang (b) that can be tolerated if the tension due to the weight is not to exceed 100 psi. Treat the overhanging portion as cantilevered where it meets the support.

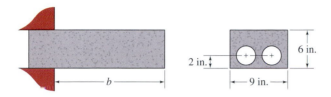

Solution

Since the cross-section is uniform, the force per length is uniform and equal to the weight per length.

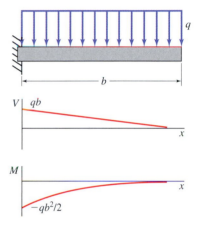

Shear force and bending moment diagrams have been found earlier for this type of problem.

To find the force per length and to compute bending stresses, we analyze the cross-section:

$$A = A_1 - A_2 = (6 \text{ in.})(9 \text{ in.}) - 2\pi(1.5 \text{ in.})^2 = 39.9 \text{ in.}^2$$

$$\bar{y} = \frac{(6 \text{ in.})(9 \text{ in.})(3 \text{ in.}) - 2\pi(1.5 \text{ in.})^2(2 \text{ in.})}{(6 \text{ in.})(9 \text{ in.}) - 2\pi(1.5 \text{ in.})^2} = 3.35 \text{ in.} \quad \text{(from the bottom)}$$

$$I = I_1 - 2I_2 = (9 \text{ in.})(6 \text{ in.})^3/12 + (6 \text{ in.})(9 \text{ in.})(3 \text{ in.} - 3.35 \text{ in.})^2$$
$$- 2[\pi(1.5 \text{ in.})^4/4 + \pi(1.5 \text{ in.})^2(2 \text{ in.} - 3.35 \text{ in.})^2] = 134.9 \text{ in.}^4$$

$$q = \frac{\text{Total Weight}}{\text{Length}} = \frac{\text{Weight}}{\text{Volume}}\frac{\text{Volume}}{\text{Length}} = \frac{\text{Weight}}{\text{Volume}}(\text{area}) = 145\frac{\text{lb}}{\text{ft}^3}\left(\frac{\text{ft}}{12 \text{ in.}}\right)^3(39.9 \text{ in.}^2) = 3.34 \text{ lb/in.}$$

$$M_{\text{max}} = -q(b^2)/2 = -(3.34 \text{ lb/in.})b^2/2$$

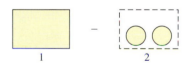

$M < 0$, so tension is on top of the beam at $y = 6$ in. $- 3.35$ in. $= 2.65$ in.

$$\sigma = \frac{-My}{I} = \frac{-(-3.34 \text{ lb/in.})(b^2/2)(2.65 \text{ in.})}{134.9 \text{ in.}^4} = 100 \text{ psi} \Rightarrow b = \sqrt{\frac{2(134.9 \text{ in.}^4)100 \text{ psi}}{(3.34 \text{ lb/in.})(2.65 \text{ in.})}} = 55.2 \text{ in.}$$

>>End Example Problem 5.22

Additional data on material properties needed to solve problems can be found in Appendix D or inside back cover.

5.111 The circular beam, with a diameter of $d = 28$ mm, is loaded as shown with $q = 400$ N/m. Determine the maximum tensile and compressive stresses at the cross-section 0.3 m from the right end of the beam. (Take the lengths to be $L_1 = 0.2$ m and $L_2 = 0.5$ m.)

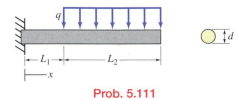

Prob. 5.111

5.112 The rectangular beam, with dimensions of $c = 3.25$ in. and $w = 0.75$ in., is loaded as shown with $M_0 = 80$ lb-ft. Determine the maximum tensile and compressive stresses (a) at the cross-section 1.5 ft from the left end of the beam and (b) at the cross-section 1 ft from the right end of the beam. (Take the lengths to be $L_1 = 3$ ft and $L_2 = 2$ ft.)

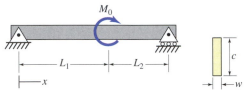

Prob. 5.112

5.113 The beam of trapezoidal cross-section, with dimensions of $a = 30$ mm, $b = 10$ mm, and $c = 15$ mm, is loaded as shown with $P = 200$ N. Determine the maximum tensile stress and maximum compressive stress in the beam. (Take the lengths to be $L_1 = 700$ mm and $L_2 = 400$ mm.)

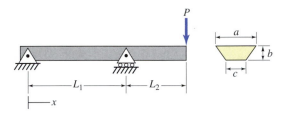

Prob. 5.113

The beam with rectangular cross-section that is loaded as shown pertains to the following problems:

5.114 Determine the bending stress (a) at the cross-section 0.5 m from the support at the point A and (b) at the cross-section 1.5 m from the support at the point B.

5.115 Determine (a) the maximum tensile stress in the beam and (b) the maximum compressive stress in the beam. In each case, indicate at what point(s) the maximum is reached, being specific as to the cross-section(s) and location(s) within the cross-section.

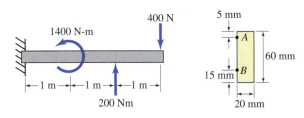

Probs. 5.114–115

The loading depicted pertains to the following two problems with beams having different cross-sections:

5.116 The beam has the T-shaped cross-section shown. Determine (a) the maximum tensile stress in the beam and (b) the maximum compressive stress in the beam. In each case, indicate at what point(s) the maximum is reached, being specific as to the cross-section(s) and location(s) within the cross-section.

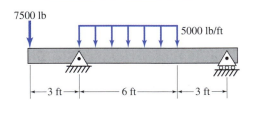

5.117 The beam has a standard C-shaped cross-section with the dimensions shown ($I = 11.0$ in.[4]). Determine (a) the maximum tensile stress in the beam and (b) the maximum compressive stress in the beam. In each case, indicate at what point(s) the maximum is reached, being specific as to the cross-section(s) and location(s) within the cross-section.

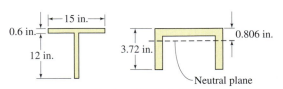

Probs. 5.116–117

5.118 The stepped circular shaft changes diameter from 1.25 in. to 0.75 in. Take the load P to be 100 lb. Determine the maximum tensile stress in the shaft, and determine the location of this maximum for two cases: (a) $L_1 = 12$ in. and $L_2 = 6$ in., and (b) $L_1 = 15$ in. and $L_2 = 3$ in. Neglect any stress concentration associated with the transition from one cross-section to the other.

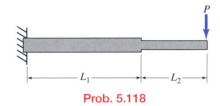

Prob. 5.118

5.119 The stepped circular shaft changes diameter. Take the load P to be 5000 N. Determine the maximum tensile stress in the shaft, and determine the location of this maximum for two cases: (a) the shaft diameter jumps from 20 mm to 28 mm, and (b) the shaft diameter jumps from 23 mm to 26 mm. Neglect any stress concentration associated with the transition from one cross-section to the other. (Take the lengths to be $L_1 = 150$ mm, $L_2 = 120$ mm, and $L_3 = 160$ mm.)

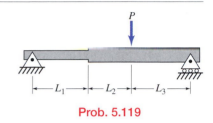

Prob. 5.119

5.120 The wood beam BCD is supported as shown and has a rectangular cross-section 5.5 in. high and 3.5 in. wide. Let the lengths be $L_1 = 6$ ft and $L_2 = 3$ ft and the angle be $\alpha = 50°$. If the bending stress is not to exceed 1200 psi, determine the maximum allowable distributed force per length q.

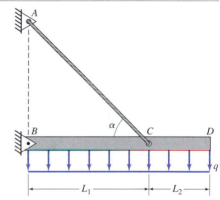

Prob. 5.120

5.121 A cable BD is used to support the post that is subjected to a lateral force F_0 at its top. Let the lengths be $L_1 = 4$ m and $L_2 = 2.5$ m and the angle be $\alpha = 35°$. Take the post to be a hollow pipe with an outer diameter of 50 mm and a wall thickness of 3 mm. If the bending stress is not to exceed 80 MPa, determine the maximum allowable force F_0.

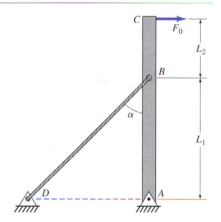

Prob. 5.121

5.122 Three plates, each of width $w = 18$ mm, are held together with an 8 mm diameter bolt. Approximate the bolt as subjected to the distributed loading shown. If the force $P = 1000$ N, determine the maximum bending stress in the bolt.

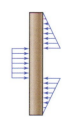

Prob. 5.122

5.123 High stresses in bicycles can occur during periods of severe braking. Let the normal force and frictional braking forces from the ground be $F_N = 180$ lb and $F_B = 90$ lb. A reasonable analysis for braking is to neglect the rotational inertia of the wheel, in which case the brake pads exert forces that balance the moment of the braking force about the wheel center. Determine the brake-pad force, the force of the axle on the wheel, and therefore the force of the axle on the fork blades. Assuming the force of the axle is divided equally between the two blades, determine the bending stress at the cross-section of the fork indicated. Approximate the fork as emanating radially from the axle at an angle of 72° from the horizontal, and the brake pad force as perpendicular to the fork (so tangentially to the wheel). The tire has a radius of 13.5 in., the brake pad is located at 12.75 in. from the axle, and the fork cross-section of interest is 14.5 in. from the axle. Approximate the fork as straight, radiating from the center of the wheel. Take the dimensions to be $a = 0.75$ in., $b = 1.25$ in., and $t = 0.1$ in.

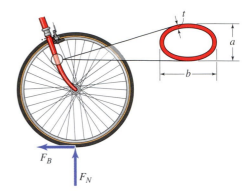

Prob. 5.123 (Appendix A1)

5.124 Pedaling produces bending and twisting of the spindle. Consider only the transverse loading (bending) and ignore the torsion. If the bending stress in the spindle is to remain below 500 MPa, what must be the maximum pedal force? There is no force applied to the other pedal, and the chain tensions act in the same plane as the bearing, so they do not contribute to the bending. The spindle is hollow with an outer diameter of 27 mm and an inner diameter of 21 mm. Take the dimensions to be $a = 70$ mm, $b = 25$ mm, $c = 9.5$ mm, $d = 80$ mm, and $L = 170$ mm.

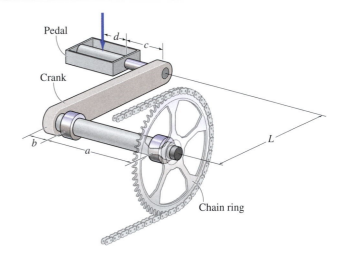

Prob. 5.124 (Appendix A1)

5.125 If held stiff, the arms can apply significant forces to the drop bars when biking on a rough road. The aluminum handlebars are tubular, with an outer diameter of $D = 25$ mm and a wall thickness of 1.7 mm. Consider bending and neglect the tendency for the forces to twist the central portion of the bar. If bending stresses are to be limited to 350 MPa, what is the maximum force each hand can apply to the drop bars? Evaluate the stress just outside the region where the handlebar tube is held by the stem. Take the dimensions to be $a = 16.8$ mm, $b = 200$ mm, $d = 29$ mm, and $R = 100$ mm.

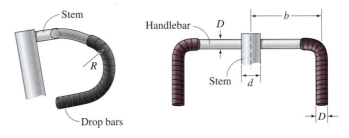

Prob. 5.125 (Appendix A1)

5.126 The downward force on the right pedal results in a complicated loading of the crank. Consider the bending stresses at the cross-section a distance x from the pedal spindle (ignore torsion). The crank is approximated as rectangular in cross-section. If the yield stress is 60 ksi, what is the factor of safety against yielding if the foot applies a force of 250 lb? Assume the force is applied in the middle of the pedal of width a. Take the dimensions to be $a = 3.625$ in., $b = 0.375$ in., $c = 1.5$ in., $w = 0.5$ in., and $x = 6$ in.

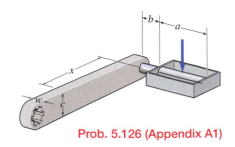

Prob. 5.126 (Appendix A1)

5.127 Determine the stress in the girder in this long center span ($L = 1400$ m), neglecting the forces due to the supporting cables. The dead load is 30 Mg/m, and I associated with bending in the vertical plane is 3.8 m^4. The centroid is located midway between the top and bottom of the 4 m high girder. Assume the span is simply supported at the pylons and that the loading due to the short side spans can be neglected. The stresses will be unbearably high—this shows that the cables are critical and the bending resistance of the girder in a long span bridge is nearly negligible compared to the support of the cables.

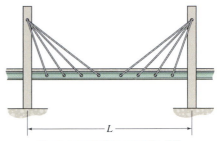

Prob. 5.127 (Appendix A2)

5.128 Sideways wind loads exert a drag force per length D on the bridge deck given by

$$D = 0.5\, \rho U^2\, h_p C_D$$

where ρ is the density of air, U is the wind velocity, h_p is the vertical projected height of the deck, and C_D is the drag coefficient. Say the relevant density of air is 1.2 kg/m^3, the vertical projected height is 3.5 m, and the drag coefficient is 0.9, assuming the wind direction is horizontal. Under a 40 m/s wind, what would be the maximum bending stress in the $L = 1400$ m span if the girder has width $w = 30$ m and an I of 200 m^4? Treat the deck as able to pivot at the pylons (pinned), and neglect the restoring forces of the pre-tensioned cables.

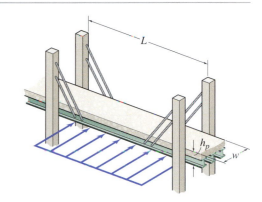

Prob. 5.128 (Appendix A2)

5.129 The cable-stayed bridge has a single plane of cables running from the tower to the middle of the roadway (only one cable is shown). The towers are balanced by the tensions on its two sides. You are to consider the effect of an accident in which the cable shown slips from its anchorage at the deck. Say the slipped cable was carrying 250 kN tension and that changes in the tensions of other cables are neglected. The tower consists of two box sections $c_1 = 12$ m $\times$ $c_2 = 10$ m of 40 mm thick steel plate. Determine the additional bending stress that is induced at the base of the tower due to the loss of cable tension. Take the dimensions to be $w = 100$ m, $h_1 = 150$ m, and $h_2 = 60$ m.

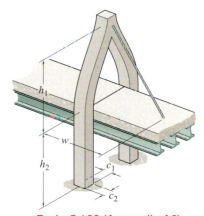

Prob. 5.129 (Appendix A2)

5.130 In using this pectoral fly machine, the hand grips and applies a force to the solid 1 in. diameter steel handle as shown. If a maximum bending stress of 15000 psi is allowed in the handle, what is the maximum allowable force from the hand? Consider the stress only in the vertical part of the handle and assume that the force of the hand is approximated as uniformly distributed over the gripped section of length w. Take the dimensions to be $w = 4$ in. and $h = 5$ in.

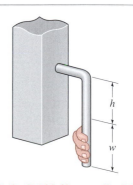

Prob. 5.130 (Appendix A4)

5.131 In using this pectoral fly machine, the hand grips and applies a force $F_0 = 40$ lb to the handle as shown. Determine the maximum bending stress at the top of the swinging arm, below the pin connection. Take the dimensions to be $w = 8$ in., $L = 25$ in., $a = 1.5$ in., $b = 2$ in., and $t = 0.25$ in.

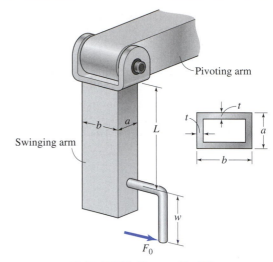

Prob. 5.131 (Appendix A4)

5.132 The simplified external fracture fixation system consists of a carbon fiber rod secured by a pin to each fragment. Say that the patient is subjected to a lateral force $F_0 = 100$ N at the level of the foot pushing the foot outward. Determine the maximum bending stress in the carbon fiber rod at (a) just above the lower pin and (b) just below the upper pin. Let the diameter of the carbon fiber rod be 10.5 mm. Take the dimensions to be $L_r = 300$ mm, $L_s = 200$ mm, and $L_p = 80$ mm.

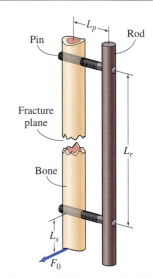

Prob. 5.132 (Appendix A5)

5.133 This type of intramedullary nail is connected to the fractured bone by screws. Say an 8 N-m torque is applied to the bone far from fracture plane. This results in bending of the screws. Approximate the bone as circular, with an outer diameter of 26 mm and an inner diameter of 20 mm. The nail itself has an outer diameter of 10 mm and an inner diameter of 5.4 mm. For a screw of 4 mm minimum diameter, determine the factor of safety with respect to a tensile yield stress of 700 MPa. Neglect any stress concentrations due to the threads. State your assumptions regarding how loads are transmitted to the screw.

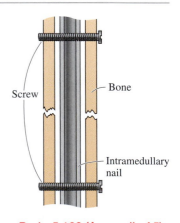

Prob. 5.133 (Appendix A5)

5.134 The wind turbine tower measures $h = 70$ m from base to the blade shaft, and consists of an annular cylinder with an outer diameter that varies from 2.134 m at the base to 1.219 m at the top. The wall thickness is 25.4 mm. The plane of the blades is located at $s = 5$ m from the centerline of the tower. One failure mode that must be designed for is the loss of a blade. Let the loss of a blade correspond to horizontal force $F_{blade} = 60$ kN in the plane of the blades perpendicular to the blade shaft. Consider only the bending of the tower due to such a load, and determine the maximum bending stress in the wall. Where in the tower is this maximum bending stress reached?

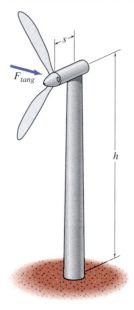

Prob. 5.134 (Appendix A6)

5.135 Let a wind turbine blade be subjected to a wind pressure $p = 800$ Pa. Say the blade is 35 m long and has a hollow cross-section that is approximated as uniform and elliptical with semi-major and semi-minor axes of 450 mm and 150 mm and a wall thickness of 25 mm. Let the air pressure act perpendicularly to the wide (450 mm) face. (Real wind turbines have blades with cross-sections that vary and twist along their length.) Estimate the bending stress that acts at the base of the blade.

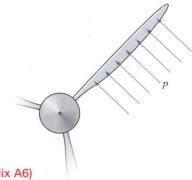

Prob. 5.135 (Appendix A6)

5.136 A steadily blowing wind tends to deflect the blades of a wind turbine out of its plane. The loading is a complex combination found from considering the drag as it depends on air-to-blade velocity and the varying blade cross-section. Let this loading vary approximately linearly from zero at the root (base) of the blade to a maximum at its tip. Say the maximum force per length at the tip of the 25 m blade is 3 kN/m. Determine the maximum stress at the root of the blade, which has a cross-section approximated as an ellipse with outer dimensions of 1.4 m and 0.7 m and a uniform wall thickness of 30 mm.

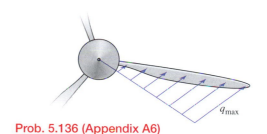

Prob. 5.136 (Appendix A6)

5.137 The stresses in a small wind turbine are believed to be significant when initially raising it up using a gin pole. The tower is a uniform steel pipe with a 2.875 in. outer diameter and a wall thickness of 0.2 in. The turbine weighs 90 lb. Including the effect of the pipe weight, the turbine weight and the cable tension, determine the maximum bending stress in the pipe. $L_g = 15$ ft, $L_1 = 37$ ft, and $L_2 = 5$ ft.

Prob. 5.137 (Appendix A6)

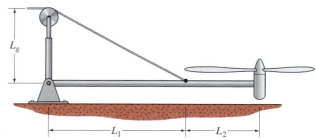

5.11 Dependence of Stiffness and Strength on Cross-Section

Bending stiffness, or the moment to produce a given (unit) amount of curvature, is equal to $M/\kappa = EI$. It depends on the material (E) and on the cross-section (I). Although the stiffness of the material (E) can vary widely, often the engineer can most strongly influence bending stiffness through I. Here we study what makes I high or low.

1. Increase I, and make the beam stiffer, by putting material farther from the neutral axis.

For a thin rectangular beam, the bending stiffness (I) is very different in the two directions. In the stiffer direction, more of the material is farther (y) from the neutral axis compared to the softer direction (z).

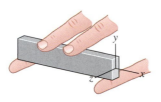

$$I_z = \int_A y^2 dA \text{ (bending in } x\text{-}y \text{ plane)}$$
(y has larger values)

$$I_y = \int_A z^2 dA \text{ (bending in } x\text{-}z \text{ plane)}$$
(z has smaller values)

$$\int_A y^2 dA \gg \int_A z^2 dA$$

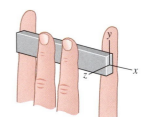

Higher stiffness   Lower stiffness

2. Compare the stiffness (I) of these beams, all with the same area. Use the logic of putting material far from the neutral axis.

A   B   C   D   E

I_z for bending in x-y plane (mm^4)

I_y for bending in x-z plane (mm^4)

If loads are primarily in one direction (e.g., only bending in the x-y plane), then make it much stiffer in that direction (choose Ⓑ). But recognize that the beam will be very compliant (soft) in the other direction.

If comparable bending loads occur in both directions, then choose cross-sections with more equal stiffnesses (Ⓒ or Ⓓ).

Solid shapes Ⓐ, Ⓑ, Ⓒ are not as stiff as hollow shapes, but are typically less expensive to manufacture.

If resistance to twisting is important, recall that torsional stiffness requires material to be moved away from the center in all directions: Ⓒ is stiffer than Ⓑ, and Ⓓ is stiffer than Ⓔ.

3. Hollow cylinders are also very stiff for their area, because material is far from the neutral axis.

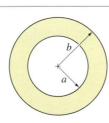

$$I = \int_A y^2 dA = \int_{\theta=0}^{y=2\pi} \int_{r=a}^{x=b} (r\sin\theta)^2 r\,dr\,d\theta = \frac{\pi(b^4 - a^4)}{4}$$

40 mm   35 mm   19.36 mm

Example: This hollow tube is 7.53 times stiffer than solid one of same area.

4. Bending strength depends on the cross-section, in particular, on I and on the height of the beam.

Section modulus, S, captures the relation between the bending moment and the maximum resulting bending stress in the beam. S depends on the shape of the cross-section, in particular, on the moment of inertia, I, and on the distance from the neutral plane to the most remote points in the cross-section.

The maximum bending stresses are at the top and bottom $y = \pm c$

$$\sigma_x = \frac{-My}{I} \qquad \sigma_{max} = \frac{Mc}{I}$$

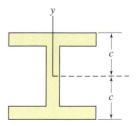

Define **section modulus**, S, as the ratio of the moment to the maximum stress.

$$S = \frac{M}{\sigma_{max}} \Rightarrow S \equiv \frac{I}{c}$$

If S of a beam is known, then the moment to reach a critical stress, σ_{crit}, is $M_{crit} = S\sigma_{crit}$.

Bending strength, M_{crit}, will depend on geometry (I and c through S) and on the material (through σ_{crit}).

Some manufacturers of complex extruded products tabulate the value of S for each shape.

---

5. Section modulus of a rectangular cross-section, like the stiffness, depends more strongly on the dimension in the plane of bending.

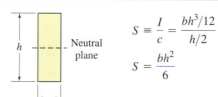

$$S \equiv \frac{I}{c} = \frac{bh^3/12}{h/2}$$

$$S = \frac{bh^2}{6}$$

By putting material farther from the neutral axis, we raise I and reduce curvature, for given M. But for a given curvature, σ_{max} is also higher because material is farther from the neutral plane.

So, while h has a stronger effect than b, its effect on $S(\sim h^2)$ is less than on $I(\sim h^3)$.

---

6. When the neutral plane is not midway between the top and bottom, there are two section moduli.

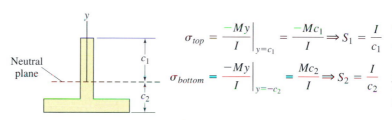

$$\sigma_{top} = \frac{-My}{I}\bigg|_{y=c_1} = \frac{-Mc_1}{I} \Rightarrow S_1 = \frac{I}{c_1}$$

$$\sigma_{bottom} = \frac{-My}{I}\bigg|_{y=-c_2} = \frac{Mc_2}{I} \Rightarrow S_2 = \frac{I}{c_2}$$

The maximum tensile stress and maximum compressive stress have different magnitudes, so the section moduli on the top and bottom differ.

---

7. If the strength (or maximum allowable stress) in tension and compression differ, then consider using a cross-section with a neutral plane not at its vertical center.

For example, let the bending moment be $M > 0$ (otherwise the opposite conclusion is reached below).

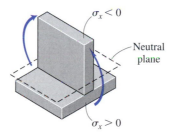

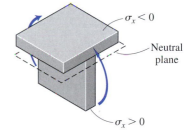

If strength in compression is greater, orient the beam this way. Since the top is farther from the neutral plane, the maximum compressive stress there is greater than the maximum tensile stress at the bottom.

If strength in tension is greater, orient the beam this way. Since the bottom is farther from the neutral plane, the maximum tensile stress there is greater than the maximum compressive stress at the bottom.

>>End 5.11

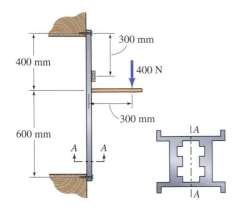

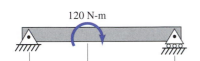

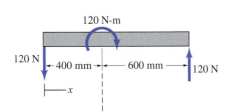

A portion of a shelving unit is constructed from an extruded aluminum profile. The shelf supports a 400 N load with center of gravity located 300 mm from the beam center. In the manufacturer's data sheets, $E = 70$ GPa and the profile is listed as having properties $I = 6.051$ cm^4 and $S = 2.689$ cm^3 for bending in the plane through A-A. Model the connections at the top and bottom as corresponding to simple supports. (a) Determine the maximum tensile bending stress in the extrusion and its location. (b) Determine the strain reading from a strain gage attached lengthwise to the front face of the profile at 300 mm from the top.

Solution

The aluminum extrusion is modeled as a simply supported beam with a concentrated moment $(400 \text{ N})(0.3 \text{ m}) = 120$ N-m, applied at the point where the shelf is attached.

Support reactions and shear force and bending moment diagrams can be found as for previous similar problems.

Note that the bending moment changes sign, and so the side that has tension or compression will change along the beam.

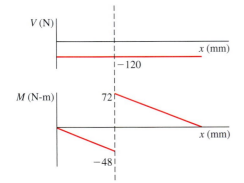

Since the profile is symmetric, both the maximum tensile and compressive stresses will occur at the cross-section where the bending moment has its maximum magnitude.

Just to the right of the applied moment, $M = 72$ N-m

$$\sigma = \frac{|M|}{S} = \frac{(72 \text{ N-m})}{2.689 \text{ cm}^3\left(\dfrac{1 \text{ m}^3}{10^6 \text{ cm}^3}\right)} = 26.8 \text{ MPa}$$

At $x = 300$ mm, $M = -(120 \text{ N})(0.3 \text{ m}) = -36$ N-m. The front face, where the strain gage is attached, is in tension.

$$\sigma = \frac{|M|}{S} = \frac{(36 \text{ N-m})}{2.689 \text{ cm}^3\left(\dfrac{1 \text{ m}^3}{10^6 \text{ cm}^3}\right)} = 13.39 \text{ MPa} \Rightarrow \varepsilon = \frac{\sigma}{E} = 1.913 \times 10^{-4}$$

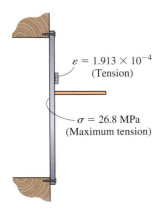

The overhanging beam consists of a built-up cross-section as shown. Determine the maximum tensile and compressive bending stresses in the beam.

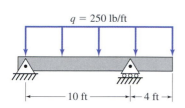

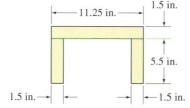

Solution

Support reactions are found and then the shear force and bending moments are drawn as in earlier problems.

There are both positive and negative bending moments. The maxima are $M(4.2 \text{ ft}) = +2205$ lb-ft and $M(10 \text{ ft}) = -2000$ lb-ft.

The cross-section is not symmetric about the *x-z* plane, so the neutral axis may not be in the center. In such cases, one must consider both maximum positive and negative bending moments.

Analyze the cross-section.

$$\bar{y} = \frac{(11.25 \text{ in.})(1.5 \text{ in.})(6.25 \text{ in.}) + 2(1.5 \text{ in.})(5.5 \text{ in.})(2.75 \text{ in.})}{(11.25 \text{ in.})(1.5 \text{ in.}) + 2(1.5 \text{ in.})(5.5 \text{ in.})} = 4.52 \text{ in.}$$

$$I = I_1 + 2I_2 = (11.25 \text{ in.})(1.5 \text{ in.})^3/12 + (11.25 \text{ in.})(1.5 \text{ in.})(6.25 \text{ in.} - 4.52 \text{ in.})^2$$
$$+ 2[(1.5 \text{ in.})(5.5 \text{ in.})^3/12 + (1.5 \text{ in.})(5.5 \text{ in.})(4.52 \text{ in.} - 2.75 \text{ in.})^2] = 147.0 \text{ in.}^4$$

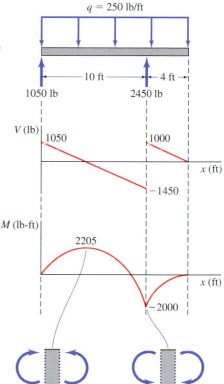

At $x = 4.2$ ft, $(M = 2205$ lb-ft):

Top: $\quad \sigma_x = \dfrac{-My}{I} = \dfrac{-(2205 \text{ lb-ft})(7 \text{ in.} - 4.52 \text{ in.})}{147.0 \text{ in.}^4} \dfrac{12 \text{ in.}}{1 \text{ ft}} = -447$ psi

Bottom: $\quad \sigma_x = \dfrac{-My}{I} = \dfrac{-(2205 \text{ lb-ft})(-4.52 \text{ in.})}{147.0 \text{ in.}^4} \dfrac{12 \text{ in.}}{1 \text{ ft}} = 814$ psi

At $x = 10$ ft, $(M = -2000$ lb-ft):

Top: $\quad \sigma_x = \dfrac{-My}{I} = \dfrac{-(-2000 \text{ lb-ft})(7 \text{ in.} - 4.52 \text{ in.})}{147.0 \text{ in.}^4} \dfrac{12 \text{ in.}}{1 \text{ ft}} = 405$ psi

Bottom: $\quad \sigma_x = \dfrac{-My}{I} = \dfrac{-(-2000 \text{ lb-ft})(-4.52 \text{ in.})}{147.0 \text{ in.}^4} \dfrac{12 \text{ in.}}{1 \text{ ft}} = -738$ psi

Notice that the maximum stresses are not necessarily at the cross-section of maximum bending moment. The tensile stress is greatest where $M = 2205$ lb-ft, but the compressive stress is greatest where $M = -2000$ lb-ft. In both cases, the maximum stress is at the bottom. When the distances of the neutral plane to the top (2.48 in.) and the bottom (4.52 in.) differ enough, a higher stress can occur at a cross-section where the bending moment is not maximum.

>>End Example Problem 5.24

Additional data on material properties needed to solve problems can be found in Appendix D or inside back cover.

5.138 (a) Derive a general expression for the section modulus of a beam with a solid circular cross-section of diameter d. (b) Determine the section modulus if $d = 30$ mm.

5.139 Determine the bending stiffness of a beam that is to bend by 1 degree over 2 m length under a bending moment of 20 N-m. Say the beam is to have a solid square cross-section. For each of these two cases, determine the beam width (equal to height) and the mass of the beam per meter (a) steel ($E = 200$ GPa, $\rho = 7830$ kg/m^3) and (b) plastic ($E = 3$ GPa, $\rho = 1000$ kg/m^3).

5.140 The manufacturer of plastic extruded parts has tabulated the following section moduli for the part shown: $S_A = 2.25$ cm^3 and $S_B = 1.85$ cm^3. Say the plastic resin used has a maximum allowable stress of 20 MPa in tension and compression. Determine the maximum allowable bending moment and the point where stress maximum is reached.

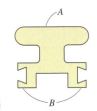

Prob. 5.140

5.141 A beam has a semicircular cross-section of diameter $d = 1.25$ in. Calculate the section modulus associated with (a) the maximum stress at the top and (b) the maximum stress at the bottom.

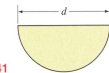

Prob. 5.141

5.142 A beam has a trapezoidal cross-section with dimensions $a = 25$ mm, $b = 20$ mm, and $c = 15$ mm. Calculate the section modulus associated with (a) the maximum stress at the top and (b) the maximum stress at the bottom.

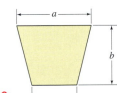

Prob. 5.142

5.143 The molded plastic beam with elastic modulus $E = 2.8$ GPa has a cross-section in the form of a triangle with a hole of 10 mm in diameter tangent to the lower surface. (a) Calculate the bending stiffness of a beam with this cross-section. (b) How long must a beam of this cross-section and material be if a 0.5 N-m moment causes a 20° angle change over the beam length?

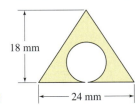

Prob. 5.143

5.144 A cast concrete form has a rectangular cross-section with a semicircular cylindrical cavity ($w = 12$ in., $h = 8$ in., and $d = 10$ in.). Bending of long beams with this cross-section can occur. The concrete has an elastic modulus $E = 3(10^6)$ psi. Calculate the bending stiffness of this concrete form.

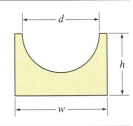

Prob. 5.144

5.145 An extruded plastic part ($E = 3$ GPa) has a rectangular cross-section (8 mm high × 4 mm wide) with a 3 mm diameter hole. Determine the bending stiffness assuming the hole is centered at: (a) the beam center and (b) 2.5 mm from the top.

5.146 An extruded plastic part ($E = 3$ GPa) has a rectangular cross-section (8 mm high × 4 mm wide) with a 3 mm diameter hole. Determine the section moduli assuming the hole is centered at: (a) the beam center and (b) 2.5 mm from the top. (For the off-center hole, find the two section moduli corresponding to the stress at the top and at the bottom.)

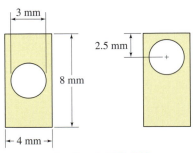

Probs. 5.145–146

5.147 The resin reinforced with chopped fibers is found to have compressive strength of 60 MPa and tensile strength of only 45 MPa. To minimize use of material, we seek to design a trapezoidal beam cross-section at which the tensile and compressive strengths are reached simultaneously when the bending moment reaches 40 N-m in the direction shown. Assume that the wider base $b = 30$ mm. (a) Determine the required section moduli corresponding to the top and bottom. (b) Find the width a at which the tensile and compressive strengths are reached simultaneously. (c) Find the beam height, to the nearest 1 mm, which just keeps the maximum stresses below the allowable values.

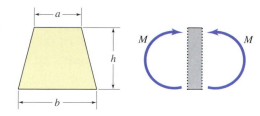

Prob. 5.147

5.148 Consider the redesign of a beam that has a solid circular cross-section with a diameter of $d = 30$ mm. As a lighter weight alternative, consider a hollow tube that has a bending stiffness equivalent to the original tube, but with 60% of the original mass. Determine the inner and outer diameters of the required tube.

5.149 Consider the redesign of a steel beam that has a rectangular cross-section with a height of 1 in. and a width of 0.5 in. The new lighter beam also has a ratio of outer height to outer width of 2 to 1 and a centered hole with a 2 to 1 height-to-width ratio. The new beam is to have 40% of the original mass, but the same stiffness as the solid bar. Determine the dimensions h_o, b_o, h_i, and b_i of the new cross-section.

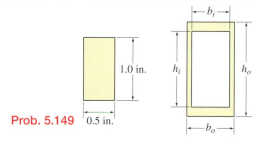

Prob. 5.149

Focused Application Problems

5.150 In designing a drill string, it is found that the bending strength of successive elements (pipes and collars) should not change too suddenly. In particular, the section moduli of adjacent elements should have a ratio of less than 5.5. In one string there is a series of drill collars, with outer and inner diameters of 7 in. and 2.81 in. Above these collars are heavy wall drill pipes with outer and inner diameters of 4.5 in. and 2.75 in. Determine the ratio of drill collar section modulus to drill pipe section modulus.

5.151 To lighten the swinging arm of this pectoral fly machine, it is proposed to reduce the wall thickness from $t = 0.25$ in. to 0.20 in., while retaining the same outer dimensions. Say the steel in the original design had a yield strength of 50 ksi and a factor of safety of 3. Consider stresses due to the bending moment shown. What should be the yield strength of the steel in the new design if the same factor of safety is to be maintained? Take the dimensions to be $a = 1.5$ in. and $b = 2$ in.

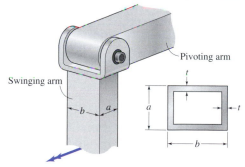

Prob. 5.151 (Appendix A4)

5.152 The padded beam is expected to sustain bending moments of up to 3000 lb-in. The beam has an outer diameter $D_1 = 1.5$ in. The beam strength is considered adequate if it offers a factor of safety of 2 against the yield strength of 50 ksi. Consider whether a hollow cylinder with a wall thickness of (a) 0.25 in., (b) 0.125 in., and (c) 0.0625 in. would be sufficiently strong.

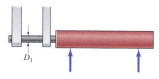

Prob. 5.152 (Appendix A4)

>>End Problems

5.12 Bending of a Beam Composed of Multiple Layers

Beams can be composed of layers of different materials bonded together. Typically, the layers all have the same width, but may have different elastic moduli and different thicknesses. Here we determine the curvature of a layered beam, and the stresses in the individual layers, given the bending moment applied to the beam. We explain the procedure assuming there are just two layers, but the approach is general.

1. **The axial strain varies linearly with *y* through the thickness of the layered beam, just as for a single material.**

A layered beam subject to pure bending (equal and opposite bending moments) bends into a circular arc with curvature κ. Following the argument in Section 5.6, the strain varies linearly through the thickness. There is one longitudinal plane where the strain is zero, the neutral plane, at a location to be determined.

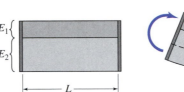

In this case, the neutral plane passes through the lower part which experiences tension and compression. The upper part is only in compression.

2. **Calculate the stress at any point within a layer from the strain at that point and the modulus of the layer.**

We approximate the stress to be uniaxial, and so it is proportional to elastic modulus and strain ($\sigma_x = E\varepsilon_x$). Since the elastic modulus changes discontinuously, the stress jumps across the plane between two layers even though the strain varies continuously. For the beam shown, the upper layer (1) has a larger elastic modulus than the lower layer (2).

$$\sigma_x = \begin{cases} -E_1\,\kappa y & \text{for } y \text{ in layer 1} \\ -E_2\,\kappa y & \text{for } y \text{ in layer 2} \end{cases}$$

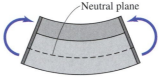

E_1 higher $\Rightarrow$ Slope σ_x vs. y higher

E_1 lower $\Rightarrow$ Slope σ_x vs. y lower

σ_x jumps

3. **Determine the position of the neutral plane by ensuring that the stresses sum to zero net force.**

Each layer has a rectangular cross-section with the same width b, but different modulus and height. The stress must be integrated over the height of each layer.

Net axial force =

$$\sum F_x = 0 \Rightarrow \int_A \sigma_x dA = -\kappa\left[\int_{A_1} E_1 y\, dA + \int_{A_2} E_2 y\, dA\right] = -\kappa\left[\int E_1 by\, dy + \int E_2 by\, dy\right] = 0$$

This equation will be used to determine the vertical position of the neutral plane.

4. Find the neutral plane by transforming the beam into one with layers of different widths, but with the same elastic modulus.

In each integral over y, E and b are constants. As long as the product $E_1 b$ has the correct value, the values of E_1 and b individually can be altered with no effect on the result.

So keep the product $E_1 b$ constant, while changing E_1 to E_2. The width b must be changed to a *transformed* value, b_t.

$$E_1 b = \frac{E_2}{E_2} E_1 b = E_2 \frac{E_1}{E_2} b = E_2 b_t \Rightarrow b_t = \frac{E_1}{E_2} b$$

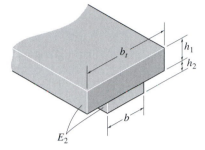

In the transformed, layered beam the upper layer has been replaced by one of modulus E_2 and width b_t. If the upper layer is stiffer $(E_1 > E_2)$, as depicted here, the transformed width b_t is larger than b.

Now the layered beam has uniform modulus E_2, but with a transformed cross-section in which the layers have different widths.

$$E_2 \left[\int_{A_1} b_t y \, dy + \int_{A_2} b y \, dy \right] = 0 \Rightarrow E_2 \int_A y \, dA = 0 \Rightarrow \int_A y \, dA = 0$$

We can find the centroid of a beam of uniform modulus, with a cross-section formed by a combination of simple shapes, by following the method of Section 5.9.

In the formula for stresses, y is measured from the centroid of this transformed cross-section.

5. Add the moments due to stresses on individual elements dA to find the total moment.

$$M = \int_A -\sigma y \, dA = \int_A E\kappa y^2 \, dA = \int_{A_1} E_1 \kappa y^2 \, dA + \int_{A_2} E_2 \kappa y^2 \, dA = \int E_1 b \kappa y^2 \, dy + \int E_2 b \kappa y^2 \, dy$$

As described above, we replace $E_1 b$ with $E_2 b_t$. Note that $dA = b_t dy$ in 1 and $dA = b \, dy$ in 2.

$$M = E_2 \kappa \left[\int y^2 b_t \, dy + \int y^2 b \, dy \right] = E_2 \kappa \left[\int_{A_t} y^2 \, dA \right] = E_2 \kappa I_t$$

where I_t is the second moment of inertia of the transformed cross-section (found using the method in Section 5.9).

The curvature could now be found from the moment and the properties $\kappa = \dfrac{M}{E_2 I_t}$

6. Substitute the formula for κ and write the stresses directly in terms of the moment.

The stresses can be expressed in terms of the curvature $\sigma_x = -E\kappa y$. Substitute the curvature in terms of the moment to relate the stresses directly to the moment.

$$\sigma_x = \begin{cases} -E_1 \dfrac{My}{E_2 I_t} = \dfrac{E_1}{E_2}\left[\dfrac{-My}{I_t}\right] & \text{for } y \text{ in layer 1} \\[3mm] -E_2 \dfrac{My}{E_2 I_t} = \dfrac{-My}{I_t} & \text{for } y \text{ in layer 2} \end{cases}$$

>>End 5.12

5.13 Bending of General (Non-Symmetric) Cross-Sections

So far, we have studied bending only for beams that are symmetric about the plane of bending. Here we see how to handle an arbitrary orientation for the plane of bending as well as non-symmetric cross-sections.

1. For a cross-section that is symmetric about two planes, resolve a general bending moment along the two symmetry planes, compute the stresses individually, and add them.

Say the beam is symmetric about two perpendicular planes, but the bending moment seeks to bend the beam in a plane other than a symmetry plane.

A moment in any direction perpendicular to the beam axis can always be resolved into two components, parallel and perpendicular to the two symmetry planes. Each moment individually causes bending as we have studied so far.

For bending in each plane, there is a different moment of inertia and a different coordinate corresponding to the distance from the neutral plane.

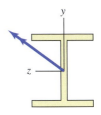

We define the respective moments of inertia as follows

$$I_y = \int_A z^2 \, dA \qquad I_z = \int_A y^2 \, dA$$

Then, the stress can be written as

$$\sigma_x = \frac{-M_z y}{I_z} + \frac{M_y z}{I_y}$$

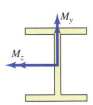

2. For a cross-section that is symmetric about one plane, resolve a general bending moment into components parallel and perpendicular to the symmetry plane, compute the stresses individually, and add them.

This cross-section is only symmetric about the vertical plane, not the horizontal. Still, we can resolve the moment into components parallel and perpendicular to the symmetry plane. The moment that causes bending in the symmetry plane (M_z) is treated as before.

Does the simple formula for bending stress above still apply when bending occurs perpendicularly to the symmetry plane (M_y)?

Consider the stress due only to M_y: $\sigma_x = \dfrac{M_y z}{I_y}$

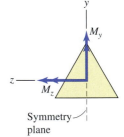

Symmetry plane

This distribution must satisfy:

$$\begin{cases} \displaystyle\int_A \sigma_x \, dA = 0 & \text{zero net force} \\[3mm] \displaystyle\int_A \sigma_x z \, dA = M_y & \text{net moment } M_y \\[3mm] \displaystyle\int_A \sigma_x y \, dA = M_z = 0 & \text{zero moment } M_z \end{cases}$$

The first condition is satisfied, since $z = 0$ at the centroid. Net moment M_y is correct because of the definition of I_y.

Consider $\quad M_z = \displaystyle\int_A \sigma_x y \, dA = \int_A \frac{M_y z y}{I_y} \, dA = \frac{M_y}{I_y} \int_A z y \, dA$

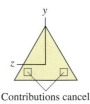

Even with a single plane of symmetry, the final integral is zero: at every plane y, there are two equivalent points with $z > 0$ and $z < 0$.

Contributions cancel

3. For a non-symmetric cross-section, bending in a single plane, *y* or *z*, results in both bending moments M_y and M_z.

Say we bend this beam with curvature κ_y about the *y*-axis and κ_z about the *z*-axis. For each, the strain is still proportional to distance from the neutral axis. So the stress is found from

$$\sigma_x = E\kappa_y z - E\kappa_z y$$

Non-symmetric cross-section

Each bending moment is found by integrating the force on each element times the perpendicular distance of that element from the neutral axis.

$$M_y = \int_A \sigma_x z\, dA = E\int_A [\kappa_y z - \kappa_z y]z\, dA = E\kappa_y \int_A z^2 dA - E\kappa_z \int_A yz\, dA$$

$$M_z = \int_A -\sigma_x y\, dA = E\int_A -[\kappa_y z - \kappa_z y]y\, dA = -E\kappa_y \int_A yz\, dA + E\kappa_z \int_A z^2 dA$$

Notice that both moments M_y and M_z result when we have only one curvature κ_y or κ_z. The two curvatures and two moments are said to be coupled. It would be simpler if curvature about one axis required only a moment about that axis.

Curvature about *y*-axis (bending in *x-z* plane)

Curvature about *z*-axis (bending in *x-y* plane)

4. Find a different pair of axes, *y′* and *z′*, rotated relative to the *y-z* axes, for which the curvature in one plane (*y′* or *z′*) results in only bending moments in that plane.

Instead of using *y-z* axes, we seek another pair of axes, *y′-z′*, which decouple the curvature and moments about the two axes. (It is critical to be familiar with the material in Appendix B, which addresses the effect of rotating axes on moments of inertia.) If we define the moments and curvatures with respect to *y′-z′*, the equations for $M_{y'}$ and $M_{z'}$ in terms of $\kappa_{y'}$ and $\kappa_{z'}$ would be identical to those above for M_y and M_z, except with *y′* and *z′* replacing *y* and *z*.

Consider what happens if we choose the axes *y′-z′* so that the mixed product of inertia is zero:

$$I_{y'z'} = \int_A y'z'\, dA = 0$$

Then the cross-terms, those that couple $M_{y'}$ to $\kappa_{z'}$ and $M_{z'}$ to $\kappa_{y'}$, would be zero. Curvature $\kappa_{y'}$ only requires moment $M_{y'}$, and similarly for *z′*. The formulas simplify to those for a cross-section symmetric about two planes *y′* and *z′*:

$$M_{y'} = E\kappa_{y'} \int_A z'^2 dA = E\kappa_{y'} I_{y'} \qquad M_{z'} = E\kappa_{z'} \int_A y'^2 dA = E\kappa_{z'} I_{z'}$$

The axes *y′-z′* on which the mixed product of inertia is zero are called the principal axes of inertia. Given the second moments of inertia about any axes, say *y-z*, one can transform axes to find the principal axes of inertia (see Appendix B).

>>End 5.13

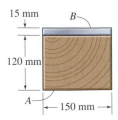

A composite beam is formed by bonding a steel plate to a wooden beam. Take the elastic moduli of steel and wood to be respectively 200 GPa and 13 GPa. The beam is subjected to a bending moment of 1500 N-m. Determine the stress at the points A and B at the bottom and top of the beam.

Solution

We treat the upper layer (steel) as material 1 and the wood as material 2.

The problem is converted to one of a single material by changing the width of material 1 in proportion to the ratio of the moduli.

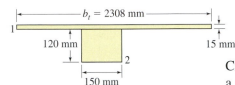

$$b_t = \frac{E_1}{E_2}b = \frac{200}{13}(150 \text{ mm}) = 2308 \text{ mm}$$

Cross-sectional properties for the transformed section can be found by treating this as a composite cross-section of uniform material properties.

$$\bar{y}_1 = 120 \text{ mm} + \frac{15}{2} \text{ mm} = 127.5 \text{ mm} \quad A_1 = (2308 \text{ mm})(15 \text{ mm}) = 3.46 \times 10^4 \text{ mm}^2$$

$$\bar{y}_2 = \frac{120}{2} \text{ mm} = 60 \text{ mm} \qquad\qquad A_2 = (150 \text{ mm})(120 \text{ mm}) = 1.80 \times 10^4 \text{ mm}^2$$

$$\bar{y} = \frac{(\bar{y}_1 A_1 + \bar{y}_2 A_2)}{A_1 + A_2} = 104.4 \text{ mm}$$

$$I = \frac{1}{12}(2308 \text{ mm})(15 \text{ mm})^3 + (3.46 \times 10^4 \text{ mm}^2)(127.5 \text{ mm} - 104.4 \text{ mm})^2$$

$$+ \frac{1}{12}(150 \text{ mm})(120 \text{ mm})^3 + (1.80 \times 10^4 \text{ mm}^2)(60 \text{ mm} - 104.4 \text{ mm})^2$$

$$= 7.62 \times 10^{-5} \text{ m}^2$$

$$\kappa = \frac{M}{E_2 I} = \frac{1500 \text{ N-m}}{(13 \text{ GPa})(7.62 \times 10^{-5} \text{ m}^4)} = 1.514 \times 10^{-3} \text{ m}^{-1}$$

The stress can now be found for any point in the cross-section using $|\sigma| = E\kappa|y|$

But the modulus must be the actual modulus corresponding to the material at the point.

Point A (in wood): $\quad y = -104.4 \text{ mm}$

$$|\sigma| = E_2\kappa|y| = (13 \text{ GPa})(1.514 \times 10^{-3} \text{ m}^{-1})(0.1044 \text{ m}) = 2.055 \text{ MPa}$$

Point B (in steel): $\quad y = 135 - 104.4 \text{ mm}$

$$|\sigma| = E_1\kappa|y| = (200 \text{ GPa})(1.514 \times 10^{-3} \text{ m}^{-1})(0.0306 \text{ m}) = 9.26 \text{ MPa}$$

>>End Example Problem 5.25

At a intermediate point in construction, an I-beam is cantilevered at one end, and subjected to a transverse force as shown. Considering the combined bending moment due to its weight and the transverse force, determine the maximum tensile stress in the beam and its location. The beam is a wide-flange section designated as W460 × 97 (see Appendix E).

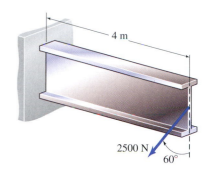

Solution

The beam cross-section is symmetric about both vertical and horizontal planes. The moment can be resolved into components parallel to each of the symmetry axes and the stress due to each moment can be added.

As seen in Appendix E, the area of this cross-section is 12300 mm, hence the force per length due to the weight is

$$q = (A)(\rho g) = (12300 \times 10^{-6}\,\mathrm{m}^2)\left(7850\,\frac{\mathrm{kg}}{\mathrm{m}^3}\right)\left(9.81\,\frac{\mathrm{m}}{\mathrm{s}^2}\right) = 947\,\frac{\mathrm{N}}{\mathrm{m}}$$

The moments of inertia about the two axes, also found from Appendix E, are shown.

The maximum moment about the z-axis is at the cantilever support. It is due to both the weight (q) and to the moment created by the 2500 N force:

$$M_z = -\frac{1}{2}\left(947\,\frac{\mathrm{N}}{\mathrm{m}}\right)(4\,\mathrm{m})^2 - (2500\,\mathrm{N})(\cos 60°)(4\,\mathrm{m}) = -12580\,\mathrm{N\text{-}m}$$

The maximum moment about the y-axis is at the cantilever support and due only to the moment created by the 2500 N force:

$$M_y = -(2500\,\mathrm{N})(\sin 60°)(4\,\mathrm{m}) = -8660\,\mathrm{N\text{-}m}$$

The moment about the z-axis is shown in its correct sense; it produces tension ($+$) and compression ($-$) as shown.

$$(\sigma)_{due\ to\ M_z} = \left|\frac{M_z y}{I}\right| = \frac{(12580\,\mathrm{N\text{-}m})\left(\dfrac{0.466\,\mathrm{m}}{2}\right)}{(445 \times 10^{-6}\,\mathrm{m}^4)} = 6.59\,\mathrm{MPa}$$

The moment about the y-axis is shown in its correct sense; it produces tension ($+$) and compression ($-$) as shown.

$$(\sigma)_{due\ to\ M_y} = \left|\frac{M_y z}{I}\right| = \frac{(8660\,\mathrm{N\text{-}m})\left(\dfrac{0.193\,\mathrm{m}}{2}\right)}{(22.8 \times 10^{-6}\,\mathrm{m}^4)} = 36.7\,\mathrm{MPa}$$

The maximum tensile stress is in the upper right corner of the cross-section, where the two maximum tensions add.

$$(\sigma)_{max} = 6.59\,\mathrm{MPa} + 36.7\,\mathrm{MPa} = 43.2\,\mathrm{MPa}$$

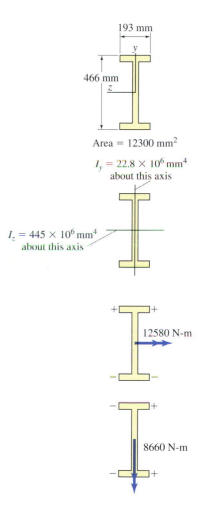

>>End Example Problem 5.26

Additional data on material properties needed to solve problems can be found in Appendix D or inside back cover.

5.153 An aluminum layer is bonded to a plastic sheet for reinforcement. Take the plastic to have an elastic modulus of 2 GPa. The width of the plastic sheet is 40 mm with a thickness of $h_p = 4$ mm. The width of the aluminum sheet is 40 mm with a thickness of $h_a = 1$ mm. Determine the maximum stresses in the plastic and aluminum if a bending moment of 5 N-m is applied.

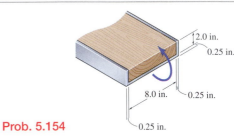

Prob. 5.153

5.154 A wood beam that is reinforced with a steel channel has the dimensions shown. Take the modulus of wood to be $1.8(10^6)$ psi. If the stress in the steel is to remain below 20 ksi and the stress in the wood is to remain below 2 ksi, determine the maximum allowable bending moment.

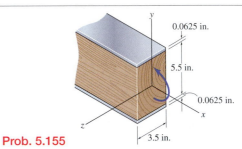

Prob. 5.154

5.155 Steel sheets 0.0625 in. thick are mounted on a 4 × 6 wood member (with finished dimensions 3.5 in. × 5.5 in.). A bending moment of 10^4 lb-in. is applied. Determine the maximum stress in the wood and in the steel. Take the wood to have an elastic modulus of $1.6(10^6)$ psi.

Prob. 5.155

5.156 An aluminum bar of square cross-section is encased in steel as shown. Determine the ratio of the bending stiffness of this composite beam with a solid aluminum bar having the same outer dimensions.

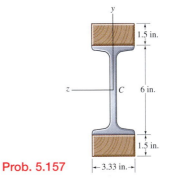

Prob. 5.156

5.157 Wood beams are secured to the top and the bottom of a steel I-beam so that they bend together with the steel beam. Determine the percentage by which the wood beams reduce the stress in the steel compared to when they are absent. Take $I = 22.1$ in.^{4} for the I-beam and the wood to have an elastic modulus of $1.6(10^6)$ psi.

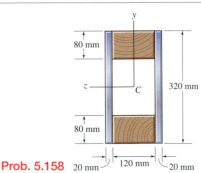

Prob. 5.157

5.158 Two steel plates are screwed to two wood beams as shown. Compare the bending stiffnesses of (a) the composite beam, (b) the steel plates alone, (c) the wood beams in their positions in the composite, and (d) a single wood beam with a height equal to the total height of the two wood beams. Treat the wood as having an elastic modulus of 10 GPa.

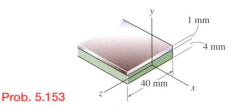

Prob. 5.158

5.159 A steel plate is screwed to a 6 × 6 wood beam (with finished dimensions 5.5 in. × 5.5 in.), and the composite beam is simply supported. The stresses in the steel and wood are not to exceed 20 ksi and 2 ksi, respectively. Take the wood to have an elastic modulus of $1.7(10^6)$ psi and the beam to have length $L = 12$ ft. Determine the maximum allowable center force P.

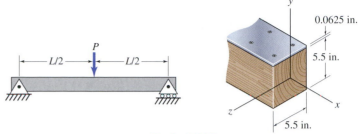

Prob. 5.159

5.160 Steel plates are screwed to both faces of the wood beam shown. The beam is cantilevered at one end and then is subjected to a uniform force per length. The stresses in the steel and wood are not to exceed 140 MPa and 14 MPa, respectively. Take the wood to have an elastic modulus of 11 GPa and the beam to have a length of $L = 3$ m. Determine the maximum allowable force per length q.

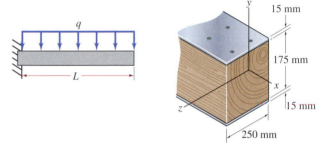

Prob. 5.160

5.161 Steel plates are screwed to both faces of a 6 × 6 wood beam (with finished dimensions 5.5 in. × 5.5 in.). The composite beam is cantilevered at one end and is subjected to a transverse force at the end. Take the wood to have an elastic modulus of $1.8(10^6)$ psi and the beam to have a length of $L = 10$ ft. If the maximum stress in the steel is 15 ksi, determine the force P.

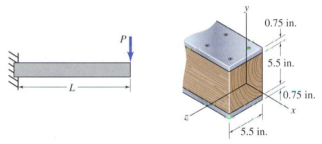

Prob. 5.161

5.162 A steel plate is screwed to one face of a wood beam with dimensions 150 mm × 100 mm. The beam is simply supported and then is subjected to a uniform force per length. Take the wood to have an elastic modulus of 11 GPa, and the beam to have length $L = 4$ m. If the maximum stress in the wood is 12 MPa, determine the force per length q.

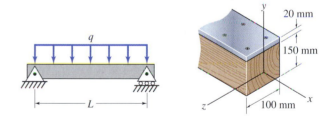

Prob. 5.162

5.163 A plastic strip is bonded to an aluminum layer and the composite functions as a spring. Take the plastic to have an elastic modulus of 2 GPa. The width of the plastic sheet is 30 mm with a thickness of $h_p = 5$ mm. The width of the aluminum sheet is 30 mm with a thickness of $h_a = 1.4$ mm. The composite strip of a length of 100 mm is supported as shown and subjected to a transverse force in the center. If the maximum stress in the aluminum is 50 MPa, determine the force P.

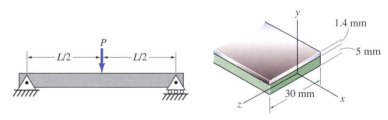

Prob. 5.163

>>End Problems

5.14 Transverse Shear Stress

So far, we have found the stresses due to the bending moment. There are also stresses associated with an internal shear force V. Stresses due to shear force may be dominant, for example, where the track transmits large forces from a rail train to the ties. Now we study how to determine the shear stresses due to V.

1. The shear force on the transverse plane (beam cross-section) equals the shear stress on each element area, times the element area, added up over all elements of the cross-section.

The distribution of shear stress due to V is complex. Two shear components, τ_{xy} and τ_{xz}, vary over the cross-section in both the y- and z-directions. Recall that each shear stress component acts on four faces of an element.

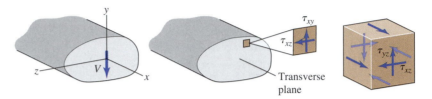

Shear stresses integrated over the cross-sectional area must give $-V$ in the y-direction and 0 in the z-direction.

$$\int_A \tau_{xy}\, dA = -V \qquad \int_A \tau_{xz}\, dA = 0$$

Therefore, the shear stress averaged over the cross-section equals the shear force divided by the area:

$$(\tau_{xy})_{\text{average}} = \frac{1}{A}\int_A \tau_{xy}\, dA = \frac{-V}{A} \qquad (\tau_{xz})_{\text{average}} = \frac{1}{A}\int_A \tau_{xz}\, dA = 0$$

Consider now only τ_{xy} and simplify the notation $\tau = |\tau_{xy}|$. We study its variation in the y-direction and neglect the variation with z.

2. Consider a beam segment of length *dx* on which shear force and bending moment act.

Here is a segment of length dx of a beam of general cross-section, symmetric about the x-y plane. Because a shear force V acts on this segment, the moment changes from M to $M + dM$. We assume that the bending stress σ_x still varies linearly with y.

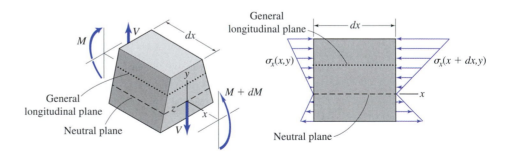

σ_x at the two transverse planes (x and $x + dx$) differ because M changes by dM. Still, the integral of σ_x over each transverse plane equals zero because the net force is zero.

We next consider a portion of this segment above a general longitudinal plane, and determine the shear stress τ that must act on this longitudinal plane to maintain equilibrium.

3. Isolate the upper portion of the segment, include σ_x on the transverse planes and τ on the longitudinal plane, and impose equilibrium in the *x*-direction.

Each end of the isolated portion is only part of the transverse plane and has area A'. The integral of σ_x over A' corresponds to a non-zero net force. Because M changes, the magnitudes of the net forces on the two transverse planes differ. The shear stress τ on the longitudinal plane balances the difference in net force.

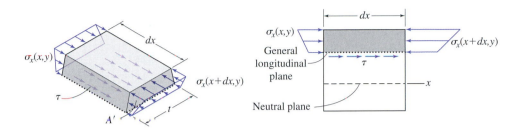

To sum forces, the normal stresses on transverse planes must be integrated over A'. Since we neglect the variation of stresses with z, we can simply multiply the shear stress τ times the area $(t)(dx)$.

$$\sum F_x = \tau(t)(dx) + \int_{A'} \frac{-(M + dM)y}{I} dA - \int_{A'} \frac{-My}{I} dA = 0 \Rightarrow \tau t dx = \int_{A'} \frac{(dM)y}{I} dA = \frac{dM}{I} \int_{A'} y dA$$

Use the relation between bending moment and shear force $\dfrac{dM}{dx} = V$ to write $\tau = \dfrac{V}{It} \displaystyle\int_{A'} y dA$

---

4. Relate the integral of y over A' to the centroid of area A' and simplify the shear stress calculation.

The integral $Q = \displaystyle\int_{A'} y dA$ is referred to as the first moment of area for area A'.

Q varies with the y-position of the longitudinal plane at the base of A', so $\tau = \dfrac{VQ}{It}$ depends on y.

Recall the definition of the y-coordinate of the centroid of area A': $\bar{y}' = \dfrac{1}{A'} \displaystyle\int_{A'} y dA$

Therefore, $Q = \displaystyle\int_{A'} y dA = \bar{y}' A'$. When the shape of A' is simple, $\bar{y}'$ and hence Q can be readily found.

---

5. The shear stress due to transverse shear is maximum at the neutral plane (centroid) and zero at free surfaces.

This example shows how Q and, hence, τ, vary with y. Q is zero at the top (because $A' = 0$), and zero at the bottom (because $\bar{y}' = 0$). This is consistent with the requirement that the top and bottom lateral surfaces have zero shear stress. The unbalanced axial force due to σ_x is greatest at the neutral plane so τ should be greatest there. Indeed, the product $A'\bar{y}'$ reaches its maximum value at the centroid.

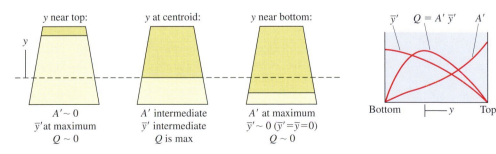

>>End 5.14

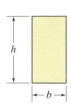 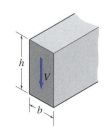

A beam of rectangular cross-section is subjected to a tranvserse force V. (a) Determine the distribution of shear stress as a function of distance from the centroid. (b) Find the location of the maximum shear stress. (c) Determine the ratio of the maximum shear stress to the average shear stress.

Solution

At any given point in the cross-section, the shear stress is found from

$$\tau = \frac{VQ}{It} = \frac{VA'\overline{y}'}{It}$$

The centroid for a rectangular cross-section is in the center. I about the centroid is:

$$I = \frac{1}{12}(b)(h)^3$$

In the formula for τ, the length t is the width at any value of y. For this rectangular cross-section $t = b$.

τ varies with y because of the term Q, which is defined as

$$Q = \int_{A'} y\, dA$$

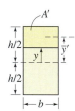

where y is defined as zero at the centroid, and A' is the area between the plane at y and the top of the cross-section. We can find Q by integration:

$$Q = \left[\int_{-b/2}^{b/2} dx\right]\left[\int_{y}^{h/2} y'\,dy'\right] = b\left.\frac{y'^2}{2}\right|_{y}^{h/2} = \frac{b}{2}\left[\left(\frac{h}{2}\right)^2 - y^2\right]$$

We can use the alternative definition: $Q = \overline{y}'A'$, where $\overline{y}'$ is the centroid of the area A'.

$$A' = b\left(\frac{h}{2} - y\right) \qquad \overline{y}' = \frac{1}{2}\left[\frac{h}{2} + y\right] \qquad Q = \overline{y}'A' = \frac{b}{2}\left[\left(\frac{h}{2}\right)^2 - y^2\right]$$

The shear stress is then found to vary with y according to:

$$\tau = \frac{VQ}{It} = \frac{V\frac{b}{2}\left[\left(\frac{h}{2}\right)^2 - y^2\right]}{\left(\frac{1}{12}bh^3\right)b} = \frac{6V\left[\left(\frac{h}{2}\right)^2 - y^2\right]}{bh^3}$$

Because of the term $-y^2$, τ is zero at the top and the bottom ($y = \pm h/2$), and decreases with distance from $y = 0$. So the maximum is at the centroid ($y = 0$):

$$\tau_{\max} = \frac{3V}{2bh}$$

The average shear stress is $V/A = V/(bh)$, so the maximum shear stress is $3/2$ times the average.

>>End Example Problem 5.27

The beam with hollow cross-section is subjected to a transverse shear force $V = 50$ kip. It is constructed from plates 1 in. thick. Determine the shear stress at the midplane of the beam.

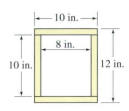

Solution

At any given point in the cross-section, the shear stress is found from

$$\tau = \frac{VQ}{It} = \frac{VA'\bar{y}'}{It}$$

where I refers to the entire cross-section.

We can analyze the beam as a composite cross-section. Since the cross-section is the difference of two rectangles with a common centroid, the centroid is in the center and I can be found from subtracting the respective values of I.

$$I = \frac{1}{12}(10 \text{ in.})(12 \text{ in.})^3 - \frac{1}{12}(8 \text{ in.})(10 \text{ in.})^3 = 773 \text{ in.}^4$$

For the midplane where we seek τ, the area A' is the upper half of the whole cross-section. We want the centroid of the upper half, which can itself be viewed as a composite cross-section. To find that centroid, we measure y from the bottom of the portion shown.

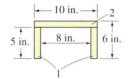

$$\bar{y}_1 = \frac{5}{2} \text{ in.} = 2.5 \text{ in.} \qquad A_1 = 2(5 \text{ in.})(1 \text{ in.}) = 10 \text{ in.}^2$$

$$\bar{y}_2 = 5 \text{ in.} + \frac{1}{2} \text{ in.} = 5.5 \text{ in.} \qquad A_2 = (10 \text{ in.})(1 \text{ in.}) = 10 \text{ in.}^2$$

$$\bar{y}' = \frac{(\bar{y}_1 A_1 + \bar{y}_2 A_2)}{A_1 + A_2} = 4 \text{ in.} \qquad A' = 10 \text{ in.}^2 + 10 \text{ in.}^2 = 20 \text{ in.}^2$$

In the shear stress formula, the centroid of area A', $\bar{y}'$, must be measured from the centroid of the entire cross-section. Since we have chosen to find the shear stress at the centroid of the entire cross-section, the value for $\bar{y}'$ is correct.

The width of the cross-section where we seek τ is $t = 2$ in.

$$\tau = \frac{VQ}{It} = \frac{VA'\bar{y}'}{It} = \frac{(50000 \text{ lb})(20 \text{ in.}^2)(4 \text{ in.})}{(773 \text{ in.}^4)(2 \text{ in.})} = 2590 \text{ psi}$$

>>End Example Problem 5.28

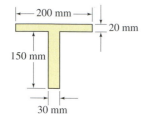

The beam of the T-shaped cross-section is subjected to a transverse shear force $V = 40$ kN. Determine the shear stress where the two portions of the cross-section meet.

Solution

At any given point in the cross-section, the shear stress is found from

$$\tau = \frac{VQ}{It} = \frac{VA'\bar{y}'}{It}$$

I refers to the entire cross-section, which is analyzed as a composite cross-section.

Each sub-area is rectangular; we find individual centroids (initially measured from the bottom) and then the centroid of entire cross-section.

$$\bar{y}_1 = 150 \text{ mm} + \frac{20}{2} \text{ mm} = 160 \text{ mm} \qquad A_1 = (200 \text{ mm})(20 \text{ mm}) = 4000 \text{ mm}^2$$

$$\bar{y}_2 = \frac{150}{2} \text{ mm} = 75 \text{ mm} \qquad\qquad A_2 = (150 \text{ mm})(30 \text{ mm}) = 4500 \text{ mm}^2$$

$$\bar{y} = \frac{(\bar{y}_1 A_1 + \bar{y}_2 A_2)}{A_1 + A_2} = 115 \text{ mm}$$

Use parallel axis theorem to find I of the entire cross-section:

$$I = \frac{1}{12}(200 \text{ mm})(20 \text{ mm})^3 + (4000 \text{ mm}^2)(160 \text{ mm} - 115 \text{ mm})^2$$

$$+ \frac{1}{12}(30 \text{ mm})(150 \text{ mm})^3 + (4500 \text{ mm}^2)(75 \text{ mm} - 115 \text{ mm})^2 = 2.39 \times 10^{-5} \text{ m}^4$$

For the cross-section where we seek τ, the highlighted area is A'. We seek the centroid $\bar{y}'$ of area A', as measured from the centroid of the whole cross-section.

$$A' = (200 \text{ mm})(20 \text{ mm}) = 4000 \text{ mm}^2 \qquad \bar{y}' = 160 - 115 \text{ mm} = 45 \text{ mm}$$

The width of the cross-section where we seek τ is $t = 30$ mm

$$\tau = \frac{VQ}{It} = \frac{VA'\bar{y}'}{It} = \frac{(40000 \text{ N})(4000 \text{ mm}^2)(45 \text{ mm})}{(2.39 \times 10^{-5} \text{ m}^4)(30 \text{ mm})} = 10.06 \text{ MPa}$$

>>End Example Problem 5.29

A 2 × 10 joist (see finished dimensions in the diagram) has a 12 ft span between supports. Take the joist to be subjected to a 40 lb/ft uniform load. (a) Determine the maximum normal stress due to bending. (b) Determine the maximum shear stress due to transverse shear.

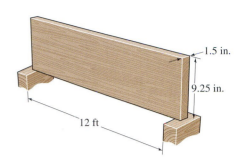

Solution

Support reactions can be found from equilibrium of the entire beam to be $qL/2$, where $q = 40$ lb/ft and $L = 12$ ft.

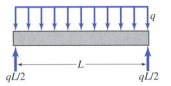

The moment M varies parabolically with distance along the beam, with the maximum bending moment at the center plane:

$$M_{max} = \frac{qL^2}{8} = \frac{(40 \text{ lb/ft})(12 \text{ ft})^2}{8} = 720 \text{ lb-ft}$$

The maximum bending stress is at the top or the bottom at the center plane:

$$\sigma_{max} = \frac{6M_{max}}{bh^2} = \frac{6(720 \text{ lb-ft})}{(1.5 \text{ in.})(9.25 \text{ in.})^2}\left(\frac{12 \text{ in.}}{1 \text{ ft}}\right) = 404 \text{ psi}$$

The shear force V varies linearly along with the beam. It has maximum values (positive and negative) equal to the force reaction at the supports:

$$V_{max} = \frac{qL}{2} = \frac{(40 \text{ lb/ft})(12 \text{ ft})}{2} = 240 \text{ lb}$$

At any cross-section, the maximum shear stress due to transverse shear is at the neutral plane. The shear stress in a beam of rectangular cross-section under transverse shear was found in Example 5.27. So, the maximum shear stress, equal to $(3/2)(V_{max}/A)$, reached at the neutral plane at each support:

$$\tau_{max} = \frac{3}{2}\frac{V_{max}}{A} = \frac{3}{2}\frac{240 \text{ lb}}{(1.5 \text{ in.})(9.25 \text{ in.})} = 25.9 \text{ psi}$$

>>End Example Problem 5.30

5.15 Shear Flow—Thin-Walled and Built-Up Cross-Sections

We consider now transverse shear of a beam with a cross-section composed of two or more portions with relatively thin walls. In such situations, the shear stress τ_{xz} can also be significant, even if V only acts in the y-direction. We use a general approach to calculating shear stress based on the concept of shear flow.

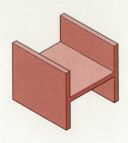

1. Simplify the shear stress distribution by recognizing stress components that are zero on free surfaces and by neglecting the variation across thin walls.

Note that the shear stress component acting on any free surface must be zero. For thin walls, the stress component that is zero at the two free surfaces is approximated as zero in between. Also, the non-zero shear stress is approximated as uniform across the thin wall.

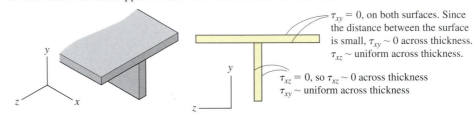

$\tau_{xy} = 0$, on both surfaces. Since the distance between the surface is small, $\tau_{xy} \sim 0$ across thickness. $\tau_{xz} \sim$ uniform across thickness.

$\tau_{xz} = 0$, so $\tau_{xz} \sim 0$ across thickness $\tau_{xy} \sim$ uniform across thickness

2. Find the shear stress needed to balance the variation of the bending stress σ_x along a segment.

We again assume the normal stresses are given by $\sigma_x = \dfrac{-My}{I}$, even though M varies with x.

We cut with an x-y or x-z plane to isolate a portion of the beam segment and draw its FBD.

In transverse shear loading of a beam we consider the loads exerted between adjacent portions that meet at planes parallel to the longitudinal axis. The **shear flow**, q, is defined as the longitudinal shear force per unit longitudinal length.

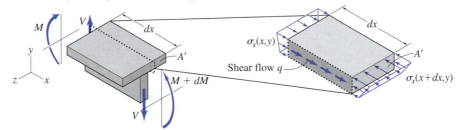

Because the bending moment varies with x and the ends of the isolated portion constitute only part of the cross-section, there is a net axial force on the isolated portion, dF_x, due to σ_x. Shear stress acts in the x-direction to balance the net axial force. **Shear flow**, q, is defined as the shear force per length in the axial direction and is equal to the average shear stress times the thickness.

$$\sum F_x = dF_x + q\,dx = \int_{A'} \frac{-(M+dM)y}{I}\,dA - \int_{A'} \frac{-My}{I}\,dA + q\,dx = 0 \implies q = \frac{dM}{dx}\frac{1}{I}\int_{A'} y\,dA = 0 \implies q = \frac{VQ}{I}$$

3. Apply the general approach to this cross-section.

The flange and web of this I-beam have thicknesses t_f and t_w, which are much smaller than b or d.

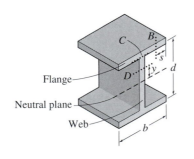

Flange

Neutral plane

Web

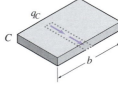

$$Q = \bar{y}'A' = \frac{d}{2}t_f s \implies \tau_B = \frac{q_B}{t_f} = \frac{Vsd}{2I}$$

$$Q = \bar{y}'A' = \frac{d}{2}t_f b \implies \tau_C = \frac{q_C}{t_w} = \frac{Vbt_f d}{2It_w}$$

$$Q = \bar{y}'A' = \frac{1}{2}\left(y + \frac{d}{2}\right)\left(\frac{d}{2} - y\right)t_w$$

$$\implies \tau_D = \frac{q_D}{t_w} = \frac{V}{2I}\left[\left(\frac{d}{2}\right)^2 - y^2\right] + \frac{Vbt_f d}{2It_w}$$

4. Transverse shear forces also affect fasteners that join portions of a beam.

Beams are sometimes built up by fastening boards together with nails or bolts across longitudinal (x-y or x-z) planes. When the beam is subjected to transverse shear V, the fastened boards usually must carry an axial force per length or shear flow q. This shear flow must be transmitted by the fasteners.

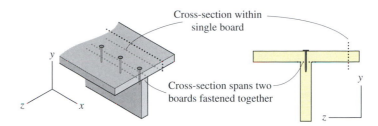

Cross-section within single board

Cross-section spans two boards fastened together

5. Find the shear flow on longitudinal planes at which the two fastened boards meet.

We find the shear flow just as we did for a thin walled cross-section.

Choose a portion of the beam of length dx that exposes the joined surface, draw the FBD, and find the net force due to σ_x.

$$dF_x = \int_{A'} \frac{-(M + dM)y}{I}\,dA - \int_{A'} \frac{-My}{I}\,dA$$

The net force, dF_x, is balanced by the shear flow q:

Find shear flow q acting across this plane spanned by fasteners

$$dF_x + q\,dx = 0 \qquad q = \frac{dM}{dx}\frac{1}{I}\int_{A'} y\,dA = 0 \Rightarrow q = \frac{VQ}{I} \qquad Q = \int_{A'} y\,dA = y'A'$$

6. When distinct fasteners, such as nails, cross a surface on which there is shear flow, each fastener exerts a shear force that depends on the shear flow and the fastener spacing.

If the boards are not bonded, but only joined by fasteners across a longitudinal plane, then the fasteners alone must transmit the shear flow q. Let each fastener carry a longitudinal force F_N.

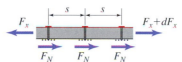

Let the fasteners be separated by a length s. Then, the axial force per length transmitted by fasteners is F_N/s, which must equal to the shear flow q.

$$F_N = (q)(s)$$

Say the maximum allowable force F_N is known (e.g., from the fastener's shear strength). Then, for a given shear flow (or V), the maximum fastener spacing, s, can be found. Or, if the fastener spacing, s, is set, then the maximum allowable shear force V can be found.

>>End 5.15

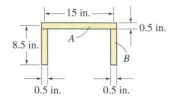

15 in.

0.5 in.

8.5 in.

A

B

0.5 in.   0.5 in.

The beam with the built-up cross-section shown is subjected to a transverse shear force $V = 15$ kN. Determine the distribution in shear flow, q, (a) in the horizontal portion and (b) in the vertical portions. Approximate the walls as thin.

Solution

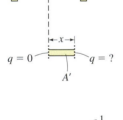

Find I for the entire cross-section, analyzing it as a composite cross-section.

Each sub-area is rectangular; we find individual centroids (initially measured from the bottom) and then the centroid of the entire cross-section.

$$\bar{y}_1 = 8.5 \text{ in.} \qquad\qquad A_1 = (15 \text{ in.})(0.5 \text{ in.}) = 7.5 \text{ in.}^2$$

$$\bar{y}_2 = \frac{8.5}{2} \text{ in.} = 4.25 \text{ in.} \qquad A_2 = 2(8.5 \text{ in.})(0.5 \text{ in.}) = 8.5 \text{ in.}^2$$

$$\bar{y} = \frac{(\bar{y}_1 A_1 + \bar{y}_2 A_2)}{A_1 + A_2} = 6.24 \text{ in.}$$

Use parallel axis theorem to find I.

$$I = \frac{1}{12}(8.5 \text{ in.})(0.5 \text{ in.})^3 + (7.5 \text{ in.}^2)(8.5 \text{ in.} - 6.24 \text{ in.})^2$$

$$+ \frac{2}{12}(0.5 \text{ in.})(8.5 \text{ in.})^3 + (8.5 \text{ in.}^2)(4.25 \text{ in.} - 6.24 \text{ in.})^2 = 123.3 \text{ in.}^4$$

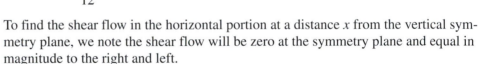

To find the shear flow in the horizontal portion at a distance x from the vertical symmetry plane, we note the shear flow will be zero at the symmetry plane and equal in magnitude to the right and left.

$$q = \frac{VQ}{I} = \frac{VA'\bar{y}'}{I} = \frac{15000(x)(0.5 \text{ in.})(8.5 \text{ in.} - 6.24 \text{ in.})}{2(123.3 \text{ in.}^4)} = (137.5)(x) \text{ lb/in.}$$

To find the shear flow in the vertical portions at a distance y from the bottom of the cross-section, we use as A' the composite cross-section shown consisting of the areas 1 and 2 (area 2 differs from above).

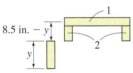

$$\bar{y}'_1 = 8.5 \text{ in.} - \bar{y} \qquad\qquad A_1 = (15 \text{ in.})(0.5 \text{ in.}) = 7.5 \text{ in.}^2$$

$$\bar{y}'_2 = \frac{y + 8.5 \text{ in.}}{2} - \bar{y} \qquad A_2 = 2(8.5 \text{ in.} - y)(0.5 \text{ in.}) = (8.5 \text{ in.} - y)(1 \text{ in.})$$

$$Q = A'\bar{y}' = A_1\bar{y}'_1 + A_2\bar{y}'_2 = (7.5 \text{ in.}^2)(8.5 \text{ in.} - \bar{y}) + (8.5 \text{ in.} - y)(1 \text{ in.})\left(\frac{y + 8.5 \text{ in.}}{2} - \bar{y}\right)$$

Substitute Q above into $q = \frac{VQ}{2I}$ using $\bar{y} = 6.24$ in. Note that Q, and so q, are functions of y.

Max q at max Q:

$$\frac{\partial Q}{\partial y} = -y + \bar{y} = 0 \Rightarrow y = \bar{y} \Rightarrow Q_{max} = 19.50 \text{ in.}^3 \quad q_{max} = 1185 \text{ lb/in.}$$

The hollow square beam is built up by nailing together four boards each with cross-sectional dimension 160 mm × 40 mm. A transverse shear force $V = 300$ N is applied. If each nail can support 120 N, determine the maximum permissible spacing of the horizontal nails and the vertical nails along the beam.

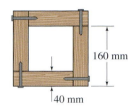

160 mm

40 mm

Solution

I for the entire cross-section is determined by treating this as a composite cross-section.

Since the cross-section is the difference of two squares with a common centroid, the centroid is in the center and I can be found from subtracting the respective values of I:

$$I = \frac{1}{12}(200 \text{ mm})(200 \text{ mm})^3 - \frac{1}{12}(120 \text{ mm})(120 \text{ mm})^3 = 1.161 \times 10^{-4} \text{ m}^4$$

We must determine the shear flow at the cross-sections through which the nails pass. By the symmetry, both horizontal nails have the same shear flow, and both vertical nails have the same shear flow. These shear flows may differ from each other because the shear force acts vertically.

The shear flow is found at a cross-section in the same way whether the cross-section is internal to the material or it is nailed or glued.

For a horizontal nail, the relevant shear flow q_H is that corresponding to the two longitudinal planes located symmetrically at the ends of the highlighted portion.

80 mm

$$A' = (120 \text{ mm})(40 \text{ mm}) = 4800 \text{ mm}^2 \qquad \bar{y}' = 80 \text{ mm}$$

$$q_H = \left(\frac{1}{2}\right)\frac{VQ}{I} = \frac{VA'\bar{y}'}{2I} = \frac{(300 \text{ N})(4800 \text{ mm}^2)(80 \text{ mm})}{2(1.161 \times 10^{-4} \text{ m}^4)} = 496 \text{ N/m}$$

The maximum nail spacing, s_H, is found from $(s_H)(q_H) = 120 \text{ N} \Rightarrow s_H = \dfrac{120 \text{ N}}{q_H} = 242 \text{ mm}$

For a vertical nail, the relevant shear flow q_V is that corresponding to the two longitudinal planes located symmetrically at the bottom of the highlighted portion.

80 mm

$$A' = (200 \text{ mm})(40 \text{ mm}) = 8000 \text{ mm}^2 \qquad \bar{y}' = 80 \text{ mm}$$

$$q_V = \left(\frac{1}{2}\right)\frac{VQ}{I} = \frac{VA'\bar{y}'}{2I} = \frac{(300 \text{ N})(8000 \text{ mm}^2)(80 \text{ mm})}{2(1.161 \times 10^{-4} \text{ m}^4)} = 827 \text{ N/m}$$

The maximum nail spacing, s_V, is found from $(s_V)(q_V) = 120 \text{ N} \Rightarrow s_V = \dfrac{120 \text{ N}}{q_V} = 145 \text{ mm}$

>>End Example Problem 5.32

Additional data on material properties needed to solve problems can be found in Appendix D or inside back cover.

5.164 Use the general formula for shear stresses due to a transverse shear force to show that the shear stress at the center plane of a beam of circular cross-section is equal to 4/3 times the average shear stress V/A.

Prob. 5.164

5.165 A simply supported beam of height h and width b is subjected to a uniformly distributed force per length. Determine the maximum shear stress due to transverse shear and where in the beam it is reached.

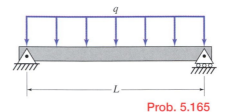

Prob. 5.165

5.166 A cantilevered beam of rectangular cross-section, h high and b wide, is subjected to an end force P. Determine the ratio of the maximum shear stress associated with transverse shear to the maximum bending stress. Calculate this ratio for a beam that is 1 m long, 30 mm high, and 10 mm wide.

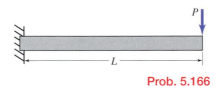

Prob. 5.166

5.167 A simply supported rod of a length of $L = 800$ mm and a diameter of $d = 16$ mm is subjected to a transverse force $P = 300$ N at the point $a = 300$ mm. (a) Determine the maximum shear stress due a transverse shear. (See Problem 5.164 for shear stress in a circular cross-section.) (b) Determine the maximum bending stress.

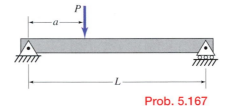

Prob. 5.167

5.168 The overhanging wood beam, $b = 3.5$ in. wide $\times$ $h = 5.5$ in. high, is loaded as shown. (a) Determine the maximum shear stress due to transverse shear. (b) Determine the maximum bending stress. Take $L_1 = 3$ ft, $L_2 = 6$ ft, $L_3 = 5$ ft, $F_1 = 5000$ lb, and $F_2 = 3000$ lb.

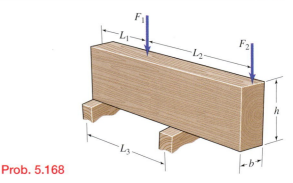

Prob. 5.168

5.169 A wood shelf, which is 200 mm wide and 16 mm high, is supported as shown on two brackets that are 1.8 m apart. A uniform downward pressure of 4000 N/m² acts on the shelf. Determine the maximum shear stress due to transverse shear.

Prob. 5.169

5.170 A simply supported beam of length L, width b, and height h is subjected to a transverse force at its center. Let the maximum allowable tensile stress due to bending be σ_{max}. What fraction of σ_{max} must the maximum allowable shear stress due to transverse shear be, if failure in shear is to occur at the same load as failure due to bending stress?

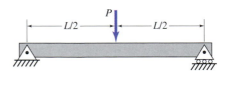

Prob. 5.170

5.171 Determine the maximum shear stress in the member (at the centroid) given a shear force of $V = 7$ kip.

5.172 If the maximum shear stress in the cross-section (at the centroid) is 1 ksi, determine the shear force.

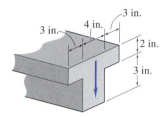

Probs. 5.171–172

5.173 Determine the maximum shear stress in the cross-section, which occurs just above the horizontal member of the cross given the shear force of $V = 30$ kip.

5.174 If the maximum shear stress in the cross-section, which occurs just above the horizontal member of the cross, is 800 psi, determine the shear force.

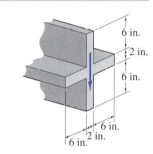

Probs. 5.173–174

5.175 Determine the maximum shear stress in the cross-section (at the centroid) given the shear force of 80 kN.

5.176 If the maximum shear stress in the cross-section (at the centroid) is 5 MPa, determine the shear force.

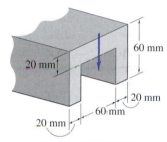

Probs. 5.175–176

5.177 Determine the maximum shear stress in the cross-section (at the centroid) given the shear force of 30 kN.

5.178 If the maximum shear stress in the cross-section (at the centroid) is 7 MPa, determine the shear force.

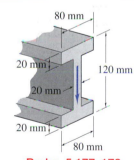

Probs. 5.177–178

5.179 Three boards are glued together to form an I-beam. The glue can resist a shear force per length of 11 N/mm. Determine the maximum shear force V, which the beam can sustain when the glue reaches its limit.

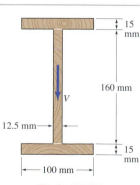

Prob. 5.179

5.180 A steel box beam, with a hollow rectangular cross-section shown, is subjected to transverse shear. If the shear stress due to transverse shear must not exceed 5000 psi, what is the maximum allowable shear force?

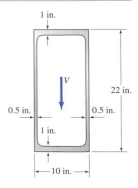

Prob. 5.180

5.181 Plates are welded together to form a beam of a T-shaped cross-section. A shear force of 15 kip is applied. Determine the maximum shear stress in the web of the beam.

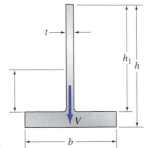

Prob. 5.181

5.182 An I-beam, 40 ft long, is built up from welded plates and is simply supported. The beam is subjected to a transverse force P in its center. Determine the maximum allowable P if (a) the allowable shear stress due to transverse shear is 7 ksi and (b) the allowable bending stress is 16 ksi.

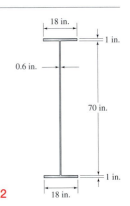

Prob. 5.182

5.183 The box beam shown is subjected to a transverse shear force of 30 kN. Determine the maximum shear stress in the web due to the transverse shear.

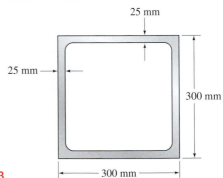

Prob. 5.183

5.184 Four boards are fastened together with nails as shown to form a box beam. Each nail can resist a force of 200 N. Assuming the top nails, spaced by 40 mm, are critical, determine the maximum shear force V that can be applied to the beam.

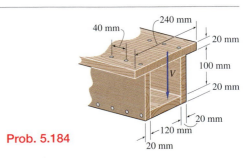

Prob. 5.184

5.185 A T-shaped beam is fabricated with 2 boards that are nailed to each other as shown. (a) If the beam has an allowable shear stress due to transverse shear of 3 MPa, determine the maximum shear force. (b) If each nail can withstand a shear force of 1500 N, what is the maximum spacing of the nails when the maximum shear force acts?

Prob. 5.185

5.186 Three boards are bolted together to form a beam, which is then subjected to a shear force $V = 8$ kip. Find the shear force in each bolt if they are regularly spaced 10 in. apart.

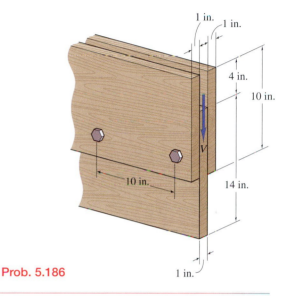

Prob. 5.186

5.187 A channel-shaped cross-section is fabricated with three boards that are nailed together as shown. If the beam is subjected to a shear force $V = 200$ lb and the nails are spaced by 5 in., what is the shear force carried by each nail?

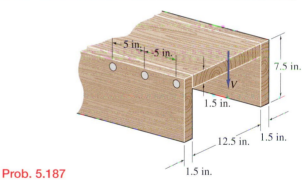

Prob. 5.187

5.188 A beam is fabricated with three boards that are nailed to each other at top and bottom. Such a beam, 6 m long and simply supported, is subjected to a distributed force of 2500 N/m. Determine the nail spacing, assumed uniform, if the nails are nowhere to exceed their maximum allowable shear force of 2000 N.

Prob. 5.188

5.16 Deflections Related to Internal Loads

We revisit the ruler from the beginning of the chapter. If the finger forces are known, we can find $V(x)$ and $M(x)$. They are plotted here.

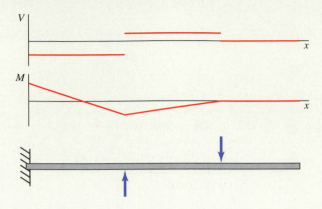

Now we learn how to find the deflection from the internal loads.

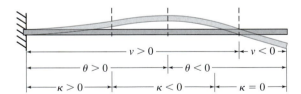

1. The proportionality between the bending moment and the curvature link the loads and the deflections.

Even though M can vary along the beam, we take the curvature, κ, at each point to be $\kappa = \dfrac{M}{EI}$

If the signs of M and κ are consistent, so will be their directions

$M > 0 \qquad \kappa > 0 \qquad\qquad M < 0 \qquad \kappa < 0$

2. The shape of any loaded beam is described by deflection, slope, and curvature.

Deflection (v): movement up ($v > 0$) or down ($v < 0$)
Slope (θ or v'): increasing or CCW rotation ($v' > 0$) or decreasing or CW rotation ($v' < 0$)
Curvature (κ): concave upward ($\kappa > 0$) or concave downward ($\kappa > 0$)

Here are the signs of the deflection, slope, and curvature for the beam above.

3. The challenge is to find the slope and deflection, when we are given only the curvature at each point.

Curvature is defined as the change in slope, θ, with arclength, s: $\kappa = d\theta/ds$. In general, it is challenging mathematically to find slope and deflection from κ. We now consider how to simplify the relations between deflection, slope, and curvature.

4. Relate the curvature and deflection through the geometry of a sloping beam.

The deflection of a beam changes by an amount dv over a distance dx. Here are general relations based on this geometry.

$$\tan \theta = dv/dx \quad \text{and} \quad ds = [dx^2 + dv^2]^{1/2}$$

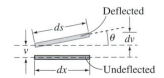

5. Find the curvature for the common situation of a small slope ($\theta \ll 1$).

A small slope implies $dv/dx = \tan\theta \approx \theta$ and $ds \approx dx$

Now recalculate the curvature: $\quad \kappa \equiv \dfrac{d\theta}{ds} \approx \dfrac{d\theta}{dx} \approx \dfrac{d(\tan\theta)}{dx} = \dfrac{d}{dx}\left(\dfrac{dv}{dx}\right) = \dfrac{d^2v}{dx^2}$

So from M at each point x along the beam, we can compute the second derivative of v: $\quad \dfrac{d^2v}{dx^2} = \dfrac{M}{EI}$

Because the slope is small, deflection in the direction parallel to the beam is usually neglected.

---

6. The loads on the beam determine M, κ (with EI), and then the deformed shape, but not the position of that shape in space.

Grip and bend a bar. Keeping it bent in a constant shape, move your hands up and down by different amounts. We say you are *rigidly* translating and rotating the bar, because you deflect it, but you do not change the deformation.

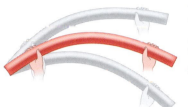

These beams each have the same $M(x)$ and so the same shape, but they are shifted and rotated.

Clearly, more information than $M(x)$ alone is needed to fully determine $v(x)$.

---

7. Integrate curvature κ twice and see that the deflection is determined to within an arbitrary rigid rotation and deflection.

Integrate $\quad \dfrac{d^2v}{dx^2} = \dfrac{M}{EI} \quad \Rightarrow \quad \dfrac{dv}{dx} = \displaystyle\int \dfrac{M}{EI}\,dx + c_1 \qquad$ Integration constant c_1

Integrate $\quad v = \displaystyle\int \dfrac{dv}{dx}\,dx = \int\left[\int\dfrac{M}{EI}\,dx\right]dx + c_1 x + c_2 \quad$ Integration constant c_2

Deflected beam shape $\qquad$ Arbitrary rotation of beam about $x = 0$ $\qquad$ Arbitrary uniform displacement of beam

$c_1 x + c_2$ is an arbitrary rotation and translation that can be added to the beam's deflection without changing the deformation (curvature).

Besides integrating M/EI twice, we must give values to c_1 and c_2 to fully specify the deflection.

---

8. Use the deflection and/or slope at supports to evaluate constants c_1 and c_2.

In addition to the loads, which determine $M(x)$ and then $\dfrac{d^2v}{dx^2}$, two pieces of information are needed to find c_1 and c_2. These are found from the deflection and/or the slope at a support.

Cantilever support
$v(0) = 0$
$v'(0) = 0$

Simply supported
$v(0) = 0$
$v(L) = 0$

x

x

L

>>**End 5.16**

Take the horizontal distance between adjacent grid lines to be 1 mm and the vertical distance to be 0.01 mm. Use a straight edge to measure and answer the following questions.

(a) Estimate the deflection and slope at D and at F (include magnitude and sign).
(b) Estimate the average curvature from B to D (assume it is uniform).
(c) Does the deflection at A have a greater or lesser magnitude than does the deflection at E? Do these deflections have the same or opposite sign?
(d) Does the slope at A have a greater or lesser magnitude than does the slope at D? Do these slopes have the same or opposite sign?
(e) Does the curvature at C have a greater or lesser magnitude than does the curvature at E? Do these curvatures have the same or opposite sign?

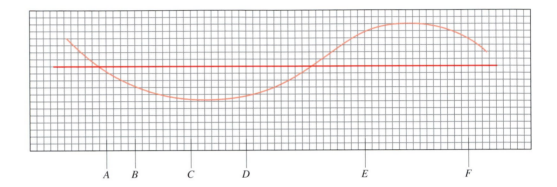

Solution

(a) Deflection at D is -0.042 mm (downward); deflection at F is 0.04 mm (upward); slope at $D = 4(0.01)/18 = 0.00222$ (slopes up, CCW rotation); slope at $F = -9(0.01)/15 = -0.006$ (slopes down, CW rotation).

(b) Assuming small slopes (which they are), the average curvature from B to D is the change in slope divided by the horizontal distance.
Slope at $B = -8(0.01)/19 = -0.00421$ (slopes down, CW rotation).
Found slope at D in part a. So, average curvature =
$[0.00222 - (-0.00421)]/16 = 4.02 \times 10^{-4}$ mm^{-1} = 0.402 m^{-1}.

(c) A moves down by a lesser distance than E moves up: $|v_A| < |v_E|$ and signs are opposite.

(d) Slope down at A is steeper than slope up at D: $|v'_A| > |v'_D|$ and signs are opposite.

(e) Curvature at C is greater than curvature at E, where curve is nearly flat. Curve is concave up at C and concave down at E. So $|\kappa_C| > |\kappa_E|$ and signs are opposite.

>>End Example Problem 5.33

Many engineering applications involve a simple cantilevered beam with a transverse force at its end. For example, this tab must be flexed outward to assemble a CD case. Determine the deflection along the entire beam and at the end point in terms of the load, the length, and EI.

Solution

Find the bending moment $M(x)$ as a function of position x along the length. Find the support reactions if convenient or necessary.

$$R_A = P, M_A = PL \qquad V(x) = P, M(x) = Px - PL$$

Calculate the curvature $\kappa(x) = M(x)/EI$

$$\kappa = \frac{d^2v}{dx^2} = \frac{M(x)}{EI} = \frac{Px - PL}{EI}$$

Integrate $\kappa(x)$ twice

$$\frac{dv}{dx}(x) = \int \left[\frac{d^2v}{dx^2}\right]dx = \int \left[\frac{Px - PL}{EI}\right]dx + c_1 = \frac{P}{EI}\left[\frac{x^2}{2} - Lx\right] + c_1$$

$$v(x) = \int \left[\frac{dv}{dx}\right]dx = \int \frac{P}{EI}\left[\frac{x^2}{2} - Lx\right]dx + c_1x + c_2 = \frac{P}{EI}\left[\frac{x^3}{6} - L\frac{x^2}{2}\right] + c_1x + c_2$$

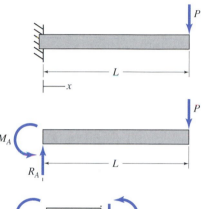
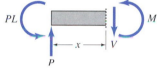

With this FBD, we don't need to find support reactions.

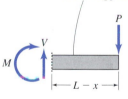

Determine the arbitrary rigid motion (c_1x and c_2) by accounting for the constraint of the cantilever support on deflection: $v(0) = 0, v'(0) = 0$

Use $v'(x)$ and $v(x)$ found from integration and evaluate c_1 and c_2:

$$v(0) = c_2, v'(0) = c_1 \Rightarrow c_2 = 0, c_1 = 0$$

$$v(x) = \frac{P}{EI}\left[\frac{x^3}{6} - L\frac{x^2}{2}\right] \qquad v'(x) = \frac{P}{EI}\left[\frac{x^2}{2} - Lx\right]$$

At the end, $\quad v(L) = \frac{-PL^3}{3EI} \qquad v'(L) = \frac{-PL^2}{2EI}$

$v(L) < 0$ means deflects down
$v'(L) < 0$ means slopes down or rotates CW

Magnitudes are shown in diagram.

$\frac{PL^3}{3EI}$

$\frac{PL^2}{2EI}$

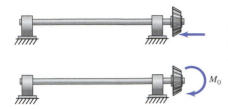

Many engineering applications involve a simply supported beam with a couple applied at one end. This occurs for example in a shaft with two bearings and a beveled gear, if we consider only the axial force on the gear, which is offset from the shaft axis. Determine the deflection along the entire beam and the rotations at the two ends in terms of the load, the length, and the beam properties. Treat the beam as simply supported.

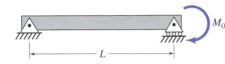

Solution

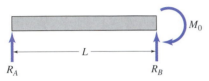

Find the bending moment $M(x)$ as a function of position x along the length. Find support reactions if convenient or necessary.

$$R_A = -M_0/L, \quad R_B = M_0/L$$
$$V(x) = -M_0/L, \quad M(x) = -M_0 x/L$$

Calculate the curvature $\kappa(x) = M(x)/EI$

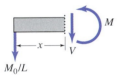

$$\kappa = \frac{d^2 v}{dx^2} = \frac{M(x)}{EI} = \frac{-M_0 x}{LEI}$$

Integrate $\kappa(x)$ twice

$$\frac{dv}{dx}(x) = \int \left[\frac{d^2 v}{dx^2}\right] dx = \int \frac{-M_0 x}{LEI} dx + c_1 = \frac{-M_0}{LEI}\left[\frac{x^2}{2}\right] + c_1$$

$$v(x) = \int \left[\frac{dv}{dx}\right] dx = \int \frac{-M_0 x^2}{2LEI} dx + c_1 x + c_2 = \frac{-M_0 x^3}{6LEI} + c_1 x + c_2$$

Determine arbitrary rigid motion ($c_1 x$ and c_2) by accounting for constraint of the supports on the deflections: $v(0) = 0, v(L) = 0$

Use $v(x)$ found from integration: $v(0) = c_2, v(L) = \frac{-M_0 L^3}{6LEI} + c_1 L + c_2$

$$\Rightarrow c_2 = 0, c_1 = \frac{M_0 L^2}{6LEI} \Rightarrow v(x) = \frac{M_0}{6LEI}\left[-x^3 + L^2 x\right], \quad v'(x) = \frac{M_0}{6LEI}\left[-3x^2 + L^2\right]$$

At the ends, $v'(0) = \frac{M_0 L}{6EI} \quad v'(L) = \frac{-M_0 L}{3 EI}$

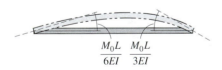

Rotations are in opposite directions ($v' > 0$ at A vs. $v' < 0$ at B).

The magnitude of the rotation is twice as great at the end where M_0 is applied.

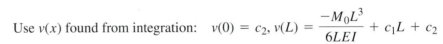

>>End Example Problem 5.35

Consider a cantilevered beam with a uniformly distributed force. Determine the deflection along the entire beam and in particular at the free end, in terms of the load, the length, and the beam properties.

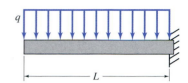

Solution

Find the bending moment as a function of position x along the length. In this instance, it is unnecessary to find the support reactions.

$$V(x) = -qx, \quad M(x) = qx^2/2$$

Calculate curvature $\kappa(x) = M(x)/EI$

$$\kappa = \frac{d^2v}{dx^2} = \frac{M(x)}{EI} = \frac{-qx^2}{2EI}$$

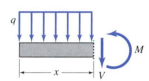

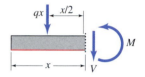

Integrate $\kappa(x)$ twice

$$\frac{dv}{dx}(x) = \int \left[\frac{d^2v}{dx^2}\right] dx = \int \frac{-qx^2}{2EI} dx + c_1 = \frac{-qx^3}{6EI} + c_1$$

$$v(x) = \int \left[\frac{dv}{dx}\right] dx = \int \frac{-qx^3}{6EI} dx + c_1 x + c_2 = \frac{-qx^4}{24EI} + c_1 x + c_2$$

Determine the arbitrary rigid motion ($c_1 x$ and c_2) by accounting for constraint of cantilever support at $x = L$: $v(L) = 0, v'(L) = 0$

Use $v'(x)$ and $v(x)$ found from integration: $v(L) = \dfrac{-qL^4}{24EI} + c_1 L + c_2 = 0, v'(L) = \dfrac{-qL^3}{6EI} + c_1 = 0$

$$\Rightarrow c_1 = \frac{qL^3}{6EI}, c_2 = \frac{-qL^3}{8EI} \Rightarrow v(x) = \frac{q}{24EI}[-x^4 + 4L^3 x - 3L^4] \quad v'(x) = \frac{q}{24EI}[-4x^3 + 4L^3]$$

At the free end ($x = 0$): $v(0) = \dfrac{-qL^4}{8EI} \quad v'(0) = \dfrac{qL^3}{6EI}$

Sign of deflection and slope indicate directions: $v(0) < 0 \Rightarrow$ deflects down, $v'(0) > 0 \Rightarrow$ slopes CCW.

One could have anticipated the correct directions of displacement and slope from the supports and the load.

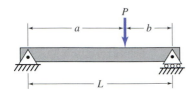

Consider a simply supported beam with a concentrated force at a general point along its length. Determine the deflection along the entire beam and the slopes at the end points in terms of the load, the lengths, and the beam properties.

Solution

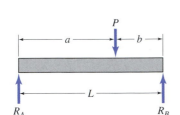

Find the bending moment as a function of position x along the length. It is convenient in this problem to find the support reactions.

$$R_A = Pb/L, R_B = Pa/L$$

Because of the concentrated force along beam, there are two different expressions for $M(x)$:

$$0 < x < a: V(x) = Pb/L, M(x) = Pbx/L$$

$$a < x < L: V(x) = -Pa/L, M(x) = Pbx/L - P(x - a)$$

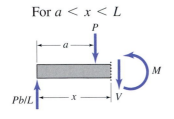

Integrate $\kappa = \dfrac{d^2v}{dx^2} = \dfrac{M(x)}{EI}$ in each region.

There are two integration constants for each region (each integration).

$$0 < x < a: \quad v'(x) = \frac{Pbx^2}{2LEI} + c_1 \qquad v(x) = \frac{Pbx^3}{6LEI} + c_1 x + c_3$$

$$a < x < L:$$

$$v'(x) = \frac{Pbx^2}{2LEI} - \frac{P(x - a)^2}{2EI} + c_2 \qquad v(x) = \frac{Pbx^3}{6LEI} - \frac{P(x - a)^3}{6EI} + c_2 x + c_4$$

Each region has only one constraint on the deflection:

$$v(0) = 0 \Rightarrow c_3 = 0 \qquad v(L) = 0 \Rightarrow c_4 = -\frac{P(bL^2 - b^3)}{6EI} - c_2 L$$

Even after the constraints of the supports are imposed, we can still:

Rotate the deflected beam in $0 < x < a$ about the pin at $x = 0$
Rotate the deflected beam in $a < x < L$ about the roller at $x = L$

These rotations correspond to adjusting the remaining constants c_1 and c_2. Consider different cases of rotating the parts of the beam:

I: Deflection and slope are discontinuous.
II: Deflection is continuous and slope is discontinuous.
III: Deflection is discontinuous and slope is continuous.
IV: Deflection and slope are continuous.

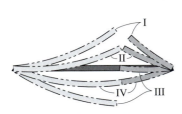

Both deflection and slope of the beam must be continuous

The formulas for the deflection and slope at $x = a$ are:

	from $0 < x < a$	from $a < x < L$
Deflection at $x = a$	$v(a) = \dfrac{Pba^3}{6LEI} + c_1a$	$v(a) = \dfrac{Pba^3}{6LEI} + c_2a - \dfrac{P(bL^2 - b^3)}{6EI} - c_2L$
Slope at $x = a$	$v'(a) = \dfrac{Pba^2}{2LEI} + c_1$	$v'(a) = \dfrac{Pba^2}{2LEI} + c_2$

Equate the expressions for $v(a)$ and $v'(a)$, and solve for the constants c_1 and c_2, to find

$$c_1 = c_2 = -\frac{Pb(L^2 - b^2)}{6LEI}, c_3 = c_4 = 0$$

After some rearrangement, the equations for the deflections and slopes are:

in $0 < x < a$ $\quad v(x) = -\dfrac{Pbx(L^2 - b^2 - x^2)}{6LEI}, \quad v'(x) = -\dfrac{Pb(L^2 - b^2 - 3x^2)}{6LEI}$

in $a < x < L$ $\quad v(x) = -\dfrac{Pbx(L^2 - b^2 - x^2)}{6LEI} - \dfrac{P(x - a)^3}{6EI},$

$$v'(x) = -\frac{Pb(L^2 - b^2 - 3x^2)}{6LEI} - \frac{P(x - a)^2}{2EI}$$

Of particular interest are the rotations at the supports, which are found to be

$$\theta_A = -v'(0) = \frac{Pb(L^2 - b^2)}{6LEI} = \frac{Pba(L + b)}{6LEI}$$

$$\theta_B = v'(L) = -\frac{Pb(-2L^2 - b^2)}{6LEI} - \frac{Pb^2}{2EI} = \frac{Pab(L + a)}{6LEI}$$

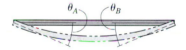

>>End Example Problem 5.37

Additional data on material properties needed to solve problems can be found in Appendix D or inside back cover.

5.189 The beam has uniform bending stiffness EI. Use the method of integration to determine the deflection and the slope as functions of x.

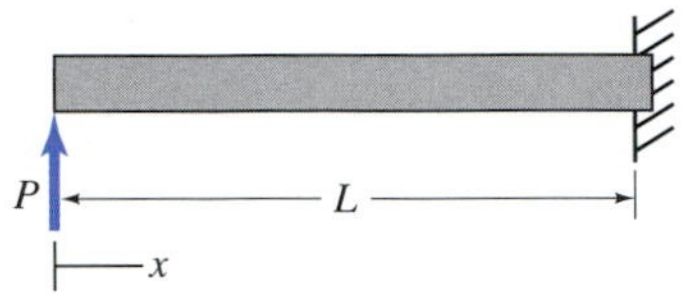

Prob. 5.189

5.190 The beam has uniform bending stiffness EI. Use the method of integration to determine the deflection and the slope as functions of x.

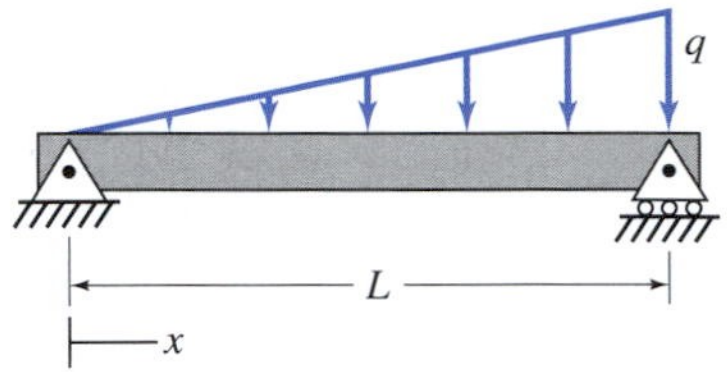

Prob. 5.190

5.191 The beam has uniform bending stiffness EI. Use the method of integration to determine the deflection and the slope as functions of x.

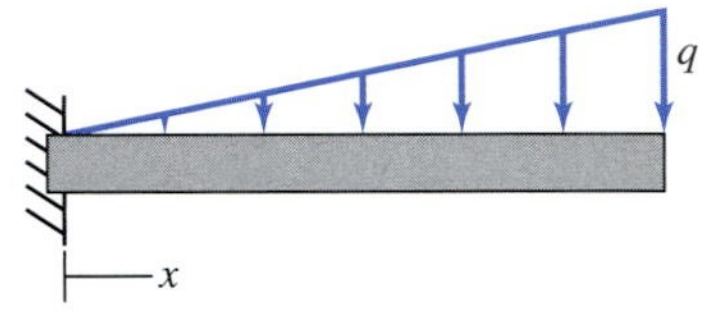

Prob. 5.191

5.192 The beam has uniform bending stiffness EI. Use the method of integration to determine the deflection and the slope as functions of x.

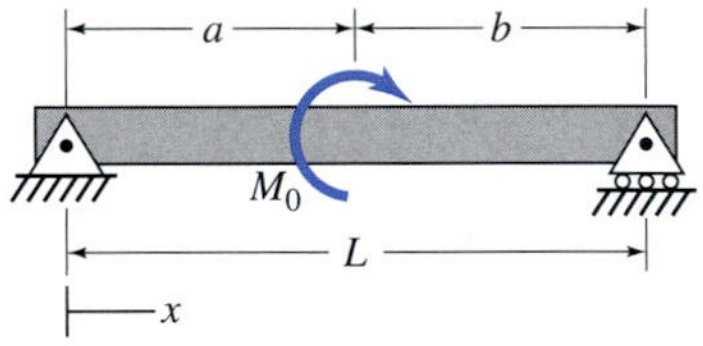

Prob. 5.192

5.193 The beam has uniform bending stiffness EI. Use the method of integration to determine the deflection and the slope as functions of x.

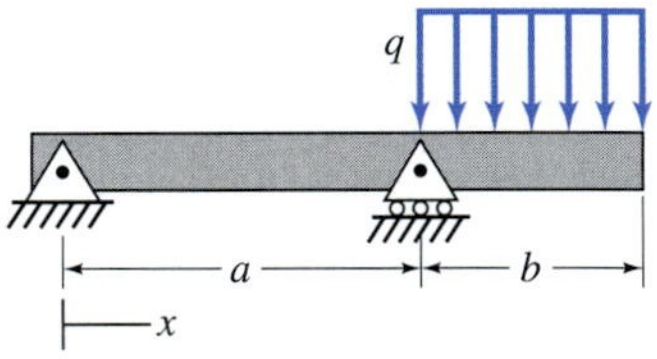

Prob. 5.193

5.194 The beam has uniform bending stiffness EI. Use the method of integration to determine the deflection and the slope as functions of x.

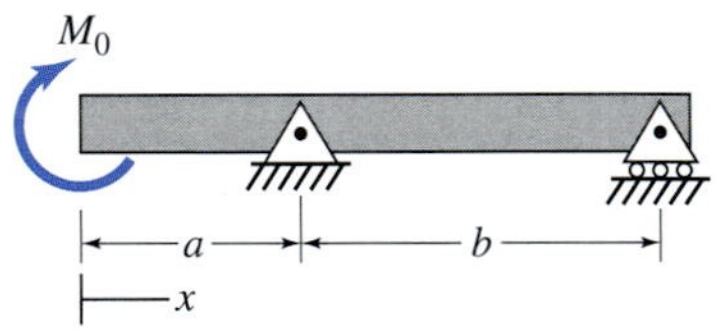

Prob. 5.194

5.195 The beam has uniform bending stiffness EI. Use the method of integration to determine the deflection and the slope as functions of x.

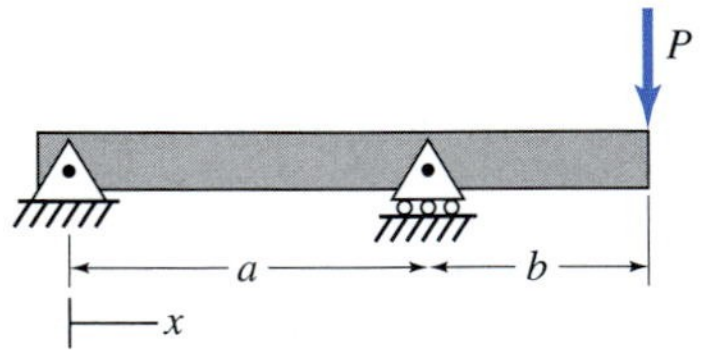

Prob. 5.195

5.196 The beam has uniform bending stiffness EI. Use the method of integration to determine the deflection and the slope as functions of x.

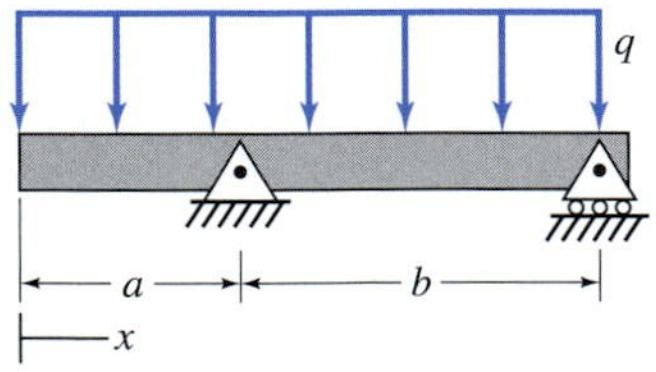

Prob. 5.196

5.197 A simply supported beam is subjected to a distribution of force per length given by the polynomial form: $q = Ax^n$.

(a) Calculate the bending moment $M(x)$ in the beam.

(b) Calculate the deflection $v(x)$ in the beam. Show that the general formula reduces to the appropriate one for a uniform load when n is set to 0.

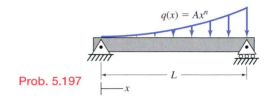

Prob. 5.197

5.198 A thin strip, of weight per length w_0 and bending stiffness EI, is laid over the round surface shown. Derive a formula for the length $2c$ over which the strip makes contact with the cylinder. Assume the deflections are small. Determine the minimum length L for which the strip contacts the cylinder at more than just the top point.

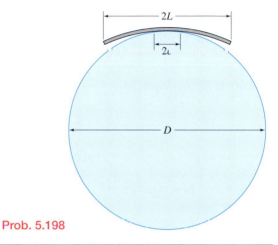

Prob. 5.198

5.199 A long length of pipe rests on the ground. The pipe has weight per length w_0 and bending stiffness EI. One end of the pipe is lifted by applying a force F_0. Determine the height h to which the end of the pipe is lifted.

Prob. 5.199

5.200 Shafts on which gears are mounted and which are supported by bearings often have steps. (For example, shaft shoulders can provide a surface against which a bearing can be located.) Excessive deflections of a gear may prevent proper engagement with the teeth of the mating gear. The shaft would be analyzed in appropriate segments to determine the deflection of the shaft at the gear at C.

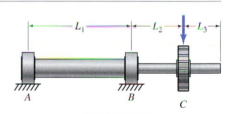

Prob. 5.200

In general circumstances, a shaft may have several changes in cross-section. In addition, bearings and mounted elements, such as gears and sprockets, can exert transverse forces and moments on the shaft. One can devise a general computer-based means of analyzing the deflection of such shafts. Such a shaft can be viewed as a series of connected shafts of uniform cross-section, with appropriate mathematical conditions at the cross-sections where they join. (These are sometimes referred to as jump conditions, since the values jump at the cross-section.)

(a) Show that along a portion of the shaft with uniform cross-section and zero applied loads, the deflection v must be of the form:
$$v = Ax^3 + Bx^2 + Cx + D$$

(b) The most general mathematical conditions at a cross-section would be where the cross-section changes and there acts an applied force and couple. (This could occur where a bevel gear is located against a shaft shoulder.) Let the deflection of the shaft on the two sides of the jump be of the form given in part (a), with constants A_1, B_1, C_1, and D_1 on the left side and A_2, B_2, C_2, and D_2 on the right side. Let the cross-section be located at $x = L$. Determine the four jump conditions that relate the constants to each other and to the force F_0 and couple M_0.

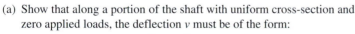

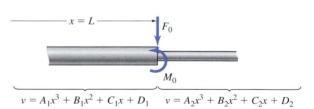

$$v = A_1x^3 + B_1x^2 + C_1x + D_1 \qquad v = A_2x^3 + B_2x^2 + C_2x + D_2$$

>>End Problems

5.17 Deflections Using Tabulated Solutions

Integrating beam deflection equations for complex loadings can be time consuming and prone to error. As an alternative, deflections of cantilevered or simply supported beams, for many simple types of loads, have been solved in variable form. The resulting formulas for beam deflections are given in tables in Appendix G. Here we explain the tables and how to use them to solve many types of problems.

1. Define the variables for tabulated cantilevered beam problems (EI = constant).

For cantilevered beams, the general variables are:

x = position (left to right)
$v(x)$ = deflection (positive in y-direction)
$v'(x)$ = slope of deflection (positive CCW)
$\delta_B = v(L)$ = deflection at end (positive down)
$\theta_B = -v'(L)$ = angle of rotation at right end (positive CW)

2. Here is a sample cantilevered beam solution from the tables.

This solution is for a uniform force per length q acting along the entire length of a beam cantilevered at the left (q positive if force acting down).

Here is the solution for this problem, using the general variables defined above:

$$v = -\frac{qx^2}{24EI}(6L^2 - 4Lx + x^2) \qquad v' = -\frac{qx}{6EI}(3L^2 - 3Lx + x^2)$$

$$\delta_B = \frac{qL^4}{8EI} \qquad \theta_B = \frac{qL^3}{6EI}$$

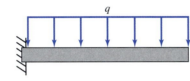

3. Define the variables for tabulated simply supported beam problems (EI = constant).

For simply supported beams, the general variables are:

x = position (left to right)
$v(x)$ = deflection (positive in y-direction)
$v'(x)$ = slope of deflection (positive CCW)
$\theta_A = -v'(0)$ = angle of rotation at left end (positive CW)
$\theta_B = v'(L)$ = angle of rotation at right end (positive CCW)
δ_c = deflection at center (positive down)

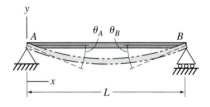

4. Here is a sample simply supported beam solution from the tables.

This solution is for a point force P applied at general point along a simply supported beam (P positive if acting down)

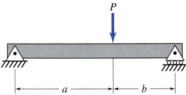

Here is the solution for this problem, using the general variables defined above:

$$v = -\frac{Pbx}{6LEI}(L^2 - b^2 - x^2) \qquad v' = -\frac{Pb}{6LEI}(L^2 - b^2 - 3x^2) \qquad (0 \le x \le a)$$

$$\theta_A = \frac{Pab(L + b)}{6LEI} \qquad \theta_B = \frac{Pab(L + a)}{6LEI}$$

$$\text{If } a \ge b, \quad \delta_C = \frac{Pb(3L^2 - 4b^2)}{48EI} \qquad \text{If } a \le b, \quad \delta_C = \frac{Pa(3L^2 - 4a^2)}{48EI}$$

$$\text{If } a < b, \quad x_1 = L - \sqrt{\frac{L^2 - a^2}{3}} \quad \text{and} \quad \delta_{max} = \frac{Pa\left(L^2 - a^2\right)^{3/2}}{9\sqrt{3}LEI}$$

5. We explain the steps to applying tabulated solutions with this example of an aluminum bracket that is attached to an uneven wall.

A bracket, formed from an aluminum extrusion, is screwed in at the top and bottom. A 40 lb force is applied at a third point to make the bracket contact the uneven wall. How far will the bracket deflect? For this cross-section, the manufacturer lists $I = 0.28$ in.[4]

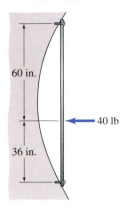

The key steps are:

1. *Draw the problem of interest as a horizontal beam problem (modeling).*

   The load is perpendicular to the bracket, so it causes bending. We turn the bracket sideways (to be horizontal) so the force acts downward, and model the screw attachments as simple supports. The remaining information, not shown in the original diagram, is given.

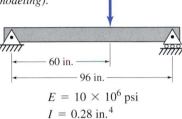

$E = 10 \times 10^6$ psi
$I = 0.28$ in.[4]

2. *Identify a tabulated problem, and the values for its variables, which is identical to the problem of interest.*

   We identify the tabulated simply supported problem with a concentrated force P as identical to our problem.

   If the variables in the tabulated problem are $a = 60$ in., $b = 36$ in., $L = 96$ in., and $P = 40$ lb, then it is identical to our problem.

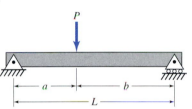

3. *Extract the desired results from the tabulated problem.*

   How far the bracket deflects corresponds to the deflection of the beam (v) at $x = a$. Here is the general formula from the tables, and the computed value given our specific problem:

$$v(x) = \frac{-Pbx}{6LEI}[L^2 - b^2 - x^2]$$

$$v(x = 60 \text{ in.}) = \frac{-(50 \text{ lb})(36 \text{ in.})(60 \text{ in.})}{6(96 \text{ in.})(10 \times 10^6 \text{ psi})(0.28 \text{ in.}^4)}[(96 \text{ in.})^2 - (36 \text{ in.})^2 - (60 \text{ in.})^2] = -0.231 \text{ in.}$$

4. *Relate the results back to the problem of interest.*

   Because the computed deflection $v(x)$ of the horizontal beam is < 0, the beam deflects down.

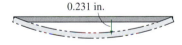

0.231 in.

Therefore, the bracket deflects as shown.

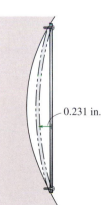

0.231 in.

>>End 5.17

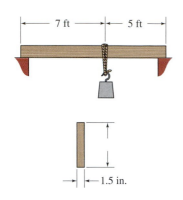

— 7 ft — — 5 ft —

—1.5 in.

A 600 lb load is hung from one of the joists in an exposed basement ceiling. The finished dimensions of a 2 × 10 joist are shown. Approximate the connection between the joist and the surfaces on which it rests as simple supports. Determine the maximum deflection of the joist due to this added load, the location of the maximum deflection and the angle at the ends of the joist. Use tabulated solutions. ($E = 1.6 \times 10^6$ psi.)

Solution

Redraw the problem as a standard beam with supports.

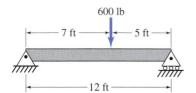

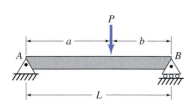

Recognize this problem as an instance of a tabulated problem: a concentrated force acting at an arbitrary position on a simply supported beam (see Appendix G-2.2).

General definitions for solutions of simply supported beams are given in Appendix G-2.

For this problem, parameters take on these particular values: $P = 600$ N, $L = 12$ ft, $a = 7$ ft, $b = 5$ ft, $E = 1.6 \times 10^6$ psi, $I = (1.5 \text{ in.})(9.25 \text{ in.})^3/12 = 98.9 \text{ in.}^4$

The maximum deflection along the beam is:

$$\delta_{max} = \frac{Pb[L^2 - b^2]^{3/2}}{9\sqrt{3}LEI} = \frac{(600 \text{ lb})(60 \text{ in.})[(144 \text{ in.})^2 - (60 \text{ in.})^2]^{3/2}}{9\sqrt{3}(144 \text{ in.})EI} = 0.227 \text{ in.}$$

δ_{max} is defined as positive downward. Since our load acts downward, the maximum deflection is downward.

δ_{max} is reached at the point: $x_1 = \sqrt{\dfrac{L^2 - b^2}{3}} = \sqrt{\dfrac{(144 \text{ in.})^2 - (60 \text{ in.})^2}{3}} = 75.6 \text{ in.}$

The rotation at A is:

$$\theta_A = \frac{Pab(L + b)}{6LEI} = \frac{(600 \text{ lb})(7 \text{ ft})(5 \text{ ft})(17 \text{ ft})}{6(12 \text{ ft})(1.6 \times 10^6 \text{ psi})(98.9 \text{ in.}^4)}\left(\frac{12 \text{ in.}}{1 \text{ ft}}\right)^2 = 5.51 \times 10^{-3} \text{ rad} = 0.259°$$

θ_A is defined as positive when the rotation is CW, so the rotation is CW.

The rotation at B is:

$$\theta_B = \frac{Pab(L + a)}{6LEI} = \frac{(600 \text{ lb})(7 \text{ ft})(5 \text{ ft})(19 \text{ ft})}{6(12 \text{ ft})(1.6 \times 10^6 \text{ psi})(98.9 \text{ in.}^4)}\left(\frac{12 \text{ in.}}{1 \text{ ft}}\right)^2 = 5.04 \times 10^{-3} \text{ rad} = 0.289°$$

θ_B is defined as positive when the rotation is CCW, so the rotation is CCW.

>>End Example Problem 5.38

An aluminum plate is built into concrete on the ground. The plate is 2 m wide and 30 mm thick. The plate is subjected to a wind load corresponding to a uniform pressure of 2.5 kPa. Determine the deflection and slope at the top of the plate. Use tabulated solutions.

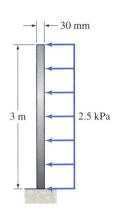

Solution

Redraw the problem as a standard beam with supports. The pressure of 2.5 kPa acts over a width 2 m (into the paper), so the force per length is 5000 N/m.

Recognize this problem as an instance of a tabulated problem: a uniformly distributed force acting over the entire length of a cantilevered beam (see Appendix G-1.5).

General definitions for solutions of cantilevered beams are given in Appendix G-1.

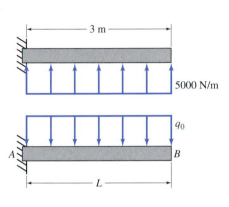

For this problem, parameters take on particular values: Load acts upward (opposite to tabulated): $q_0 = -5000$ N/m, $L = 3$ m, $E = 70$ GPa, $I = (2\text{ m})(0.03\text{ m})^3/12 = 4.50 \times 10^{-6}\text{ m}^4$

The deflection at the free end of the beam is:

$$\delta_{max} = \frac{q_0 L^4}{8EI} = \frac{(-5000\text{ N/m})(3\text{ m})^4}{8(70\text{ GPa})(4.5 \times 10^{-6})\text{ m}^4} = -0.161\text{ m}$$

From the general definitions for cantilever beam solutions $\delta_{max} > 0$ means downward. The upward pressure creates $\delta_{max} < 0$ or an upward deflection.

The rotation at the free end of the beam is:

$$\theta_B = \frac{q_0 L^3}{6EI} = \frac{(-5000\text{ Pa})(3\text{ m})^3}{6(70\text{ GPa})(4.5 \times 10^{-6}\text{ m}^4)} = -0.0714\text{ rad} = -4.09\text{ deg}$$

θ_B is defined as positive when the rotation at B is CW. The upward pressure creates $\theta_B < 0$ and, thus, a CCW rotation.

The end deflection and slope of the (horizontal) beam problem we analyzed is:

Convert this deflection and slope to those pertaining to the vertical plate (see figure at right):

5.18 Simple Generalizations of Tabulated Solutions _____

One can use the deflection tables and some creativity to solve problems with different support conditions and complex loadings. We illustrate several specific techniques for expanded use of the tables.

1. Consider this problem, which cannot be immediately solved by using a tabulated solution.

Toothbrushes given out for free at hotels are sometimes too flexible to use comfortably. For a given toothbrush design, we want to quantify the flexibility, defined here as the end deflection with a 1 N force on bristles. The elastic modulus is $E = 2$ GPa, and the dimensions are shown.

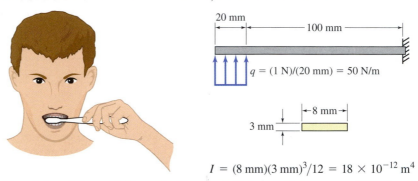

$q = (1 \text{ N})/(20 \text{ mm}) = 50 \text{ N/m}$

$I = (8 \text{ mm})(3 \text{ mm})^3/12 = 18 \times 10^{-12} \text{ m}^4$

2. Change how you look at the beam.

The cantilever support is at the right in this toothbrush problem, but at the left in the tables. But if you looked at the beam through the back of the page, it would look like this.

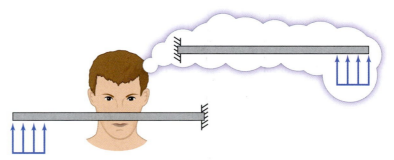

Solve the imagined problem cantilevered at its left and reinterpret the results for our problem.

3. Add and subtract distributions to get new ones.

Because the deflection is found from $d^2v/dx^2 = M/EI$, the deflection due to a sum of loadings, $M_1(x) + M_2(x)$, is equal to the sum of their deflections. So tabulated solutions, corresponding to the same supports, can be superposed (added). The solution for a uniformly distributed force that is zero over some distance next to the support is found from adding the two tabulated solutions shown.

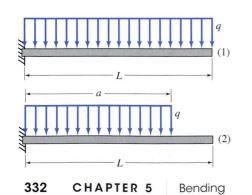

(1) $q = -50$ N/m, $L = 0.12$ m

$$(\delta_B)_1 = \frac{qL^4}{8EI} = \frac{-(50 \text{ N/m})(0.120 \text{ m})^4}{8(2 \times 10^9 \text{ Pa})(18 \times 10^{-12} \text{ m}^4)} = -0.036 \text{ m}$$

(2) $q = 50$ N/m, $L = 0.12$ m, $a = 0.1$ m

$$(\delta_B)_2 = \frac{qa^3(4L - a)}{24EI} = \frac{(50 \text{ N/m})(0.1 \text{ m})^3[4(0.12 \text{ m}) - 0.1 \text{ m})]}{24(2 \times 10^9 \text{ Pa})(18 \times 10^{-12} \text{ m}^4)} = 0.022 \text{ m}$$

$(\delta_B) = (\delta_B)_1 + (\delta_B)_2 = -0.036 + 0.022 = -0.014 \text{ m}$ $\delta_B < 0$, so end deflects up

4. With this problem we demonstrate the technique of finding a tabulated problem that has the same loads as the problem of interest and then adding a rigid motion to the tabulated solution.

A moment, M_0, is applied to the center of this beam. Determine the rotation of the center point of the beam, due to extension of the springs and bending of the beam.

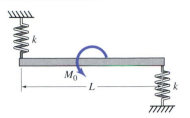

5. Try to identify standard supports that would result in the same loads on the beam.

The linear springs are in tension when the moment is applied; the forces on the beam are as shown. A simply supported beam would produce the same reactions as the springs, because the supports exert only forces. The beam with springs has the same $M(x)$ (and curvature) as the simply supported beam. So they differ only by a rigid motion.

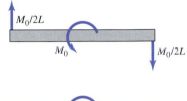

6. Determine the rigid motion by which the actual beam differs from the beam with standard supports.

Rather than $v = 0$ at pin supports, the ends of the beam actually displace because the springs stretch. From the spring force and spring constant, k, $\delta_{spring} = M_0/2Lk$. The rigid motion consists of the left end moving down and the right moving up, both by $M_0/2Lk$.

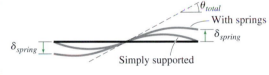

7. Combine the supported beam deflection with the rigid motion.

The total deflection equals that of the simply supported beam, plus the rigid motion.

In particular, the total rotation, v' or θ, at the center equals the sum of the rotations of the simply supported beam and the rigid motion

$$\theta_{total} = \theta_{s.s.} + \theta_{rigid}$$

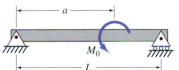

8. Compute deflections and combine.

Use this tabulated solution to find the rotation at the center of the simply supported beam (v' at $x = L/2$, $a = L/2$).

Special case: $a = L/2$

$$\theta_{s.s.} = (v'|_{x=L/2})_{beam} = \frac{M_0 L}{12EI}$$

The rigid motion of the beam is given by $v = c_1 x + c_1$ and satisfies $v(0) = -M_0/2Lk$ and $v(L) = M_0/2Lk$.

$$\theta_{rigid} = \frac{v(L) - v(0)}{L} = \frac{\dfrac{M_0}{2Lk} - \left(-\dfrac{M_0}{2Lk}\right)}{L} = \frac{M_0}{L^2 k}$$

The total rotation, θ_{total}, is given by

$$\theta_{total} = \theta_{s.s.} + \theta_{rigid} = \frac{M_0 L}{12EI} + \frac{M_0}{L^2 k}$$

>>End 5.18

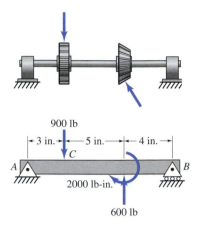

900 lb

← 3 in. → | ← 5 in. → | ← 4 in. →

C

A ▲ ⟋⟋⟋   2000 lb-in.   B ▲ ⟋⟋⟋

600 lb

The steel shaft shown (diameter = 0.875 in.) is part of a transmission system. The loads and bearings are modeled as shown. Often, we determine the rotation of a shaft at bearing supports because a bearing functions properly only if the tilting of the shaft is kept with acceptable limits. Determine the rotation of the shaft at bearing A (a) under the loads shown and (b) if only the load C were to act.

Solution

Recognize this problem as a combination of 3 tabulated solutions of simply supported beams (see Appendix G-2).

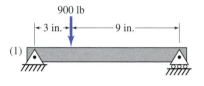

(1) 900 lb, ← 3 in. → | ← 9 in. →

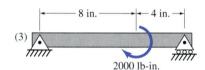

(2) ← 8 in. → | ← 4 in. → ; 600 lb

(3) ← 8 in. → | ← 4 in. → ; 2000 lb-in.

1. Appendix G-2.2: $P = 900$ lb, $a = 3$ in., $L = 12$ in.
2. Appendix G-2.2: $P = -600$ lb, $a = 8$ in., $L = 12$ in.
3. Appendix G-2.5: $M_0 = -2000$ lb-in., $a = 8$ in., $L = 12$ in.

In all cases,

$$E = 30 \times 10^6 \text{ psi}, \ I = \pi (0.875 \text{ in.})^4/4 \Rightarrow EI = 8.63 \times 10^5 \text{ lb-in.}^2$$

Loading 1: $(\theta_A)_1 = \dfrac{Pab(L + b)}{6LEI} = \dfrac{(900 \text{ lb})(3 \text{ in.})(9 \text{ in.})(21 \text{ in.})}{6(12 \text{ in.})(8.63 \times 10^5 \text{ lb-in.})} \text{ rad}$

Loading 2: $(\theta_A)_2 = \dfrac{Pab(L + b)}{6LEI} = \dfrac{(-600 \text{ lb})(8 \text{ in.})(4 \text{ in.})(16 \text{ in.})}{6(12 \text{ in.})(8.63 \times 10^5 \text{ lb-in.})} \text{ rad}$

Loading 3:

$$(\theta_A)_3 = \frac{M_0[6aL - 3a^2 - 2L^2]}{6LEI} = \frac{(-2000 \text{ lb-in.})[(6 \text{ in.})(8 \text{ in.})(12 \text{ in.}) - 3(8 \text{ in.})^2 - 2(12 \text{ in.})^2]}{6(12 \text{ in.})(8.63 \times 10^5 \text{ lb-in.})} \text{ rad}$$

The total rotation due to all three loads is:

$$\theta_A = (\theta_A)_1 + (\theta_A)_2 + (\theta_A)_3 = 1.787 \times 10^{-4} \text{ rad} = 0.01023°$$

The total rotation due to just the 900 lb force is:

$$(\theta_A)_1 = 8.21 \times 10^{-3} \text{ rad} = 0.471°$$

Note θ_A is positive so rotation is CW (shaft slopes down) at A. Loads 2 and 3 act to rotate the end A oppositely to load 1, which is why the total rotation is much smaller with all three loads than with load 1 alone.

>>End Example Problem 5.40

The motor is bolted to a cantilevered steel beam made of a C10 × 30 channel section (see Appendix E). The motor weighs 150 lb, and is delivering a torque of 600 lb-ft. Determine the deflection at the end of the beam in two cases: (a) the torque exerted by the motor acts CCW (as shown) and (b) the torque exerted by the motor acts CW.

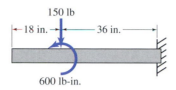

Solution

Because the support is on the right, we should carefully adapt tabulated solutions. Imagine looking from the other side of the page, and define distances and directions of the loads appropriately.

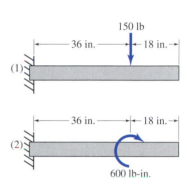

To solve this problem we need a combination of 2 tabulated solutions of cantilevered beams (see Appendix G-1). For both of these problems, the end deflection should be down.

1. Appendix G-1.3: $P = 150$ lb, $a = 36$ in., and $L = 54$ in.
2. Appendix G-1.4: $M_0 = 600$ lb-ft, $a = 36$ in., and $L = 54$ in.

In both cases, $E = 30 \times 10^6$ psi, $I = 3.94$ in.^{4} (from Appendix E).

Loading 1: $(\delta_B)_1 = \dfrac{Pa^2(3L - a)}{6EI} = \dfrac{(150 \text{ lb})(36 \text{ in.})^2[3(54 \text{ in.}) - (36 \text{ in.})]}{6(54 \text{ in.})(30 \times 10^6 \text{ psi})(3.94 \text{ in.}^4)} = 3.45 \times 10^{-2}$ in.

Loading 2: $(\delta_B)_2 = \dfrac{M_0(a)(2L - a)}{2EI} = \dfrac{(600 \text{ lb-ft})\left[\dfrac{12 \text{ in.}}{\text{ft}}\right](36 \text{ in.})[2(54 \text{ in.}) - (36 \text{ in.})]}{2(54 \text{ in.})(30 \times 10^6 \text{ psi})(3.94 \text{ in.}^4)} = 7.89 \times 10^{-2}$ in.

Deflection δ_B is defined as positive when downward.

Case (a): Loads are in the directions drawn, so both $(\delta_B)_1$ and $(\delta_B)_2$ are downward

$\delta_B = (\delta_B)_1 + (\delta_B)_2 = 3.45 \times 10^{-2}$ in. $+ 7.89 \times 10^{-2}$ in. $= 0.1135$ in. $\Rightarrow 0.1135$ in. down.

Case (b): Force is in the direction shown, but the moment is opposite: $(\delta_B)_1$ is down, $(\delta_B)_2$ is up

$\delta_B = (\delta_B)_1 + (\delta_B)_2 = 3.45 \times 10^{-2}$ in. $- 7.89 \times 10^{-2}$ in. $= -0.0444$ in. $\Rightarrow 0.0444$ in. up.

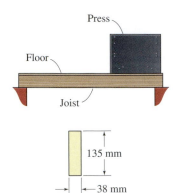

Press

Floor

Joist

135 mm

38 mm

The joist, 4 m in length, with cross-section shown carries a typical uniform loading from the floor above of 2400 N/m. In addition, a heavy press is also to be placed on the floor above, resulting in an additional 200 kg of mass carried by the joist over 1.5 m. Determine the total deflection of the joist at its midpoint. Model the joist as simply supported. Take the modulus of wood to be $E = 12$ GPa.

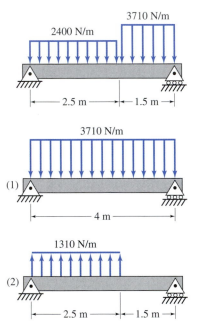

3710 N/m

2400 N/m

— 2.5 m — 1.5 m —

3710 N/m

(1)

— 4 m —

1310 N/m

(2)

— 2.5 m — 1.5 m —

Solution

The additional loading due to the press is $(200 \text{ kg})(9.81 \text{ m/s}^2)/(1.5 \text{ m}) = 1308$ N/m. The total force per length acting on the portion of the joist under the press is 3710 N/m.

To create a piecewise uniform force per length, we can use tabulated solutions for uniform distributed force on the entire beam, and uniform load acting only over $0 < x < a$.

We can produce the desired loading by applying a uniform load of 3710 N/m on the entire beam and removing 1310 N/m over $0 < x < 2.5$ m.

1. Appendix G-2.6: $q = 3710$ N/m, $L = 4$ m
2. Appendix G-2.7: $q = -1310$ N/m, $a = 2.5$ m, $L = 4$ m

In both cases,

$$E = 12 \text{ GPa}, \quad I = (38 \text{ mm})(135 \text{ mm})^3/12 = 4.11 \times 10^{-5} \text{ m}^2$$

Loading 1 (tables give deflection at the center for this load)

$$v_1 = (\delta_c)_1 = \frac{5qL^4}{384EI} = \frac{5(3710 \text{ N/m})(4 \text{ m})^4}{384(12 \text{ GPa})(4.111 \times 10^{-5} \text{ m}^4)} = 0.0251 \text{ m (down)}$$

Loading 2 use $v(x)$ at $x = 2$ m (positive v means deflection is up)

$$v_2 = \frac{qx[a^4 - 4a^3L + 4a^2L^2 + 2a^2x^2 - 4aLx^2 + Lx^3]}{24LEI} = 0.00614 \text{ m (up)}$$

The total downward deflection is that associated with the sum of two loads:

$$\delta = v_1 + v_2 = 0.0251 \text{ m} - 0.00614 \text{ m} = 18.93 \text{ mm (down)}$$

>>End Example Problem 5.42

The steel beam, with $I = 0.1$ in.^{4}, is supported at its ends by compression springs with spring constant 250 lb/in. A person weighing 200 lb stands at the point shown. Determine the deflection of the beam where the person stands, and determine the rotation of the beam at the right end B.

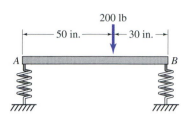

Solution

From equilibrium, the forces on the left and right springs are 75 lb and 125 lb. From the spring compressions the ends of the beam deflect down by:

$$v_A = \frac{P}{k} = \frac{75 \text{ lb}}{250 \text{ lb/in.}} = 0.3 \text{ in.} \qquad v_B = \frac{P}{k} = \frac{125 \text{ lb}}{250 \text{ lb/in.}} = 0.5 \text{ in.}$$

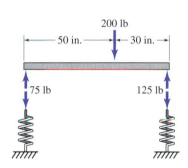

The deflection of the beam involves displacements at the ends consistent with the spring deflections, and the bending of the beam.

Since the springs apply only vertical forces to the beam, the loads on the beam are the same as if it were simply supported. The beam bends into the same shape as would a simply supported beam under the same load. To the deflection of the simply supported beam we add a rigid motion ($c_1 x$ and c_2) that gives the correct deflections at the ends.

Total deflection is the sum of two deflections:

1. Appendix G-2.2: $P = 200$ lb, $a = 50$ in., $b = 30$ in., $L = 80$ in.
2. Rigid deflection $v(x) = -0.3$ in. $- (0.5$ in. $- 0.3$ in.$)x/80$ in.

$$v_1(a) = \frac{-Pba(L^2 - b^2 - a^2)}{6LEI} = \frac{-(200 \text{ lb})(30 \text{ in.})(50 \text{ in.})((80 \text{ in.})^2 - (30 \text{ in.})^2 - (50 \text{ in.})^2)}{6(80 \text{ in.})(30 \times 10^6 \text{ psi})(0.1 \text{ in.}^4)} = -0.625 \text{ in.}$$

$$v_2(a) = -0.3 \text{ in.} - (0.5 \text{ in.} - 0.3 \text{ in.})(50 \text{ in.})/(80 \text{ in.}) = -0.425 \text{ in.}$$

The total downward deflection is due to the sum of two deflections: $v(50) = -1.05$ in. (down)

The rotation at B is found for each of the two deflections

$$(\theta_B)_1 = \frac{Pab(L + a)}{6LEI} = \frac{(200 \text{ lb})(50 \text{ in.})(30 \text{ in.})(80 \text{ in.})}{6(80 \text{ in.})(30 \times 10^6 \text{ psi})(0.1 \text{ in.}^4)} = 0.0271 \text{ rad CCW}$$

$$(\theta_B)_2 = -(0.5 \text{ in.} - 0.3 \text{ in.})/(80 \text{ in.}) = -0.0025 \text{ rad (CW)}$$

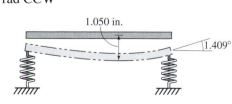

The net rotation at B is $= 0.0246$ rad $= 1.409°$ CCW.

>>**End Example Problem 5.43**

Additional data on material properties needed to solve problems can be found in Appendix D or inside back cover.

5.201 Use the tabulated solutions to determine the deflection at 300 mm from the support. Express the deflection in terms of the bending stiffness EI, and indicate whether the deflection is up or down.

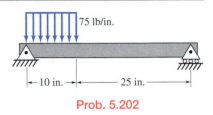

Prob. 5.201

5.202 Use the tabulated solutions to determine the slope at 8 in. from the left support. Express the slope in terms of the bending stiffness EI, and indicate whether the rotation at that point is CW or CCW.

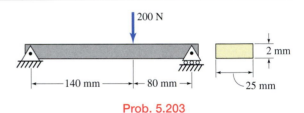

Prob. 5.202

5.203 The loading is applied to an aluminum strip that is 25 mm wide and 2 mm thick. Use the tabulated solutions to determine the deflection and slope at 100 mm from the left support.

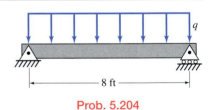

Prob. 5.203

5.204 A maximum deflection of 0.25 in. due to the distributed load q can be tolerated. The beam is steel with wide-flange cross-section W18 × 71 (see Appendix E). Use the tabulated solutions to determine the maximum allowable force per length q.

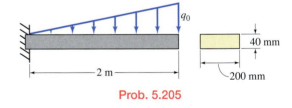

Prob. 5.204

5.205 The stress in the wood beam must not exceed 2 MPa. The cross-section is 40 mm high and 200 mm wide. The beam is loaded with a linearly varying distributed force as shown, for which the allowable stress of 2 MPa is reached at one point. Use the tabulated solutions to determine the resulting maximum allowable end deflection. Take $E = 11$ GPa.

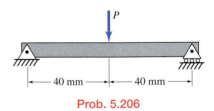

Prob. 5.205

5.206 If the 2 mm diameter steel wire shown is viewed as a spring, use the tabulated solutions to determine the spring constant, which is defined as the force applied per unit deflection at the point of application.

Prob. 5.206

5.207 The steel strip, 12 mm wide × 3 mm thick, acts like a torsional spring because it rotates at its center due to the concentrated moment. Use the tabulated solutions to determine the spring constant, which is defined as moment applied per unit rotation (in radians) at the point of application.

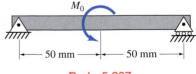

Prob. 5.207

5.208 Use the tabulated solutions to determine the deflection at 30 in. from the left support. Indicate whether the deflection is up or down. ($EI = 10^6$ lb-in.^{2}.)

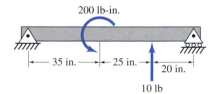

200 lb-in.

35 in. — 25 in. — 20 in.

10 lb

Prob. 5.208

5.209 Use the tabulated solutions to determine the slope at 1 m from the support. The beam is steel with wide-flange section W410 × 85 (see Appendix E). Indicate whether the rotation is CW or CCW. (Take $I = 315 (10)^6$ mm^4.)

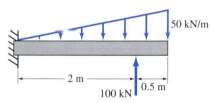

50 kN/m

2 m

100 kN 0.5 m

Prob. 5.209

5.210 Use the tabulated solutions to determine the deflection and the slope at the center of the beam. Express the results in terms of the bending stiffness EI, and indicate whether the deflection is up or down and the rotation is CW or CCW.

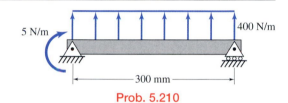

5 N/m

400 N/m

300 mm

Prob. 5.210

5.211 Use the tabulated solutions to determine the deflection and the slope at the end of the beam. Express the results in terms of the bending stiffness EI, and indicate whether the deflection is up or down and the rotation is CW or CCW.

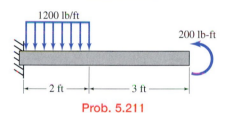

1200 lb/ft

200 lb-ft

2 ft — 3 ft

Prob. 5.211

5.212 Consider the distributed load q as given. Use the tabulated solutions to determine the force P necessary to produce a zero net deflection at the center of the beam. P will be expressed in terms of q, the dimension a, and the bending stiffness EI.

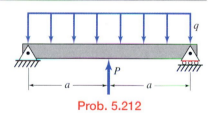

q

a P a

Prob. 5.212

5.213 Consider the force P as given. Use the tabulated solutions to determine the end moments M_0 necessary to produce a zero net rotation at the supports. M_0 will be expressed in terms of P, the dimension b, and the bending stiffness EI.

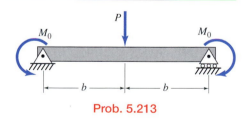

P

M_0 M_0

b b

Prob. 5.213

5.214 The distributed load is given. Use the tabulated solutions to determine the moment M_0 which results in zero net rotation at the right end of the beam.

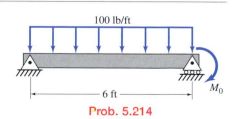

100 lb/ft

6 ft

M_0

Prob. 5.214

5.215 The distributed load is given. The force P is not known, but it results in an end deflection that is 1 mm downward. Use the tabulated solutions to determine P, and then determine the maximum stress in the beam. The cross-section is 2 mm high × 10 mm wide. $E = 70$ GPa.

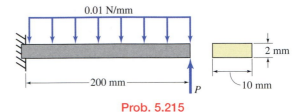

0.01 N/mm

200 mm

P

2 mm

10 mm

Prob. 5.215

5.216 The beam has a uniform bending stiffness EI. Use the tabulated solutions to derive expressions for the deflection and slope of the beam at the free end B.

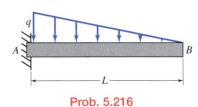

q

A

B

L

Prob. 5.216

5.217 The beam has a uniform bending stiffness EI. Use the tabulated solutions to derive an expression for the slope of the beam at the supports.

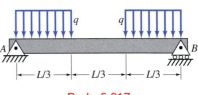

q q

A B

$L/3$ $L/3$ $L/3$

Prob. 5.217

5.218 The beam has a uniform bending stiffness EI. Use the tabulated solutions to derive expressions for the slope of the beam at the (a) left (A) and (b) right (B) supports. Indicate whether the direction of the slopes at each is CW or CCW.

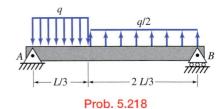

q $q/2$

A B

$L/3$ $2L/3$

Prob. 5.218

5.219 Use the tabulated solutions to derive expressions for the deflection and slope of the beam at the free end A. Indicate whether the beam deflects up or down and whether the slope is CW or CCW. The bending stiffness is EI.

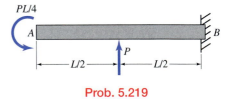

$PL/4$

A B

P

$L/2$ $L/2$

Prob. 5.219

5.220 Determine the deflection and slope at the point $x = 3L/4$. Indicate whether the slope is CW or CCW. (Use the tabulated solutions.) The bending stiffness is EI.

A B

L

M_0

Prob. 5.220

5.221 Use the tabulated solutions to derive expressions for the slope of the beam at the (a) left (A) and (b) right (B) supports. Indicate whether the direction of the slopes at each is CW or CCW. The bending stiffness is EI.

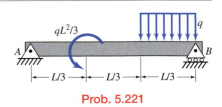

$qL^2/3$ q

A B

$L/3$ $L/3$ $L/3$

Prob. 5.221

5.222 The plates of a toenail clipper must be designed with just the right level of stiffness. A simple model for a clipper consists of two cantilevered beams. Determine the force P_3 to bring together the cutting edges, which are initially 0.125 in. apart. Take $L_1 = 2.25$ in., $L_2 = 2.5$ in., $L_3 = 0.4$ in., $P_1/P_2 = 0.94$, and $P_3/P_2 = 0.06$. Let the thickness and width of the two beams be 0.08 in. and 0.55 in., respectively. The material is steel.

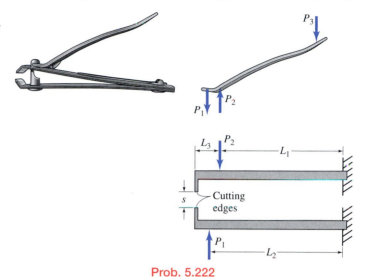

Prob. 5.222

5.223 A pen clip is helpful in keeping the pen clamped to a shirt pocket or a stack of paper. Let there be zero gap between the pen and the steel clip under zero load. The pen has a mass of 28 gm. The clip cross-section is 5 mm wide × 1.2 mm thick. Determine the minimum thickness a clamped object must have if the pen is not to fall under its own weight. Take the friction coefficient between the pen, clip and the clamped part to be 0.2. Treat the clip as cantilevered to the top of the pen.

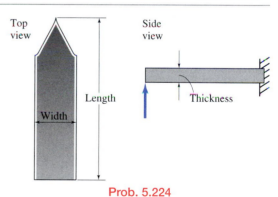

44 mm

Prob. 5.223

5.224 Atomic force microscope (AFM) probes are cantilevered beams, which are designed to have certain stiffnesses. A silicon probe, with an elastic modulus of 90 GPa, has a rectangular cross-section. For a particular model to be used in non-contacting tapping mode, the manufacturer quotes the range in values as width (40–50 μm), thickness (3.6–5.6 μm) and length (150–170 μm). Given this range in parameters, determine the maximum and minimum possible values of stiffness, that is, end force per unit end deflection. Neglect the change in cross-section at the tip.

Top view

Side view

Length

Width

Thickness

Prob. 5.224

5.225 The cantilevered wood chopsticks are subjected to repeating loads, and so the stress is of concern. Say the chopsticks are used to pick up food that is 0.375 in. wide and that the fingers are applying forces as shown. Assume the picked up food exerts negligible force. Determine the maximum stress experienced by the wood. The cross-section is rectangular, with dimensions shown. ($E = 1.5 \times 10^6$ psi.)

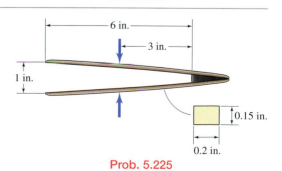

6 in.

3 in.

1 in.

0.15 in.

0.2 in.

Prob. 5.225

5.226 Some of the flexibility of a bicycle comes from deflections of the fork. Severe deflections are associated with descending suddenly from a sidewalk to the road. Let an upward 300 lb force on the tire be transmitted to the fork tips through the axle. Assume that the force is shared equally by the two blades of the fork. Consider the deflection of the fork blades from the crown to the tips due to bending and neglect the axial deformation of the fork. We simplify the cross-section to be uniform with the cross-section shown (in practice it is usually tapered). Take the cross-section at a distance h from the fork tips to be cantilevered. The bending deflection at the fork tip is in the direction perpendicular to the blade. Once the bending deflection is determined, resolve that deflection to determine how much the crown moves vertically relative to the fork tip under such a loading. Assume the fork is carbon fiber composite with $E = 30(10)^6$ psi. Approximate the fork as straight, radiating from the center of the wheel. Take the dimensions to be $h = 14$ in., $\theta = 72°$, $a = 0.75$ in., $b = 1.25$ in., and $t = 0.1$ in.

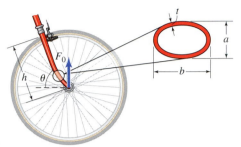

Prob. 5.226 (Appendix A1)

5.227 Flexibility due to deflection of the pedals is not desirable. In the pedal shown, the pedal contacts the pedal spindle at two points. To lower the weight, spindles made of titanium have been considered, although their greater flexibility is viewed as a potential drawback. One detrimental consequence of deflection is that the rider may sense the rotation of the spindle and pedal due to bending. To estimate this effect, the spindle is approximated as having a uniform cross-section. Let a downward force of 250 lb on the pedal be divided equally between the two contact points. Determine the angle of rotation of the spindle midway between the contact points. Neglect any twisting of the crank. Let the titanium spindle have an outer diameter of 0.4 in. and an inner diameter of 0.2 in. Take the dimensions to be $a = 0.5$ in., $b = 1.0$ in., and $c = 1.5$ in.

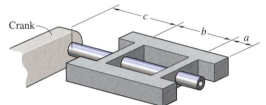

Prob. 5.227 (Appendix A1)

5.228 The cable-stayed bridge has one cable running to the center-line of the roadway. The A-shaped steel tower consists of two separate box sections ($c_1 = 12$ m $\times$ $c_2 = 10$ m of 40 mm thick steel plate) that join at a certain point. Determine the increase in the tension in the cable on one side of the tower that will deflect the top of the pylon perpendicularly to the height by 10 mm. Approximate the pylon as two straight beams of length $h_1 + h_2$, each cantilevered at its base, with the cable force evenly split between the two beams. Take the dimensions to be $w = 100$ m, $h_1 = 150$ m, and $h_2 = 60$ m.

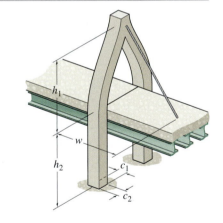

Prob. 5.228 (Appendix A2)

5.229 The short cable-stayed bridge has a single pylon with a single cable running from the tower to the center of the span. The span is simply supported and is not connected to the pylon, except via the cable. Say a 10 Mg truck that passes over the bridge is located at the midpoint of the span. For bending in the plane, the steel structure under the deck has a moment of inertia $I = 0.8$ m^4. The cable shown has an area of 0.02 m^2 and an effective elastic modulus of 165 GPa. Calculate separately and compare the deflections that would occur if (a) the deck bends and there were no cable and (b) the deck bending stiffness is neglected and only tension in the cable provides resistance. Judging from the results of (a) and (b), if both the cable and the understructure bending stiffness were included, which would most determine the deflection? Take the dimensions to be $w_1 = 50$ m, $w_2 = 20$ m, $h_1 = 40$ m, and $h_2 = 30$ m.

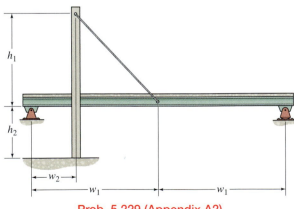

Prob. 5.229 (Appendix A2)

5.230 In this cable-stayed bridge the roadway is not supported by the pylons directly, but by cables running to the center-line of the deck. The tension is to be increased in the two cables supporting the deck. Assuming the pylons themselves do not deflect, by how many mm is the middle of the span lifted for every kN increase in both of the cable tensions? The bending resistance of the roadway comes from the steel girder, with $I = 0.9$ m^4. Neglect the elongation of the cables. Take the dimensions to be $w_1 = 10$ m, $w_2 = 20$ m, $w_3 = 30$ m, $h_1 = 30$ m, and $h_2 = 20$ m.

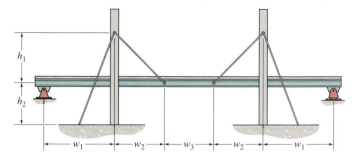

Prob. 5.230 (Appendix A2)

5.231 Lift off of the deck from an end support due to cable tensions overcoming the force of gravity on the deck is a concern in some cable-stayed bridges. A pair of cables is attached symmetrically to the two edges of the deck (one is shown). Let the deck be reinforced concrete with a cross-section of 2.4 m^2 and a density of 2400 kg/m^3. Determine the tension in each of the two cables at which the end of the deck would first lose contact with the support (not shown). The deck has an elastic modulus of 40 GPa and $I = 0.2$ m^4. Simplify the side span as being cantilevered at the pylon. Take the dimensions to be $w = 40$ m and $h = 60$ m.

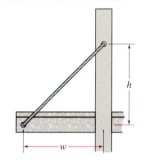

Prob. 5.231 (Appendix A2)

5.232 The simplified external fracture fixation system consists of a carbon fiber rod secured by a pin to each fragment. Say that the patient is subjected to a lateral force $F_0 = 100$ N at the level of the foot pushing the foot outward. Determine (a) the outward deflection and (b) the rotation of the carbon fiber rod where it is connected to the lower pin. Let the diameter of the solid carbon fiber bar be 10.5 mm. Take the carbon fiber to have an elastic modulus of 200 GPa and the dimensions to be $L_r = 300$ mm, $L_s = 200$ mm, and $L_p = 80$ mm. Approximate the upper pin as being rigid and acting as a cantilever support for the rod.

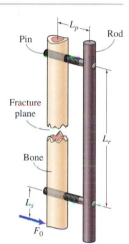

Prob. 5.232 (Appendix A5)

5.233 Consider the external fracture fixation system under the condition that the foot is caught while the leg is trying to move up. This results in a downward 100 N force on the foot. Take the rod to be solid carbon fiber with 10.5 mm diameter and with a 200 GPa elastic modulus. Approximate the upper end of the rod where it is attached to the pin as cantilevered, neglect deformation due to axial loading, and determine the bending deflection of the carbon fiber rod where it connects to the lower pin. Which way does the rod deflect? Let the dimensions be $L_r = 300$ mm and $L_p = 50$ mm.

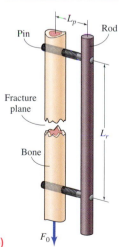

Prob. 5.233 (Appendix A5)

>>End Problems

5.19 Complex Generalizations of Tabulated Solutions

Here we show how to find deflections of more complex structures, if they can be decomposed into beams that have tabulated solutions.

1. Analysis of complex structures using tabulated solutions relies on two ideas.

This example of a diving board subjected to an end load contains the major ideas in analyzing complex structures:

- When a structure is separated into adjacent parts, the internal loads at their connections are treated as external loads on the separated parts.
- If a beam has the same $M(x)$ as a tabulated solution (and hence the same deformed shape), we can add a rigid motion to the tabulated deflection to give the beam the correct total deflection.

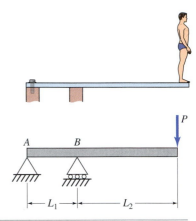

2. Break the structure into a *base*, which is simply supported or cantilevered, and the additional portions that are straight beams.

Because there is a pin and roller, we can identify a simply supported base portion AB, by separating just to the right of B.

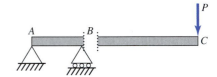

3. Apply equilibrium to determine the force and moment where the structure has been separated into portions.

From equilibrium of BC, we can find the loads that must act on BC at the cut.

By Newton's 3rd Law, the equal and opposite loads act on the base AB at the cut.

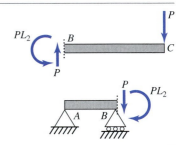

4. Use the tables to find the deflection of the base.

Deflection of the base AB is found from the superposition of two simply supported beams, one with a force at the end and one with a moment.

A force acting through a roller or pin support (the cut is essentially at the support) often appears when dismembering structures into beams. This loading produces zero deflection of the beam, so it can be disregarded.

$M(x) = 0 \Rightarrow \kappa(x) = 0 \Rightarrow v(x) = 0$
since $v = 0$ at supports

As we see later, we need the deflection and slope of the base at B where it connects to BC due to the end moment PL_2. The deflection at the roller support is zero. To find the slope use this tabulated problem, with $M_0 = PL_2$ and $L = L_1$. We want the slope at the end where the moment is applied (θ_A from Appendix G-2.3).

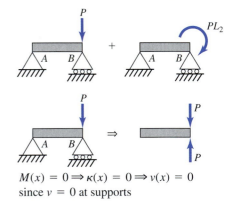

$$\theta_A = \frac{M_0 L}{3EI}$$

So for our problem

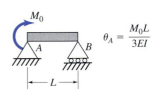

$$\theta_B \qquad \theta_B = \frac{M_0 L}{3EI} = \frac{PL_2 L_1}{3EI}$$

5. Analyze the remaining portion as a cantilever with an added rigid motion.

The remaining portion (or portions if the structure needed to be dismembered into more than two beams) has loads that include a force and moment at the cut. Such a portion would have the same $M(x)$ and deformed shape if the portion were instead cantilevered at the cut. To the deflection of the cantilever, we must add the correct rigid motion, since BC is not truly cantilevered at B.

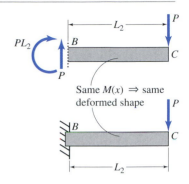

Same $M(x) \Rightarrow$ same deformed shape

---

6. Find the deflections of the additional portions assuming they are cantilevered at the cut.

If BC were cantilevered at B, its deflection and slope at C are found from Appendix G-1.1:

$$(\delta_C)_{cant} = \frac{PL_2^3}{3EI}$$

$$(\theta_C)_{cant} = \frac{PL_2^2}{2EI}$$

---

7. Find the rigid motion that must be added to a portion so it connects with continuous deflection and slope to the base (or to the adjacent portion which is closer to the base).

Here is the deflection of AB, and here is the deflection of BC, assuming it is cantilevered at B. Because AB has a non-zero slope at B, there would be a kink (jump in slope) if BC truly deflected as if it were cantilevered at B. To avoid the kink and have continuous slope at B, we must add to BC a CW rigid rotation about B equal to $PL_1L_2/3EI$, the rotation we found from analyzing AB.

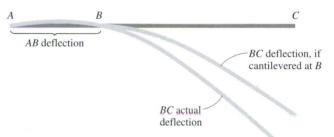

AB deflection

BC deflection, if cantilevered at B

BC actual deflection

The rigid rotation of BC by angle $PL_1L_2/3EI$ about point B adds a deflection $v = -(PL_1L_2/3EI)\,x$, where $x = 0$ is at B. This rigid motion has a slope $PL_1L_2/3EI$ CW everywhere and a downward deflection of $(PL_1L_2/3EI)(L_2)$ at point C.

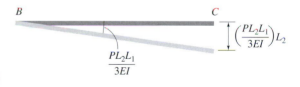

$$(\delta_C)_{rigid} = \frac{PL_2L_1}{3EI}L_2 \qquad (\theta_C)_{rigid} = \frac{PL_2L_1}{3EI}$$

---

8. Add the rigid motions to the cantilever solution to get the total deflection and slope.

The total deflection and slope at C, due to the cantilever deflection of BC and the rigid rotation of the base at B, are:

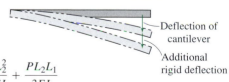

Deflection of cantilever

Additional rigid deflection

$$(\delta_C)_{total} = \frac{PL_2^3}{3EI} + \frac{PL_2^2L_1}{3EI} \qquad (\theta_C)_{total} = \frac{PL_2^2}{2EI} + \frac{PL_2L_1}{3EI}$$

>>End 5.19

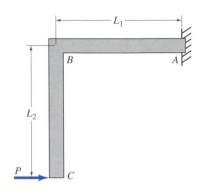

Determine the horizontal and vertical deflections at point C and the rotation there. Neglect deflections due to axial deformations.

Solution

Break the beam into straight segments and recognize the loads on each, as shown.

Start to determine deflections with the base portion that has standard supports (here AB). Let u be the horizontal deflection (positive to the right), and let v be vertical deflection (positive down).

AB

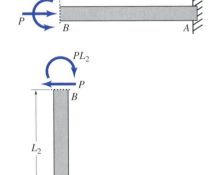

Force P produces axial deformation, so we ignore its influence on the displacement of B.

The moment PL_2 on AB causes B to deflect down and rotate CCW.

To get the magnitudes of the deflection and rotation, recognize that this problem is equivalent to the tabulated problem (see Appendix G-1.2).

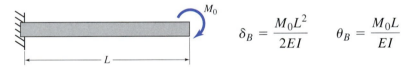

$$\delta_B = \frac{M_0 L^2}{2EI} \qquad \theta_B = \frac{M_0 L}{EI}$$

Apply to our problem $\quad L = L_1 \quad M_0 = PL_2$

Note that deflection parallel to beam, u_B, is taken to be zero, as usual.

BC

BC is attached to AB.

The right angle remains a right angle even when the beams deflect. If the angle ceased to be a right angle, it would be as if the loading produced a kink or sudden change in angle in the middle of a beam.

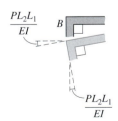

Since AB rotates CCW at B, BC must also rotate CCW at B by the same angle.

The displacements at B must also be the same, whether one considers AB or BC.

The segment BC has the same bending moment distribution as a cantilever with an end force.

BC differs from a cantilever beam, though, because the deflection and rotation at B are not zero.

We need to add a rigid motion to the cantilever deformation, so that the motion at the upper end (B) is consistent with what we determined from analyzing AB.

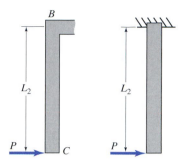

Consider first the deflections that BC would have if it were cantilevered at B.

Note: $(u_C)_{cant}$ is to the right, $(v_C)_{cant}$ is zero, and $(\theta_C)_{cant}$ is CCW.

Use tabulated solution (see Appendix G-1.1).

Displaces down by $\dfrac{PL^3}{3EI}$

Rotates CW by $\dfrac{PL^2}{2EI}$

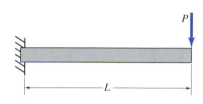

Use this tabulated solution, except with $L = L_2$.

$$(u_C)_{cant} = \frac{PL_2^3}{3EI} \text{ to the right}, (v_C)_{cant} = 0, (\theta_C)_{cant} \text{ is } \frac{PL_2^2}{2EI} \text{ CCW}$$

Consider now the motion of BC if it instead moved rigidly (without deformation) along with the end B in AB.

$$(\theta_C)_{rigid} = (\theta_B) = \frac{PL_2L_1}{EI}, (u_C)_{rigid} = (\theta_B)L_2 = \frac{PL_2^2L_1}{EI}$$

$$(v_C)_{rigid} = (v_B) = \frac{PL_2L_1^2}{2EI}$$

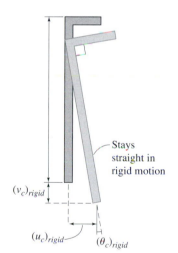

Stays straight in rigid motion

The total deflection and rotation of point C is the summation of deflections and rotations due to the cantilever deformation of BC and the rigid motion. Keep track of directions and add or subtract as necessary.

Horizontal
Both to the right: $\quad u_C = (u_C)_{cant} + (u_C)_{rigid} = \dfrac{PL_2^3}{3EI} + \dfrac{PL_2^2L_1}{EI}$

Vertical
Rigid motion down: $\quad v_C = (v_C)_{rigid} = \dfrac{PL_2L_1^2}{2EI}$

Rotation
Both CCW: $\quad \theta_C = (\theta_C)_{cant} + (\theta_C)_{rigid} = \dfrac{PL_2^2}{2EI} + \dfrac{PL_2L_1}{EI}$

>>End Example Problem 5.44

Additional data on material properties needed to solve problems can be found in Appendix D or inside back cover.

In the problems on this page, neglect axial deformations (relative to bending), and assume that rotations are all small.

5.234 Split the compound beam into two or more beams, one of which can be analyzed using the tables. Find the internal loads between the split beams and use the tabulated solutions to derive an expression for the rotation of the beam at the support A. The bending stiffness is EI.

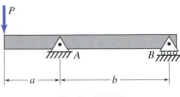

Prob. 5.234

5.235 Split the compound beam into two or more beams, one of which can be analyzed using the tables. Find the internal loads between the split beams and use the tabulated solutions to derive expressions for the horizontal and vertical deflections and rotation at the point B. The bending stiffness is EI.

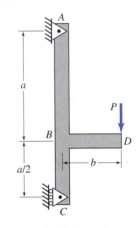

Prob. 5.235

5.236 Split the compound beam into two or more beams, one of which can be analyzed using the tables. Find the internal loads between the split beams and use the tabulated solutions to derive expressions for the horizontal and vertical deflections and rotation at the point B. The bending stiffness is EI.

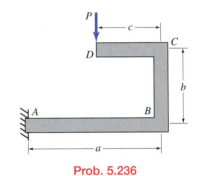

Prob. 5.236

5.237 Split the compound beam into two or more beams, one of which can be analyzed using the tables. Find the internal loads between the split beams and use the tabulated solutions to derive expressions for the rotation at the point B. Indicate if the rotation is CW or CCW. The bending stiffness is EI.

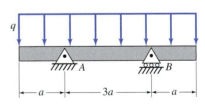

Prob. 5.237

5.238 Split the compound beam into two or more beams, one of which can be analyzed using the tables. Find the internal loads between the split beams and use the tabulated solutions to derive expressions for the horizontal and vertical deflections and rotation at the point B. The bending stiffness is EI.

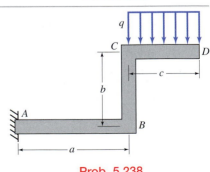

Prob. 5.238

In the problems on this page, neglect axial deformations (relative to bending) and assume that rotations are all small.

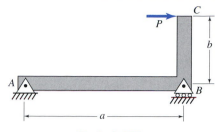

5.239 Imagine the compound beam has been split and analyzed, and the rotation at B was found to be CW with magnitude θ_0. (Deflections are zero at B.) Recognize that segment BC has the same internal loads as a cantilevered beam with an end load. Use the tabulated solutions and the rotation at B to derive expressions for the horizontal and vertical deflections at point C. The bending stiffness is EI.

Prob. 5.239

5.240 Imagine the compound beam has been split and analyzed, and the rotation at B was found to be CCW with magnitude θ_0. (Deflections are zero at B.) Recognize that segment BC has the same internal loads as a cantilevered beam with a distributed load. Use the tabulated solutions and the rotation at B to derive expressions for the horizontal and vertical deflections at point C. The bending stiffness is EI.

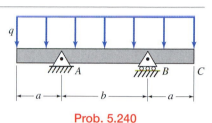

Prob. 5.240

5.241 Imagine the compound beam has been split and analyzed, and the rotation at C was found to be CCW with magnitude θ_0. (Let the deflections at C be defined to be zero.) Recognize that segment CD has the same internal loads as a cantilevered beam with the same distributed force. Use the tabulated solutions and the rotation at C to derive expressions for the horizontal and vertical deflections at point D. The bending stiffness is EI.

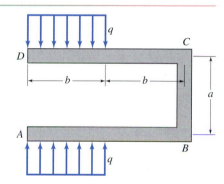

Prob. 5.241

5.242 Imagine the compound beam has been split and analyzed; the rotation at B was found to be CW with magnitude θ_0 and the horizontal deflection was found to be u_0 to the right. (The vertical deflection was zero at B.) Recognize that segment BC has the same internal loads as a cantilevered beam with appropriate loads. Use the tabulated solutions and the deflections and rotation at B to derive expressions for the horizontal and vertical deflections at point C. The bending stiffness is EI.

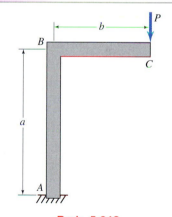

Prob. 5.242

5.243 Imagine the compound beam has been split and analyzed; the rotation at B was found to be CW with magnitude θ_0 and the vertical deflection was found to be v_0 downward. (The horizontal deflection was zero at B.) Recognize that segment BC has the same internal loads as a cantilevered beam with appropriate loads. Use the tabulated solutions and the deflections and rotation at B to derive an expression for the horizontal and vertical deflections and rotation at point C. The bending stiffness is EI.

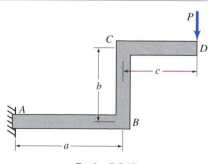

Prob. 5.243

5.244 Split up the structure shown as appropriate and use the tabulated solutions to derive expressions for the horizontal and vertical deflections at point C. The bending stiffness is EI.

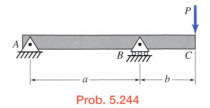

Prob. 5.244

5.245 Split up the structure shown as appropriate and use the tabulated solutions to derive expressions for the horizontal and vertical deflections at point C. The bending stiffness is EI.

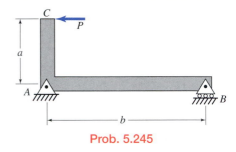

Prob. 5.245

5.246 Split up the structure shown as appropriate and use the tabulated solutions to derive expressions for the horizontal and vertical deflections at point D. The bending stiffness is EI.

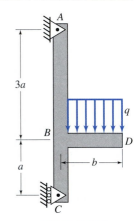

Prob. 5.246

5.247 Split up the structure shown as appropriate and use the tabulated solutions to derive expressions for the horizontal and vertical deflections at point C. The bending stiffness is EI.

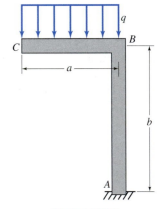

Prob. 5.247

5.248 Split up the structure shown as appropriate and use the tabulated solutions to derive expressions for the vertical deflections at (a) the center of the beam and (b) at point C. The bending stiffness is EI.

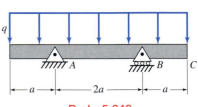

Prob. 5.248

5.249 Split up the structure shown as appropriate and use the tabulated solutions to derive expressions for the horizontal and vertical deflections at point C. The bending stiffness is *EI*.

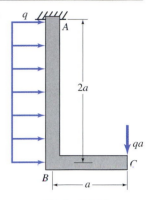

Prob. 5.249

5.250 Split up the structure shown as appropriate and use the tabulated solutions to derive expressions for the horizontal and vertical deflections at point D. The bending stiffness is *EI*.

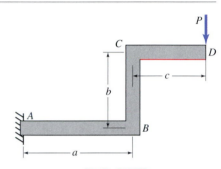

Prob. 5.250

5.251 The problem shown models a beam with supports at A and B that are compliant. Determine the deflection at the loaded point C. (Hint: you can solve this by treating it as one simply supported beam with a single load, which is then rigidly displaced consistent with the spring deflections.)

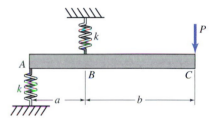

Prob. 5.251

5.252 The moment applied causes twisting of the solid circular rod and bends the simply supported beam. Determine the rotation where the moment *T* is applied. Take the beam to have height *h*, width *b*, and Young's modulus *E*. Take the rod to have radius *R* and shear modulus *G*.

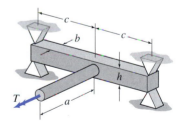

Prob. 5.252

5.253 A weight is to be hung at point C from the long wire that is attached to the bracket AB. The steel wire BC has length 200 ft and diameter 0.0625. The steel bracket is 0.25 in. high, 1 in. wide, and 3 ft long. Determine the downward deflection at C when a weight of 100 lb is hung, and the fraction of the deflection that is due to the bracket bending.

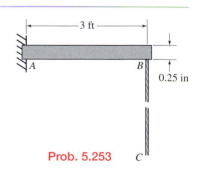

Prob. 5.253

5.254 Deformation in so-called torsional springs, such as in a mousetrap, is due primarily to bending of the wire in the coil. Let the force F be 0.5 lb and the distance L be 1 in. Let there be $N = 20$ complete loops, each of diameter 0.25 in.; the wire itself has a diameter of 0.06 in. Estimate the deflection δ. Neglect the contribution of the short length of the wire at the other end of the coil. Hint: approximate the bending moment as constant in the coil and determine the change in rotation angle along the coil.

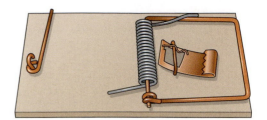

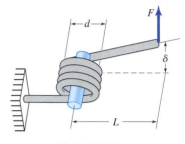

Prob. 5.254

5.255 When a heavily weighted barbell deflects, the bar rotates where it is gripped. The solid steel barbell is 7 ft long and 1 in. in diameter. The gripping hands are 2 ft apart. The barbell holds two sets of four weight plates stacked tightly one next to the other. Each plate weighs 50 lb and is 2 in. wide. The inner surfaces of the inner plates are 5 ft apart. Determine the equal and opposite rotations of the bar at the hands.

Prob. 5.255

5.256 This clamp is used to hold up a broom. Clamping action comes from having adequate stiffness. An estimate of the squeezing action of the clamp can be found from this simple model. Say the material is steel, the width is 20 mm, and the thickness is 2 mm. If the arms are widened by 4 mm, what are the squeezing forces P? Take the corners A and B to have zero deflections. Take the dimensions to be $a = 30$ mm and $b = 80$ mm.

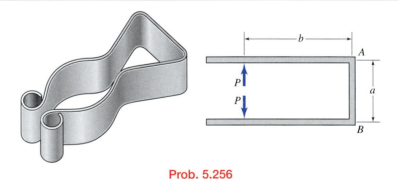

Prob. 5.256

5.257 Hand grippers serve to strengthen the muscles of the hand. Say there are two and one-half loops of wire between the handles. Estimate the deformation in the loops by taking the moment there to be constant; this produces a constant angle change per length. Take the wire diameter to be 0.15 in., $R = 0.5$ in., and $L = 3$ in. If the gripping forces are $P = 20$ lb, by how much does the handle deflect inward on each side where the forces are applied? Approximate the handles as parallel.

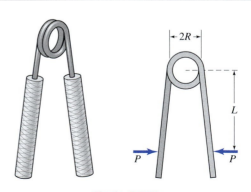

Prob. 5.257

5.258 The simplified external fracture fixation system consists of a carbon fiber rod secured by a stainless steel pin to each fragment. Say that the patient is subjected to a lateral force $F_0 = 50$ N at the level of the foot pushing the foot outward. Determine the transverse and axial displacements and rotation of the lower bone fragment relative to the upper bone fragment at the fracture plane. Account for the deflections of both pins and the carbon fiber rod. Note the bone fragments are not loaded between the pins, and so deflect rigidly. Let the diameter of the solid carbon fiber rod be 10.5 mm and the diameter of the stainless steel pins be 5 mm. Take the carbon fiber and the steel both to have elastic moduli of 200 GPa and the dimensions to be $L_r = 300$ mm, $L_s = 120$ mm, and $L_p = 60$ mm. Take each of the two fractured bone segments to have length $L_r/2$.

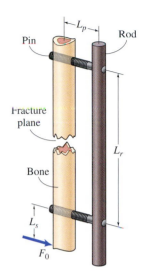

Prob. 5.258 (Appendix A5)

5.259 Consider the external fracture fixation system under the condition in which the foot is caught while the leg is trying to move up. This results in a downward 70 N force on the foot. Determine the transverse and axial displacements and rotation of the lower bone fragment relative to the upper bone fragment at the fracture plane. Account for the deflections of the pins and the carbon fiber rod. Note the bone fragments are not loaded between the pins, and so deflect rigidly. Let the diameter of the solid carbon fiber bar be 10.5 mm and the diameter of the stainless steel pins be 5 mm. Take the carbon fiber and the steel both to have elastic moduli of 200 GPa and the dimensions to be $L_r = 300$ mm and $L_p = 60$ mm. Take each of the two fractured bone segments to have length $L_r/2$.

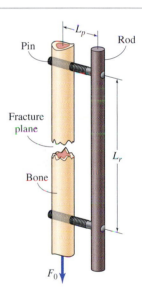

Prob. 5.259 (Appendix A5)

>>End Problems

5.20 Statically Indeterminate Structures Subjected to Bending

Beam problems considered so far always had only two constraint conditions due to the supports. The two unknown support reactions could always be found from the two equilibrium conditions. Sometimes, however, beams are *statically indeterminate*: they have more than two constraint conditions, and their reactions cannot be found from equilibrium alone. Here we show how to superpose tabulated solutions to solve statically indeterminate beam problems.

1. Use the same principles and equations involving equilibrium and deformation to solve statically indeterminate problems, but combine the equations rather than solve them sequentially.

For any mode of deformation, axial (Section 3.4), torsion (Section 4.14), or bending, we must relate four types of **quantities** using three types of **relations**, as depicted in this schematic.

Here are those quantities and relations for the case of bending.

Tabulated solutions satisfy all these relations; we just need to choose the right tabulated solutions.

2. Statically indeterminate beams have more than two support reactions, but enough extra constraints.

In this typical problem there are 3 unknown reactions (R_A, R_B, M_A), and only 2 equilibrium equations $(\Sigma F_y = 0$ and $\Sigma M = 0)$.

But there are 3 constraints $(v(0) = 0; v'(0) = 0; v(L) = 0)$, which is one more than the two needed to evaluate constants c_1 and c_2.

In statically indeterminate beam problems, there are always extra constraint conditions to allow one to solve for the extra support reactions.

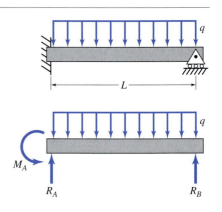

3. Superpose tabulated solutions for either only cantilevered or only simply supported beams, and automatically satisfy two constraints of the original problem.

Generally, we can take advantage of tabulated solutions and automatically satisfy two constraints: $v(0) = 0$ and $v(L) = 0$ (with simply supported solutions) or $v(0) = 0$ and $v'(0) = 0$ (with cantilevered solutions). Any constraints not automatically satisfied are treated as unknown loads. In the solutions shown here, we can automatically satisfy the cantilever support and treat the support at B as an additional force R_B. Superpose solutions for a cantilevered beam with (1) the original load (q) and (2) the unknown force R_B.

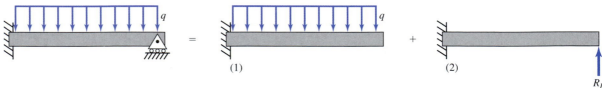

4. Find the superposed unknown load by making it satisfy the only deflection constraint not satisfied by the supports in the standard tabulated solutions.

Since we are using tabulated solutions, equilibrium, the moment–curvature relation, and the relation between deflection and curvature are automatically satisfied, no matter the value of R_B. Also, because both superposed solutions are for cantilever beams, any superposition of them also satisfies the constraints $v(0) = 0$, $v'(0) = 0$, no matter the value of R_B. Only the constraint $v(L) = 0$ is not automatically satisfied.

But, since R_B is still not determined, we can adjust R_B to make $v(L) = 0$.

Here are the deflected shapes for different values of R_B.

We need to find the value of R_B that is just right.

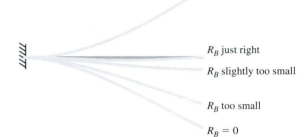

5. Write down formulas for the deflections due to the applied load q and the unknown R_B, and choose R_B so $v(L) = 0$.

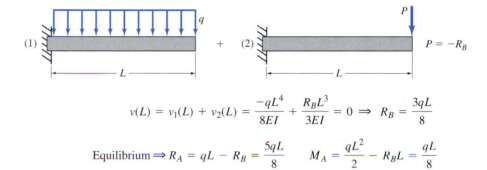

$$v(L) = v_1(L) + v_2(L) = \frac{-qL^4}{8EI} + \frac{R_B L^3}{3EI} = 0 \Rightarrow R_B = \frac{3qL}{8}$$

$$\text{Equilibrium} \Rightarrow R_A = qL - R_B = \frac{5qL}{8} \qquad M_A = \frac{qL^2}{2} - R_B L = \frac{qL}{8}$$

6. As an alternative approach to this same problem, superpose simply supported beams with a moment at A as the unknown superposed load.

If we superpose two simply supported problems, $v(0) = 0$ and $v(L) = 0$ are automatically satisfied.

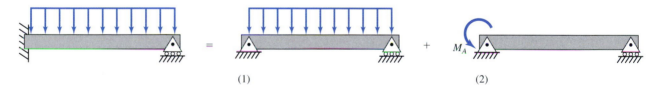

Adjust the unknown moment reaction M_A to satisfy the remaining unsatisfied condition $v'(0) = 0$

$$v'(0) = v_1'(0) + v_2'(0) = \frac{-qL^3}{24EI} + \frac{M_A L^3}{3EI} = 0 \Rightarrow M_A = \frac{qL}{8} \quad \text{Agrees with above!}$$

>>End 5.20

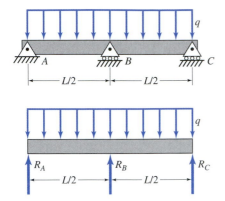

The beam has three pin supports and is subjected to a uniform distributed load. (a) Determine the support reactions at A, B, and C. (b) Compare the bending moment at the center of the beam with the bending moment for the same beam and loading if there were no center pin support.

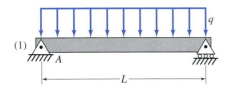

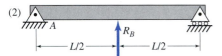

Solution

We have defined pin support reactions (with assumed senses that can be verified by the signs of the results). From equilibrium we find:

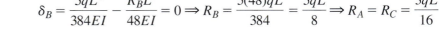

$$\sum F_y = -qL + R_A + R_B + R_C = 0$$

$$\sum M|_B = -R_A(L/2) + R_C(L/2) = 0 \Rightarrow R_A = R_c$$

This is statically indeterminate: there is one extra (redundant) reaction beyond what can be found from equilibrium alone.

The constraints of the pin supports at the ends can be satisfied by superposing problems that are simply supported. The total loading is then a superposition of a uniformly distributed force and the unknown concentrated upward force exerted by the center support. The total deflection must be zero at B.

1. Appendix G-2.6
2. Appendix G-2.1: $P = -R_B$, $a = L/2$

$$\delta_B = \frac{5qL^4}{384EI} - \frac{R_B L^3}{48EI} = 0 \Rightarrow R_B = \frac{5(48)qL}{384} = \frac{5qL}{8} \Rightarrow R_A = R_C = \frac{3qL}{16}$$

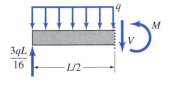

Determine the moment at the center by isolating the segment from the left support to just before the center support.

$$\sum M|_{x=L/2} = -\frac{3qL}{16}\left(\frac{L}{2}\right) + \frac{qL}{2}\left(\frac{L}{4}\right) + M = 0 \Rightarrow M = \frac{-qL^2}{32}$$

If there were no center support, then the reactions at the ends would be

$$R_A = R_C = qL/2$$

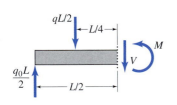

The bending moment at the center would be: $M = \dfrac{qL^2}{8}$

By adding a center support, the magnitude of the moment at the center is reduced by a factor of 4.

>>End Example Problem 5.45

Determine the support reactions for the doubly cantilevered beam with concentrated applied force.

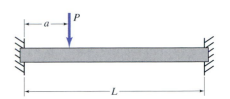

Solution

We have defined support reactions (with assumed senses that can be verified by the signs of the results). From equilibrium we find:

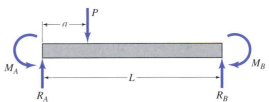

$$\sum F_y = -P + R_A + R_B = 0$$

$$\sum M \mid_A = -P(a) + R_B(L) + M_A - M_B = 0$$

This is statically indeterminate: there are two extra (redundant) reactions beyond what can be found from equilibrium alone.

The constraints at the left end can be satisfied by combining solutions for cantilevered beams.

The loading is then a superposition of the concentrated force P and the unknown supporting force and moment reactions at B. The unknown reactions must be chosen so the total deflection and the total slope at B are zero.

1. Appendix G-1.3
2. Appendix G-1.1: $P = -R_B$
3. Appendix G-1.2: $M_0 = M_B$

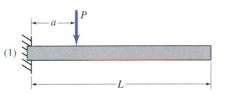

$$\delta_B = \frac{Pa^2(3L - a)}{6EI} - \frac{R_B L^3}{3EI} + \frac{M_B L^2}{2EI} = 0$$

$$\theta_B = \frac{Pa^2}{2EI} - \frac{R_B L^2}{2EI} + \frac{M_B L}{EI} = 0$$

Combine equations as follows: $[\delta_B = 0] - 2L[\theta_B = 0] \Rightarrow M_B = \dfrac{Pa^2(L - a)}{L^2}$

Substitute back into equation $[\theta_B = 0] \Rightarrow R_B = \dfrac{Pa^2(3L - 2a)}{L^3}$

Substitute back into equation $\left[\sum F_y = 0\right] \Rightarrow R_A = \dfrac{P(L^3 - 3a^2L + 2a^3)}{L^3}$

Substitute back into equation $\left[\sum M \mid_A = 0\right] \Rightarrow M_A = \dfrac{Pa(L - a)^2}{L^2}$

One check on the solution is for the case when the force is applied at the center: $a = L/2$. When we substitute into the general results, we find that $R_A = R_B = P/2$ and $M_A = M_B$, which are consistent with symmetry.

>>End Example Problem 5.46

Additional data on material properties needed to solve problems can be found in Appendix D or inside back cover.

5.260 For the beam shown, determine the reactions at the supports.

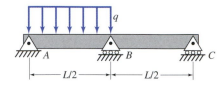

Prob. 5.260

5.261 For the beam shown, determine the reactions at the support and the tension in the bar. Take the bar to have elastic modulus E_b, length L_b, and area A_b and the beam to have bending stiffness EI.

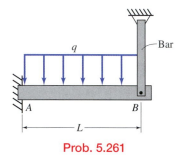

Prob. 5.261

5.262 For the beam shown, determine the reactions at the support and the deflection at the end A. Take the spring to have a spring constant k and the beam to have bending stiffness EI.

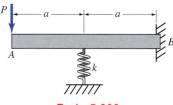

Prob. 5.262

5.263 For the beam shown, determine the reactions at the supports.

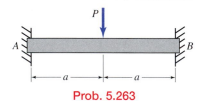

Prob. 5.263

5.264 For the beam shown, determine the reactions at the supports.

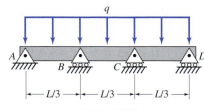

Prob. 5.264

5.265 For the beam shown, determine the reactions at the supports and the deflection in the center.

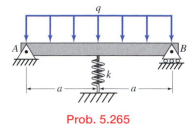

Prob. 5.265

5.266 The center of a simply supported beam rests on the end of a cantilevered beam. The cantilever has a bending stiffness of E_1I_1, and the simply supported beam has a bending stiffness of E_2I_2. Determine the reactions at the supports of both beams.

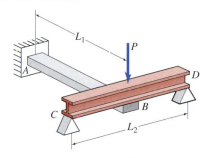

Prob. 5.266

5.267 The wood beam is 3.5 in. wide, 5.5 in. high, and 14 ft long. The elastic modulus of the wood is $1.7(10^6)$ psi. A steel rod 8 ft long is attached to the center of the beam. Determine what the diameter of the steel rod must be if the deflection is to be half of its value without the rod.

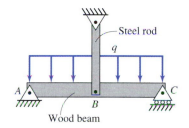

Prob. 5.267

5.268 The beam AB has bending stiffness E_1I_1 and is attached at its end to a rod with properties E_2, A_2, and α_2. Determine the deflection of the beam at B and the stress in the rod if its temperature is reduced by ΔT.

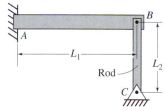

Prob. 5.268

5.269 Two simply supported beams are coupled so their centers have equal deflections. The upper beam has length L_1, and the lower beam has length L_2. Determine the deflection of the center of the beams.

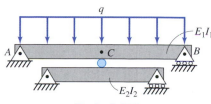

Prob. 5.269

5.270 Beams sometimes lie upon a compliant material, which exerts upward forces on the beam. Two examples are a concrete footer on earth and a ski on snow. The deflections and stresses in the beam may be of interest in such circumstances.

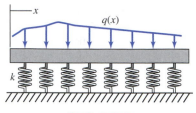

One simple model for the compliant material (the foundation) is to assume that the force per length exerted by the foundation on the beam is proportional to the deflection (with constant k). The force of the foundation acts to oppose the direction of the beam deflection.

Using equations from beam theory, derive the following differential equation governing the deflection satisfied by the beam on a compliant foundation. Take $v > 0$, if upward.

$$EI\frac{d^4v}{dx^4} + kv + q(x) = 0$$

Prob. 5.270

5.271 Consider a single concentrated force applied to the center of a compliantly supported beam (see Problem 5.270). Since there is no applied load on any of the beam except at the central point, the solution for the beam deflection can be found by combining solutions to the homogeneous version of the equation derived in Problem 5.270, namely with $q(x) = 0$. A general solution to the homogeneous equation is given by:

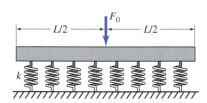

$v = A\cosh(\lambda x)\cos(\lambda x) + B\cosh(\lambda x)\sin(\lambda x) + C\sinh(\lambda x)\cos(\lambda x) + D\sinh(\lambda x)\sin(\lambda x)$

where A, B, C, and D are arbitrary constants.

Determine the constant λ in these equations.

Prob. 5.271

5.272 A tube with a thin wall is reinforced with 8 rods, equally spaced around the circumference, that are bonded to its interior wall. Say that the tube together with the rods deform as a single beam, and that each rod is so thin that its strain is uniform. The rods have area A and elastic modulus E_r. The tube has thickness t, elastic modulus E_t, and radius R.

(a) Say the tube is bent with B-B as the neutral plane and given a curvature κ. Determine the moment.
(b) Say the tube is bent with C-C as the neutral plane and given a curvature κ. Determine the moment.

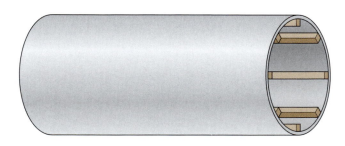

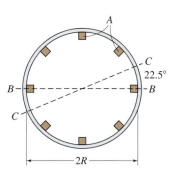

Prob. 5.272

5.273 A construction toy for children features wooden rods that are inserted tightly into holes in thick disks. One particular arrangement is shown below. Say the two disks are rotated in opposite directions, each through an angle ϕ, by the application of twisting moments T. The rods are circular with a radius of R_r and an elastic modulus of E_r. Assume the disks are rigid and that the rods remain straight within each disk. Estimate the angle ϕ.

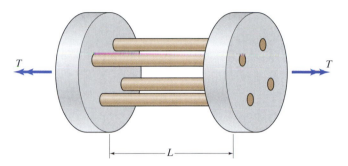
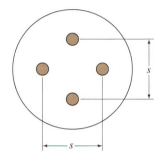

Prob. 5.273

5.274 A tube consists of two layers of separate materials, designated A and B, with elastic moduli E_A and E_B. Let the tube be subjected to a bending moment M. In analyzing the bending, ensure that bending strains are continuous from one material to another. (However, adopt the simplification of neglecting continuity of transverse strains.)

(a) Determine the curvature κ.
(b) Determine the maximum stress in each of the materials.

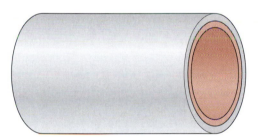
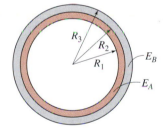

Prob. 5.274

>>End Problems

Chapter Summary

5.1 A **beam** is a relatively long, slender member that bends. Points on the beam displace mostly in the direction perpendicular to the member's length.

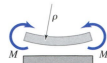

Deformation of the beam due to a bending moment M is quantified by **radius of curvature** ρ or **curvature** $\kappa = 1/\rho$.

5.2–5.3 External transverse loads produce internal loads at each cross-section: **shear force** V and **bending moment** M. Internal loads vary along the beam.

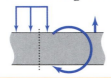

Represent internal loads by drawing them on a short beam segment. By convention, the signs of V and M signal their senses.

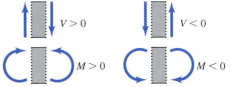

5.4–5.5 Determine V and M in one of two ways.

To determine V and M at one cross-section, apply equilibrium to a segment from one end of the beam to that cross-section.

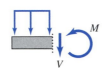

To determine V and M along an entire beam, plot V and M vs. x directly based on the variation due to external loads.

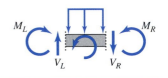

5.6 The primary strain due to bending is normal strain ε_x parallel to the beam's length. Strain only varies through thickness of the beam (linearly with y) from tension on the convex side to compression on the concave side.

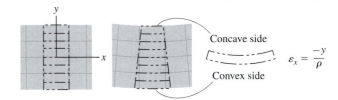

$$\varepsilon_x = \frac{-y}{\rho}$$

5.7 Bending causes only normal stress σ_x parallel to the beam's length. Like strain, it is uniform with z and varies linearly with y, being tensile on the convex side and compressive on the concave side.

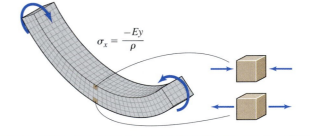

$$\sigma_x = \frac{-Ey}{\rho}$$

5.8 Linear variation of stress across the beam must be equivalent to zero net force and moment M.

Forces due to tensile and compressive stresses balance to give zero net force. This implies:

Neutral plane (plane of zero stress at $y = 0$) must coincide with the centroid of the cross-section (at $\overline{y}$).

Total moment about the z-axis due to stresses must equal bending moment M. This implies:

Moment–curvature relation: $\kappa = \dfrac{1}{\rho} = \dfrac{M}{EI}$

Moment–stress relation: $\sigma_x = \dfrac{-My}{I}$

I is **second moment of inertia** about centroid: $I = \displaystyle\int y^2 \, dA$

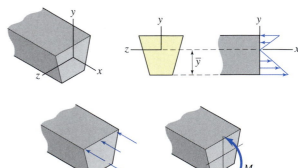

5.9 Composite cross-sections consist of simple shapes that are added or subtracted.

$\bar{y}$ and I for simple shapes are tabulated (Appendix C)

Centroid of Composite: Area-weighted average of centroids (subtract if holes)
$$\bar{y} = \frac{[A_1\bar{y}_1 + A_2\bar{y}_2 - A_3\bar{y}_3 \ldots]}{[A_1 + A_2 - A_3 \ldots]}$$

I of Composite: Sum I of simple shapes (subtract if holes) $\quad I = I_1 + I_2 - I_3$

For each shape, use I about centroid of whole cross-section. If shape centroid is different from that of whole cross-section, use parallel axis theorem: $I = I_c + Ad^2$

5.10–5.11 Bending moment M varies along the beam. Stress distribution is the same as for pure bending (uniform M), but use $M(x)$ that pertains to cross-section at x.

$$\sigma_x = \frac{-M(x)y}{I}$$

Bending stiffness EI depends on elastic modulus, E, and moment of inertia I.

To increase I, move material farther from the neutral plane. **Bending strength** depends on **section modulus** $S = I/y_{max}$ of cross-section and on material strength.

5.12–5.13 Analyze beams with layers of different E by giving all layers the same E but different widths.

For non-symmetric cross-sections, resolve M into components along principal axes of inertia.

5.14–5.15 Shear stress τ_{xy} due to shear force V varies principally with y: $\quad \tau = |\tau_{xy}| = \dfrac{VQ}{It}$

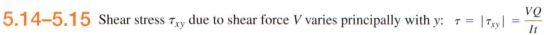

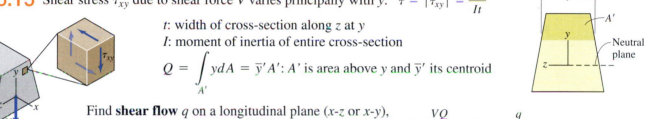

t: width of cross-section along z at y
I: moment of inertia of entire cross-section

$$Q = \int_{A'} y\,dA = \bar{y}'A' : A' \text{ is area above } y \text{ and } \bar{y}' \text{ its centroid}$$

Find **shear flow** q on a longitudinal plane (x-z or x-y), and from it shear stress τ_{xy} or τ_{xz}:

$$q = \frac{VQ}{I} \qquad \tau = \frac{q}{thickness}$$

5.16

Deflection v is related to curvature by $\kappa = \dfrac{d^2v}{dx^2}$, if slope $\left|\dfrac{dv}{dx}\right|$ is small

To determine v: find $M(x)$, then $\kappa = M(x)/EI$, then integrate $\dfrac{d^2v}{dx^2}$ twice.

Evaluate integration constants using known values of v or v' at supports.

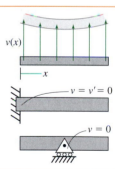

5.17–5.19 Deflections are tabulated for many common loadings of cantilevered and simply supported beams. Solutions are expressed in terms of general parameters (e.g., a, P) that are adapted to specific problems.

Combine tabulated solutions to solve more complex beam problems.

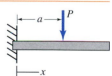

5.20 Statically indeterminate beam problems have more support reactions than can be found by equilibrium alone. They can often be solved by superposition, with an unknown reaction load treated as one of the superposed solutions.

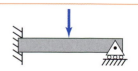

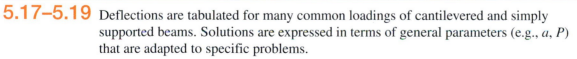

>>**End Chapter Summary**

Combined Loads

1. **Recognize** distinct internal load types

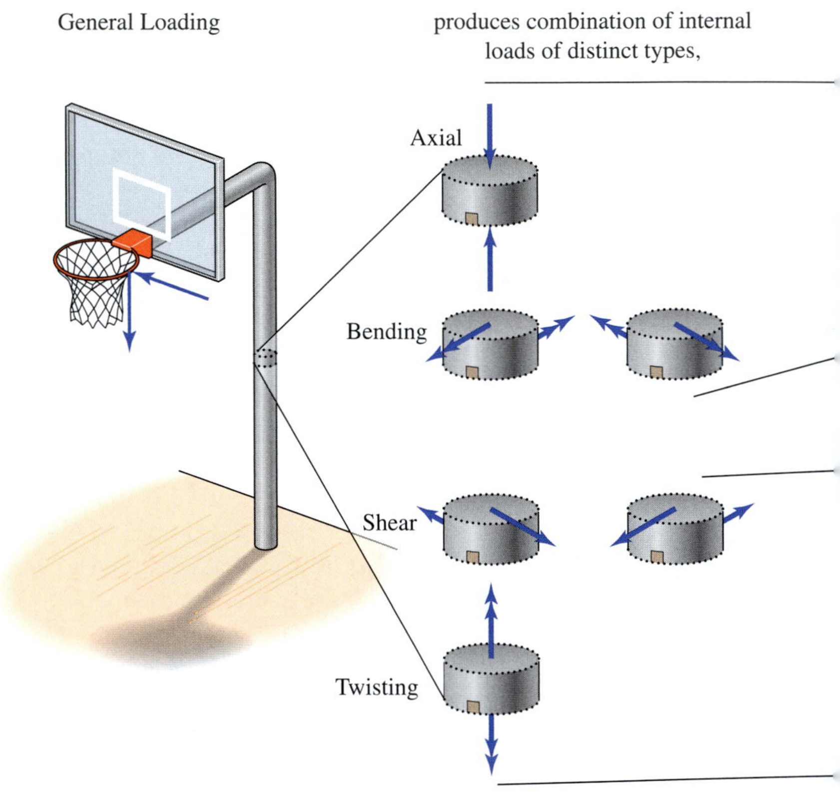

General Loading

produces combination of internal loads of distinct types,

Axial

Bending

Shear

Twisting

2. **Find** deformation due to combination

Element subjected to multiple stresses due to combined loads

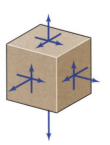

Determine strains due to simultaneous action of multiple stresses

and their stresses

each with its own stress, so a
combination of stresses results

or

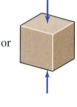

or

Combined stresses also arise in pressure
vessels: hollow bodies with thin walls
subjected to internal pressure

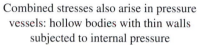

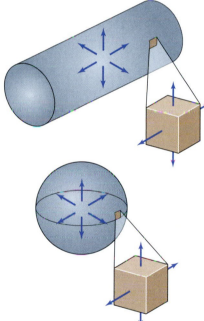

of load types

Combination of internal
loads vary along structure

Determine deflection due to
combined internal loads

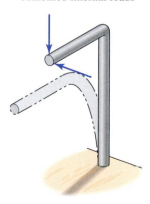

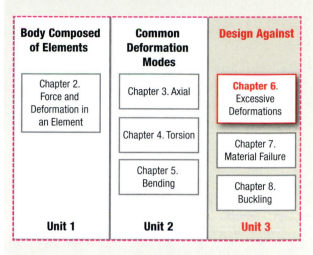

Body Composed of Elements	Common Deformation Modes	Design Against
Chapter 2. Force and Deformation in an Element	Chapter 3. Axial	**Chapter 6.** Excessive Deformations
	Chapter 4. Torsion	Chapter 7. Material Failure
	Chapter 5. Bending	Chapter 8. Buckling
Unit 1	**Unit 2**	**Unit 3**

Chapter Outline

The internal loads at a cross-section of an elongated member, regardless of the complexity of the overall structure or machine, can be resolved into a combination of axial force, shear force, bending moment, and twisting moment (**6.1**). Assuming the member responds elastically, each internal load produces its own normal stress or shear stress, which may vary from point to point, as studied in previous chapters, just as if that internal load acted alone. The different stresses at any point can be represented on the faces of a cubic element (**6.2**). Internal or external pressure acting on a thin-walled vessel can also produce multiple (normal) stresses on each element (**6.3**). One can determine the strains due to a combination of stresses: the normal strain along any one direction depends both on the normal stress in the same direction and on normal stresses in the transverse direction because of the Poisson ratio effect (**6.4**). To determine the deflection of a structure composed of multiple segments in which combinations of internal loads may act, one must carefully track the deflections of various segments and ensure that they are continuous where they meet (**6.5**). An alternative method of determining the deflection in a structure can be devised using the principle of conservation of energy (**6.6–6.7**).

6.1 Determining Internal Loads

We have studied different types of loading, axial, shear, torsion, and bending when each occurs individually. Often, however, these loadings occur together. In this chapter we will study how to capture stress, deformation, and stiffness when a body simultaneously undergoes multiple loading types.

1. A single straight member can have combinations of internal loads because multiple loads are applied directly to it or, for example, because a single force is applied off axis.

Sometimes a single straight body is subjected directly to two or more loadings. For example, when drilling a hole, this drill bit feels both a force along its length, as the workpiece pressing on it, and a moment about the bit axis, as the workpiece resists the turning of the bit.

The loads at an internal cross-section must keep the bit in equilibrium (if it is spinning at constant speed). So the internal loads are axial compression (P) and a twisting moment (T).

Sometimes a single force acts on the body, but because of the direction of the force and its line of action relative to the body, the force induces multiple internal loadings. The internal loads are found from isolating a portion of the body. In the shaft of this golf club, there is a shear force (V), a twisting moment (T), and a bending moment (M).

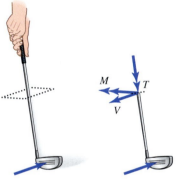

2. Resolve forces and moments into components parallel and perpendicular to a member's axis to decouple the types of internal loads.

We distinguish internal forces, axial and shear, by the direction of the force relative to the long direction of a body. Torsion and bending moments are similarly distinguished.

Depending on the direction of the load relative to the body, the internal force might have to be resolved into components parallel and perpendicular to the member before it can be interpreted as axial vs. shear. Likewise, a moment vector might need to be resolved to separate torsion and bending. At the cross-section shown, there is a bending moment and an internal force. But, to compute the stresses produced by the force, it first needs to be resolved into components corresponding to normal and shear.

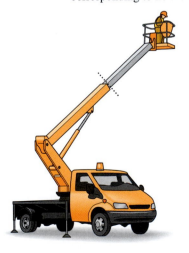

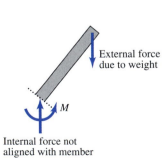

External force due to weight

M

Internal force not aligned with member

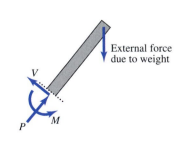

External force due to weight

V

P

M

3. Recognize portions of a complex structure that can be approximated as straight and find internal loads in them.

Engineering components or systems rarely consist of a single straight member. However, an individual portion of a component might be at least approximately straight. Often the internal loads in such a straight portion involve combinations of the basic load types. One must learn to dismember a component or system and detect the prevailing internal loads.

This simple C-clamp has been tightened up to clamp the block. What are the internal loads at different cross-sections of the clamp due the forces applied by the squeezed block?

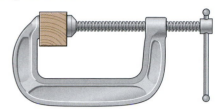

Notice that the clamp consists of distinct portions, each of which is approximately straight. Within each portion, we can find internal loads. Depending on their directions relative to the cut surface, we interpret internal loadings differently. Notice that different portions experience, respectively, simple axial compression, shear plus bending, and axial tension plus bending.

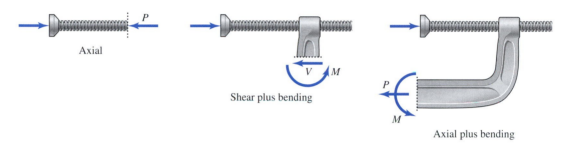

Axial

Shear plus bending

Axial plus bending

---

4. Even a relatively simple loading of a simple structure can lead to combinations of all possible internal load types: axial, shear, bending, and twisting.

This basketball backboard and hoop are supported by a post. Given the height and weight of players, the hoop can be subjected to significant forces acting in any direction. Here the player approaches at high speed from the side, jumps, stuffs the ball, and hangs on the hoop. The forces on the hoop shown produce a wide range of internal loadings that depend on the location of the cross-section of interest.

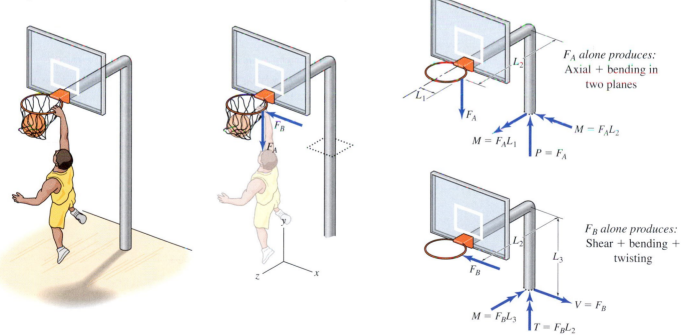

F_A alone produces:
Axial + bending in two planes

$M = F_A L_1$

$M = F_A L_2$

$P = F_A$

F_B alone produces:
Shear + bending + twisting

$M = F_B L_3$

$V = F_B$

$T = F_B L_2$

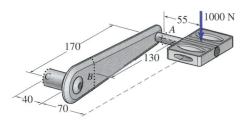

A bicycle pedal and crank are subjected to a 1000 N pedaling force. Determine the internal loads at three cross-sections: A, B, and C. Identify each load type and draw the equal and opposite loads on an infinitesimal segment. Dimensions are in millimeters.

Solution

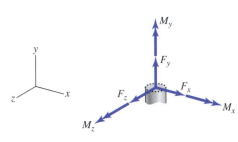

Note that there can, in general, be three components of internal force and three components of internal moment at a cross-section. However, there is no need to include components that are clearly not present (must obviously be zero by equilibrium).

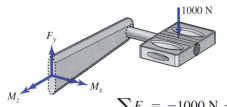

A: Of the 6 possible components, only F_y and M_z are necessary to maintain this portion of the structure in equilibrium.

$$\sum F_y = -1000 \text{ N} + F_y = 0 \qquad \Rightarrow F_y = 1000 \text{ N (shear force)}$$

$$\sum M|_{cut,z} = -1000 \text{ N (55 mm)} + M_z = 0 \Rightarrow M_z = 55 \text{ N-m (bending moment)}$$

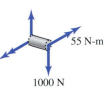

B: Of the 6 components, only F_y, M_x, and M_z are necessary to maintain this portion of the structure in equilibrium.

$$\sum F_y = -1000 \text{ N} + F_y = 0 \qquad \Rightarrow F_y = 1000 \text{ N (shear force)}$$

$$\sum M|_{cut,x} = -1000 \text{ N (130 mm)} + M_x = 0 \Rightarrow M_x = 130 \text{ N-m (bending moment)}$$

$$\sum M|_{cut,z} = -1000 \text{ N (70 mm)} + M_z = 0 \Rightarrow M_z = 70 \text{ N-m (twisting moment)}$$

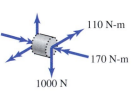

C: Of the 6 components, only F_y, M_x, and M_z are necessary to maintain this portion of the structure in equilibrium.

$$\sum F_y = -1000 \text{ N lb} + F_y = 0 \qquad \Rightarrow F_y = 1000 \text{ N (shear force)}$$

$$\sum M|_{cut,x} = -1000 \text{ N (170 mm)} + M_x = 0 \Rightarrow M_z = 170 \text{ N-m (twisting moment)}$$

$$\sum M|_{cut,z} = -1000 \text{ N (110 mm)} + M_z = 0 \Rightarrow M_z = 110 \text{ N-m (bending moment)}$$

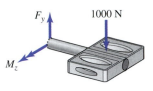
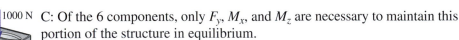

>>End Example Problem 6.1

We often need to consider the combined loadings in shafts that transmit power. Shafts are subject to loads from gears and other elements that are mounted on them and from bearings that constrain them. The shaft shown has bearings at A and B, and tangential and normal forces acting on the gears. The diameters of gears G_1 and G_2 are $D_1 = 4.5$ in. and $D_2 = 6$ in. respectively. (a) Determine the internal loads at bearing A, just toward gear G_1. (b) Determine the internal loads at gear G_2, just toward bearing B. Treat the bearings as exerting only forces at their mid-planes perpendicular to the shaft.

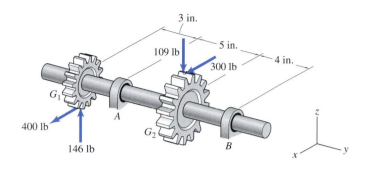

Solution

Notice that the tangential components of the gear forces satisfy $\Sigma M|_y = 0$. Use equilibrium to find the reactions at the bearings in the x and z directions.

$$\Sigma F_x = 0, \ \Sigma M|_z = 0 \Rightarrow R_{Ax} = 667 \text{ lb} \qquad R_{Bx} = 33.3 \text{ lb}$$

$$\Sigma F_z = 0, \ \Sigma M|_x = 0 \Rightarrow R_{Az} = -146.2 \text{ lb} \qquad R_{Bz} = 109.2 \text{ lb}$$

We can anticipate the positive directions of internal forces and moments and determine the magnitudes.

For the cross-section near bearing A:

$$\Sigma M|_{cut,x} = -146.2 \text{ lb (3 in.)} + M_x = 0 \Rightarrow M_x = 439 \text{ lb-in.}$$

$$\Sigma M|_{cut,y} = -400 \text{ lb (2.25 in.)} + M_y = 0 \Rightarrow M_y = 900 \text{ lb-in.}$$

$$\Sigma M|_{cut,z} = 400 \text{ lb (3 in.)} - M_z = 0 \Rightarrow M_z = 1200 \text{ lb-in.}$$

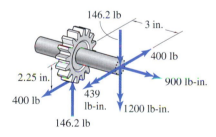

Shear Forces: 146 lb, 400 lb; Twisting Moment: 900 lb-in.; Bending Moments: 439 lb-in., 1200 lb-in.

For cross-section near gear G_2:

$$\Sigma M|_{cut,x} = 109.2 \text{ lb (4 in.)} - M_x = 0 \Rightarrow M_x = 437 \text{ lb-in.}$$

$$\Sigma M|_{cut,z} = 33.3 \text{ lb (4 in.)} - M_z = 0 \Rightarrow M_z = 133.2 \text{ lb-in.}$$

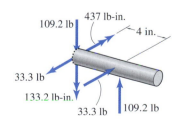

Shear Forces: 109.2 lb, 33.3 lb
Bending Moments: 437 lb-in., 133.2 lb-in.

>>End Example Problem 6.2

Additional data on material properties needed to solve problems can be found in Appendix D or inside back cover.

6.1 Force $F_1 = 20$ kN ($F_2 = 0$) is applied to the structure shown. Determine the magnitudes of the internal loads along the x-y-z axes at the base of the structure (A), and describe each force as axial or shear and each moment as bending or twisting. Neglect the dimensions of the cross-section relative to the lengths of the members, which are $a = 2$ m, $b = 1.5$ m, and $c = 1.75$ m.

6.2 Forces $F_1 = 500$ lb and $F_2 = 750$ lb are applied to the structure shown. Determine the magnitudes of the internal loads along the x-y-z axes at the base (A), and describe each force as axial or shear and each moment as bending or twisting. Neglect the dimensions of the cross-section relative to the lengths of the members, which are $a = 8$ ft, $b = 6$ ft, and $c = 7$ ft.

6.3 The two forces F_1 and F_2 act in the senses shown. Midway along AB they produce a twisting moment of 4000 N-m, and midway along BC they produce a bending moment of 3000 N-m about the x-axis. Determine the magnitudes of F_1 and F_2, and find the twisting moment midway along BC and the bending moment about the y-axis midway along BC. Neglect the dimensions of the cross-section relative to the lengths of the members, which are $a = 1.5$ m, $b = 2$ m, and $c = 1.25$ m.

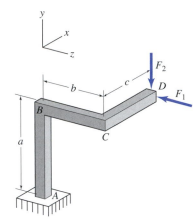

Probs. 6.1–3

6.4 Forces $F_1 = 4$ kN and $F_2 = 6$ N ($F_3 = 0$) are applied to the structure shown. On a sketch of a segment comprising the lower half of the vertical member AB, draw the internal loads along the x-y-z directions exerted on that cut surface by the upper half of AB. Draw the loads in the correct senses and give their magnitudes. Neglect the dimensions of the cross-section relative to the lengths of the members, which are $a = 0.5$ m, $b = 1.25$ m, and $c = 2$ m.

6.5 Forces $F_1 = 2$ kip and $F_2 = 5$ kip ($F_3 = 0$) are applied to the structure shown. On a sketch of the member AB and the connected half of BC, draw the internal loads along the x-y-z directions exerted on that cut surface by the other half of BC. Draw the loads in the correct senses and give their magnitudes. Neglect the dimensions of the cross-section relative to the lengths of the members, which are $a = 2$ ft, $b = 5$ ft, and $c = 7$ ft.

6.6 The loads on the structure result in no twisting of any portion. At the cross-section A, there are internal loads, which include a bending moment of 20 kN about the y-axis. (a) Determine the magnitudes of the forces F_1, F_2, and F_3. (b) Determine and describe each of the internal loads (normal, shear, bending, or twisting) acting through the cross-section midway along BC. Neglect the dimensions of the cross-section relative to the lengths of the members, which are $a = 1.25$ m, $b = 2$ m, and $c = 2.5$ m.

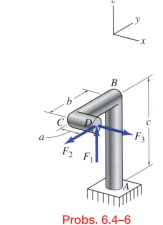

Probs. 6.4–6

In Problems 6.7 to 6.10, the sign is attached to the post that is built-in at the ground. Treat the sign as having uniform mass. Neglect the contribution of the post weight. The length d is from the center plane of the sign to the center of the post.

6.7 If the sign weighs 500 kg and the wind pressure is 2 kPa, determine the internal loads acting along the x-y-z directions at the base of the post. Take the sign dimensions to be $h_1 = 8.5$ m, $h_2 = 1.5$ m, $w = 3.5$ m, and $d = 400$ mm.

6.8 If the sign weighs 1000 lb and there is no wind pressure blowing, determine the internal loads acting along the x-y-z directions at the base of the post. Take the sign dimensions to be $h_1 = 26$ ft, $h_2 = 4$ ft, $w = 10$ ft, and $d = 16$ in.

6.9 The sign weighs 500 kg. Because of a concern for fatigue, the cycles of loading at the base of the post due to intermittent wind of 1.75 kPa are of interest. Consider each of the two bending moments acting along the x-y-z axes. Determine the range over which each moment varies as the wind blows and doesn't blow. Indicate whether or not each bending moment changes direction and magnitude. Take the sign dimensions to be $h_1 = 8.5$ m, $h_2 = 1.5$ m, $w = 3.5$ m, and $d = 400$ mm.

6.10 Consider moments at the base of the post due only to the wind pressure. By what factor must the width of the sign (w) be reduced if all moments at the base (along the x-y-z axes) are to be reduced by a factor of *at least* two? Which moment gets reduced by the factor of two and by what factor will the other moment be reduced?

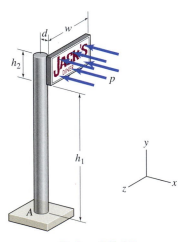

Probs. 6.7–10

In Problems 6.11 to 6.14, the shaft from a transmission has two gears B and C and is supported by bearings at A and D. Assume that the bearings offer no rotational resistance about the x-axis.

6.11 Consider only the effects of the tangential forces $F_{t1} = 600$ lb and $F_{t2} = 1000$ lb (not the normal forces F_{n1} and F_{n2}). Calculate and describe the internal loads along each of the x-y-z axes at the cross-section midway between gears B and C. Take the dimensions to be $L_1 = 4$ in., $L_2 = 8$ in., and $L_3 = 3$ in., and gears at B and C to have diameters $d_B = 10$ in. and $d_C = 6$ in.

6.12 Consider only the effects of the tangential forces $F_{t1} = 2400$ N and $F_{t2} = 4000$ N (not the normal forces F_{n1} and F_{n2}). Calculate and describe the internal loads along each of the x-y-z axes at the cross-section midway between gears B and C. Take the dimensions to be $L_1 = 100$ mm, $L_2 = 200$ mm, and $L_3 = 75$ mm, and gears at B and C to have diameters $d_B = 200$ mm and $d_C = 120$ mm.

6.13 Let the tangential forces be $F_{t1} = 600$ lb and $F_{t2} = 1000$ lb and the normal forces be $F_{n1} = 218$ lb and $F_{n2} = 364$ lb. Calculate and describe the internal loads along each of the x-y-z axes at the cross-section midway between gears B and C. Take the dimensions to be $L_1 = 4$ in., $L_2 = 8$ in., and $L_3 = 3$ in., and gears at B and C to have diameters $d_B = 10$ in. and $d_C = 6$ in.

6.14 Let the tangential forces be 2400 N and $F_{t2} = 4000$ N and the normal forces be $F_{n1} = 874$ N and $F_{n2} = 1456$ N. Calculate and describe the internal loads along each of the x-y-z axes at the cross-section midway between gears B and C. Take the dimensions to be $L_1 = 100$ mm, $L_2 = 200$ mm, and $L_3 = 75$ mm, and gears at B and C to have diameters $d_B = 200$ mm and $d_C = 120$ mm.

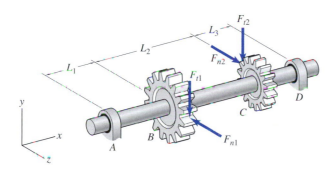

Probs. 6.11–14

>>End Problems

6.2 Drawing Stresses on 3-D Elements

As seen in Chapters 3–5, the different internal load types produce different distributions of different types of stress. In general, when multiple internal loads act, their stresses cannot simply be added. The first step in combining stresses properly is to draw them on a small element that represents a tiny portion of the material taken from one point.

1. Recall the interaction represented by each stress drawn on an element.

The stresses drawn on one face of an element (pink) are the force components per area exerted by the element (blue) that contacts it.

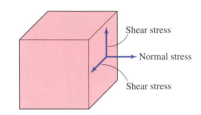

Shear stress

Normal stress

Shear stress

2. Recall the stress distribution for each type of internal load, and draw the stresses due to each internal load on an element aligned with the axes of the member.

The forces on the faces of an element are most easily determined if the sides of the element are aligned with the surfaces of the body from which the element is extracted.

Here we review the stresses on an element for the basic types of internal loads:

Axial Force

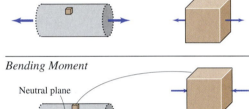

Normal stress; parallel to member axis; uniform in cross-section
(Chapter 2)

Bending Moment

Neutral plane

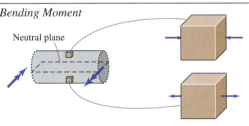

Normal stress; parallel to member axis; varies linearly with distance from neutral plane, tension on one side and compression on the other
(Chapter 5)

Shear Force

No stress

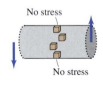

No stress

Shear stress; varies over cross-section
(Chapter 5)

Twisting Moment

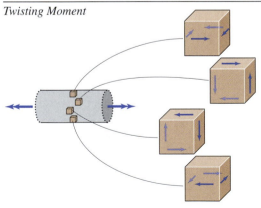

Shear stress; on face normal to axis, shear stress varies linearly with radial distance and acts circumferentially to contribute to torque; complementary shear stresses on longitudinal planes
(Chapter 4)

No forces act on the lateral surfaces of members for any of the four load types. Therefore, no stresses act on element faces that coincide with the lateral surfaces of the member.

3. Focus on a particular point in a member, draw stresses due to each internal load, and add stress components only if they act on the same face of the element and in the same direction.

When multiple internal loads act, consider the stresses they produce at the same point, and draw them on identically oriented cubic elements.

Under this loading, the basketball post is subjected to both axial compression and bending about a single axis. Both internal loads produce normal stress in the vertical direction.

The stresses due to the axial compression are uniform over the cross-section, and the stresses due to bending vary from point A to B. Compare the total normal stress for points A and B.

Point A: Axial and bending both produce compression

To illustrate how to add normal stresses in the same direction, we give them arbitrary magnitudes, 10 MPa compression due to the axial force and a maximum of 70 MPa due to the bending moment. At point A, where the bending stress is compressive, the stresses together produce a compressive stress equal to the sum of the stress magnitudes (80 MPa).

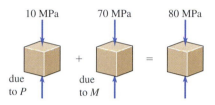

Point B: Axial produces compression and bending produces tension

At point B, the stresses have opposite sign. The forces on, say, the upper face of an element are both vertical, although they have opposite senses. We can add a downward force and an upward force. The net force acts in the sense of the force with the larger magnitude (in this case, upward or in tension) and has a magnitude equal to the difference between the two magnitudes (60 MPa).

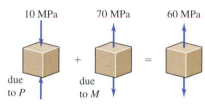

4. Follow this example of drawing multiple stresses, now with combined shear, torsion, and bending.

This loading produces shear, torsion, and bending. At each point, A, B, C, and D, draw the stresses associated with the different loads on an element. Add stresses that act on the same face in the same direction. If shear stress due to T has greater magnitude than that due to V, as is typical, the net shear stress will be in the direction dictated by T.

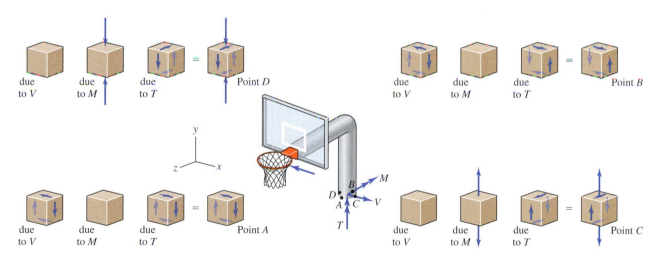

Note that we do not add normal stresses to shear stresses, nor do we add normal stresses that act on different faces. In Chapter 7, we study the combined effect of such stresses acting simultaneously.

>>End 6.2

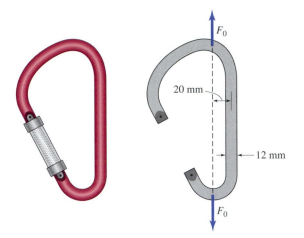

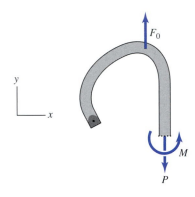

The carabiner used for rock climbing can withstand a maximum force of $F_0 = 5$ kN. The cross-section is circular with a diameter of 12 mm. What are the maximum tensile and compressive stresses? Assume that no stresses are transmitted at the opening.

Solution

The internal loads along the side opposite to the opening can be found from isolating the portion shown.

$$\sum F_y = 5000 \text{ N} - P = 0 \qquad\qquad \Rightarrow P = 5000 \text{ N (normal force)}$$

$$\sum M|_{cut,z} = -5000 \text{ N (20 mm)} + M = 0 \Rightarrow M = 100 \text{ N-m (bending moment)}$$

Cross-section:

12 mm

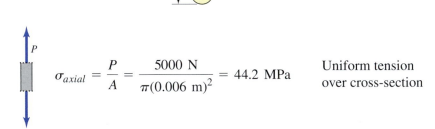

$$\sigma_{axial} = \frac{P}{A} = \frac{5000 \text{ N}}{\pi(0.006 \text{ m})^2} = 44.2 \text{ MPa}$$

Uniform tension over cross-section

44.2 MPa

$$\sigma_{bend} = \frac{Mr}{I} = \frac{(100 \text{ N-m})(0.006 \text{ m})}{\frac{\pi}{4}(0.006 \text{ m})^4} = 589 \text{ MPa}$$

589 MPa

589 MPa

Combine stresses due to axial and due to bending. One finds tension on one side and compression on the other, with different magnitudes.

$589 + 44.2 = 633$ MPa

$589 - 44.2 = 545$ MPa

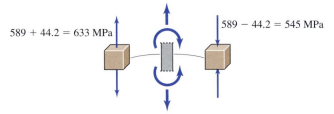

Notice that the stresses due to axial and bending add at the critical point, giving a maximum stress of 633 MPa.

>>End Example Problem 6.3

A tungsten carbide drill bit 3.5 mm in diameter is used to drill aluminum. The drill bit sustains an axial compressive force of 50 N and a torque of 0.06 N-m. Determine stresses at the two points A and B shown.

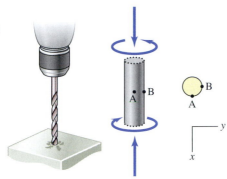

Solution

Draw stresses on a cubic element oriented relative to the x-y-z axes.

The stresses due to the axial force and torsion are found separately.

Axial

$$\sigma_{axial} = \frac{P}{A} = \frac{-50 \text{ N}}{\pi \left(\dfrac{3.5 \text{ mm}}{2}\right)^2} = -5.20 \text{ MPa}$$

Uniform compression over cross-section

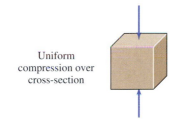

Torsion

$$\tau_{torsion} = \frac{Tr}{I_p} = \frac{(0.06 \text{ N-m})\left(\dfrac{3.5 \text{ mm}}{2}\right)}{\dfrac{\pi}{2}\left(\dfrac{3.5 \text{ mm}}{2}\right)^4} = 7.13 \text{ MPa}$$

Point A

Point B

For each point A and B, draw stresses together on an element.

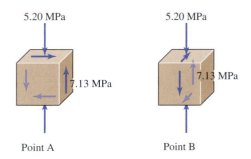

5.20 MPa

7.13 MPa

Point A

5.20 MPa

7.13 MPa

Point B

6.15 The cable under a tension of 500 lb is attached to the structure shown. The steel post has a diameter of 4 in. Determine the stresses acting on a cubic element aligned with the x-y-z axes located at the point A due to (a) each internal moment, (b) the axial force, and (c) each shear force. Dimensions are $a = 12$ ft, $b = 8$ ft, $c = 7$ ft, and $d = 5$ ft. (Note: the maximum shear stress due to a shear force on a circular cross-section is 4/3 times the average shear stress.)

6.16 The cable under a tension of 500 lb is attached to the structure shown. The steel post has a diameter of 4 in. Determine the stresses acting on a cubic element aligned with the x-y-z axes located at the point B due to (a) each internal moment, (b) the axial force, and (c) each shear force. Dimensions are $a = 12$ ft, $b = 8$ ft, $c = 7$ ft, and $d = 5$ ft. (Note: the maximum shear stress due to a shear force on a circular cross-section is 4/3 times the average shear stress.)

6.17 The cable under a tension of 2000 N is attached to the structure shown. The steel post has a diameter of 110 mm. Determine the stresses acting on a cubic element aligned with the x-y-z axes located at the point C due to (a) each internal moment, (b) the axial force, and (c) each shear force. Dimensions are $a = 4$ m, $b = 3$ m, $c = 2.5$ m, and $d = 2$ m. (Note: the maximum shear stress due to a shear force on a circular cross-section is 4/3 times the average shear stress.)

6.18 The cable under a tension of 2000 N is attached to the structure shown. The steel post has a diameter of 110 mm. Determine the stresses acting on a cubic element aligned with the x-y-z axes located at the point D due to (a) each internal moment, (b) the axial force, and (c) each shear force. Dimensions are $a = 4$ m, $b = 3$ m, $c = 2.5$ m, and $d = 2$ m. (Note: the maximum shear stress due to a shear force on a circular cross-section is 4/3 times the average shear stress.)

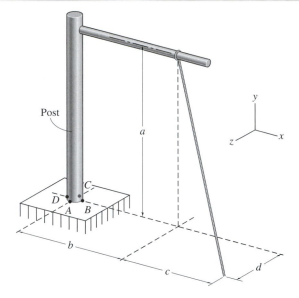

Probs. 6.15–18

6.19 The vertical column with a rectangular cross-section is subjected to an axial load. Determine the maximum offset c of the load from the center line if there is to be no tensile stress in the cross-section.

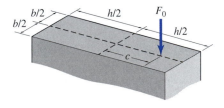

Prob. 6.19

6.20 The vertical column with a rectangular cross-section is subjected to an axial load. The load is located 20 mm from the one symmetry plane as shown. Determine the maximum offset c of the load from the other symmetry plane if there is to be no tensile stress in the cross-section. Dimensions are in millimeters.

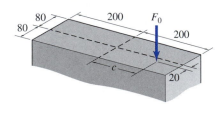

Prob. 6.20

6.21 The carabiner used for rock climbing has a cross-section that can be approximated as having a T shape. If the axial load is $F_0 = 300$ lb and the offset $d = 0.5$ in., determine the maximum tensile and compressive stresses in the cross-section.

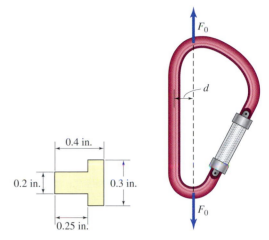

0.4 in.

0.2 in.   0.3 in.

0.25 in.

Prob. 6.21

6.22 Chain links consist of incomplete steel loops. For the chain shown of width $w = 1.5$ in. and wire diameter $d = 0.25$ in., determine the maximum allowable axial force F_0, if the maximum tensile stress is not to exceed 5000 psi.

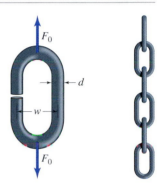

Prob. 6.22

6.23 In contrast to spur gears which only have tangential and normal forces, helical gears experience forces at their teeth that act in the axial direction as well. Consider the effect only of an axial force $F_a = 2000$ N acting on the helical gear at C (ignore all tangential and normal force components applied to the gears). The bearing at A can support the resulting thrust (axial force) on the shaft. Determine the maximum tensile and compressive stress σ_z in the shaft due to F_a. Gears at B and C have diameters $d_B = 200$ mm and $d_C = 120$ mm, the shaft has diameter $d_s = 28$ mm, and other dimensions are $L_1 = 100$ mm, $L_2 = 200$ mm, and $L_3 = 75$ mm.

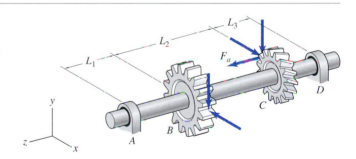

Prob. 6.23

6.24 Proper functioning of a hacksaw requires pre-tensioning of the blade. The blade must have a high enough pre-tension that it does not buckle in use, but not so high that the frame becomes permanently deformed (yields plastically).

(a) Determine the maximum blade pre-tension if the frame is not to reach a yield stress of 50 ksi.

(b) Say the maximum transverse force on the blade while cutting is 75 lb. Estimate the minimum blade pre-tension if no part of the blade is to go into compression due to bending during cutting. Model the 10 in. long blade as simply supported, and take the force to act in the center of the blade.

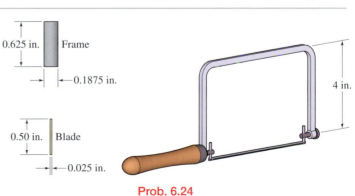

0.625 in.   Frame

0.1875 in.

0.50 in.   Blade

0.025 in.

4 in.

Prob. 6.24

6.25 When descending suddenly from a sidewalk to the road, a large upward force is transmitted from the road through the tire to the fork ends. Say the cross-section is approximated as elliptical with a constant wall thickness. For the fork cross-section $h = 14.5$ in. from the axle, determine the ratio of the maximum bending stress to the normal stress due to the axial loading. The fork is oriented at $\theta = 72°$ from the vertical. Approximate the fork as straight, radiating from the center of the wheel. Take the dimensions to be $a = 0.75$ in., $b = 1.25$ in., and $t = 0.1$ in.

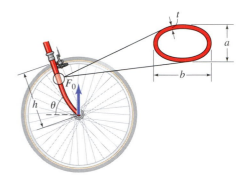

Prob. 6.25 (Appendix A1)

6.26 Pedaling produces bending and twisting of the spindle. Let the pedal force be 1000 N. Determine the maximum stresses in the spindle separately due to torsion and to bending. Indicate where on the spindle the maximum stresses would be found, and draw the stresses on an element. There is no force applied to the other pedal, and the chain tensions act in the same plane as the bearing, so they do not contribute to the bending. The spindle is hollow with outer and inner diameters of 27 mm and 21 mm. Take the dimensions to be $a = 70$ mm, $b = 25$ mm, $c = 9.5$ mm, $d = 80$ mm, and $L = 170$ mm.

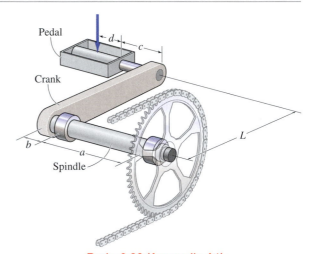

Prob. 6.26 (Appendix A1)

6.27 A drill pipe changes direction in a dog leg, below which the rest of the drill string extends to the bottom of the well bore. In the dog leg, the drill pipe has an outer diameter of 130 mm and an inner diameter of 110 mm. Below the dog leg the string weighs $W_s = 800$ kN, which is fully taken up as axial loading by the pipe at the dog leg. If the total stress in the drill pipe in the dog leg (including axial and bending) is to be 400 MPa, over what length L_d should a $\theta_0 = 20°$ change in direction be accomplished?

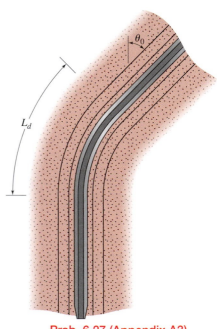

Prob. 6.27 (Appendix A3)

6.28 In using this pectoral fly machine, the hand grips and applies a 40 lb force to the handle as shown. The pivoting arm is approximated here as horizontal. Determine the stresses at the far end of the pivoting arm, where the bending stresses are maximum. Draw the stresses on an element. Use the approximation of a thin wall for the torsion of the pivoting arm. Take the dimensions to be $w = 6$ in., $L = 24$ in., $a = 1.5$ in., $b = 2$ in., $t = 0.25$ in., and $c = 20$ in.

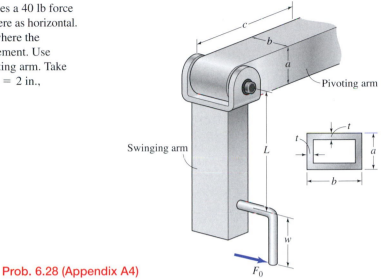

Prob. 6.28 (Appendix A4)

6.29 Consider the external fracture fixation system under a 100 N forward load on the foot. Determine the stress components due to moments in the rod where they reach their maximum values near the upper pin. Take the solid carbon fiber rod to have a diameter of 10.5 mm, an elastic modulus of 200 GPa, and a shear modulus of 90 GPa. Let the dimensions be $L_r = 300$ mm, $L_s = 120$ mm, and $L_p = 60$ mm.

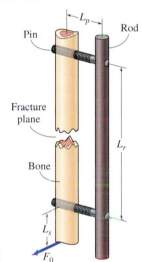

Prob. 6.29 (Appendix A5)

6.30 Consider the external fracture fixation system under the condition in which the foot is caught while the leg is trying to move up. This results in a downward 80 N force on the foot. Determine the maximum tensile stress in the rod. What fraction of the stress is due to axial loading? Take the rod to be solid carbon fiber with a 10.5 mm diameter. Let the dimensions to be $L_r = 300$ mm and $L_p = 50$ mm.

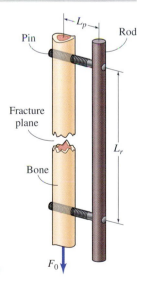

Prob. 6.30 (Appendix A5)

>>End Problems

6.3 Pressure Vessels

We now consider a common situation in which normal stresses are produced on more than one pair of faces of an element. An inflated balloon exhibits such stresses and will help us picture the deformation and stresses. However, this stress state appears notably, for example, in pressure vessels, an important engineering application. This stress state also appears in many other situations, including hyperbaric chambers for medical treatment, submarines, and inflatable structures.

1. Increasing the internal pressure in a hollow cylinder causes both its length and diameter to increase.

Consider an elongated balloon, which is not inflated. Draw on its surface a small square midway along the length with the sides of the square parallel and perpendicular to the long direction of the balloon. Inflate the balloon, and the square becomes larger. Also, it does not remain a square, but becomes rectangular with the longer dimension aligned in the circumferential direction of the balloon.

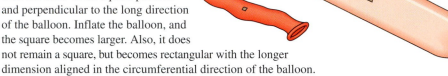

2. Picture the stresses that would cause the deformation observed on elements of the inflated cylinder.

Think about the small piece of balloon making up the square. How would forces be applied to produce the strains observed in the balloon?

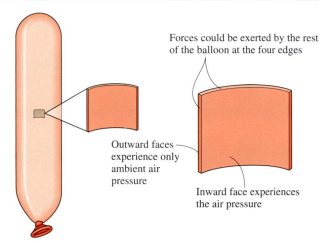

Forces could be exerted by the rest of the balloon at the four edges

Outward faces experience only ambient air pressure

Inward face experiences the air pressure

To produce the observed expansion of the element, the tensile stress would be greater in the circumferential direction than in the axial direction.

3. Recall the circumferential tension in a wrapping band due to an outward force per length (see Section 2.6).

Loads on headband and segment of headband

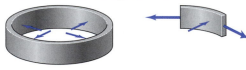

Loads on segment of balloon

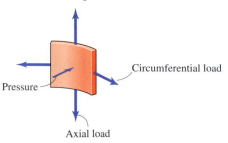

Circumferential load

Pressure

Axial load

From the chapter on axial loading, we saw that a circular band (e.g., a headband) expanded due to pressure from the head. The tension in the headband was in the circumferential direction and tended to increase the circumference. We will use similar arguments here to relate the pressure to the circumferential tension.

The headband is open at the top and bottom, and the pressure of the head on the band is only radial. By contrast, the balloon is closed at its ends, and the air pressure inside is hydrostatic; that is, it acts equally in all directions. Therefore, the air presses against the closed ends of the balloon and causes tensile stresses in the balloon wall in the axial direction, as well as in the circumferential direction (like the headband).

4. Approximate the wall of the cylinder as thin relative to the radius.

Often, the wall is very thin relative to the radius. With this assumption, we consider only the mean radius, rather than inner and outer radii. The stresses we find will be interpreted as averaged over the thickness.

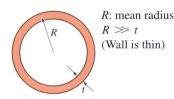

R: mean radius
$R \gg t$
(Wall is thin)

---

5. Find the two different stresses in a cylindrical vessel by isolating two distinct portions of the vessel.

Cut the cylinder perpendicularly to its axis to reveal the surface on which the axial stress acts. Include in the FBD the volume of fluid (air) up to the cut, so the pressure acts on a flat surface. Similarly, cut the cylinder through the axis to reveal the surface on which the circumferential or hoop stress acts.

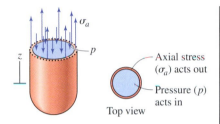

σ_a
p
Top view

Axial stress (σ_a) acts out
Pressure (p) acts in

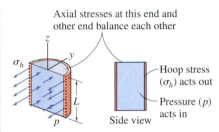

Axial stresses at this end and other end balance each other

σ_h
L
p
Side view

Hoop stress (σ_h) acts out
Pressure (p) acts in

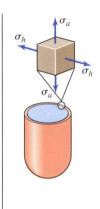

σ_a
σ_h
σ_h
σ_a

$$\sum F_z = -p(\pi R^2) + \sigma_a(2\pi R t) = 0$$

Axial Stress $\quad \sigma_a = \dfrac{pR}{2t}$

$$\sum F_y = p(2RL) - \sigma_h(2Lt) = 0$$

Hoop Stress $\quad \sigma_h = \dfrac{pR}{t}$

---

6. Find the single stress in a spherical vessel by isolating any half of the vessel.

Draw a FBD for half of the sphere plus the fluid in that half. No matter the plane that cuts the sphere in half, there must be the same tensile stress in the wall. It is termed the circumferential or hoop stress σ_h.

σ_h
p
σ_h

p

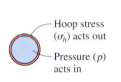

Hoop stress (σ_h) acts out
Pressure (p) acts in

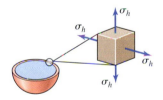

σ_h
σ_h
σ_h
σ_h

$$\sum F = -p(\pi R^2) + \sigma_h(2\pi R t) = 0$$

Hoop Stress $\quad \sigma_h = \dfrac{pR}{2t}$

---

7. Tensile stresses in the wall are greatly magnified relative to the pressure by a factor of R/t.

All wall stresses in a pressure vessel are proportional to pressure p, and to the ratio R/t. Since $R \gg t$ (relatively thin wall), and hence $R/t \gg 1$, the tensile stresses in the vessel wall are much greater than p. When representing the stresses on an element taken from the wall, we do not usually include p, since it is relatively small. It acts in the third direction.

>>End 6.3

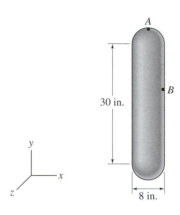

30 in.

8 in.

The external aluminum fuel tank on a hobbyist rocket is subject to an internal pressure of 500 psi. The end caps are spherical. The cylinder has a length of 30 in., a diameter of 8 in., and a wall thickness of 0.125 in. Determine the stresses in the wall at two points A and B. For each, draw the stresses on a 3-D element. Treat the upper cap as a spherical vessel and the sidewalls as a cylindrical vessel.

Solution

Point A is from the spherical end cap. The tensile stress in the wall of a spherical pressure vessel is the same in all directions in the wall. The radius is 4 in. and the thickness 0.125 in.

$$\sigma_h = \frac{pR}{2t} = \frac{(500 \text{ psi})(4 \text{ in.})}{2(0.125 \text{ in.})} = 8000 \text{ psi}$$

The directions of stresses are parallel to the wall surface, not perpendicular to it. To reinforce the picture of the directions of stress, stresses on three elements from different points on the spherical shell are shown.

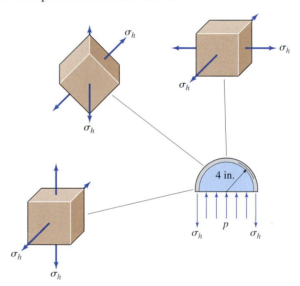

Point B is from the cylindrical portion.

The axial stress acts parallel to the length of the cylinder. The hoop stress acts in the circumferential direction.

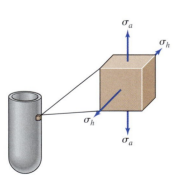

$$\sigma_a = \frac{pR}{2t} = \frac{(500 \text{ psi})(4 \text{ in.})}{2(0.125 \text{ in.})} = 8000 \text{ psi} \qquad \sigma_h = \frac{pR}{t} = \frac{(500 \text{ psi})(4 \text{ in.})}{(0.125 \text{ in.})} = 16000 \text{ psi}$$

>>End Example Problem 6.5

A stainless steel tank, supported at its two ends, stores liquid nitrogen under a pressure of 2.4 MPa. Determine the stresses at the top and bottom of the tank at its center plane, accounting for both pressure vessel stresses and bending stresses. The liquid nitrogen weighs 820 kg/m³, and the steel weighs 7860 kg/m³. The tank has an inner diameter of 3 m and a wall thickness of 30 mm. Model the tank as a beam to estimate the stress due to the weight.

15 m

Solution

The stresses due to internal pressure are:

$$\sigma_a = \frac{pR}{2t} = \frac{(2.4 \text{ MPa})(1.515 \text{ m})}{2(0.03 \text{ m})} = 60.6 \text{ MPa} \qquad \sigma_h = \frac{pR}{t} = \frac{(2.4 \text{ MPa})(1.515 \text{ m})}{0.03 \text{ m}} = 121.2 \text{ MPa}$$

Stresses due to bending are determined as follows.

Since the end caps are over the supports, they will not contribute to the bending. The force per length due to the weight is otherwise uniform and includes the cylindrical steel cross-section and the contained liquid nitrogen.

Weight over a length of 1 m:

$$q = \left[\pi \left[(1.53 \text{ m})^2 - (1.5 \text{ m})^2 \right] \left(7860 \frac{\text{kg}}{\text{m}^3} \right) + \pi (1.5 \text{ m})^2 \left(820 \frac{\text{kg}}{\text{m}^3} \right) \right] \left(9.81 \frac{\text{m}}{\text{s}^2} \right) = 78880 \text{ N/m}$$

The bending moment is maximum in the center plane:

$$M_{max} = \frac{qL^2}{8} = \frac{(78880 \text{ N/m}) (15 \text{ m})^2}{8} = 2.22 \times 10^6 \text{ N-m}$$

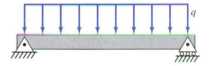

The moment of inertia of the steel tank is: $I = \frac{\pi}{4} \left[(1.53 \text{ m})^4 - (1.5 \text{ m})^4 \right] = 0.328 \text{ m}^4$

$$|\sigma_{max}| = \frac{M_{max} r}{I} = \frac{(2.20 \times 10^6 \text{ N-m})(1.53 \text{ m})}{0.328 \text{ m}^4} = 10.36 \text{ MPa}$$

The bending stress acts parallel to the long dimension—tension on the bottom, compression on the top. The tensile bending stress at the bottom adds to the axial stress σ_a. On the top, it subtracts from σ_a. The bending adds nothing to the hoop stress (121.2 MPa).

(60.6 + 10.36) = 71.0 MPa

121.2 MPa

(60.6 − 10.36) = 50.2 MPa

121.2 MPa

>>End Example Problem 6.6

Additional data on material properties needed to solve problems can be found in Appendix D or inside back cover.

6.31 The pressure vessel with a wall thickness of 15 mm, a diameter of 700 mm, and a length of 2 m is subjected to an internal pressure of 2.2 MPa. Determine the stresses acting on the faces of the element oriented as shown taken from (a) point A and (b) point B.

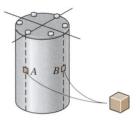

Prob. 6.31

6.32 The cylindrical vessel with an outer diameter of 4 ft and a wall thickness of 0.625 in. is allowed to have a maximum normal stress in its wall of 4000 psi. Determine the maximum allowable pressure.

4 ft

Prob. 6.32

6.33 The spherical vessel 2.5 m in diameter is to contain gas under pressure of 2 MPa without the tensile stress in its wall exceeding 60 MPa. Determine the minimum required wall thickness.

Prob. 6.33

In order to raise the pressure capacity of a pressure vessel with closed ends, it is proposed to wrap steel bands around the vessel and then tighten the bands with bolts. This gives the vessel initial compressive stresses. Assume that the friction between the bands and the vessel does not affect the axial expansion. Also, assume that pressurizing the vessel does not affect the force in the bands, so all the pressure is taken up by the vessel wall.

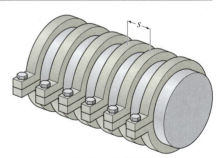

6.34 The vessel has a mean diameter of 50 in. and a wall thickness of 0.75 in. Let every wrapped band be tightened to carry a tensile force of $12(10^3)$ lb. If the maximum normal stress in the vessel wall is not to exceed 4000 psi, determine the maximum allowable internal pressure. Consider two cases: (a) the band spacing is $s = 3$ in. and (b) the band spacing is $s = 6$ in.

Probs. 6.34–35

6.35 The vessel has a mean diameter of 1.2 m and a wall thickness of 19 mm. The bands are spaced apart a distance of $s = 100$ mm. Say the maximum normal stress in the vessel wall is not to exceed 30 MPa. Determine the optimum band tension that raises the allowable pressure to its maximum value, and determine that maximum allowable pressure.

6.36 The vessel shown is formed by bending a plate to nearly a complete circle and then riveting a second plate to close it. The ends are then closed. The vessel has a diameter of $D = 300$ mm, a thickness of $t = 5$ mm, and a length of 800 mm. The rivets are 10 mm in diameter, spaced a distance of $s = 25$ mm apart, and have a shear strength of 35 MPa. Determine the maximum internal pressure if the rivets are not to shear, and determine the resulting maximum normal stress in the wall.

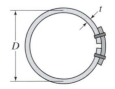

Prob. 6.36

6.37 A steel scuba tank is 24 in. long and has an external diameter of 7.25 in. and an internal diameter of 6.75 in. Say the tank is to carry a pressure of 3000 psi. Determine the stresses in the tank wall.

6.38 The manufacturer of aluminum scuba tanks wants to reduce the wall thickness of its flagship tank from the standard 11 mm down to 10 mm. The tank is 660 mm long, has a 180 mm outer diameter, and typically carries 17 MPa pressure. Will the hoop stress increase or decrease due to the thickness change and by how much?

6.39 An aluminum scuba tank that is 26 in. long with an external diameter of 7 in. and a wall thickness 0.43 in. is pressurized to 2400 psi. A diver takes the tank in the ocean to 100 ft below the surface. Take the density of the water to be 62.4 lb/ft^3. Assume the pressure of the contained air does not change. Does the hoop stress in the tank increase or decrease relative to its value on the surface and by how much? Why is it appropriate to assume that the pressure of the contained air does not change?

Probs. 6.37–39

6.40 Submarines that descend to great depths (up to 7 miles or 10000 m), must withstand very high external pressures. Since their interiors have to be at atmospheric pressure to accommodate human life, they are like a pressure vessel except with external pressure rather than internal pressure. While there is an external hull that is hydrodynamic, it does not carry the pressure difference. Instead, a cylinder shell carries the pressure difference. If seawater has a density of 1027 kg/m^3, what should be the wall thickness of the cylindrical shell with a 13 m internal diameter when diving to 400 m, if the normal stress in the steel is limited to 200 MPa?

Prob. 6.40

6.41 Submersible vehicles are being designed to travel down to 37000 ft below sea level. Take the liquid density to be 64.1 lb/ft^3, typical of salt water. A transparent ceramic is being considered for the hemispherical viewport. The ceramic can be obtained only up to 1 in. thick and can withstand compressive stresses only as high as 300 ksi. What is the largest diameter window that can safely withstand these depths?

Prob. 6.41

6.42 A toroidal shaped habitat is proposed for human occupation in space. Let the inner diameter be denoted by D_i, the outer diameter by D_o, and the wall thickness by t. The pressure in space is zero, while the interior is to be pressurized to atmospheric pressure on earth. Derive an estimate of the stresses in terms of p_{atm}, D_i, D_o, and t using formulas for spherical or cylindrical pressure vessels. Indicate the relative sizes for the variables D_i and D_o at which this estimate is valid.

6.43 A toroidal space habitat has an inner diameter of 200 ft, and an outer diameter of 280 ft. An internal pressure of 15 psi (comparable to atmospheric pressure) is to be contained. Use formulas from the cylindrical pressure vessel and estimate the necessary thickness of the wall, if it is made of plexiglass (acrylic) and limited to a stress of 1500 psi.

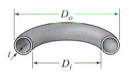

Probs. 6.42–43

6.44 A hyperbaric chamber is used to study the effects of pressure changes on humans and animals. One such chamber is 9 ft in diameter and 26 ft long and is designed to handle up to 30 atmospheres of pressure (30 × 14.7 psi) above the external atmospheric pressure. If the steel wall is to be kept below 10000 psi (to allow for cut-outs), determine the acceptable wall thickness.

Prob. 6.44

6.45 A pressure vessel is stainless steel on the inside to provide corrosion resistance and fiber-reinforced composite on the outside to provide strength at lower weight. Let the fibers be wound on the outside so that the composite has elastic properties E_c, G_c, and ν_c. The steel has properties E, G, and ν.

Let the vessel have an internal pressure of p. Using the same assumption of a relatively thin wall, determine the axial and hoop stresses in the steel and composite, assuming that the materials are perfectly bonded to each other.

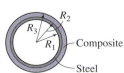

Prob. 6.45

>>End Problems

6.4 Elastic Stress–Strain Relations

In preceding chapters, we only needed strains due to a single stress component. Now, we determine strains when multiple stresses act. We need a notation to distinguish the different stresses.

1. Consider the full set of stresses that can act on the faces of an element.

Here are the full set of stresses, normal and shear, that can act on an element oriented with the *x-y-z* axes

Sometimes, to avoid clutter, stresses are only drawn on the three visible faces, not on the back faces.

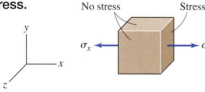

2. Assumptions about the material response simplify the combining of strains at a point due to different stress components acting at that point.

> A material is defined as **isotropic** if it has the same mechanical properties regardless of the orientation of the material when tested mechanically. For an isotropic, elastic material, *E*, *G*, and ν have unique values, regardless of orientation.

We only combine strains of the same type (normal or shear) that act in the same direction.

Let the strains be small and material response elastic. Then, not only are the stress and strain proportional to one another, the strain of a given type due to two stress components that act simultaneously equals the sum of the strains when the stresses act independently.

Many common materials are *isotropic*: they have no intrinsic directionality. For **isotropic** materials, normal stress produces only normal strain, not shear. Also, no matter the direction of the stress, a tensile stress causes the same tensile strain in that direction and identical contractions in directions perpendicular to the tension.

3. Recognize the strains associated with each normal stress.

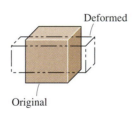

No stress   Stress   Deformed   Original

Signs are significant:

If $\sigma_x > 0$ (tensile stress), then

$$\varepsilon_x = \frac{\sigma_x}{E} \qquad \text{Strain along } x: \quad \varepsilon_x > 0 \text{ (tensile)}$$

$$\varepsilon_y = -\nu\frac{\sigma_x}{E} \qquad \text{Strain along } y: \quad \varepsilon_y < 0 \text{ (compressive)}$$

$$\varepsilon_z = -\nu\frac{\sigma_x}{E} \qquad \text{Strain along } z: \quad \varepsilon_z < 0 \text{ (compressive)}$$

If $\sigma_x < 0$ (compressive), all signs reverse

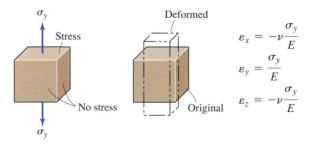

Stress   Deformed   No stress   Original

$$\varepsilon_x = -\nu\frac{\sigma_y}{E}$$
$$\varepsilon_y = \frac{\sigma_y}{E}$$
$$\varepsilon_z = -\nu\frac{\sigma_y}{E}$$

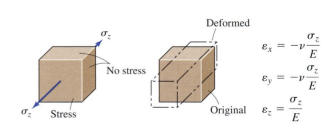

Deformed   No stress   Stress   Original

$$\varepsilon_x = -\nu\frac{\sigma_z}{E}$$
$$\varepsilon_y = -\nu\frac{\sigma_z}{E}$$
$$\varepsilon_z = \frac{\sigma_z}{E}$$

4. Add only strains in the same direction.

Superpose the effects of the three normal stresses, but only add strains in the same direction; that is, separately add the contributions of the three stresses to ε_x, to ε_y, and to ε_z.

$$\varepsilon_x = \frac{\sigma_x}{E} - \nu\frac{\sigma_y}{E} - \nu\frac{\sigma_z}{E} \qquad \varepsilon_y = -\nu\frac{\sigma_x}{E} + \frac{\sigma_y}{E} - \nu\frac{\sigma_z}{E} \qquad \varepsilon_z = -\nu\frac{\sigma_x}{E} - \nu\frac{\sigma_y}{E} + \frac{\sigma_z}{E}$$

Note that σ_x, σ_y, and σ_z in these formulas can be tensile (> 0) or compressive (< 0). The associated positive or negative signs allow the contributions to the strains to combine properly.

5. Recognize deformed elements that are superficially different, but correspond to the same shear strain.

For isotropic materials, shear stress produces only shear (not normal) strain, and the shear strain is only in the direction of the particular shear stress. As we have seen, shear strain captures element deformation in which edges are no longer mutually perpendicular.

Here is one shear stress component, τ_{xy}, and several different configurations of the deformed element. All have the same shear strain ($\gamma_{xy} > 0$), and they can be changed into one another by translating and rotating (not deforming). For all cases, the lower left corner, from which the edges emerge as do the positive x- and y-axes, deforms into an angle $< 90°$. The shear strain is the angle change in radians.

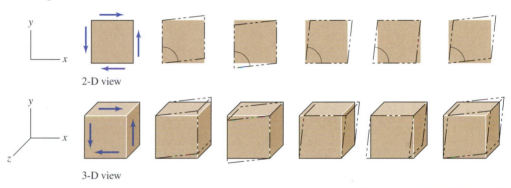

2-D view

3-D view

6. Use the sign of the shear strain to distinguish increase or decrease in angle.

Here we show the shear stress $\tau_{xy} < 0$. For $\gamma_{xy} < 0$, the lower left corner deforms into an angle greater than 90° (not all cases from above are shown).

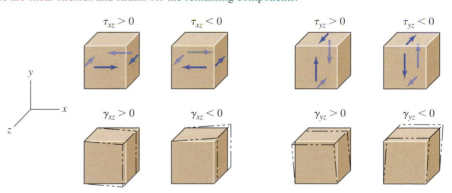

7. Define shear strains in the x-z and y-z planes in similar ways.

Here are shear stresses and strains for the remaining components.

| $\tau_{xz} > 0$ | $\tau_{xz} < 0$ | $\tau_{yz} > 0$ | $\tau_{yz} < 0$ |

| $\gamma_{xz} > 0$ | $\gamma_{xz} < 0$ | $\gamma_{yz} > 0$ | $\gamma_{yz} < 0$ |

8. Each shear strain is proportional only to the shear stress that acts along the same pair of axes.

For an isotropic material, each shear stress causes only its corresponding strain. The proportionality is the same shear modulus for all three components.

$$\gamma_{xy} = \frac{\tau_{xy}}{G} \qquad \gamma_{xz} = \frac{\tau_{xz}}{G} \qquad \gamma_{yz} = \frac{\tau_{yz}}{G}$$

>>End 6.4

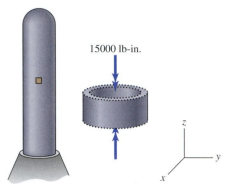

15000 lb-in.

A pressurized tank that is 20 in. long, with 5 in. mean diameter and 0.1 in. wall thickness, is under a pressure of 300 psi. As part of a processing operation, the tank is suddenly spun by the motor to which it is attached. A dynamic analysis (not carried out here) suggests that the sudden rotation produces stresses that are consistent with a torque (twisting moment) of 15000 lb-in. Determine the x-y-z stresses at the indicated point (at the front of the tank), and the change in shape of a 0.2 in. × 0.2 in. square. ($E = 29 \times 10^6$ psi, $v = 0.3$, and $G = 11 \times 10^6$ psi.)

Solution

Stresses due to internal pressure are:

$$\sigma_a = \frac{pR}{2t} = \frac{(300\ \text{psi})(2.5\ \text{in.})}{2(0.1\ \text{in.})} = 3750\ \text{psi} \qquad \sigma_h = \frac{pR}{t} = \frac{(300\ \text{psi})(2.5\ \text{in.})}{(0.1\ \text{in.})} = 7500\ \text{psi}$$

Stresses due to twisting are determined as follows.

Polar Moment of Inertia: $\quad I_p = \dfrac{\pi}{2}\left[(2.55\ \text{in.})^4 - (2.45\ \text{in.})^4\right] = 9.82\ \text{in.}^4$

$$|\tau_{max}| = \frac{Tr}{I_p} = \frac{(15000\ \text{lb-in.})\ (2.55\ \text{in.})}{9.82\ \text{in.}^4} = 3890\ \text{psi}$$

Stresses on the element are:

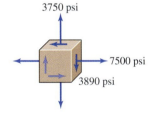

3750 psi

7500 psi

3890 psi

$$\sigma_x = 0,\ \sigma_y = 7500\ \text{psi},\ \sigma_z = 3750\ \text{psi}$$
$$\tau_{xy} = 0,\ \tau_{xz} = 0,\ \tau_{yz} = -3890\ \text{psi}$$

To determine the change in shape of the element, we want the strains in y-z plane:

$$\varepsilon_y = -v\frac{\sigma_x}{E} + \frac{\sigma_y}{E} - v\frac{\sigma_z}{E} = \frac{7500\ \text{psi}}{29 \times 10^6\ \text{psi}} - 0.3\frac{3750\ \text{psi}}{29 \times 10^6\ \text{psi}} = 2.20 \times 10^{-4}$$

$$\varepsilon_z = -v\frac{\sigma_x}{E} - v\frac{\sigma_y}{E} + \frac{\sigma_z}{E} = -0.3\frac{7500\ \text{psi}}{29 \times 10^6\ \text{psi}} + \frac{3750\ \text{psi}}{29 \times 10^6\ \text{psi}} = 5.17 \times 10^{-5}$$

$$\gamma_{yz} = \frac{\tau_{yz}}{G} = \frac{-3890\ \text{psi}}{11 \times 10^6\ \text{psi}} = -3.54 \times 10^{-4}$$

Changes in lengths and angle of a 0.2 in. × 0.2 in. square are as follows:

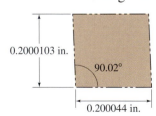

0.2 in.

0.2 in.

0.2000103 in.

90.02°

0.200044 in.

$$\delta_y = (\varepsilon_y)(L_y) = (2.20 \times 10^{-4})(0.2\ \text{in.}) = 4.40 \times 10^{-5}$$
$$\delta_z = (\varepsilon_z)(L_z) = (5.17 \times 10^{-5})(0.2\ \text{in.}) = 1.034 \times 10^{-5}$$

Angle change $= -3.54 \times 10^{-4}(180°/\pi) = -0.0203°$ (increases from 90°).

>>End Example Problem 6.7

The reinforced-rubber structure is part of an inflatable evacuation system. Each individual tube, which has a diameter of 200 mm and a thickness of 2 mm, is inflated to a pressure of 150 kPa. To test the structure, it is placed on supports and loaded in the center, which results in 500 N on each tube. Determine the stresses and strains in the top and bottom of a tube in the plane of the load. ($E = 150$ MPa and $v = 0.4$.)

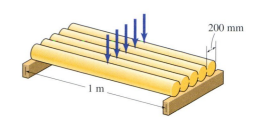

200 mm

1 m

Solution

Stresses due to internal pressure are:

$$\sigma_a = \frac{pR}{2t} = \frac{(150 \text{ kPa})(0.1 \text{ m})}{2(0.002 \text{ m})} = 3.75 \text{ MPa} \qquad \sigma_h = \frac{pR}{t} = \frac{(150 \text{ kPa})(0.1 \text{ m})}{(0.002 \text{ m})} = 7.50 \text{ MPa}$$

Stresses due to bending of one tube are determined as follows:

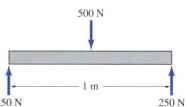

500 N

250 N   1 m   250 N

Maximum moment at center: $M_{max} = (250 \text{ N})(0.5) = 125$ N-m

Moment of Inertia: $I = \frac{\pi}{4}[(0.101 \text{ m})^4 - (0.099 \text{ m})^4] = 6.28 \times 10^{-6} \text{ m}^4$

$$|\sigma_{max}| = \frac{M_{max} r}{I} = \frac{(125 \text{ N-m})(0.101)}{6.28 \times 10^{-6} \text{ m}^4} = 2.01 \text{ MPa}$$

Compression on top
Tension on bottom

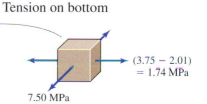

(3.75 + 2.01) = 5.76 MPa

7.50 MPa

(3.75 − 2.01) = 1.74 MPa

7.50 MPa

We use this coordinate system

y

x

z

Top:
$\sigma_x = 1.74$ MPa, $\sigma_y = 0$, $\sigma_z = 7.5$ MPa

$$\varepsilon_x = \frac{\sigma_x}{E} - v\frac{\sigma_y}{E} - v\frac{\sigma_z}{E} = \frac{1.74 \text{ MPa}}{150 \text{ MPa}} - 0.4\frac{7.5 \text{ MPa}}{150 \text{ MPa}} = 0.0084$$

$$\varepsilon_z = -v\frac{\sigma_x}{E} - v\frac{\sigma_y}{E} + \frac{\sigma_z}{E} = -0.4\frac{1.74 \text{ MPa}}{150 \text{ MPa}} + \frac{7.5 \text{ MPa}}{150 \text{ MPa}} = 0.0454$$

Bottom:
$\sigma_x = 5.76$ MPa, $\sigma_y = 0$, $\sigma_z = 7.5$ MPa

$$\varepsilon_x = \frac{\sigma_x}{E} - v\frac{\sigma_y}{E} - v\frac{\sigma_z}{E} = \frac{5.76 \text{ MPa}}{150 \text{ MPa}} - 0.4\frac{7.5 \text{ MPa}}{150 \text{ MPa}} = 0.0184$$

$$\varepsilon_z = -v\frac{\sigma_x}{E} - v\frac{\sigma_y}{E} + \frac{\sigma_z}{E} = -0.4\frac{5.76 \text{ MPa}}{150 \text{ MPa}} + \frac{7.5 \text{ MPa}}{150 \text{ MPa}} = 0.0346$$

>>End Example Problem 6.8

PROBLEMS

Additional data on material properties needed to solve problems can be found in Appendix D or inside back cover.

6.46 A strain gage oriented in the axial direction is attached to the outer surface of a steel pressure vessel. The vessel has a mean diameter of 3 ft and a wall thickness of 0.75 in. If the strain measured by the gage increases by $3(10^{-5})$, determine the increase in the internal pressure. Neglect the effect of the radial stress in a pressure vessel in comparison with the hoop and axial stresses.

Prob. 6.46

6.47 A flat aluminum plate has planar dimensions 6 in. $\times$ 4 in. and 0.02 in. thickness when no stresses act. Stresses in the x- and y-directions of 4000 psi and 9000 psi are then applied. Determine the dimensions of the plate when the stresses act.

Prob. 6.47

6.48 A plastic cube 20 mm on a side with elastic properties $E = 3$ GPa, $v = 0.4$ is to be placed in a container with fluid under pressure. Fluid pressure results in all three normal stresses in the cube being compressive, equal in magnitude to the pressure. If the cube reduces by 0.01 mm on each side upon being placed in the fluid, determine the magnitude of the pressure.

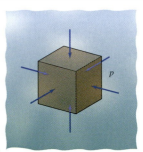

Prob. 6.48

6.49 During a processing operation, a steel sheet is subjected to tensile stresses $\sigma = 75$ ksi in the plane and a squeezing pressure $p = 10$ ksi. Consider a portion of the sheet that is 2 in. $\times$ 2 in. and 0.04 in. thick when the loads are acting. Determine the in-plane dimensions and thickness of this portion of the sheet if these loads are released and the sheet springs back elastically.

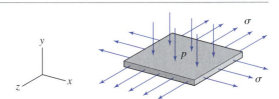

Prob. 6.49

6.50 A long steel pressure vessel supported at its ends contains a liquid under pressure. The vessel has an outer diameter of 2.5 m and a thickness of 40 mm. Due to the downward load of the weight of the liquid and the vessel weight, there is a bending moment of $5(10)^6$ N-m at its central cross-section. Four strain gages are attached to the vessel: one in each of the axial and circumferential directions, at two points, on the top and on the side. The vessel is also subject to an internal pressure of 2.2 MPa. Determine the strains in the four gages.

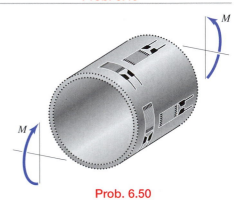

Prob. 6.50

6.51 A plastic sheet that is initially 200 mm $\times$ 300 mm in the x- and y-direction and 4 mm thick is subjected to tensile forces at its edges that result in uniform in-plane stresses. With the forces acting, the in-plane dimensions are 199 mm and 304 mm. Determine the tensile forces on the sheet in the x- and y-directions, if the elastic modulus and Poisson ratio are known to be 3.7 GPa and $v = 0.4$.

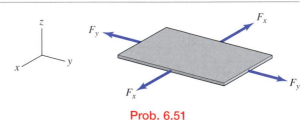

Prob. 6.51

6.52 A balloon is inflated and currently has the form of a sphere with an 8 in. diameter and a wall thickness of 0.04 in. In the current state, the elastomeric material has an elastic modulus of 120 psi and Poisson ratio of 0.45. If the diameter is to be increased to 8.25 in., by how much must the pressure be increased?

Prob. 6.52

6.53 An aluminum tank is originally 600 mm long with a mean diameter of 240 mm and a wall thickness of 3 mm. A gas having a pressure $p = 1.6$ MPa is added, and the tank is also subjected to a compressive axial force of $F_0 = 30$ kN. Determine by how much the tank length and diameter change under the combined action of the pressure and axial force.

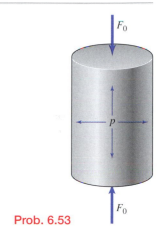

Prob. 6.53

6.54 A rubber disk initially has a diameter of 8 in. and a thickness of 0.25 in. The disk is observed to expand to a diameter of 8.05 in. when the disk is squeezed by compressive forces $F_0 = 500$ lb in the z-direction. Despite efforts to lubricate the disk, it is known that the observed expansion is affected by some frictional resistance. The effect of friction is modeled as inducing uniform normal stresses that are equal in the x- and y-directions. The elastic modulus and Poisson ratio of the disk are $E = 300$ psi and $\nu = 0.43$. (a) Deduce the equal x-y normal stresses that are consistent with the expansion and (b) the resulting reduction in thickness of the disk.

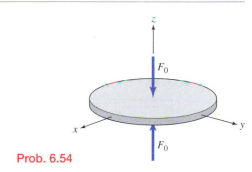

Prob. 6.54

6.55 A steel cylindrical pressure vessel is 4 m long, 1.75 m in diameter, and has a wall thickness of 30 mm. The vessel is externally unconstrained and is subjected to an internal pressure of 2 MPa while the temperature rises by 50°C. Determine the stresses in the vessel wall and the change in length and diameter of the vessel. Take $\alpha = 12(10)^{-6}(°C)^{-1}$.

6.56 An inflatable cylindrical tube, capped at its ends, is also subjected to bending loads. The tube is 1.3 m long and has an outer diameter of 400 mm and a wall thickness of 6 mm. Its elastic properties are $E = 4$ MPa and $\nu = 0.4$. The pressure must be high enough so that the normal stress and normal strains are both positive. Say a bending moment of up to 110 N-m is expected. Determine the minimum internal pressure to ensure that (a) the normal stress is not compressive and (b) the normal strain is not compressive.

6.5 Deflections Under Combined Internal Loads

Engineering structures sometimes consist of combinations of connected straight segments. Each segment of the body is subjected to a combination of internal loads. We can determine deflections by suitably combining results based on individual internal loads. The goal is to determine the motion of the member center-line in three dimensions, both its deflection and its rotation. When segments are long compared to the cross-sectional dimensions, we often neglect axial and shear deformation, and consider only bending or twisting.

1. The method of calculating deflections of complex structures is illustrated by a simple structure of two segments.

The legs E and F of this waterstrider robot provide propulsion. The force of the leg pushing on the water causes complex deflection of the leg.

This loaded structure is typical of the problems in which we seek the deflection due to deformation of multiple straight segments.

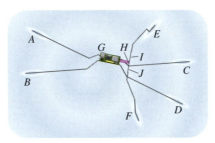

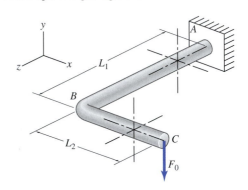

2. Track the 3-D deflection and rotation of each cross-section.

When the member has a distinct axis (long direction) and a cross-section, as this problem does, we follow the motion of each cross-section. We imagine three mutually perpendicular lines, a triad, one of which is aligned with the member's axis. The other two axes are chosen as convenient, for example, aligned with other coordinate directions that have been defined.

The motion of the cross-section is described by a 3-D displacement and a 3-D rotation. The displacement is at the point where the axis pierces the cross-section, and it is usually defined with components u, v, and w along the x-y-z axes. If we neglect deformation due to shear, then a segment's cross-section stays perpendicular to its center-line. The triad rotates through angles θ_x, θ_y, and θ_z about the x-y-z axes, usually defined in the right hand rule sense. θ_x, θ_y, and θ_z shown here are > 0.

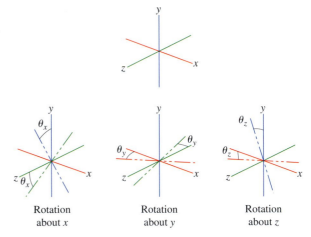

Rotation about x Rotation about y Rotation about z

We have studied these angles before, although we used a different notation. The angle of rotation about the member axis (z on AB, and x on BC, above) is the angle of rotation in torsion; we used the symbol ϕ. The other angles of rotation are due to bending; there we used the symbol θ. The twisting and bending moments acting on any segment produce rotations measured by angles.

Often a structure is fully fixed (cantilevered, built-in) at one end, that is, it is rigidly connected to another structure that is assumed not to move. Then, we can find the deflection and rotation of portions of the structure beginning with the segment that is attached to the fixed cross-section.

3. Separate the structure into straight segments and find the loads acting on each.

As we did with beam deflections, we divide the member into straight segments, and find the internal loads between the segments. These loads can then constitute the external loads when we analyze only one of the two segments.

For example, we can isolate BC and find the loads ($F_y = F_0$ and $M_z = F_0 L_2$) where the members are separated. The equal and opposite loads, F_y and M_z, act on the built-in segment AB.

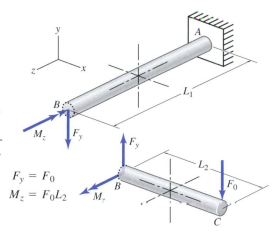

$$F_y = F_0$$
$$M_z = F_0 L_2$$

4. Consider the member where some deflections and/or rotations are known (one cross-section is fixed), and find the deflection and rotation where that member connects to the next member.

The segment AB is subjected to a force (F_y) that causes bending and a moment (M_z) that causes twisting. F_y causes y-deflection at B, and rotation about the x-axis, which can be found from Appendix G-1. The moment M_z causes rotation about the z-axis (θ_z), but no deflection. Notice the signs of v_B, θ_{xB}, and θ_{zB}.

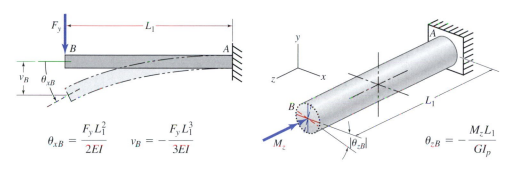

$$\theta_{xB} = \frac{F_y L_1^2}{2EI} \qquad v_B = -\frac{F_y L_1^3}{3EI}$$

$$\theta_{zB} = -\frac{M_z L_1}{GI_p}$$

5. Next, find the deflection of the connected segment, taking it to have zero deflection and rotation where it connects to the solved member, and then adding the rigid deflection and rotation from the connection.

For segment BC, we now know the motion at end B, where it attaches to AB; our goal is to find the motion at C. We use an idea from beam deflections: we first calculate the deflection at C, given the loads on BC, assuming the end B is fixed. Then, we add a rigid motion to BC, which would give the end B its correct deflection and rotation, as computed from the analysis of AB.

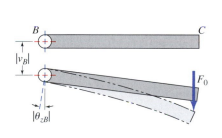

Bending of BC as if cantilevered at B

Rigid motion of BC corresponding to motion at B found from AB

$$v_C = -\frac{F_0 L_2^3}{3EI} - \frac{F_0 L_1^3}{3EI} - \left(\frac{F_0 L_2 L_1}{GI_p}\right) L_2$$

>>End 6.5

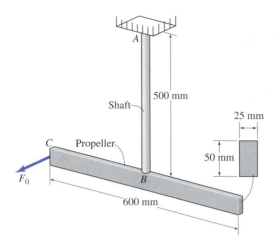

500 mm

Shaft

25 mm

C   Propeller

50 mm

F_0

B

600 mm

The molded reinforced plastic propeller system is subject to vibration. One aspect of analyzing a structure for vibration is determining the stiffness of the structure. Consider the force F_0 applied to the tip of the blade as shown. Determine the magnitude of the force that would cause a deflection of 1 mm. The shaft is 25 mm in diameter. Approximate the blade as having a rectangular cross-section with dimensions given. ($E = 50$ GPa, $v = 0.38$, and $G = 18$ GPa.)

Solution

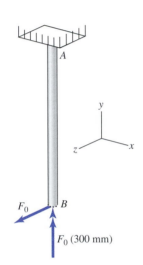

F_0 B

F_0 (300 mm)

The shaft is under a combination of bending and twisting. The loaded half of the propeller is subjected only to bending.

Determine the deflection of the shaft as follows:

Transverse force: the deflection u_z is the same as for a cantilever with an end force (see Appendix G-1.1).

$$(u_z)_B = \frac{F_0 L_{AB}^3}{3EI_{AB}} = \frac{F_0(0.5 \text{ m})^3}{3(50 \text{ GPa})\frac{\pi}{4}(0.0125 \text{ m})^4}$$

Twisting moment: rotation θ_y is that of a shaft fixed at one end with a torque applied to the other end.

$$(\theta_y)_B = \frac{TL_{AB}}{GI_p} = \frac{F_0(0.3 \text{ m})(0.5 \text{ m})}{(18 \text{ GPa})\frac{\pi}{2}(0.0125 \text{ m})^4}$$

Determine the deflection of the blade by summing the deflection if BC were cantilevered at B, plus the deflection due to motion of B from analyzing AB.

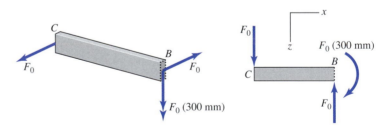

If the blade were cantilevered at B (see Appendix G-1.1):

$$(u_z)_{cant.} = \frac{F_0 L_{BC}^3}{3EI_{BC}} = \frac{F_0(0.3 \text{ m})^3}{3(50 \text{ GPa})\frac{1}{12}(0.05 \text{ m})(0.025 \text{ m})^3}$$

Additional rigid motion of the blade is due to u_z and θ_x of the shaft: $(u_z)_B + (\theta_y)_B(0.3 \text{ m})$

If $(u_z)_{cant.} + (u_z)_B + (\theta_y)_B(0.3 \text{ m}) = 0.001 \text{ m} \Rightarrow F_0 = 3.79 \text{ N} \Rightarrow$ Stiffness $= 3.79$ N/mm

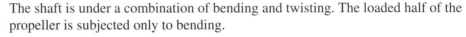

The supporting structure shown is loaded at the center of BC. Determine the deflection at the load point. The cross-section is uniform with cross-sectional properties I and I_p and moduli E and G.

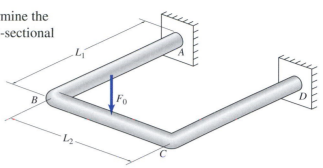

Solution

Think of the structure as broken into three straight segments. Notice that the loading and structure are symmetric. Consider all possible internal loads at cross-sections B and C. Assume they have equivalent magnitudes and act in senses that are consistent with symmetry. Now, recognize that certain internal loads are zero.

$F_x = 0$ (otherwise AB and CD would bend outward in x-z plane)

$F_z = 0$ (loading of BC not symmetric if F_z in opposite directions, and not in equilibrium if in same direction)

$M_x = 0$ (symmetric loading cannot cause BC to twist)

$M_y = 0$ (loading of BC not symmetric if M_y in opposite directions, and not in equilibrium if in same direction)

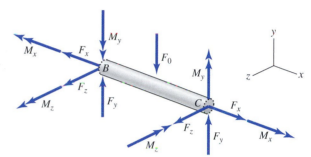

Re-draw members with remaining non-zero loads.

Cantilevers AB and CD deflect down due to $F_0/2$ and twist due to M_z.

Internal loads in BC are those of a simply supported beam with center force and moments at ends.

Rotation at B must be the same as calculated from twist of AB (with symmetric rotation at C).

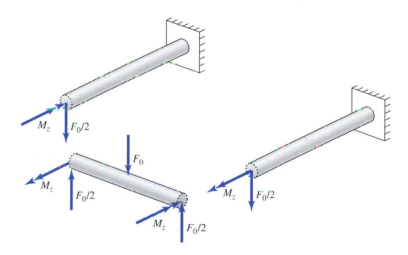

Use Appendix G-1.1, G-2.2, and G-2.3.

Deflection of AB	Twisting of AB	Beam deflection of BC

$$(u_z)_B = \frac{F_0 L_1^3}{3EI} \qquad (\theta_z)_B = -\frac{M_z L_1}{GI_p} \qquad (\theta_z)_B = -\frac{F_0 L_2^2}{16EI} + \frac{M_z L_2}{3EI} + \frac{M_z L_2}{6EI}$$

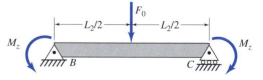

Equate $(\theta_z)_B \Rightarrow$

$$M_z = \frac{F_0 L_2^2}{16EI\left[\dfrac{L_1}{GI_p} + \dfrac{L_2}{2EI}\right]}$$

Then, for u_y at load point, add deflection of cantilever AB and simple beam BC

$$u_y = \frac{F_0 L_1^3}{3EI} + \frac{F_0 L_2^3}{48EI} - \frac{M_z L_2^2}{8EI} = \frac{F_0}{EI}\left[\frac{L_1^3}{3} + \frac{L_2^3}{48} - \frac{L_2^4}{128\left[L_1\left(\dfrac{EI}{GI_p}\right) + \dfrac{L_2}{2}\right]}\right]$$

>>End Example Problem 6.10

Additional data on material properties needed to solve problems can be found in Appendix D or inside back cover.

6.57 A tray rests on an extendible arm consisting essentially of two steel tubes of different outer diameters. Deflecting and tilting of the tray is undesirable and needs to be quantified. The downward force at the tray center is $F_0 = 1000$ N. Determine the downward deflection of the tray center and the tilt angles of the tray about each of the coordinate axes. Both members have wall thickness of 4 mm. Take the dimensions to be $d_1 = 40$ mm, $d_2 = 34$ mm, $a = 700$ mm, and $b = 500$ mm.

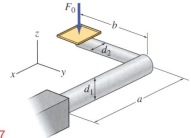

Prob. 6.57

6.58 It is necessary to quantify the tilting of a mirror that is attached to a space structure when the force F_0 acts. The structure consists of aluminum tubular members, each with an outer diameter of 50 mm and a wall thickness of 6 mm. If the force is $F_0 = 150$ N, determine the tilting of the mirror about each of the coordinate axes. The gravitational force is zero. Take the dimensions to be $a = 3$ m, $b = 2$ m, and $c = 1$ m.

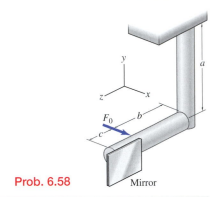

Prob. 6.58

Mirror

6.59 For increased stability, it is proposed to locate aircraft landing wheels outward from the center of the plane (one such wheel is shown). The deflections of the wheel upon landing and braking are of interest. Consider the tilting of the tire away from straight ahead (rotation about the y-axis) due only to the braking force $F_2 = 30$ kip, which is taken as acting through the center of the tire. Take the members to have dimensions $a = 30$ in. and $b = 40$ in. These members are hollow steel shafts with an outer diameter of 6.0 in. and an inner diameter of 5.0 in.

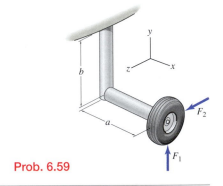

Prob. 6.59

6.60 Copper pipe (mass density $= 8960$ kg/m^3, $E = 120$ GPa, $G = 48$ GPa) is supported by suspenders distributed along its length. The final hanger that supported the pipe at the end C has become detached. Water flows out from the open pipe at the end C. However, because the end C is unsupported, the distributed load due to the pipe weight deflects the pipe and the water exits at C not horizontally (not parallel to the x-axis) but at some angle. Assume that the pipe was fully supported up to the cross-section A. Take the pipe to have outer diameter 12.7 mm and inner diameter 10.9 mm, and the unsupported segments to have lengths $a = 1.9$ m and $b = 2.6$ m. Determine the angle below the x-axis at which water exits at C.

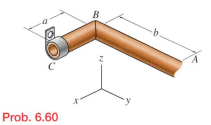

Prob. 6.60

6.61 A photographer captures the deflection of a highway sign during a severe storm. When the image is analyzed, the sign is estimated to have tilted by $\alpha = 15°$. The steel post AB has an outer diameter of 16 in., a wall thickness of 0.75 in. and a length of 25 ft. The steel pole BC has an outer diameter of 12 in., a wall thickness of 0.625 in., and a length of 40 ft to the midpoint where the sign is attached. If the sign is 15 ft wide × 10 ft high, determine the apparent pressure.

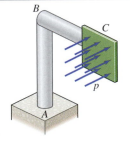

Prob. 6.61

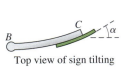

Top view of sign tilting

6.62 A decorative lighting system consists of a lamp at the end of a solid bent steel rod 15 mm in diameter. The rod deflects a negligible amount due to its own weight, but does deflect due to the weight of the lamp at the end. The installer supports the weight of the lamp by hand and attaches it at the support A so that BC initially tilts in the y-z plane making an angle of θ_0 with the horizontal. Determine the value for θ_0 if the lamp at the end of BC is to appear horizontal once the installation is complete and the weight is no longer supported by hand. Take the lengths to be $a = 1.2$ m and $b = 0.9$ m and the lamp to weigh 3 kg.

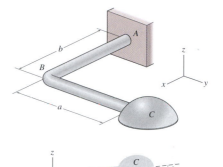

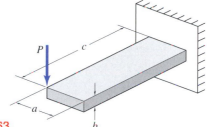

Prob. 6.62

6.63 Determine the downward deflection of the corner of the plate due to the applied force, accounting for both bending and torsion. The elastic moduli are E, G, and ν. Approximate the plate as thin compared to its width ($b \ll a$).

Prob. 6.63

Focused Application Problems

6.64 In using this pectoral fly machine, the hand grips and applies a 40 lb force to the handle as shown. The pivoting arm is approximated here as horizontal. Stiffness of the machine is important in that tilting at the handle is undesirable. Considering both the torsion of the pivoting arm and the bending of the swinging arm, determine the angle at which the handle tilts. Use the approximation of a thin wall for the torsion of the pivoting arm, and assume the pivoting arm is fixed at the far end. Neglect the deflection of the short handle itself. Take the dimensions to be $w = 6$ in., $L = 24$ in., $a = 1.5$ in., $b = 2$ in., $t = 0.25$ in., and $c = 20$ in.

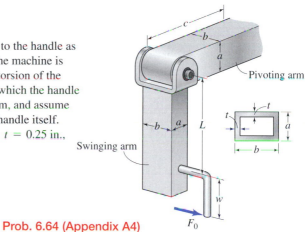

Prob. 6.64 (Appendix A4)

6.65 Consider the external fracture fixation system under a 100 N forward load on the foot. Determine the forward displacement of the lower bone fragment relative to the upper fragment at the fracture plane. Take the solid carbon fiber rod to have a diameter of 10.5 mm, an elastic modulus of 200 GPa, and a shear modulus of 90 GPa. Let the dimensions be $L_r = 300$ mm, $L_s = 120$ mm, and $L_p = 60$ mm. Treat the carbon fiber rod as fully fixed by the pins to each of the bone fragments. Consider only deformation of the carbon rod, and neglect any deformation of the pins.

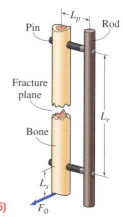

Prob. 6.65 (Appendix A5)

>>End Problems

6.6 Strain Energy

In Newtonian mechanics, problems can be solved with more than one method, for example by directly applying $F = ma$ or by the work–energy theorem. So far in this book we have used only direct methods that use equilibrium, the material law, and geometric compatibility to relate force, stress, strain, and displacement. Now we consider energy methods for mechanics of materials.

1. **The principal of conservation of energy simplifies when applied to elastic bodies under load.**

All physical systems observe the principle of conservation of energy. In many practical systems, to a reasonable approximation, only mechanical work and energy are involved. There is negligible heat generated, and no other forms of stored energy, such as electromagnetic or chemical energy, are relevant. For elastic systems, any work that is done on the system is stored as elastic energy that is fully recoverable. Often, we can devise efficient means of computing overall deformations by equating the work done to the stored energy.

2. **Recall the example of the energy in a spring.**

Springs are studied physics. One method of solving problems with springs involves accounting for the energy stored in the deformed spring. For a linear spring with $P = k\delta$,

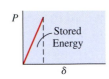

$$\text{External Work} = \int P d\delta = \int (k\delta) d\delta = \frac{1}{2}k\delta^2 = \text{Stored Energy}$$

To solve problems in mechanics of materials using energy, it is useful to be able to determine the energy stored in an element that is strained.

3. **Compute the strain energy of an element under uniaxial stress.**

In Chapter 3, we considered the analogy between a bar under axial loading and a spring. For a bar of length L, cross-sectional area A, and elastic modulus E, the axial stiffness of the bar is $k = EA/L$, because $P = EA\delta/L$. When such a bar is elongated by δ, the stored energy $= k\delta^2/2 = EA\delta^2/2L$.

Strain energy density is defined as the elastic energy stored in a deformed element divided by the element volume.

The stress and strain in an axially deformed bar are uniform, so each element of the bar stores an identical amount of energy. We define the **strain energy density** as the energy stored in a unit volume of the bar. Since the volume of the bar is AL, for an element under uniaxial stress,

$$\text{Strain Energy Density} = \frac{\text{Stored Energy}}{\text{Volume}} = \frac{\frac{1}{2}\left(\frac{EA}{L}\right)\delta^2}{AL} = \frac{1}{2}E\left(\frac{\delta}{L}\right)^2 = \frac{1}{2}E\varepsilon^2 = \frac{1}{2E}\sigma^2$$

4. **Compute the strain energy of an element under pure shear stress.**

Consider a rectangular block that is subjected to loads that cause a uniform shear stress and strain. If the block displaces as shown, and only shear stresses act, only the external force V on the top does work, acting through the displacement u.

Since no energy is lost, the work done on the block equals the stored energy. Because the shear stress and strain are uniform, the energy stored per volume is uniform. So, for an elastic element under pure shear ($\tau = G\gamma$),

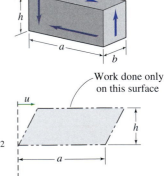

Work done only on this surface

$$\text{Strain Energy Density} = \frac{\text{Stored Energy}}{\text{Volume}} = \frac{\frac{1}{2}(V)(u)}{abh} = \frac{\frac{1}{2}(\tau ab)(\gamma h)}{abh} = \frac{1}{2}G\gamma^2 = \frac{1}{2G}\tau^2$$

5. Compute the strain energy in a unit length of a shaft subjected to a twisting moment T.

Recall that the shear stress in a shaft in torsion varies linearly with radial distance ρ from the center of the shaft: $\tau = T\rho/I_p$. Since the strain energy density varies radially with ρ, it is instead convenient to find the strain energy per unit length of the twisted shaft, which includes the entire cross-section. Consider a unit length and integrate the strain energy density per volume over the cross-sectional area:

$$\text{Strain Energy per Length} = \int_A \left(\frac{1}{2G}\tau^2\right)dA = \int \frac{1}{2G}\left(\frac{T\rho}{I_p}\right)^2 dA = \frac{1}{2G}\left(\frac{T}{I_p}\right)^2 \int_A \rho^2 dA = \frac{T^2}{2GI_p}$$

Alternatively, consider a shaft of length L with left end fixed, subjected to a torque at the right end. Gradually, increase the torque until rotation reaches $\phi = TL/GI_p$. The strain energy per length equals the work done on the shaft divided by the length L:

$$\text{Strain Energy per Length} = \frac{1}{L}\int T d\phi = \frac{1}{L}\int\left(\frac{GI_p\phi}{L}\right)d\phi = \frac{1}{L}\left(\frac{GI_p\phi^2}{2L}\right) = \frac{T^2}{2GI_p}$$

6. Compute the strain energy in a unit length of a beam subjected to a bending moment M.

Recall that the normal stress in a beam in bending varies linearly with distance y from the neutral plane. Since the strain energy varies spatially, it is instead convenient to find the strain energy per unit length of the bent beam, which includes the entire cross-section. Consider a unit length and integrate the strain energy density per volume over the cross-sectional area:

$$\text{Strain Energy per Length} = \int_A \left(\frac{1}{2E}\sigma^2\right)dA = \int \frac{1}{2E}\left(\frac{-My}{I}\right)^2 dA = \frac{1}{2E}\left(\frac{M}{I}\right)^2 \int_A y^2 dA = \frac{M^2}{2EI}$$

Alternatively, consider a beam of length L with left end fixed, subjected to a bending moment at the right end. Gradually, increase the moment until the rotation reaches $\theta = ML/EI$. The strain energy per length equals the work done on the shaft divided by the length L:

$$\text{Strain Energy per Length} = \frac{1}{L}\int M d\theta = \frac{1}{L}\int\left(\frac{EI\theta}{L}\right)d\theta = \frac{1}{L}\left(\frac{EI\theta^2}{2L}\right) = \frac{M^2}{2EI}$$

7. Compute the strain energy in a member subjected to a combination of axial, torsional, and bending loads by integrating the respective strain energies per length over the member length.

The total strain energy per length can be shown to be the sum of the strain energies, due to axial loading (which equals $A\sigma^2/2E = P^2/2EA$), due to twisting and due to bending. The total strain energy can be found by integrating the strain energy per length over the length of the member:

$$\text{Strain Energy} = \int_{Length} \left[\frac{P^2}{2EA} + \frac{T^2}{2GI_p} + \frac{M^2}{2EI}\right]ds$$

Recall that for long, slender members, we often neglect axial deformation in comparison with deformation due to twisting and bending.

>>End 6.6

6.7 Solving Problems Using Conservation of Energy ──────

There are a number of different methods for solving problems using energy. Here we consider one method: Conservation of energy.

1. For energy to be conserved when an elastic body is loaded, the work done by external loads deforming the body from the unloaded state to the loaded state must equal the strain energy stored in the body.

External work is put into a body when an applied load acts through a displacement or rotation. Since the displacements increase together with the applied loads, one would need to integrate the load at each point of application over the increment in displacement, for example $F_x \, du_x$. Moments do work when they produce a rotation. Support loads do not do work, since there are no displacements associated with the non-zero loads. The total energy stored is found from integrating the Strain Energy Density, which can vary at each point, over the volume. To illustrate, we write down conservation of energy for a supported body subjected to two external loads:

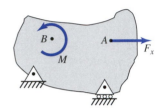

$$\text{External Work} = \text{Stored Energy}$$

$$\int_{at\ A} F_x \, du_x + \int_{at\ B} M d\theta = \int_{entire\ body} (\text{Strain Energy Density}) \, dV$$

2. When only a single external load does work, then the displacement or rotation associated with that load is proportional to only that load, and it appears in the expression for external work.

When there are multiple external loads, the displacement at each point is a linear function of the external loads (not the support loads). When there is a single external force, F_{ext}, its associated displacement, u_{ext}, is proportional to F_{ext}. Then, the external work integral simplifies to:

$$\text{External Work} = \int F du = \frac{1}{2} F_{ext} u_{ext}$$

If the applied load is a single moment M_{ext}, with its associated rotation θ_{ext}, then the External Work simplifies to

$$\text{External Work} = \int M d\theta = \frac{1}{2} M_{ext} \theta_{ext}$$

3. If the total strain energy in the loaded body can be determined, because the distribution of stresses or internal loads are known, then the displacement associated with the single external load can be calculated.

We insert the External Work expression for the case of a single external load F_{ext} into the equation of conservation of energy and rearrange to find:

$$u_{ext} = \frac{2}{F_{ext}} \int (\text{Strain Energy Density}) \, dV$$

Often the stored energy, that is the integral of the Strain Energy Density, can be written down as a function of the external load, F_{ext}, because all the stresses or internal loads can be expressed as functions of the single external load, F_{ext}. Since each stress or internal load will be proportional to F_{ext}, the stored energy will be proportional to $(F_{ext})^2$. Using this last equation, the deflection u_{ext} can then be found, and it will be a linear function of F_{ext}.

4. Apply conservation of energy if these criteria are satisfied: (i) there is a single external load, (ii) the desired unknown is the displacement through which that load does work, and (iii) the stored energy can be calculated in terms of the applied load.

The beam shown is an example of a body that is supported and has a single external load, F_0. The end deflection, v_0, is the deflection through which the external load does work. (Note that the supporting reaction force and moment do no work because the displacement and rotation at the support are zero.) The internal loads (V and M) at each cross-section can be found and are proportional to the applied force F_0, as are the stresses at each point. So we can use conservation of energy to find v_0.

$$\text{External Work} = \frac{1}{2}F_{ext}\,u_{ext} = \frac{1}{2}F_0 v_0$$

$$V(x) = -F_0 \qquad M(x) = -F_0(L - x)$$

5. Determine the *Stored Energy* in terms of the external load, and then equate it to the *External Work*.

In this case of bending, the stored energy can be found by integrating the *Strain Energy per Length* ($M^2/2EI$) over the length. We consider only the energy in bending, just as we neglected the shear deformation in calculating the displacement due to bending in Chapter 5.

$$\text{Stored Energy} = \int(\text{Strain Energy per Length})ds = \int_0^L \left[\frac{M^2}{2EI}\right]dx = \frac{F_0^2}{2EI}\int_0^L (L - x)^2 dx = \frac{F_0^2}{2EI}\left(\frac{L^3}{3}\right)$$

Equate the External Work to the stored energy, and solve for the desired deflection. Note the result agrees with what we found in Chapter 5.

$$\frac{1}{2}F_0 v_0 = \frac{F_0^2}{2EI}\left(\frac{L^3}{3}\right) \Rightarrow v_0 = \frac{2}{F_0}\left[\frac{F_0^2}{2EI}\left(\frac{L^3}{3}\right)\right] = \frac{F_0 L^3}{3EI}$$

6. Follow the same procedure when seeking an angle corresponding to a single applied moment.

Conservation of energy can be used to find the angle of rotation ϕ_0 at the end C of the shaft shown. It is the angle through which the single applied moment T does work, and internal load (twisting) can be found at all cross-sections.

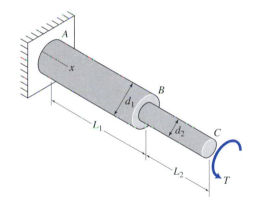

$$\frac{1}{2}T\phi_0 = \int\left[\frac{T^2}{2GI_p}\right]dx = \frac{T^2 L_1}{2G\left[\frac{\pi}{2}\left(\frac{d_1}{2}\right)^4\right]} + \frac{T^2 L_2}{2G\left[\frac{\pi}{2}\left(\frac{d_2}{2}\right)^4\right]} \qquad \phi_0 = \frac{32T}{\pi G}\left[\frac{L_1}{(d_1)^4} + \frac{L_2}{(d_2)^4}\right]$$

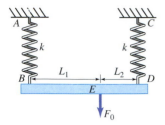

The rigid horizontal bar is suspended from two springs and is subjected to a vertical force F_0. Each spring has stiffness k. Use conservation of energy to determine the deflection of the load at point E. Neglect the bar's weight in comparison with F_0.

Solution

The conditions for using conservation of energy hold:

1. There is a single external load F_0 (the support loads do not do work since the displacements there are zero).
2. The deflection at the load point E, v_E, is desired, and that is the deflection through which the external load does work.
3. Energy is stored in the springs only, and the energy in each spring can be expressed in terms of the force in that spring. Using equilibrium, the forces in each spring can be found in terms of the external load. So, the stored energy can be expressed in terms of the external load.

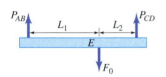

The free body diagram is as shown. From equilibrium, the internal forces can be found.

$$P_{AB} = \frac{L_2}{L_1 + L_2}F_0 \qquad P_{CD} = \frac{L_1}{L_1 + L_2}F_0$$

The expression of External Work = Stored Energy is:

$$\frac{1}{2}F_0 v_E = \frac{1}{2}\frac{P_{AB}^2}{k} + \frac{1}{2}\frac{P_{CD}^2}{k} = \frac{1}{2}\frac{F_0^2}{k}\left[\left(\frac{L_2}{L_1 + L_2}\right)^2 + \left(\frac{L_1}{L_1 + L_2}\right)^2\right]$$

Solve for the desired deflection, v_E, and find:

$$v_E = \frac{F_0}{k}\left[\left(\frac{L_2}{L_1 + L_2}\right)^2 + \left(\frac{L_1}{L_1 + L_2}\right)^2\right]$$

The same problem was considered in Section 3.3. The method in Chapter 3 allowed one to compute the deflection of any point on the horizontal bar, although the specific result presented was for the load point E.

$$v_E = v_B + \frac{L_1}{L_1 + L_2}(v_D - v_B) = \delta_{AB} + \frac{L_1}{L_1 + L_2}(\delta_{CD} - \delta_{AB})$$

We recover the formula found by conservation of energy by substituting in expressions for δ_{AB} and δ_{CD} in terms of F_0.

>>**End Example Problem 6.11**

The slender, curved beam is subjected to the horizontal force F_0. Use conservation of energy to determine the horizontal deflection at the load point. Consider only the bending of the beam, and neglect axial and shear deformation.

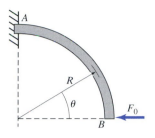

Solution

The conditions for using conservation of energy hold:

1. There is a single external load F_0 (the loads at the fixed support do not do work since the displacement and rotation there are zero).
2. The horizontal deflection at the load point B, u_B, is desired, and that is the deflection through which the external load does work.
3. Energy is stored in the bending deformation only (since axial and shear are neglected). Using equilibrium, the bending moment at each point can be found in terms of the external load. So, the stored energy can be expressed in terms of the external load by integration.

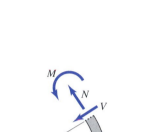

The bending moment can be found from a free body diagram of the portion of the beam shown.

$$M = F_0 R \sin \theta$$

The expression of External Work = Stored Energy for this problem is:

$$\frac{1}{2} F_0 u_B = \int \frac{M^2}{2EI} ds = \int_0^{\pi/2} \frac{(F_0 R \sin \theta)^2}{2EI} R d\theta = \frac{(F_0^2 R^3)}{2EI} \int_0^{\pi/2} (\sin \theta)^2 d\theta$$

Use the trig identity to evaluate the integral:

$$\int_0^{\pi/2} (\sin \theta)^2 d\theta = \int_0^{\pi/2} \frac{1}{2}(1 - \cos 2\theta) d\theta = \frac{1}{2}\left[\theta - \frac{1}{2}\sin 2\theta\right]_0^{\pi/2} = \frac{\pi}{4}$$

Solve for the desired deflection: $u_B = \dfrac{\pi F_0 R^3}{4EI}$

An interesting comparison is with a straight beam with length equal to that of the quarter circle: $L = \pi R/2$.

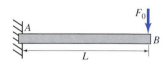

$$v_B \Big|_{straight} = \frac{F_0 L^3}{3EI} = \frac{F_0}{3EI}\left(\frac{\pi R}{2}\right)^3 = \frac{\pi^3 F_0 R^3}{24EI} = \left(\frac{\pi^2}{6}\right)\frac{\pi F_0 R^3}{4EI}$$

As seen from the result, the straight beam has a larger deflection by a factor $\pi^2/6 = 1.64$. The straight beam plausibly has a larger deflection because the bending moment reaches greater values in some portions of the beam.

>>End Example Problem 6.12

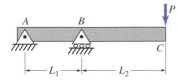

The overhanging beam shown is subjected to a concentrated force P. Use conservation of energy to determine the end deflection and compare with the solution from Chapter 5.

Solution

The conditions for using conservation of energy hold. A single external force P does work, we desire the end deflection through which P does work, and the energy can be found by integrating the strain energy due to bending over the beam.

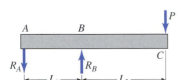

The bending moment in each part of the beam can be found in terms of the force P, if the support loads are expressed in terms of P.

$$R_A = \frac{L_2}{L_1}P \qquad R_B = \frac{L_1 + L_2}{L_1}P$$

Sometimes one can simplify the expression for strain energy $M^2/2EI$ that needs to be integrated, and one can do separate integrations over different parts of a body. Also, notice that one only needs the magnitude of M, not the sign, because we integrate M^2.

$$|M| = R_A x = \frac{L_2}{L_1}Px \quad (0 < x < L_1) \qquad |M| = Px_2 \quad (0 < x_2 < L_2)$$

Now integrate separately over the lengths AB and BC. While it was not necessary to use the distance, x_2, the moment M was then easy to find and the integration is very simple.

$$\text{Stored Energy} = \int_{AB} \frac{M^2}{2EI}ds + \int_{BC} \frac{M^2}{2EI}ds = \int_0^{L_1} \frac{\left(\frac{L_2}{L_1}Px\right)^2}{2EI}dx + \int_0^{L_2} \frac{(Px_2)^2}{2EI}dx_2 = \frac{P^2L_2^2L_1}{6EI} + \frac{P^2L_2^3}{6EI}$$

The expression of external work = stored energy is then:
$$\frac{1}{2}Pv_C = \frac{P^2L_2^2L_1}{6EI} + \frac{P^2L_2^3}{6EI}$$

The desired deflection, v_C, is found to be:
$$v_C = \frac{PL_2^2L_1}{3EI} + \frac{PL_2^3}{3EI}$$

The same problem was considered in Chapter 5, and the same end deflection was found.

>>End Example Problem 6.13

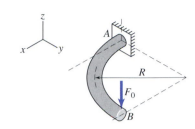

A slender member in the form of a quarter circle is rigidly supported at A and subjected to a force F_0 acting perpendicularly to its plane. The loading produces both bending and twisting of the member. Use conservation of energy to determine the vertical deflection w_0 (parallel to F_0) at the load point. Neglect transverse shear deformation. The beam has uniform bending stiffness EI, and torsional stiffness GI_p.

Solution

The conditions for using conservation of energy hold. A single external force F_0 does work, we desire the end deflection through which F_0 does work, and the energy can be found by integrating the strain energy, due to bending and twisting, over the beam.

The internal moment varies along the length of the member. We want the bending and twisting moments, M and T, because their strain energies per length are found separately and then added. M and T are defined as the components of the moment that are perpendicular and parallel to the outward normal to the internal cross-section. But the orientation of the internal cross-section varies with θ. The external force F_0 acts into and out of the plane of the paper and the member and is represented with $\otimes$ and $\odot$ signs.

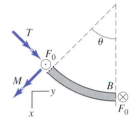

One approach to determining M and T is first to find the components of the moment along x- and y-axes, M_x and M_y. The respective perpendicular distances are:

$$d_x = R(1 - \cos\theta) \qquad d_y = R \sin\theta$$

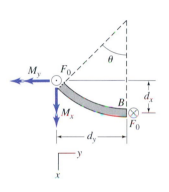

The moment components are then:

$$M_x = F_0 d_y = F_0 R \sin\theta \qquad M_y = F_0 d_x = F_0 R(1 - \cos\theta)$$

We drew the components M_x and M_y to be in the correct senses so they are both positive.

Compute the bending and twisting components by resolving the x-y moment components.

$$M = M_y \sin\theta + M_x \cos\theta = F_0 R \sin\theta \qquad T = -M_y \cos\theta + M_x \sin\theta = -F_0 R(1 - \cos\theta)$$

(Check that you have the correct combinations of $\sin\theta$ and $\cos\theta$ by inspecting whether the signs and values are correct for θ approaching $0°$ and $90°$.)

Integrate and use trig identities to find:

$$\text{Stored Energy} = \int_0^{90°} \frac{M^2}{2EI} d\theta + \int_0^{90°} \frac{T^2}{2GI_p} d\theta = \frac{F_0^2 R^3}{2EI}\left[\left(\frac{1}{2}\right)\left(\frac{\pi}{2}\right)\right] + \frac{F_0^2 R^3}{2GI_p}\left[\left(\frac{3\pi}{4} - 2\right)\right]$$

Equate the external work with the stored energy and find the desired deflection, w_0, to be:

$$w_0 = F_0 R^3\left[\frac{\pi}{4}\left(\frac{1}{EI}\right) + \left(\frac{3\pi}{4} - 2\right)\left(\frac{1}{GI_p}\right)\right]$$

>>End Example Problem 6.14

Additional data on material properties needed to solve problems can be found in Appendix D or inside back cover.

6.66 The rigid bar ABC is pinned at A and supported by a spring. Use conservation of energy to determine the vertical deflection v_C at point C.

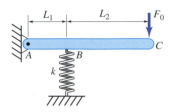

Prob. 6.66

6.67 The rigid bar $ABCD$ is pinned at A and supported by two springs. Use conservation of energy to determine the vertical deflection v_D at point D. (Hint: relate the deflections at points B and C to the desired deflection v_D.)

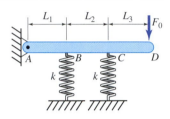

Prob. 6.67

6.68 The shaft is supported at both ends A and C and is subjected to an external torque T_0 at the midpoint B. Use conservation of energy to determine the rotation at the midpoint. (Hint: use symmetry to recognize that the reaction torques are the same at both ends.) The shaft has a uniform torsional stiffness GI_p.

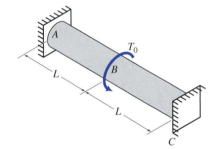

Prob. 6.68

6.69 Shaft AB is supported at A and has a gear (diameter of d_B) mounted on the end B. A second shaft CD has a gear at C (diameter of d_C), which meshes with the gear at B. The gears are supported by frictionless bearings (not shown). A torque T_0 is applied to end D of the shaft CD. Use conservation of energy to determine the rotation at D. Shaft AB has a uniform torsional stiffness G_1I_{p1}, and shaft BC has a uniform torsional stiffness G_2I_{p2}.

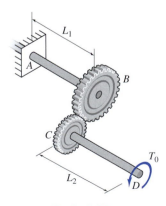

Prob. 6.69

6.70 The simply supported beam is subjected to moment M_0 at the right end. Decide which quantity can be found using conservation of energy, and then use conservation of energy to find it. The beam has uniform bending stiffness EI.

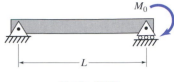

Prob. 6.70

6.71 The simply supported beam is subjected to a concentrated force P at the center. Determine the deflection at the center of the beam using conservation of energy. The beam has uniform bending stiffness EI.

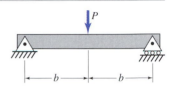

Prob. 6.71

6.72 The simply supported beam is subjected to concentrated moment M_0 at the center. Determine the rotation at the center of the beam using conservation of energy. The beam has uniform bending stiffness EI.

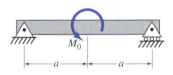

Prob. 6.72

6.73 The simply supported beam is subjected to concentrated force F_0 at some point along the beam. Use conservation of energy to determine the deflection at the point where the force is applied. Compare with the result from Appendix G-2. The beam has uniform bending stiffness EI.

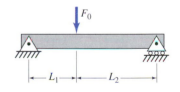

Prob. 6.73

6.74 The cantilevered beam, subjected to a force F_0 at its end, consists of two portions with different moments of inertia I_1 and I_2. Determine the deflection at the end C using conservation of energy. The beam has uniform elastic modulus E.

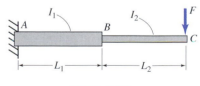

Prob. 6.74

6.75 The simply supported beam is non-uniform, but symmetric about the center plane through point C. Portions AB and DE have the same moment of inertia I_1, while the center portion has moment of inertia I_2. Use conservation of energy to determine the deflection at the center point where force F_0 is applied. The beam has uniform elastic modulus E.

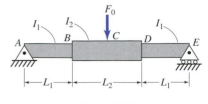

Prob. 6.75

6.76 The beam shown is subjected to symmetric loading as shown. Assume that the deflections are also symmetric, so that points A and D each deflect inwardly (parallel to the load P) by v_0. Use conservation of energy to determine the deflection v_0. Note that both forces do work through the deflection v_0. The beam has uniform bending stiffness EI.

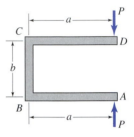

Prob. 6.76

6.77 The beam shown is subjected to the horizontal force P. Use conservation of energy to determine the horizontal deflection u_C. The beam has uniform bending stiffness EI.

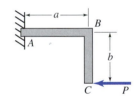

Prob. 6.77

6.78 The curved beam is subjected to a horizontal force F_0 at point B. Use conservation of energy to determine the horizontal deflection at B. The beam has uniform bending stiffness EI.

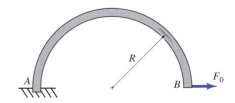

Prob. 6.78

6.79 The curved beam is subjected to a vertical force F_0 at point B. Use conservation of energy to determine the vertical deflection at B. The beam has uniform bending stiffness EI.

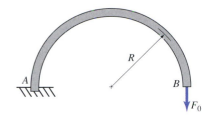

Prob. 6.79

6.80 The slender member in the form of a half circle is rigidly supported at one end and subjected to a force F_0 acting perpendicularly to its plane. The loading produces both bending and twisting of the member. Use conservation of energy to determine the vertical deflection (parallel to F_0) at the load point. Neglect transverse shear deformation. The member has uniform bending stiffness EI, and torsional stiffness GI_p.

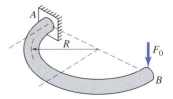

Prob. 6.80

6.81 The slender member in the form of a three-quarter circle is rigidly supported at one end and subjected to a force F_0 acting perpendicularly to its plane. The loading produces both bending and twisting of the member. Use conservation of energy to determine the vertical deflection (parallel to F_0) at the load point. Neglect transverse shear deformation. The member has uniform bending stiffness EI, and torsional stiffness GI_p.

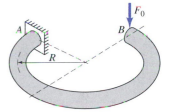

Prob. 6.81

>>End Problems

Chapter Summary

6.1   External loads can produce multiple internal loads at a cross-section.

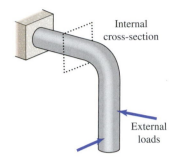

Internal
cross-section

External
loads

Resolve internal loads into
components normal to and
parallel to the member axis.

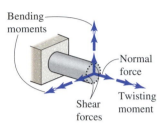

Bending
moments

Normal
force

Twisting
moment

Shear
forces

Categorize each internal
load as one of:

Normal force P
Shear force V
Bending moment M
Twisting moment T

6.2   Different internal loads produce different stresses in an element. Therefore, a combination of internal loads
will cause a combination of stresses.

Stresses can depend on the position of the element in a cross-section, as seen in Chapters 3, 4, and 5.
Stresses shown are for an element at the upper point of the cross-section.

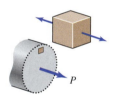

P

Zero stress

V

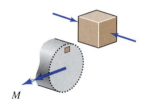

M

T

6.3   Internal pressure in a thin-walled **pressure vessel** causes tensile stress in both directions of the plane
of the wall.

Cylindrical Pressure Vessel

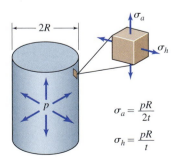

$2R$

σ_a

σ_h

p

$$\sigma_a = \frac{pR}{2t}$$

$$\sigma_h = \frac{pR}{t}$$

Spherical Pressure Vessel

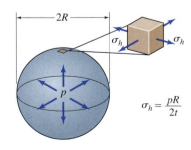

$2R$

σ_h

σ_h

p

$$\sigma_h = \frac{pR}{2t}$$

6.4   The **generalized elastic stress–strain relations** are the equations that relate each strain component to the full set of stress components. An **isotropic** material responds identically to stress regardless of how the material is oriented. For an isotropic material, the strains are related to the stresses as follows:

To calculate each normal strain, add the contributions (positive or negative) of each of the normal stress components.

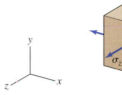

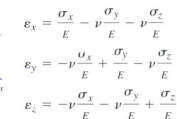

$$\varepsilon_x = \frac{\sigma_x}{E} - \nu\frac{\sigma_y}{E} - \nu\frac{\sigma_z}{E}$$

$$\varepsilon_y = -\nu\frac{\sigma_x}{E} + \frac{\sigma_y}{E} - \nu\frac{\sigma_z}{E}$$

$$\varepsilon_z = -\nu\frac{\sigma_x}{E} - \nu\frac{\sigma_y}{E} + \frac{\sigma_z}{E}$$

Each shear stress component is proportional to its corresponding shear strain component and related to no other.

$$\tau_{xy} = G\gamma_{xy} \qquad \tau_{xz} = G\gamma_{xz} \qquad \tau_{yz} = G\gamma_{yz}$$

6.5   For a body composed of multiple segments, some portions can undergo combinations of deformations (e.g., bending and twisting). Account carefully for continuity of deflection and rotation at the connections between segments.

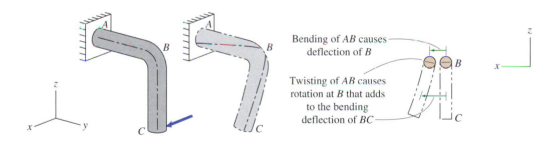

Bending of *AB* causes deflection of *B*

Twisting of *AB* causes rotation at *B* that adds to the bending deflection of *BC*

6.6–6.7   If there is a single external load (besides supports), **Conservation of Energy** can sometimes be used to determine the deflection. The work done by the external load through the to-be-determined deflection is equal to the strain energy in the body. The strain energy may be determined by integrating the strain energy density over the volume of the body, which can be expressed in terms of local stresses or internal loads.

Stress Transformations and Failure

1. Define stresses on inclined surface

Focus on one point at which a combination of stresses, σ_x, σ_y, and τ_{xy}, is found to act.

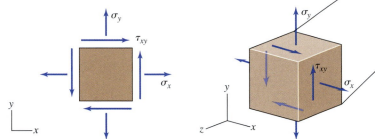

To predict failure under a combination of stresses, we cannot simply add them.

Follow the elongations and angle changes for an element inclined at angle θ from x-axis. For illustration, consider only one x-y stress: σ_x.

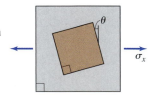

The inclined element deforms into a parallelogram, signaling that normal and shear stresses act on its surfaces

On a surface of any orientation θ, σ_x, σ_y, and τ_{xy} each produce normal stress σ and shear stress τ. We add the normal stress due to all three components σ_x, σ_y, and τ_{xy} and shear stress due to all three.

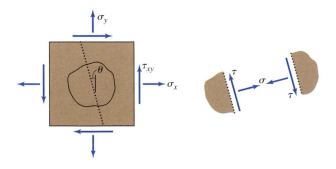

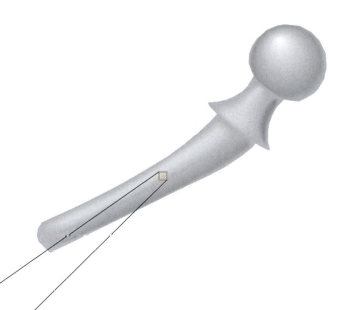

2. Determine maximum stresses and relate to failure

Compare the normal or shear stress acting
on differently oriented surfaces θ.

Find the orientation θ with the
maximum shear stress, and find
the magnitude of the maximum
shear stress.

Find the orientation θ with the
maximum normal stress, and find
the magnitude of the maximum
normal stress.

Maximum normal and shear stresses and their orientations depend only
on the combination of σ_x, σ_y, and τ_{xy}.

Each major failure mode tends to be promoted by primarily
shear stress or normal stress

Ductile failure

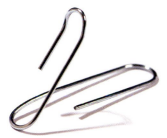

Ductile failure (plastic deformation)
is sensitive to shear stress

Predict ductile failure based
on maximum shear stress

Brittle failure

Brittle failure is sensitive
to normal stress

Predict brittle failure based
on maximum normal stress

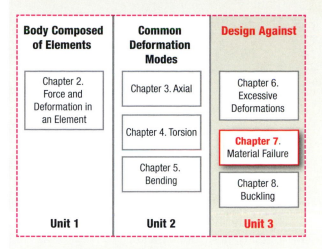

Body Composed of Elements	Common Deformation Modes	Design Against
Chapter 2. Force and Deformation in an Element	Chapter 3. Axial	Chapter 6. Excessive Deformations
	Chapter 4. Torsion	**Chapter 7.** Material Failure
	Chapter 5. Bending	Chapter 8. Buckling
Unit 1	**Unit 2**	**Unit 3**

Chapter Outline

When multiple stresses act at the same point in a member, judging failure requires special attention to how stresses combine. To prepare for studying the effect of a combination of stresses, we recognize that deformation, which clearly corresponds to a normal strain when viewed from one perspective (x-y axes), involves both shear and normal strain when viewed from a different perspective (**7.1**). We consequently define normal and shear stresses acting across a plane of general orientation θ, not just across x-y planes (**7.2**). The shear and normal stresses acting on a general plane depend on its orientation and on the stresses on x-y planes (**7.3**). When planes of all orientations are considered, the shear or normal stresses on certain planes are maximum or minimum (**7.4**). An alternative, graphical representation allows one to picture stresses on all planes, as well as maximum and minimum stresses (**7.5**). Brittle and ductile failure modes correspond to whether the normal or the shear stress, maximized over all planes, reaches a critical value (**7.6–7.7**). Strains in directions other than x-y directions can also be defined, and experimental measurements of strains acting along three directions are used to infer multiple components of stress (**7.8**). Fatigue failure can occur due to stresses that are cycling (applied and released multiple times), even if the maximum stress would not cause failure when applied steadily (**7.9**). Stresses can be elevated relative to the levels studied in previous chapters at sudden changes in cross-section. These stress concentrations, which are particularly likely to lead to failure under fatigue loading, can often be computed from tables of stress concentration factors (**7.10**).

7.1 Goal of Chapter, and Strain is in the Eye of the Beholder

As we have seen in the previous chapter, loads in many practical situations produce a combination of stresses. Normal and shear stresses, or stresses acting on different planes or in different directions, cannot simply be added. Still we need to determine if failure occurs when these stresses act simultaneously.

1. No matter how complex the loading, it produces a set of three (2-D) or six (3-D) stress components at each point of the body.

Here are examples of loads that produce multiple stress components: a shaft under bending and twisting, and a hip implant subject to a complex loading that is analyzed with finite elements. In any event, we focus on a point in the loaded body. In general, multiple stress components act at that point. Here, we show stresses on x-y planes. We want to know if these stresses cause failure.

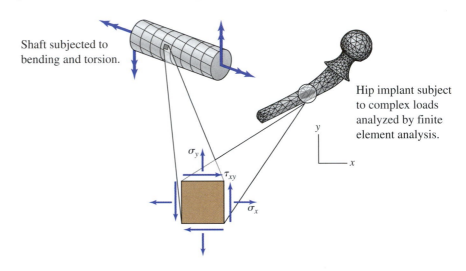

Shaft subjected to bending and torsion.

Hip implant subject to complex loads analyzed by finite element analysis.

2. Even when there are only three stress components, it is not practical to determine failure experimentally for all possible combinations.

It is infeasible to conduct measurements of the stresses to cause failure under multiple combinations of normal and shear stress.

In fact, it is common only to measure failure properties in one state, typically uniaxial tension. How can we predict the tendency for failure under combinations of stresses from only measurements under uniaxial tension?

3. While forces at a point can be added, normal and shear stresses in an element cannot.

Even though stress has dimensions of force per area $\left(\text{e.g., N/m}^2\right)$, stress is very different from a force, beyond the normalization by area. Stress components give information on the different forces (per area) that act on different faces.

Consider this example in which the stress components σ_x and τ_{xy} are positive in an element, but $\sigma_y = 0$. Each of σ_x and τ_{xy} corresponds to a collection of forces. σ_x corresponds to opposite normal forces acting on the left and right faces. τ_{xy} corresponds to shear forces acting on all four faces; those forces act horizontally or vertically depending on the face.

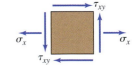

Here is the problem that arises when merely adding: if we were to add the numbers σ_x and τ_{xy}, which stress component would that sum correspond to? What would be the directions of the resulting forces, and on which faces would they act? In fact, we cannot make sense of that sum.

So, we do not add different stress components. Instead, we take a more general view of internal force per area that also provides a means of properly combining stress components.

To motivate a more general view of internal force per area, we first show that the same deformation can appear to be either normal or shear strain depending entirely on one's perspective.

4. Describe the deformation of a body where the motions of four points are tracked.

Consider a body under load, and focus on a small region that deforms in the plane. Four points, A, B, C, and D are labeled on the body, and these points displace as the body deforms. Lines joining the points will be drawn, and the deformations of the lines will be considered.

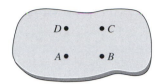

5. For two distinct deformations of this region, follow the displacements of these four points and what happens to various line segments connecting them.

	Deformation 1	Deformation 2
How would you describe these two deformation patterns, based only on the displacements of the points?	D C ... A B	D C ... A B
Draw line segments AB, BC, CD, and DA, aligned with the x-y axes, and follow them with each deformation.	D C / A B / y x	D C / A B / y x
Draw line segments AC and BD, oriented at 45° from x-y axes, and follow them with each deformation.	D C / A B / y x	D C / A B / y x

Look at the four cases with line segments and ask: Which segments elongate (normal strain)? Does the angle between a pair of line segments change (shear strain)?

6. For the same deformation, we observe lines to change length or shear or both, depending on the orientations of the lines.

For each of the two deformations, how we characterize the strain depends on which segments are followed.

Segments aligned with x-y: Deformation 1 involves only normal strain (and only along AB and CD). Deformation 2 involves only shear strain; no segment elongates (for small strain).

Segments at 45° from x-y: Deformation 1 involves shear strain and normal strain (AC and BD get longer). Deformation 2 involves no shear strain; BD shortens, AC elongates.

Neither way of drawing segments is better or preferred. But, for the same deformation pattern, there may or may not be shear strain or normal strain, depending on how the segment is drawn. We need a way of describing strain and stress that accounts more fully for orientation.

>>End 7.1

7.2 Defining Stresses on General Surfaces

We saw in the previous chapter that, for isotropic materials, normal strains along *x-y* axes are related to σ_x and σ_y, and shear strain is related to shear stress τ_{xy}. We just found that there are normal and shear strains involving segments not aligned with *x-y* axes. Here we discuss the stresses that must be associated with those strains. All these different stress components correspond to different ways of viewing forces on an element at the same point.

1. **Each stress component represents a force (per area) between two bodies across some internal surface.**

Stress is force per area, but between exactly which two bodies does the force act?

The *x-y* coordinate axes define one set of natural surfaces on which we can separate the region of interest into two bodies.

For example, we can talk about the stress between these two bodies:

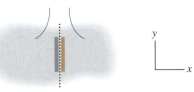

The surface separating the bodies is a *y-z* plane and has a normal in the *x*-direction.

Separate these two bodies and draw the equal and opposite force (per area) between them (resolved into *x-y* components).

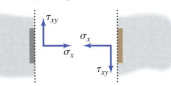

Or, we can think of the stress between bodies formed by a different division of the same small region:

The surface separating the bodies is a *x-z* plane and has a normal in the *y*-direction.

Separate these two bodies and draw the equal and opposite force (per area) between them (resolved into *x-y* components).

The forces per area across surfaces with normals in the *x*- and *y*-directions are different.

2. **Represent the stresses across two perpendicular planes at the same time by drawing them on a square element extracted from the central point.**

It is common to use a single drawing to show the stresses on two perpendicular surfaces of separation.

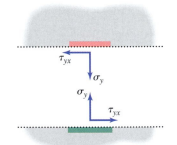

σ_x: acts on the *x*-face in the *x*-direction

σ_y: acts on the *y*-face in the *y*-direction

τ_{xy}: acts on the *x*-face in the *y*-direction

τ_{yx}: acts on the *y*-face in the *x*-direction

As we have seen, the signs give the sense of the stress component. The senses shown are for positive values.

The components shown are independent, except that $\tau_{xy} = \tau_{yx}$ for moments to be in equilibrium.

3. Consider now the interaction between two portions of a body that meet at an internal surface that is inclined with respect to *x-y* axes.

The *x-y* coordinate axes are not the only surfaces on which to separate a small region into two bodies. We can separate along any plane.

The small region is now separated into two bodies at an internal surface of arbitrarily chosen orientation.

Here are the two bodies. The force per area on their shared internal surface is of interest.

4. Define the orientation of the inclined surface by the angle between its outward normal vector and the *x*-axis.

We need a means of describing the orientation of any internal surface. We draw a vector pointing outward from the internal surface, and then use an angle to describe the orientation of the vector.

For each internal surface, there is an outward normal vector. To describe the orientation of the outward normal, rotate the positive *x*-axis counterclockwise through some angle to make it align with the outward normal. That angle of rotation, θ, defines the orientation of the outward normal. Surfaces 1 and 2 correspond to opposite faces so they satisfy $(\theta)_2 = (\theta)_1 + 180°$.

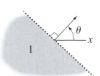

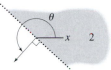

5. On an inclined surface, define the normal stress as positive if outward (tension), and define the shear stress as positive when it is 90° CCW from the outward normal.

The force per area between the two bodies is equal and opposite. We resolve the force on the surface into components, not along *x-y* axes, but in directions perpendicular and parallel to the surface. These force components per area, normal and shear, depend on θ, and so are written as functions of θ.

The normal stress $\sigma(\theta)$ is defined as positive if it is in tension (outward). The shear stress $\tau(\theta)$ is defined as positive if it points 90° counterclockwise from the outward normal. Since the forces must be equal and opposite on opposing faces on which θ differs by 180°, $\sigma(\theta) = \sigma(\theta + 180°)$ and $\tau(\theta) = \tau(\theta + 180°)$.

Our goal is to determine $\sigma(\theta)$ and $\tau(\theta)$ in terms of the *x-y* components σ_x, σ_y, and τ_{xy} and the angle θ.

6. Stresses $\sigma(\theta)$ and $\tau(\theta)$ are related to the normal and shear strains of inclined elements.

In the previous section, we recognized that there could be normal and shear strains when viewing segments inclined relative to *x-y* axes. The stresses $\sigma(\theta)$ and $\tau(\theta)$ are related to these strains along directions other than *x-y* axes.

 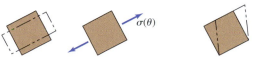

>>End 7.2

Determine the angle θ that describes each of the internal surfaces shown.

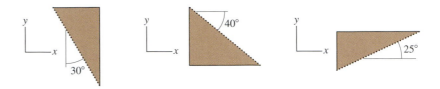

Solution

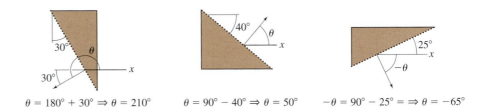

$$\theta = 180° + 30° \Rightarrow \theta = 210°$$ $$\theta = 90° - 40° \Rightarrow \theta = 50°$$ $$-\theta = 90° - 25° = \Rightarrow \theta = -65°$$

>>End Example Problem 7.1

Draw internal surfaces corresponding to the angles $\theta = 105°$, $-160°$, and $330°$.

Solution

Draw the outward normal that is at the angle θ clockwise from the x-axis.

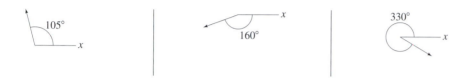

Draw internal surfaces so that they are perpendicular to the outward normals just found.

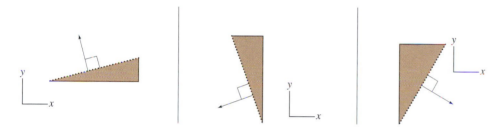

>>**End Example Problem 7.2**

For each of the surfaces and stresses shown, indicate if σ and τ are positive or negative.

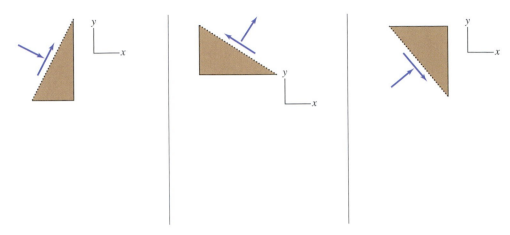

Solution

For each surface, draw the outward normal, and from it draw the directions of positive σ and positive τ. Recall that $\sigma > 0$ if outward from the surface (tension). Recall that $\tau > 0$ if it acts in the direction 90° counterclockwise from $\sigma > 0$.

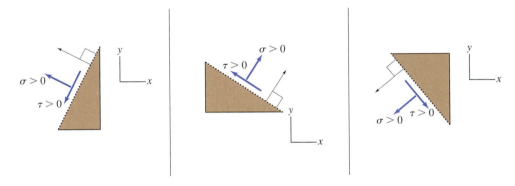

Compare given stress directions with positive directions:

$$\Rightarrow \sigma < 0 \text{ and } \tau < 0 \quad \Big| \quad \Rightarrow \sigma > 0 \text{ and } \tau < 0 \quad \Big| \quad \Rightarrow \sigma < 0 \text{ and } \tau > 0$$

>>End Example Problem 7.3

Draw internal surfaces and stresses corresponding to (a) $\theta = -75°$, $\sigma = 80$ MPa, and $\tau = -60$ MPa and (b) $\theta = 230°$, $\sigma = -40$ MPa, and $\tau = -70$ MPa.

Solution

Draw the internal surface and directions of positive stresses.

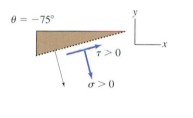

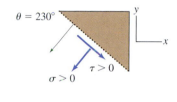

Now, draw the stresses in correct directions and magnitudes labeled on internal surfaces.

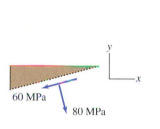

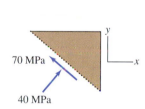

>>End Example Problem 7.4

Additional data on material properties needed to solve problems can be found in Appendix D or inside back cover.

7.1 We are interested in the stresses that act on the internal surface shown. What θ describes this surface?

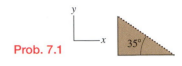

Prob. 7.1

7.2 We are interested in the stresses that act on the internal surface shown. What θ describes this surface?

Prob. 7.2

7.3 Draw the right triangular element that includes the internal surface described by angle $\theta = 170°$.

7.4 Draw the right triangular element that includes the internal surface described by angle $\theta = -30°$.

7.5 Given a polygon, with surfaces a, b, c, and d, that make angles with the horizontal or vertical as shown, determine the angle θ for each of the surfaces.

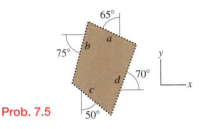

Prob. 7.5

7.6 Four surfaces a, b, c, and d, are described by angles $\theta_a = 55°$, $\theta_b = 120°$, $\theta_c = 210°$, and $\theta_d = 325°$, respectively. Draw a closed polygonal figure having these surfaces, and label the angle (between 0 and 90) that each surface makes with the horizontal.

7.7 Shear and normal stresses are drawn in particular senses on the internal surface shown. Identify whether each of the normal (σ) and shear (τ) stresses is > 0 or < 0.

Prob. 7.7

7.8 Shear and normal stresses are drawn in particular senses on the internal surface shown. Identify whether each of the normal (σ) and shear (τ) stresses is > 0 or < 0.

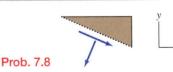

Prob. 7.8

7.9 On the internal surface shown, $\sigma = -150$ MPa and $\tau = -90$ MPa. Draw the shear and normal stresses in the correct senses.

Prob. 7.9

7.10 On the internal surface shown, $\sigma = 8000$ psi and $\tau = -5000$ psi. Draw the shear and normal stresses in the correct senses.

Prob. 7.10

7.11 Four internal surfaces, a, b, c, and d, are shown on the polygon taken from a small region of the material. The shear and normal stresses are drawn in particular senses on each surface. For each surface, identify whether each of the normal (σ) and shear (τ) stresses is > 0 or < 0.

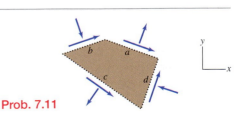

Prob. 7.11

7.12 Four internal surfaces, a, b, c, and d, are shown on the polygon taken from a small region of the material. The shear and normal stresses are drawn in particular senses on each surface. For each surface, identify whether each of the normal (σ) and shear (τ) stresses is > 0 or < 0.

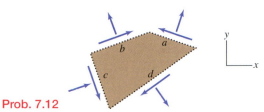

Prob. 7.12

7.13 Four internal surfaces, a, b, c, and d, are shown on the polygon taken from a small region of the material. The shear and normal stresses have been determined for each surface. Draw the stresses on each surface in the correct senses.

a: $\sigma = -44.6$ MPa and $\tau = 162$ MPa
b: $\sigma = -185$ MPa and $\tau = -118$ MPa
c: $\sigma = 101$ MPa and $\tau = -46.9$ MPa
d: $\sigma = -176$ MPa and $\tau = -126$ MPa

Prob. 7.13

7.14 The shear and normal stresses are drawn in particular senses on the internal surface as shown. Identify the angle θ, and whether the stresses σ and τ are each > 0 or < 0.

Prob. 7.14

7.15 The shear and normal stresses are drawn in particular senses on the internal surface as shown. Identify the angle θ, and whether the stresses σ and τ are each > 0 or < 0.

Prob. 7.15

7.16 Draw the right triangular element containing the surface with angle $\theta = 120°$. Assuming the stresses on the surface are $\sigma = 9000$ psi and $\tau = -7500$ psi, draw the stresses in the correct senses on the surface.

7.17 Draw the right triangular element containing the surface with angle $\theta = 210°$. Assuming the stresses on the surface are $\sigma = -200$ MPa and $\tau = 150$ MPa, draw the stresses in the correct senses on the surface.

7.18 Given a polygon, with surfaces a, b, c, and d, and the shear and normal stresses drawn in particular senses on each surface, identify the angle θ of each face and whether stresses σ and τ on each are > 0 or < 0.

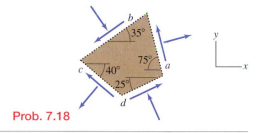

Prob. 7.18

7.19 Consider a polygon with four surfaces a, b, c, and d, with the following angles and stresses:

a: $\theta = 50°$, $\sigma = -12.9$ ksi, $\tau = 7.18$ ksi
b: $\theta = 110°$, $\sigma = 11.2$ ksi, $\tau = 6.74$ ksi
c: $\theta = 210°$, $\sigma = -14.8$ ksi, $\tau = -2.17$ ksi
d: $\theta = 305°$, $\sigma = 12.9$ ksi, $\tau = -0.25$ ksi

Draw a closed polygonal figure having these surfaces, label the angle (between 0 and 90°) that each surface makes with the horizontal, and draw the stresses in the correct senses on the faces.

>>End Problems

7.3 Stress Transformation Formulas

We are considering a planar stress state, in which the stresses act in the x-y plane on surfaces with normals in the x-y plane. The stresses on just two perpendicular surfaces through a point, for example σ_x, σ_y, and τ_{xy}, fully determine the *state of stress* at that point. Here we derive formulas to determine the normal and shear stress $\sigma(\theta)$ and $\tau(\theta)$ acting on any other surface with normal in the x-y plane through the same point.

1. **Isolate a portion of material on which all the stresses, σ_x, σ_y, τ_{xy}, $\sigma(\theta)$, and $\tau(\theta)$, act.**

Here again is a small region in the body. The stresses are uniform from point to point within this region, but they do depend on the orientation of the plane on which they act. Say the stresses on x-y planes, σ_x, σ_y, and τ_{xy}, are known. We can find stresses on a surface oriented at θ by drawing an FBD of a region on which all stresses act.

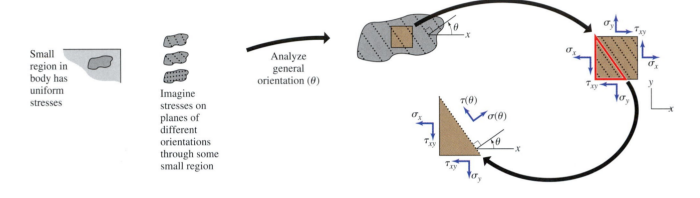

Small region in body has uniform stresses

Imagine stresses on planes of different orientations through some small region

Analyze general orientation (θ)

2. **Multiply the stress on each face times the face area and sum the resulting forces in the x- and y-directions.**

Choose the triangular element to have hypotenuse h and out-of-plane thickness t. If the side of length h is the surface oriented at θ according to the earlier definition, then θ is also the angle between the vertical side and the hypotenuse. Multiply each stress component by the area on which it acts to get its force. Then add forces in each of the x- and y-directions.

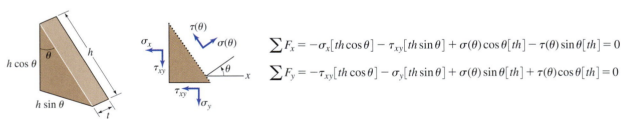

$$\sum F_x = -\sigma_x[th\cos\theta] - \tau_{xy}[th\sin\theta] + \sigma(\theta)\cos\theta[th] - \tau(\theta)\sin\theta[th] = 0$$

$$\sum F_y = -\tau_{xy}[th\cos\theta] - \sigma_y[th\sin\theta] + \sigma(\theta)\sin\theta[th] + \tau(\theta)\cos\theta[th] = 0$$

3. **Combine equations, use trig identities, and find formulas for $\sigma(\theta)$ and $\tau(\theta)$.**

Combine the force summation equations above as follows:

$$\left(\sum F_x\right)\cos\theta + \left(\sum F_y\right)\sin\theta = 0 \qquad -\left(\sum F_x\right)\sin\theta + \left(\sum F_y\right)\cos\theta = 0$$

Use these trigonometric identities: $2\sin\theta\cos\theta = \sin 2\theta \qquad \cos^2\theta - \sin^2\theta = \cos 2\theta$

Rearrange the equations, solve for $\sigma(\theta)$ and $\tau(\theta)$, and arrive at these final transformation equations (more details in Appendix I-2):

$$\sigma(\theta) = \frac{1}{2}(\sigma_x + \sigma_y) + \frac{1}{2}(\sigma_x - \sigma_y)\cos 2\theta + \tau_{xy}\sin 2\theta$$

$$\tau(\theta) = \tau_{xy}\cos 2\theta + \frac{1}{2}(\sigma_y - \sigma_x)\sin 2\theta$$

4. Confirm that the transformation formulas give the correct results when the inclined plane is perpendicular to the *x*- or *y*-axis.

$\sigma(\theta)$ and $\tau(\theta)$ given by the transformation formulas should have the correct values for $\theta = 0$ (normal is *x*-axis) and $\theta = 90°$ (normal is *y*-axis).

$$\theta = 0°: \quad \sigma(0) = \frac{1}{2}(\sigma_x + \sigma_y) + \frac{1}{2}(\sigma_x - \sigma_y) = \sigma_x \quad \tau(0) = \tau_{xy}$$

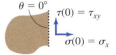
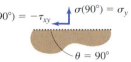

$$\theta = 90°: \quad \sigma(90°) = \frac{1}{2}(\sigma_x + \sigma_y) - \frac{1}{2}(\sigma_x - \sigma_y) = \sigma_y \quad \tau(90°) = -\tau_{xy}$$

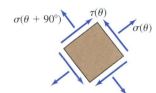

---

5. Draw the stresses on an inclined square element by finding the stresses on its four faces.

Pick some θ in $0 < \theta < 90°$. Look at four surfaces $\theta, \theta + 90°, \theta + 180°$, and $\theta + 270°$. These are faces of a square element rotated by angle θ relative to the *x* axis. Use these trig relations:

From these trig identities $\quad \sin 2(\theta + 90°) = -\sin 2\theta \quad \cos 2(\theta + 90°) = -\cos 2\theta$

we find $\quad \tau(\theta + 90°) = -\tau(\theta)$

From these trig identities $\quad \sin 2(\theta + 180°) = \sin 2\theta \quad \cos 2(\theta + 180°) = \cos 2\theta$

we find $\quad \sigma(\theta + 180°) = \sigma(\theta) \quad \tau(\theta + 180°) = \tau(\theta)$

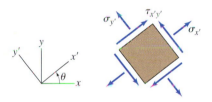

Therefore, on the element at θ, there are only three independent components, $\sigma(\theta), \sigma(\theta + 90°)$, and $\tau(\theta)$, just like there are only three components on *x-y* axes, σ_x, σ_y, and τ_{xy}.

---

6. Use rotated axes *x'-y'* as an alternative notation for stresses on inclined planes.

Here is a common alternative notation for stresses on an element oriented at θ. Define additional axes *x'-y'*, with *x'* oriented at θ from the *x*-axis. Then, the normal and shear stresses on the *x'-y'* axes are $\sigma_{x'} = \sigma(\theta), \sigma_{y'} = \sigma(\theta + 90°)$, and $\tau_{x'y'} = \tau(\theta)$.

---

7. Assuming isotropic elastic behavior, relate stresses on rotated *x'-y'* axes to the strains on those axes.

Just as the *x-y* stresses are related to the *x-y* strains by the material law, the stresses on the *x'-y'* axes are similarly related to their respective strains.

To see these strains, draw a square element with sides oriented along the *x'-y'* axes. The changes in length of the edges, $\delta_{x'}$ and $\delta_{y'}$, normalized by their original lengths, give the normal strains $\varepsilon_{x'}$ and $\varepsilon_{y'}$. The change in the included angle gives the shear strain $\gamma_{x'y'}$.

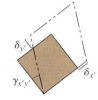

Recall that an isotropic material strains in the direction of the stress by the same amount, regardless of how that stress is oriented. For an isotropic material, the strains along $\varepsilon_{x'}, \varepsilon_{y'}$, and $\gamma_{x'y'}$ are related to the normal stresses $\sigma_{x'}$ and $\sigma_{y'}$ and shear stress $\tau_{x'y'}$ by the same elastic constants, *E*, ν, and *G*, as are *x-y* components.

$$\varepsilon_{x'} = \frac{\sigma_{x'}}{E} - \nu\frac{\sigma_{y'}}{E} \qquad \varepsilon_{y'} = \frac{\sigma_{y'}}{E} - \nu\frac{\sigma_{x'}}{E} \qquad \gamma_{x'y'} = \frac{\tau_{x'y'}}{G}$$

>>End 7.3

For this state of stress with the given x-y components, determine the stress acting on the surface shown.

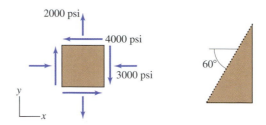

Solution

Note the directions of the stresses on the x-y faces:

$$\sigma_x = -3000 \text{ psi}, \quad \sigma_y = 2000 \text{ psi}, \text{ and } \tau_{xy} = -4000 \text{ psi}$$

The surface of interest is $\theta = 180° - 30°$ or $90° + 60° = 150°$.

Evaluate the transformed stresses $\sigma(\theta)$ and $\tau(\theta)$ for the specific values of σ_x, σ_y, and τ_{xy} and the surface θ.

$$\sigma(\theta) = \frac{1}{2}(\sigma_x + \sigma_y) + \frac{1}{2}(\sigma_x - \sigma_y)\cos 2\theta + \tau_{xy} \sin 2\theta$$

$$\sigma(150°) = \frac{1}{2}(-3000 + 2000) + \frac{1}{2}(-3000 - 2000)\cos 2(150°) - 4000 \sin 2(150°) = 1714 \text{ psi}$$

$$\tau(\theta) = \tau_{xy} \cos 2\theta + \frac{1}{2}(\sigma_y - \sigma_x)\sin 2\theta$$

$$\tau(150°) = -4000 \cos 2(150°) + \frac{1}{2}(2000 + 3000)\sin 2(150°) = -4160 \text{ psi}$$

The directions of stresses if $\sigma > 0$ and $\tau > 0$ would be

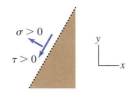

Since $\sigma = 1714$ psi and $\tau = -4160$ psi, the correct directions of stresses are

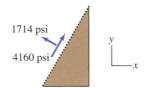

>>End Example Problem 7.5

Given the state of stress shown, determine the change in length of the 2 mm segment. Take the elastic moduli to be $E = 70$ GPa and $\nu = 0.33$.

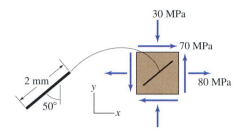

Solution

Note the directions of the stresses on the x-y faces:

$$\sigma_x = 80 \text{ MPa}, \ \sigma_y = -30 \text{ MPa}, \ \tau_{xy} = 70 \text{ MPa}$$

To find the strain of the element shown, we need the normal stress acting parallel to the line segment and the normal stress perpendicular to the line segment (because of the Poisson effect).

Define the direction parallel to the segment as x'.

The stress parallel to the segment, $\sigma_{x'}$, corresponds to $\sigma(40°)$.
The stress perpendicular to the segment, $\sigma_{y'}$, corresponds to $\sigma(130°)$.

Evaluate the transformed stresses $\sigma(\theta)$ for the specific values of σ_x, σ_y, and τ_{xy}, and angles $\theta = 40°$ and $130°$.

$$\sigma(\theta) = \frac{1}{2}(\sigma_x + \sigma_y) + \frac{1}{2}(\sigma_x - \sigma_y)\cos 2\theta + \tau_{xy}\sin 2\theta$$

$$\sigma_{x'} = \sigma(40°) = \frac{1}{2}(80 - 30) + \frac{1}{2}(80 + 30)\cos 2(40°) + 70 \sin 2(40°) = 103.5 \text{ MPa}$$

$$\sigma_{y'} = \sigma(130°) = \frac{1}{2}(80 - 30) + \frac{1}{2}(80 + 30)\cos 2(130°) + 70 \sin 2(130°) = -53.5 \text{ MPa}$$

$$\text{Strain along segment} = \varepsilon_{x'} = \frac{\sigma_{x'}}{E} - \nu\frac{\sigma_{y'}}{E} = \frac{103.5 \text{ MPa}}{70 \text{ GPa}} - \nu\frac{-53.5 \text{ MPa}}{70 \text{ GPa}} = 1.731 \times 10^{-3}$$

The change in length of the segment δ is related to the strain by $\varepsilon_{x'} = \dfrac{\delta}{2 \text{ mm}}$.

The change in length is, therefore, $\delta = (2 \text{ mm})(1.731 \times 10^{-3}) = 3.46 \times 10^{-3}$ mm.

>>End Example Problem 7.6

Additional data on material properties needed to solve problems can be found in Appendix D or inside back cover.

7.20 The stresses on the x-y axes are $\sigma_x = -3$ ksi, $\sigma_y = 5$ ksi, and $\tau_{xy} = 6$ ksi. Calculate the normal and shear stresses acting on the inclined surface shown, and then draw them in the correct senses, labeling their magnitudes.

Prob. 7.20

7.21 The stresses on the x-y axes are $\sigma_x = 100$ MPa, $\sigma_y = -200$ MPa, and $\tau_{xy} = -150$ MPa. Calculate the normal and shear stresses acting on the inclined surface shown, and then draw them in the correct senses, labeling their magnitudes.

Prob. 7.21

7.22 The stresses are depicted on the square element aligned with the x-y axes. Calculate and then draw and label the normal and shear stresses acting on the inclined surface shown.

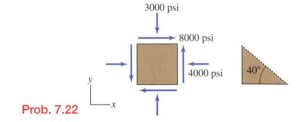

Prob. 7.22

7.23 The stresses are depicted on the square element aligned with the x-y axes. Calculate and then draw and label the normal and shear stresses acting on the inclined surface shown.

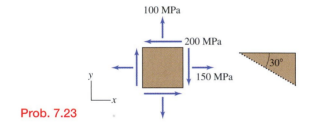

Prob. 7.23

7.24 The stresses on the x-y axes are $\sigma_x = -80$ MPa, $\sigma_y = -40$ MPa, and $\tau_{xy} = 50$ MPa. Calculate the normal and shear stresses acting on the four faces of the inclined element shown, and then draw them in the correct senses, labeling their magnitudes.

Prob. 7.24

7.25 The stresses on the x-y axes are $\sigma_x = -5$ ksi, $\sigma_y = 4$ ksi, and $\tau_{xy} = 3$ ksi. Calculate the normal and shear stresses acting on the four faces of the inclined element shown, and then draw them in the correct senses, labeling their magnitudes.

Prob. 7.25

7.26 The stresses are drawn on an element oriented with respect to the x-y axes. Calculate the normal and shear stresses acting on the four faces of the inclined element shown, and then draw them in the correct senses, labeling their magnitudes.

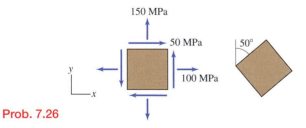

Prob. 7.26

7.27 The stresses are drawn on an element oriented with respect to the x-y axes. Calculate the normal and shear stresses acting on the four faces of the inclined element shown, and then draw them in the correct senses, labeling their magnitudes.

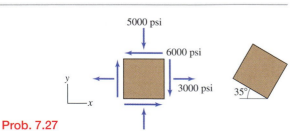

Prob. 7.27

7.28 The stresses are drawn on the inclined element oriented as shown. Determine the normal and shear stresses σ_x, σ_y, and τ_{xy}.

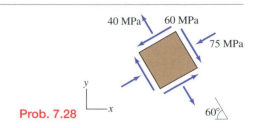

Prob. 7.28

7.29 The stresses are drawn on the inclined element oriented as shown. Determine the normal and shear stresses σ_x, σ_y, and τ_{xy}.

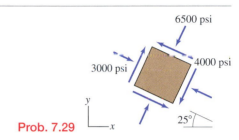

Prob. 7.29

7.30 The normal and shear stresses are drawn on one surface, while only the normal stress is known on the second surface. Assuming these are consistent with a uniform state of stress, determine the stresses σ_x, σ_y, and τ_{xy}, and then calculate and draw the shear stress on the second surface.

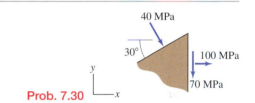

Prob. 7.30

7.31 The normal and shear stresses are drawn on one surface, while only the shear stress is known on the second surface. Assuming these are consistent with a uniform state of stress, determine the stresses σ_x, σ_y, and τ_{xy}, and then calculate and draw the normal stress on the second surface.

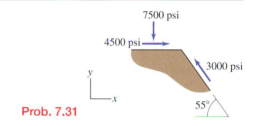

Prob. 7.31

7.32 From the general stress transformation formulas, prove that τ has the same value at θ and at $\theta + 180°$. Prove that τ has the same magnitude, but opposite signs, at θ and at $\theta + 90°$.

7.33 From the general stress transformation formulas, prove that σ has the same value at θ and at $\theta + 180°$.

7.34 From the general stress transformation formulas, prove that $\sigma(\theta) + \sigma(\theta + 90°)$ has the same value, equal to $\sigma_x + \sigma_y$, independent of the angle θ.

7.35 The stresses on the x-y axes are $\sigma_x = 80$ MPa, $\sigma_y = 50$ MPa, and $\tau_{xy} = 60$ MPa. Determine the normal strain parallel to the line segment shown. Include both the direct stress and the transverse stress (Poisson effect). Take $E = 70$ GPa and $\nu = 0.33$.

Prob. 7.35

7.36 The stresses on the x-y axes are shown. Determine the normal strain parallel to the line segment shown. Include both the direct stress and the transverse stress (Poisson effect). Take $E = 300$ ksi and $\nu = 0.4$.

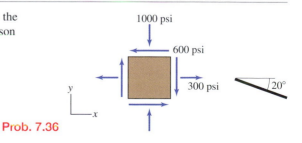

Prob. 7.36

7.37 The two line segments shown are perpendicular under zero stress. The stresses on the x-y axes are $\sigma_x = 10$ ksi, $\sigma_y = 6$ ksi, and $\tau_{xy} = 5$ ksi. What angle do these two segments make with respect to each other when the stresses act? Take $G = 11.5\left(10^6\right)$ psi.

Prob. 7.37

7.38 The two line segments shown are perpendicular under zero stress. The stresses on the x-y axes are shown. What angle do these two segments make with respect to each other when the stresses act? Take $G = 2.5$ GPa.

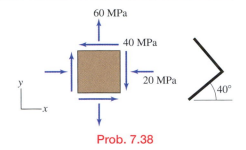

Prob. 7.38

7.39 The mutually perpendicular line segments, AB and AC, each having length 1 mm when zero stresses act. Under the action of stresses on the x-y axes, $\sigma_x = 100$ MPa, $\sigma_y = 200$ MPa, and $\tau_{xy} = 150$ MPa, AB and AC are, respectively, $2.481\left(10^{-3}\right)$ mm longer and $7.592\left(10^{-4}\right)$ mm shorter. Determine the Young's modulus and the Poisson ratio that are consistent with this observation.

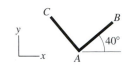

Prob. 7.39

7.40 The mutually perpendicular line segments, AB and AC, each having length 1 mm when zero stresses act. Under the action of stresses on the x-y axes, $\sigma_x = -120$ MPa, $\sigma_y = -140$ MPa, and $\tau_{xy} = 100$ MPa, AB and AC are, respectively, $1.044\left(10^{-3}\right)$ mm shorter and $1.336\left(10^{-4}\right)$ mm longer. Determine the Young's modulus and the Poisson ratio that are consistent with this observation.

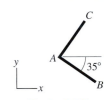

Prob. 7.40

7.41 The stresses on the x-y axes are $\sigma_x = -5$ MPa, $\sigma_y = 8$ MPa, and $\tau_{xy} = -4$ MPa. The angle between the two lines shown, which was 90° when unstressed, is 89.370° when the stresses act. Determine the elastic shear modulus consistent with this observation.

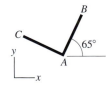

Prob. 7.41

7.42 The stresses on the x-y axes are $\sigma_x = 60$ MPa, $\sigma_y = 80$ MPa, and $\tau_{xy} = 90$ MPa. The angle between the two lines shown, which was 90° when unstressed, is 90.055° when the stresses act. Determine the elastic shear modulus consistent with this observation.

Prob. 7.42

7.43 Without any stresses acting, the element shown was a 1×1 mm square oriented at 40° with respect to the horizontal axis. The elastic moduli are $E = 2$ GPa, $\nu = 0.4$, and $G = 0.714$ GPa. Determine the normal and shear stresses that must act on the x-y axes to cause the element to deform into the parallelogram shown.

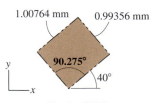

Prob. 7.43

7.44 A bone fracture is oriented at $\theta = 30°$ from the axial direction. The bone is required to withstand a loading due to weight on the foot that produces a compressive stress of 5 MPa along the bone axis. The fracture can withstand these forces if there is sufficient friction to resist slippage at the fracture plane due to shear stress. The friction coefficient, μ, gives the maximum shear stress that the surfaces can sustain without slipping as a proportionality of the normal stress ($\mu \sigma_n$). In order to prevent slippage, a screw is inserted perpendicular to the fracture and tightened to give some initial compressive normal stress (and no shear stress) squeezing the fragments together prior to any loading. If μ is assumed to be 0.2, determine the average level of normal stress across the fracture plane that the screw must induce, if, together with the applied axial stress, no slippage between fragments is to occur.

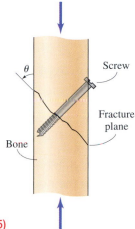

Prob. 7.44 (Appendix A5)

7.45 The fracture plane of orientation $\theta = 30°$ from the axial direction has been pre-compressed to a level of 10 MPa by the insertion and tightening of a screw. The friction coefficient, μ, gives the maximum shear stress that the surfaces can sustain without slipping as a proportionality of the normal stress ($\mu \sigma_n$). Let the friction coefficient at the fracture plane be 0.3. Determine the maximum bending moment that can be applied before slippage occurs on the tensile side of the bone. Take the bone to have an outer diameter of 28 mm and an inner diameter of 14 mm.

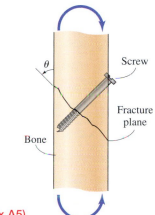

Prob. 7.45 (Appendix A5)

7.46 Assume there is no pre-compression at a fracture plane via screws. Under axial compression along the bone, there will still not be slippage for fracture planes at suitable orientations because the shear stress increases more slowly than the normal stress. The friction coefficient, μ, gives the maximum shear stress that the surfaces can sustain without slipping as a proportionality of the normal stress ($\mu \sigma_n$). Determine the range of stable orientations of the fracture plane α assuming a friction coefficient of $\mu = 0.3$.

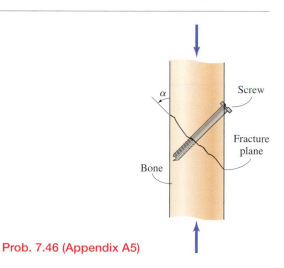

Prob. 7.46 (Appendix A5)

>>End Problems

7.4 Maximum and Minimum Stresses

For a given state of stress at a point, specified by one set of components, σ_x, σ_y, and τ_{xy}, the normal and shear stresses vary with the orientation of the plane θ. Failure, we will see later, is often driven primarily by only normal stress or by only shear stress, depending on the material. If, for example, normal stress drives failure, we will judge whether failure occurs by looking at the normal stress on different planes θ and find that plane on which the normal stress is greatest. Here we show how the maximum normal and shear stress are found.

1. Follow this example, which shows a single stress state and how the normal and shear stresses $\sigma(\theta)$ and $\tau(\theta)$ vary depending on the orientation of the plane.

We illustrate maximum stresses for one specific case: $\sigma_x = -10$ MPa, $\sigma_y = 30$ MPa, and $\tau_{xy} = 40$ MPa. $\sigma(\theta)$ and $\tau(\theta)$ which are sinusoidal functions given by the transformation equations, are plotted vs. θ for $0 < \theta < 180°$. These smooth curves have maxima and minima. Stress elements are drawn for $\theta = 0°$, 22.5°, 45°, and 67.5° to show the general variation, and for $\theta = 13.28°$ and 58.28° to show the maximum shear and normal stress.

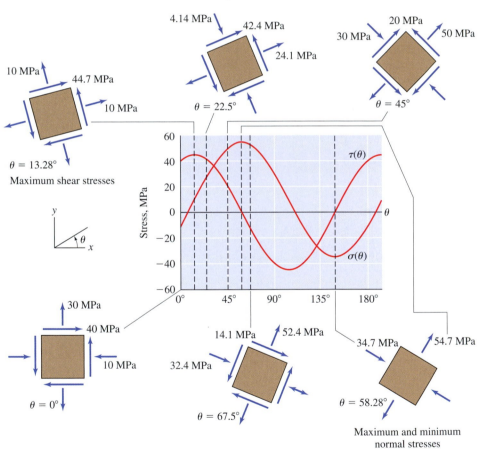

Maximum and minimum normal stresses

When considering the stresses acting on elemental surfaces of different orientations passing through the same point, the **Maximum Shear Stress**, τ_{max}, is defined as the magnitude of the shear stress that is maximum from among all possible surface orientations.

When considering the stresses acting on elemental surfaces of different orientations passing through the same point, the **Principal Stresses**, σ_{max} and σ_{min}, are defined as the maximum and minimum normal stresses from among all possible surface orientations.

For these particular values of σ_x, σ_y, and τ_{xy}, the shear stress is maximum on the element at $\theta = 13.28°$. That stress is referred to as the **maximum shear stress**. An element drawn in this orientation along with its stresses is called the maximum shear stress element. On the element at $\theta = 58.28°$, the normal stresses are maximum and minimum; they are also termed the **principal stresses**. An element drawn in this orientation along with its stresses is called the principal stress element.

We can find general formulas for the maximum and minimum normal and shear stresses and the planes on which they act using calculus $\dfrac{d\sigma(\theta)}{d\theta} = \dfrac{d\tau(\theta)}{d\theta} = 0$. The formulas and the procedure for drawing the maximum shear stress and principal stress elements are derived in Appendix I-2.

2. From the stresses σ_x, σ_y, **and** τ_{xy} **evaluate the quantities** τ_{max}, σ_m, θ_s $(-90° < \theta_s < 90°)$, σ_{max}, **and** σ_{min}.

These quantities, computed from σ_x, σ_y, and τ_{xy}, are used to determine the maximum shear and principal stress elements.

Maximum Shear Stress	**Maximum Shear Angle**	**Mean Stress**
$\tau_{max} = \sqrt{\frac{1}{4}(\sigma_y - \sigma_x)^2 + \tau_{xy}^2}$	$\theta_s = \frac{1}{2}\tan^{-1}\left[\frac{(\sigma_y - \sigma_x)/2}{\tau_{xy}}\right]$	$\sigma_m = \frac{1}{2}(\sigma_x + \sigma_y)$

The maximum and minimum normal stresses (principal stresses) have values of

$$\sigma_{max} = \sigma_m + \tau_{max} \text{ and } \sigma_{min} = \sigma_m - \tau_{max}$$

3. Draw the maximum shear stress element.

- Draw an element at angle θ_s (one face is at θ_s).
- Evaluate $\tau(\theta_s) = \tau_{xy}\cos 2\theta_s + \frac{1}{2}(\sigma_y - \sigma_x)\sin 2\theta_s$.
  ($\tau(\theta_s)$ should equal $\pm\tau_{max}$; if not, check for calculation error.)
- Draw shear stress τ_{max} on face θ_s consistent with sign of $\tau(\theta_s)$.
- Draw consistent shear stresses on the remaining three faces.
- Draw normal stress σ_m on all four faces (sense consistent with sign of σ_m).

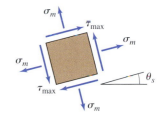

4. Draw the principal stress element.

- Draw an element at the principal stress angle $\theta_p = \theta_s - 45°$.
- Evaluate $\sigma(\theta_p) = \frac{1}{2}(\sigma_x + \sigma_y) + \frac{1}{2}(\sigma_x - \sigma_y)\cos 2\theta_p + \tau_{xy}\sin 2\theta_p$.
  ($\sigma(\theta_p)$ should equal σ_{max} or σ_{min}; if not, check for calculation error.)
- Draw the normal stress, σ_{max} or σ_{min}, on face θ_p consistent with $\sigma(\theta_p)$.
- Draw the normal stress on face $\theta_p + 90°$ (equal to the other principal stress σ_{max} or σ_{min}).
- Shear stress is zero on all faces of this element.

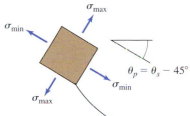

5. Alternatively, determine the planes of the principal stress element on which σ_{max} **and** σ_{min} **act by using the directions of shear of the maximum shear element.**

σ_{max} and σ_{min} always act on the element oriented at $\theta_s - 45°$. Instead of finding the faces on which σ_{max} and σ_{min} act by evaluating $\sigma(\theta_p)$, note that σ_{max} acts on faces near τ_{max} arrow heads, and σ_{min} acts on faces near τ_{max} arrow tails.

See this from the drawn triangles, on which the senses of τ_{max} differ.

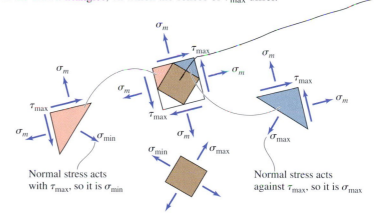

Normal stress acts with τ_{max}, so it is σ_{min}

Normal stress acts against τ_{max}, so it is σ_{max}

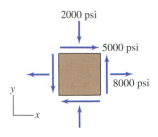

Given the state of stress shown, draw the maximum shear stress element and the principal stress element.

Solution

Note the directions of the stresses on x-y faces:

$$\sigma_x = 8000 \text{ psi}, \; \sigma_y = -2000 \text{ psi, and } \tau_{xy} = 5000 \text{ psi}$$

To find the maximum shear stress and principal stresses, the key parameters that we need to calculate are

$$\tau_{max} = \sqrt{\frac{1}{4}(\sigma_y - \sigma_x)^2 + \tau_{xy}^2} = \sqrt{\frac{1}{4}(-2000 - 8000)^2 + (5000)^2} = 6400 \text{ psi}$$

$$\theta_s = \frac{1}{2} \tan^{-1}\left[\frac{(\sigma_y - \sigma_x)/2}{\tau_{xy}}\right] = \frac{1}{2}\tan^{-1}\left[\frac{(-2000 - 8000)/2}{4000}\right] = -25.67°$$

$$\sigma_m = \frac{1}{2}(\sigma_x + \sigma_y) = \frac{1}{2}(8000 - 2000) = 3000 \text{ psi}$$

Test the direction of $\tau(\theta)$ on $\theta = \theta_s = -25.67°$.

$$\tau(-25.67°) = 5000 \cos 2(-25.67°) + \frac{1}{2}(-2000 - 8000)\sin 2(-25.67°) = +6400 \text{ psi}$$

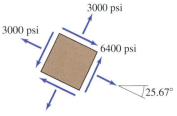

Maximum shear stress element:
Note $\tau(\theta) = +6400$ psi on $\theta = -25.67°$.
So the shear stress is drawn in the positive sense on that face, and the remaining shear stresses are drawn consistently. Both normal stresses on this element equal σ_m.

Principal stress element is 45° from maximum shear stress element.
Draw element at $-25.67° + 45° = 19.32°$. Evaluate $\sigma(\theta)$ at $\theta = 19.32°$.

$$\sigma(19.32°) = \frac{1}{2}(8000 - 2000) + \frac{1}{2}(8000 + 2000)\cos 2(19.32°) + 5000\sin 2(19.32°) = 9400 \text{ psi}$$

Notice that $\sigma(19.32°)$ equals $\sigma_{max} = \sigma_m + \tau_{max}$. The stress on $\theta = 19.32° + 90°$ must be $\sigma_{min} = \sigma_m - \tau_{max} = 3000 \text{ psi} - 6400 \text{ psi} = -3400 \text{ psi}$.

Here is the principal stress element. Notice that σ_{max} is on the face that is the closest to the arrowheads of τ_{max} on the maximum shear element.

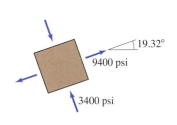

>>End Example Problem 7.7

A thin-walled cylindrical steel tube under internal pressure of $p = 3$ MPa is also subjected to torsion $T = 800$ N-m. The tube has a mean diameter of 80 mm and a wall thickness of 2 mm. Determine the maximum shear stress element and the principal stress element at a point on the visible side of the tube.

Solution

Calculate the stresses on the x-y axes and find them to have these magnitudes:

$$\sigma_h = 2\sigma_a = \frac{pR}{t} = \frac{3 \text{ MPa}(40 \text{ mm})}{2 \text{ mm}} = 60 \text{ MPa} \qquad \tau = \frac{T\rho}{I_p} = \frac{(800 \text{ N-m})(41 \text{ mm})}{\frac{\pi}{2}\left[(41 \text{ mm})^4 - (39 \text{ mm})^4\right]} = 40.8 \text{ MPa}$$

Note the directions of the stresses on the x-y faces: $\sigma_x = \sigma_a = 30$ MPa, $\sigma_y = \sigma_h = 60$ MPa, and $\tau_{xy} = -40.8$ MPa.

The key parameters that we need to calculate are:

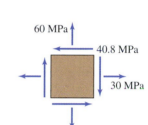

$$\tau_{max} = \sqrt{\frac{1}{4}(\sigma_y - \sigma_x)^2 + \tau_{xy}^2} = \sqrt{\frac{1}{4}(60 - 30)^2 + (40.8)^2} = 43.5 \text{ MPa}$$

$$\theta_s = \frac{1}{2}\tan^{-1}\left[\frac{(\sigma_y - \sigma_x)/2}{\tau_{xy}}\right] = \frac{1}{2}\tan^{-1}\left[\frac{(60 - 30)/2}{-40.8}\right] = -10.09°$$

$$\sigma_m = \frac{1}{2}(\sigma_x + \sigma_y) = \frac{1}{2}(30 + 60) = 45 \text{ MPa}$$

Maximum shear stress element is at $-10.09°$.
Test the direction of $\tau(\theta)$ on $\theta = -10.09°$.

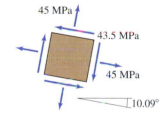

$$\tau(-10.09°) = -40.8\cos 2(-10.09°) + \frac{1}{2}(60 - 30)\sin 2(-10.09°) = -43.5 \text{ MPa}$$

Draw τ_{max} and σ_m in the correct senses on the maximum shear element.
Principal stress element is at $-10.09° + 45° = 34.91°$.

$$\sigma(34.91°) = \frac{1}{2}(30 + 60) + \frac{1}{2}(30 - 60)\cos 2(34.91°) - 40.8\sin 2(34.91°) = 1.53 \text{ MPa}$$

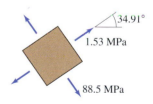

Notice that $\sigma(34.91°)$ equals $\sigma_{min} = \sigma_m - \tau_{max}$.
So $\sigma(34.91° + 90°)$ must be $\sigma_{max} = \sigma_m + \tau_{max} = 45 \text{ MPa} + 43.5 \text{ MPa} = 88.5 \text{ MPa}$.

Notice that σ_{max} is on the face that is closest to the arrowheads of τ_{max}.

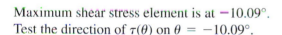

>>End Example Problem 7.8

Additional data on material properties needed to solve problems can be found in Appendix D or inside back cover.

7.47 An element has stresses $\sigma_x = 5400$ psi, $\sigma_y = -3000$ psi, and $\tau_{xy} = -2800$ psi. Draw the principal stress element, showing its orientation relative to the x-y axes and all the stresses acting on it.

7.48 An element has stresses $\sigma_x = 80$ MPa, $\sigma_y = -110$ MPa, and $\tau_{xy} = 60$ MPa. Draw the maximum shear stress element, showing its orientation relative to the x-y axes and all the stresses acting on it.

7.49 An element has stresses $\sigma_x = 130$ MPa, $\sigma_y = 60$ MPa, and $\tau_{xy} = 90$ MPa. Draw the principal stress and maximum shear stress elements, showing their orientations relative to the x-y axes and all the stresses acting on each.

7.50 An element has stresses $\sigma_x = -5$ ksi, $\sigma_y = 6$ ksi, and $\tau_{xy} = 3$ ksi. Draw the principal stress and maximum shear stress elements, showing their orientations relative to the x-y axes and all the stresses acting on each.

7.51 The stresses on the x-y axes are $\sigma_x = 80$ MPa, $\sigma_y = -50$ MPa, and $\tau_{xy} = 70$ MPa. The elastic moduli are $E = 200$ GPa, $\nu = 0.3$, and $G = 80$ GPa. Find the orientation of the line segment that extends the most. What is the strain of this segment?

7.52 The stresses on the x-y axes are $\sigma_x = 2$ ksi, $\sigma_y = 6$ ksi, and $\tau_{xy} = -4$ ksi. The elastic moduli are $E = 10(10^6)$ psi, $\nu = 0.33$, and $G = 3.75(10^6)$ psi. Find the orientation of the line segment that extends the most. What is the strain of this segment?

7.53 The stresses on the x-y axes are $\sigma_x = -7$ MPa, $\sigma_y = 4$ MPa, and $\tau_{xy} = -3$ MPa. The elastic moduli are $E = 3$ GPa, $\nu = 0.4$, and $G = 1.1$ GPa. Find the orientation of the line segment that contracts the most. What is the strain of this segment?

7.54 The stresses on the x-y axes are $\sigma_x = -6$ ksi, $\sigma_y = -8$ ksi, and $\tau_{xy} = 5$ ksi. The elastic moduli are $E = 58(10^6)$ psi, $\nu = 0.2$, and $G = 24(10^6)$ psi. Find the orientation of the line segment that contracts the most. What is the strain of this segment?

7.55 The stresses on the x-y axes are $\sigma_x = -30$ MPa, $\sigma_y = -60$ MPa, and $\tau_{xy} = 40$ MPa. The elastic moduli are $E = 70$ GPa, $\nu = 0.33$, and $G = 26$ GPa. Find the orientation of an initially 1 mm × 1 mm square which remains a rectangle. Determine the dimensions of the sides of the rectangle into which the square deforms.

7.56 The stresses on the x-y axes are $\sigma_x = 6$ ksi, $\sigma_y = -10$ ksi, and $\tau_{xy} = -7$ ksi. The elastic moduli are $E = 30(10^6)$ psi, $\nu = 0.3$, and $G = 11(10^6)$ psi. Find the orientation of an initially 1 in. × 1 in. square which remains a rectangle. Determine the dimensions of the sides of the rectangle into which the square deforms.

7.57 The stresses on the x-y axes are $\sigma_x = -30$ MPa, $\sigma_y = -60$ MPa, and $\tau_{xy} = -50$ MPa. The elastic moduli are $E = 400$ GPa, $\nu = 0.2$, and $G = 167$ GPa. Find the orientation of an initially 1 mm $\times$ 1 mm square which shears the most. Draw the deformed element. Determine the shear strain of this element, and the lengths of the sides of the element when the stresses are acting.

7.58 The stresses on the x-y axes are $\sigma_x = 2$ ksi, $\sigma_y = -4$ ksi, and $\tau_{xy} = 1$ ksi. The elastic moduli are $E = 350$ ksi, $\nu = 0.4$, and $G = 125$ ksi. Find the orientation of an initially 1 in. $\times$ 1 in. square which shears the most. Draw the deformed element. Determine the shear strain of this element, and the lengths of the sides of the element when the stresses are acting.

7.59 A hollow shaft with a 14 mm outer diameter and a wall thickness of 2 mm is subjected to a twisting moment of $T_0 = 20$ N-m and a bending moment of $M_0 = 30$ N-m. Determine the stresses at point A (where x is maximum), and then compute and draw the maximum shear stress element. Describe its orientation relative to the shaft axis.

7.60 A hollow shaft with a 1.6 in. outer diameter and a wall thickness of 0.125 in. is subjected to a twisting moment of $T_0 = 2500$ lb-in. and a bending moment of $M_0 = 2000$ lb-in. Determine the stresses at point A (where x is maximum), and then compute and draw the maximum shear stress element. Describe its orientation relative to the shaft axis.

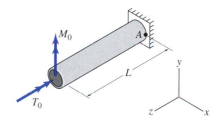

Probs. 7.59–60

7.61 The solid bar of circular cross-section (diameter $= 20$ mm) is fixed at one end, and then the other end is rotated as shown by $\phi_0 = 1°$ and given an extension $\delta_0 = 0.2$ mm. Determine the stresses at point A (where x is maximum), and then compute and draw the maximum shear stress element. Describe its orientation relative to the shaft axis. The elastic moduli are $E = 200$ GPa, $\nu = 0.3$, and $G = 80$ GPa. Take the length $L = 1.2$ m.

7.62 A hollow shaft (outer diameter $= 1$ in., inner diameter $= 0.5$ in.) is fixed at one end, and then the other end is rotated as shown by $\phi_0 = 1°$ and given an extension $\delta_0 = 0.01$ in. Determine the stresses at point A (where x is maximum), and then compute and draw the principal stress element. Describe its orientation relative to the shaft axis. The elastic moduli are $E = 10(10^6)$ psi, $\nu = 0.33$, and $G = 3.75(10^6)$ psi. Take the length $L = 50$ in.

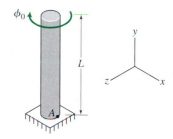

Probs. 7.61–62

7.63 A solid circular shaft with a 0.25 in. diameter is fixed at one end. The other end is rotated as shown by $\phi_0 = 0.3°$ and then subjected to a transverse force which causes a deflection of 0.007 in. in the negative y-direction. Determine the stresses at point A (where y is maximum), and then compute and draw the maximum shear stress element. Describe its orientation relative to the shaft axis. The elastic moduli are $E = 58(10^6)$ psi, $\nu = 0.2$, and $G = 24(10^6)$ psi. Take the length $L = 6$ in.

7.64 A solid circular shaft with 6 mm diameter is fixed at one end. The other end is rotated as shown by $\phi_0 = 0.4°$ and then subjected to a transverse force which causes a deflection of 0.15 mm in the negative y-direction. Determine the stresses at point A (where y is maximum), and then compute and draw the principal stress element. Describe its orientation relative to the shaft axis. The elastic moduli are $E = 3$ GPa, $\nu = 0.4$, and $G = 1.1$ GPa. Take the length $L = 150$ mm.

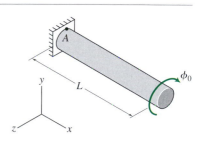

Probs. 7.63–64

7.65 The aluminum member shown is subjected to a transverse force $F_0 = 30$ N at its end. Determine the stresses at point A (where z is maximum), and then compute and draw the maximum shear and principal stress elements. Describe the orientation of these elements relative to the axis of the portion of length L_1. Take the member to have a circular cross-section of 12 mm in diameter and the dimensions to be $L_1 = 200$ mm and $L_2 = 300$ mm.

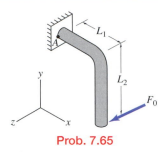

Prob. 7.65

7.66 The steel member shown is subjected to a transverse force F_0. The loading results in a deflection of 0.25 in. at the load point. Determine the stresses at point A (where y is maximum), and then compute and draw the maximum shear and principal stress elements. Describe the orientation of these elements relative to the axis of the portion of length L_1. The elastic moduli are $E = 30(10^6)$ psi, $\nu = 0.3$, and $G = 11(10^6)$ psi. Take the member to have a hollow circular cross-section with 3 in. outer diameter and a wall thickness of 0.09 in., and take the dimensions to be $L_1 = 36$ in. and $L_2 = 24$ in.

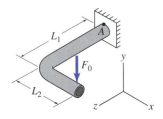

Prob. 7.66

7.67 The steel post supporting the sign has a cross-section with an outer diameter of 350 mm and a wall thickness of 8 mm. Consider the loading due to only the wind pressure of 2.3 kPa. Determine the stresses at point A (where x is maximum), and then compute and draw the maximum shear and principal stress elements. Describe the orientation of these elements relative to the post axis. Take the sign dimensions to be $h_1 = 8.5$ m, $h_2 = 1.5$ m, and $w = 3.5$ m.

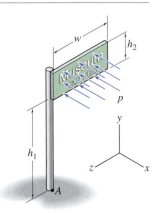

Prob. 7.67

7.68 A cylindrical pressure vessel with an outer diameter of 40 in. and a wall thickness of 0.6 in. is subjected to an internal pressure of 125 psi. Determine the stresses at a point on the surface and then compute and draw the maximum shear and principal stress elements. Describe the orientation of these elements relative to the vessel axis.

7.69 A cylindrical vessel with capped end is pressurized to 1.8 MPa. The structure on which the vessel rests becomes damaged, resulting in uneven support that causes a twisting moment T_0 of 10^7 N-m on the vessel. The vessel has an outer diameter of 2.5 m and a wall thickness of 50 mm. Determine the stresses at a point on the outer surface midway along the length of the vessel, and then compute and draw the maximum shear and principal stress elements. Describe the orientation of these elements relative to the vessel axis.

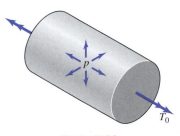

Prob. 7.69

7.70 A large downward force on the right pedal results in a moment about the forward axis of a bike. This moment can be balanced by an opposite moment of the hands pulling up on right handle bar and pushing down on the left. Thus, two equal and opposite moments have to be absorbed by the frame. Assume that the down tube, rather than the top tube, takes most of this moment. Some of the moment acts parallel to the tube, producing torsion and some of it perpendicularly, producing bending. Let the total moment from the hands be 1800 lb-in. Combining the stresses due to the twisting and bending, determine the maximum normal stress and maximum shear stress in the down tube. Take the down tube to be of aluminum 24 in. long, oriented at an angle $\theta = 48°$, with hollow circular cross-section having an outer diameter of 1.75 in. and a wall thickness of 0.125 in.

Prob. 7.70 (Appendix A1)

7.71 In using this pectoral fly machine, the hand grips and applies a 50 lb force to the handle as shown. The pivoting arm is approximated here as horizontal. Determine the stresses at the far end of the pivoting arm, where the bending stresses are maximum. Find the maximum normal stress and maximum shear stress at this point. Use the approximation of a thin wall for the torsion of the pivoting arm. Take the dimensions to be $w = 6$ in., $L = 24$ in., $a = 2$ in., $b = 1.5$ in., $t = 0.25$ in., and $c = 20$ in.

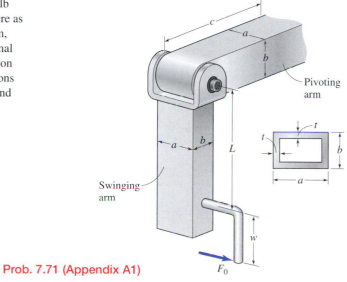

Prob. 7.71 (Appendix A1)

7.72 Say that each leg exerts on upward force of 50 lb. Determine the maximum shear stress at two cross-sections in the shaft connecting the pivot beam to the cord plate: (a) at the cord plate and (b) at the pivot beam. Account for both the twisting and the bending of the shaft that occurs at these cross-sections. Note that the shaft bends in two planes, due to the vertical force and moment applied by the pivot beam and due to the horizontal force applied by the cord. Take the dimensions to be $a = 4$ in., $b = 16$ in., $c = 4$ in., $d = 3.5$ in., $D_2 = 1$ in., $p = 2$ in., $q = 6$ in., $R = 9$ in., $s = 3$ in., and $w = 17$ in.

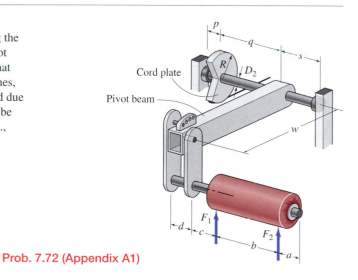

Prob. 7.72 (Appendix A1)

>>**End Problems**

7.5 Mohr's Circle

Formulas for the transformation of *x-y* stresses to stresses acting on surfaces at other angles can be represented graphically with Mohr's circle, named after Otto Mohr. You may find it easier to picture principal stresses and maximum shear stress with Mohr's circle.

1. Recall the quantities needed for the maximum stresses, and define the angle of the principal stresses.

Here again are the key quantities one determines from σ_x, σ_y, and τ_{xy}:

Maximum Shear Stress	Maximum Shear Angle	Mean Stress
$\tau_{max} = \sqrt{\dfrac{1}{4}(\sigma_y - \sigma_x)^2 + \tau_{xy}^2}$	$\theta_s = \dfrac{1}{2}\tan^{-1}\left[\dfrac{(\sigma_y - \sigma_x)}{2\tau_{xy}}\right]$	$\sigma_m = \dfrac{1}{2}(\sigma_x + \sigma_y)$

With the trig identity $\tan(\alpha - 90°) = \dfrac{-1}{\tan \alpha}$ for $\alpha = 2\theta_s$, and the relation $\theta_p = \theta_s - 45°$,

one can write the angle of the principal stresses, θ_p, as $\theta_p = \dfrac{1}{2}\tan^{-1}\left[\dfrac{\tau_{xy}}{(\sigma_x - \sigma_y)/2}\right]$.

2. Interpret the relations between θ_p, the *x-y* stresses, and τ_{max} in terms of the sides of a triangle, and re-write the transformation formulas.

Think of the formula for τ_{max} and θ_p as defining a right triangle:

$$\tan 2\theta_p = \frac{\tau_{xy}}{(\sigma_x - \sigma_y)/2} \implies$$

$$\implies \frac{\sigma_x - \sigma_y}{2} = \tau_{max} \cos 2\theta_p$$

$$\tau_{xy} = \tau_{max} \sin 2\theta_p$$

Use the quantities τ_{max}, σ_m, and θ_p to rewrite the formulas for transformed stresses $\sigma(\theta)$ and $\tau(\theta)$ as

$$\sigma(\theta) - \sigma_m = \frac{1}{2}(\sigma_x - \sigma_y)\cos 2\theta + \tau_{xy}\sin 2\theta$$

$$= \tau_{max}(\cos 2\theta_p)\cos 2\theta + \tau_{max}(\sin 2\theta_p)\sin 2\theta = \tau_{max}\cos 2(\theta_p - \theta)$$

$$\tau(\theta) = \tau_{xy}\cos 2\theta + \frac{1}{2}(\sigma_y - \sigma_x)\sin 2\theta$$

$$= \tau_{max}(\sin 2\theta_p)\cos 2\theta - \tau_{max}(\cos 2\theta_p)\sin 2\theta = \tau_{max}\sin 2(\theta_p - \theta)$$

3. Recognize $\sigma(\theta) - \sigma_m$ and $\tau(\theta)$ as legs of a triangle with hypotenuse τ_{max} and angle $2\theta_p - 2\theta$.

Note that the quantities $\sigma(\theta) - \sigma_m$ and $\tau(\theta)$ satisfy the equation for a circle of radius τ_{max}:

$$[\sigma(\theta) - \sigma_m]^2 + [\tau(\theta)]^2 = [\tau_{max}\cos 2(\theta_p - \theta)]^2 + [\tau_{max}\sin 2(\theta_p - \theta)]^2 = \tau_{max}^2$$

The quantities $\sigma(\theta) - \sigma_m$ and $\tau(\theta)$ can be viewed as horizontal (cos) and vertical (sin) components of a vector of magnitude τ_{max} and angle $(2\theta_p - 2\theta)$ relative to the horizontal. As θ varies over 180°, the vector tip sweeps out a circle, which is called *Mohr's circle*. Because the angle θ appears in the expressions for $\sigma(\theta) - \sigma_m$ and $\tau(\theta)$ in the form -2θ, it will be convenient to plot the stress $\tau(\theta)$ downward.

4. Draw Mohr's circle following these steps.

- Draw horizontal and vertical axes $[\sigma, \tau]$ with τ positive down.
- Draw points A at $[\sigma, \tau] = [\sigma_m, 0]$, B at $[\sigma, \tau] = [\sigma_x, \tau_{xy}]$, and C at $[\sigma, \tau] = [\sigma_y, -\tau_{xy}]$.
- Draw the circle centered at point A, with points B and C forming a diameter. (The angle from the σ-axis to point B should be $2\theta_p$.)

$$B: [\sigma_x, \tau_{xy}] = [\sigma_m + \tau_{max} \cos 2\theta_p, \tau_{max} \sin 2\theta_p]$$
$$C: [\sigma_y, -\tau_{xy}] = [\sigma_m - \tau_{max} \cos 2\theta_p, -\tau_{max} \sin 2\theta_p]$$

5. Use Mohr's circle as follows to find the stresses at angle θ, and the maximum stresses.

Rotate the diameter BC counterclockwise by angle 2θ to $B'C'$. The coordinates of B' are:

$$[\sigma_m + \tau_{max} \cos 2(\theta_p - \theta), \tau_{max} \sin 2(\theta_p - \theta)] = [\sigma(\theta), \tau(\theta)]$$

This means that the point B' on the circle, located at an angle 2θ from B, has $[\sigma, \tau]$ coordinates that correspond to $\sigma(\theta)$ and $\tau(\theta)$, which are the stresses that act on the surface oriented at θ relative to the x-axis. So the stresses that are found by rotating around Mohr's circle through an angle 2θ correspond to the stresses on an element rotated by θ with respect to the x-y axes.

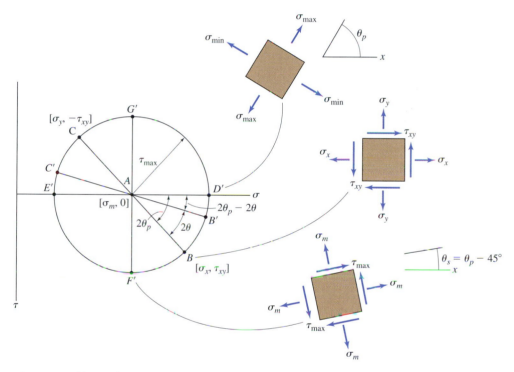

As one considers points around the circle from point B, the horizontal and vertical components change cyclically (like the curves for $\sigma(\theta)$ and $\tau(\theta)$ vs. θ).

The normal stress reaches a maximum value $\sigma_{max} = \sigma_m + \tau_{max}$ at point D', corresponding to a surface θ_p from the x-axis, and a minimum value $\sigma_{min} = \sigma_m - \tau_{max}$ at point E', corresponding to a surface $(180/2)°$ from θ_p.

The shear stress reaches a maximum value τ_{max} at point F', corresponding to a surface $(90/2)°$ CW from θ_p, and a minimum value $-\tau_{max}$ at point G', corresponding to a surface $(90/2)°$ CCW from θ_p.

So Mohr's circle confirms that the maximum shear stress and principal stress elements are rotated by $90°/2 = 45°$ with respect to each other. Mohr's circle is also another way of finding the relations between the maximum and minimum stresses: $\sigma_{max} = \sigma_m + \tau_{max}$ and $\sigma_{min} = \sigma_m - \tau_{max}$.

>>End 7.5

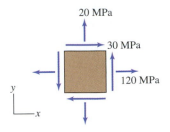

For the stress state shown, draw Mohr's circle, and from the circle determine the stress on an element 25° counterclockwise from the x-y axes, and the principal stress element.

Solution

Stresses are $\sigma_x = 120$ MPa, $\sigma_y = 20$ MPa, and $\tau_{xy} = 30$ MPa.

$$\tau_{max} = \sqrt{\frac{1}{4}(\sigma_y - \sigma_x)^2 + \tau_{xy}^2} = \sqrt{\frac{1}{4}(20 - 120)^2 + (30)^2} = 58.3 \text{ MPa}$$

$$\sigma_m = \frac{1}{2}(\sigma_x + \sigma_y) = \frac{1}{2}(120 + 20) = 70 \text{ MPa}$$

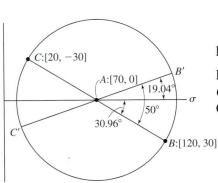

Draw the horizontal and vertical axes $[\sigma, \tau]$ with τ positive down.

Draw points A at $[\sigma, \tau] = [70, 0]$, B at $[\sigma, \tau] = [120, 30]$, and C at $[\sigma, \tau] = [20, -30]$. The circle is centered at A with radius $\tau_{max} = 58.3$. Calculate angle $2\theta_p$ between AB and the σ-axis.

$$2\theta_p = \tan^{-1}\left[\frac{30}{(120-70)}\right] = 30.96°$$

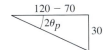

Determine stresses on an element oriented at 25° counterclockwise from the x-y axes.

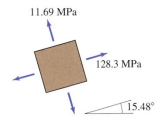

We rotate BC by $2(25°) = 50°$ to get $B'C'$, which is at $50° - 30.96° = 19.04°$ CCW from the σ-axis (so $\tau < 0$ at B').

The coordinates of B' and C' are

$$B': [\sigma_{x'}, \tau_{x'y'}] = [70 + 58.3 \cos 19.04°, -58.3 \sin 19.04°] = [125.1, -19.0]$$

$$C': [\sigma_{y'}, -\tau_{x'y'}] = [70 - 58.3 \cos 19.04°, 58.3 \sin 19.04°] = [14.9, 19.0]$$

Draw the element at this angle with $\sigma_{x'}$, $\sigma_{y'}$, and $\tau_{x'y'}$ in correct directions.

The principal stress element is found by rotating the diameter BC counterclockwise by 30.96° so that it lies on the σ-axis. $\tau = 0$ and σ is at maximum (B') and minimum (C') values. The element is oriented at $30.96°/2 = 15.48°$ counterclockwise from the x-y axes.

$$\sigma_{max} = 70 + 58.3 = 128.3 \text{ MPa} \qquad \sigma_{min} = 70 - 58.3 = 11.69 \text{ MPa}$$

>>End Example Problem 7.9

For the stress state shown, draw Mohr's circle, and from the circle determine the principal stress element and the maximum shear element.

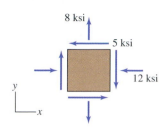

Solution

Stresses are $\sigma_x = -12$ ksi, $\sigma_y = 8$ ksi, and $\tau_{xy} = -5$ ksi.

$$\tau_{max} = \sqrt{\frac{1}{4}(\sigma_y - \sigma_x)^2 + \tau_{xy}^2} = \sqrt{\frac{1}{4}(8 + 12)^2 + (5)^2} = 11.18 \text{ ksi}$$

$$\sigma_m = \frac{1}{2}(\sigma_x + \sigma_y) = \frac{1}{2}(-12 + 8) = -2 \text{ ksi}$$

Draw the horizontal and vertical axes $[\sigma, \tau]$ with τ positive down.

Draw points A at $[\sigma, \tau] = [-2, 0]$, B at $[\sigma, \tau] = [-12, -5]$, and C at $[\sigma, \tau] = [8, 5]$. The circle is centered at A with radius $\tau_{max} = 11.18$ ksi.

Calculate the angle $2\theta_p$ between AB and the σ-axis.

$$2\theta_p = \tan^{-1}\left[\frac{5}{12 - 2}\right] = 26.57°$$

Rotate BC counterclockwise by $26.57°$ to get to the principal stress element. B' is at $\sigma_{min} = -2 - 11.18 = -13.18$ ksi, and C' is at $\sigma_{max} = -2 + 11.18 = 9.18$ ksi.

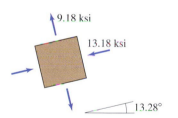

Rotate BC clockwise by $90 - 26.56° = 63.44°$ to get to the maximum shear element. This element is oriented at $63.44°/2 = 31.72°$ clockwise from x-y. B'' corresponds to $[\sigma_{x'}, \tau_{x'y'}]$. $\tau_{x'y'}$ is -11.18 at B'' because τ is positive downward. So the shear stress on face $\theta = -31.72°$ is drawn in negative sense. The remaining shear stresses are drawn to be consistent with the shear stress drawn on the face $\theta = -31.72°$, and a normal stress equal to $\sigma_m = -2$ ksi is drawn on all four faces.

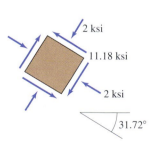

>>End Example Problem 7.10

Additional data on material properties needed to solve problems can be found in Appendix D or inside back cover.

7.73 The stresses on the x-y axes are $\sigma_x = 80$ MPa, $\sigma_y = 110$ MPa, and $\tau_{xy} = 50$ MPa. Draw the Mohr's circle for this stress state and determine the normal and shear stresses on a surface with normal oriented at 20° CCW from the x-axis.

7.74 The stresses on the x-y axes are $\sigma_x = 5$ ksi, $\sigma_y = -3$ ksi, and $\tau_{xy} = 3$ ksi. Draw the Mohr's circle for this stress state and determine the normal and shear stresses on a surface with normal oriented at 30° CCW from the x-axis.

7.75 The stresses on the x-y axes are $\sigma_x = -50$ MPa, $\sigma_y = 60$ MPa, and $\tau_{xy} = -70$ MPa. Draw the Mohr's circle for this stress state. Identify the principal stresses from the Mohr's circle, and draw the principal stress element, specifying its orientation.

7.76 The stresses on the x-y axes are $\sigma_x = 8$ ksi, $\sigma_y = 12$ ksi, and $\tau_{xy} = -5$ ksi. Draw the Mohr's circle for this stress state. Identify the maximum shear stress from the Mohr's circle, and draw the maximum shear stress element, specifying its orientation.

7.77 A solid shaft with 1.25 in. diameter is subjected to a bending moment of 2500 lb-in. and a twisting moment of 3200 lb-in. Determine stresses at the point where the bending and torsional stresses are greatest. Draw the Mohr's circle for that stress state.

7.78 A tube with an outer diameter of 14 mm and a wall thickness of 1.5 mm is subjected to a bending moment of 8 N-m and a twisting moment of 6 N-m. Determine stresses at the point where the bending and torsional stresses are greatest. Draw the Mohr's circle for that stress state.

7.79 A tray rests on an extendible arm consisting essentially of two steel tubes of different outer diameters. The downward force is applied at the tray center $F_0 = 1000$ N. Determine the stresses at the point of tube 1 where the bending and torsional stresses are greatest. Draw the Mohr's circle for that stress state. Both members have a wall thickness of 4 mm. $d_1 = 40$ mm, $d_2 = 34$ mm, $a = 700$ mm, and $b = 500$ mm.

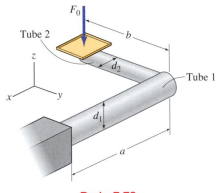

Prob. 7.79

7.80 The cylindrical pressure vessel with end caps, an outer diameter of 4 ft and a wall thickness of 0.625 in., contains gas under a pressure of 200 psi. Determine the stresses in the wall and draw the Mohr's circle for the stress state.

4 ft

Prob. 7.80

7.81 A cylindrical pressure vessel with end caps has a mean diameter of 800 mm and a wall thickness of 20 mm. In addition to being subjected to an internal pressure of 2 MPa, a disruption in its supports leads the vessel also to be twisted with a twisting moment of $3(10^5)$ N-m. Determine the stresses on the outer surface of the vessel and draw the Mohr's circle for that stress state.

7.02 A road sign consists of a steel post AB that has an outer diameter of 16 in., a wall thickness of 0.75 in., and a length of 25 ft. The pole BC has an outer diameter of 12 in., a wall thickness of 0.625 in., and a length of 40 ft to the center of the sign. The sign is 15 ft wide by 10 ft high and is subjected to a wind pressure of 0.2 psi. Determine the largest stresses in the vertical post AB due to the wind pressure, and draw Mohr's circle for that stress state.

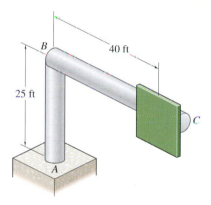

B   40 ft

25 ft

C

A

Prob. 7.82

7.6 Failure Criteria

The major purpose in looking at stresses on different planes is to determine when a stress state produces failure. Failure of materials is complex, and a solid understanding involves knowledge of both materials science and mechanics. Materials scientists would be most concerned, for example, with how the microstructure or the processing affects the stress at which a material yields plastically or fails. However, the materials scientists generally considers yield and failure only in, say, uniaxial tension. Let's say that the critical stress for failure in uniaxial tension is known from measurements of the material. Mechanics of materials, and stress transformations in particular, are needed to determine the combinations of normal and shear stress at which failure in the same material would occur.

1. Model failure in a simple, approximate way, as being either ductile or brittle, and note their respective dependence on shear and normal stress.

In judging failure for combinations of stresses, we simplify significantly actual observations of failure. In this simplest view, failure occurs in one of two general forms and depends on stresses as follows:

Ductile	**Brittle**
Material exhibits significant plastic flow before finally breaking (like a paper clip).	Material breaks with no evidence of plastic deformation (like a glass).

Microscopic Mechanism:
Deformation by motion of atomistic defects

Microscopic Mechanism:
Opening up and propagation of small cracks

Dependence on Stress:
Ductile failure depends on shear stress

Dependence on Stress:
Brittle failure depends on normal stress

2. Recall that at any point the maximum normal and shear stresses can act on inclined planes.

We have learned that the same stress state will have different stresses acting on surfaces of different orientations. For a given stress state, say described by σ_x, σ_y, and τ_{xy}, the shear stress is maximum on a surface of one orientation, and the normal stress is maximum on another orientation.

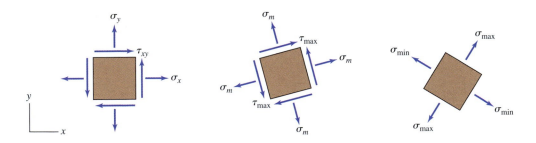

Does a normal stress or shear stress of a given magnitude have the same tendency to cause failure regardless of the plane on which it acts?

3. Approximate the material as isotropic: namely, a normal (or shear) stress has the same effect regardless of the plane on which it acts.

Most engineering materials are approximately isotropic. All directions in an isotropic material are the same: a normal stress or shear stress will have the same effect, including the tendency to cause failure regardless of the plane on which it acts. Anisotropic materials respond differently depending on the plane on which the stresses act. Anisotropic materials, such as fiber-reinforced composites and wood, sometimes require more complex failure criteria.

4. For an isotropic material that fails in a ductile fashion, predict failure to occur if the maximum shear stress exceeds a critical value.

For a ductile material, we often consider failure to correspond to first yield. That is, the part is not reusable if it has yielded. When a ductile material is tested in tension, we measure the tensile yield strength σ_Y. (Do not confuse the yield strength of a material σ_Y, with the stress σ_y, which is the normal stress acting in the y-direction.) What then is the shear yield strength τ_Y?

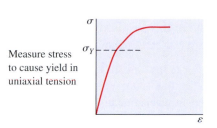  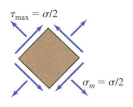

Measure stress to cause yield in uniaxial tension

Maximum shear element

Under uniaxial tension σ, the maximum shear stress is $\sigma/2$. So, if a ductile material yields in uniaxial tension when $\sigma = \sigma_Y$, then $\tau_Y = \sigma_Y/2$. We take yield to occur whenever the maximum shear stress reaches $\sigma_Y/2$, regardless of the particular combination of stresses.

For general stresses σ_x, σ_y, and τ_{xy}, using this criterion of maximum shear stress in the plane, we predict yield to occur if:

$$\tau_{max} = \sqrt{\frac{1}{4}(\sigma_y - \sigma_x)^2 + \tau_{xy}^2} = \frac{\sigma_Y}{2}$$

We will show in the next section that yield can sometimes occur even before the above condition is reached, if we consider the maximum shear stress on any 3-D plane.

5. For a material that fails in a brittle fashion, predict failure to occur if the maximum tensile stress exceeds a critical value.

When a material is tested in tension, we measure the tensile failure strength. What is the maximum normal stress at failure?

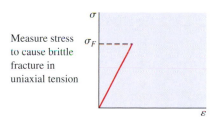  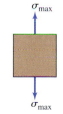

Measure stress to cause brittle fracture in uniaxial tension

Principal stress element

Under uniaxial tension σ, the maximum normal stress is equal to σ. So, if brittle fracture occurs when $\sigma = \sigma_F$, we take brittle fracture to occur whenever $\sigma_{max} = \sigma_F$, regardless of the particular combination of stresses.

For general stresses σ_x, σ_y, and τ_{xy}, using a criterion of maximum normal stress, we predict brittle failure to occur if:

$$\sigma_{max} = \sigma_m + \tau_{max} = \frac{\sigma_x + \sigma_y}{2} + \sqrt{\frac{1}{4}(\sigma_y - \sigma_x)^2 + \tau_{xy}^2} = \sigma_F$$

>>End 7.6

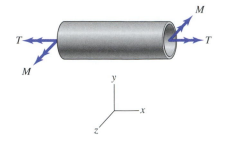

The cast iron pipe shown has a 6 in. outer diameter and a 5 in. inner diameter. It is subjected to simultaneous bending and twisting. The bending moment is $250(10^3)$ lb-in. Cast iron is generally treated as a brittle material; its strength, measured in uniaxial tension, is 30 ksi. (a) Determine the points on the pipe that failure under combined bending and twisting would occur. (b) Determine the torque that, together with bending, will cause failure. (c) Draw Mohr's circle for this element, draw the principal stress element, and show the orientation of the failure plane.

Solution

Because the material is modeled as brittle, failure is determined by the maximum tensile stress. For this cast iron, failure occurs when the maximum tension due to the combination of bending and torsion must equal 30 ksi.

Bending produces normal stress in the x-direction; the maximum is at the top of the pipe (where y is maximum). The other normal stresses at this point are zero. Torsion produces shear stresses that are maximum at the outer diameter. So failure must occur on the top of the pipe.

$$\sigma_x = \frac{My}{I} = \frac{(250(10^3) \text{ lb-in.})(3 \text{ in.})}{\frac{\pi}{4}\left[(3 \text{ in.})^4 - (2.5 \text{ in.})^4\right]} = 22.8 \text{ ksi} \qquad \sigma_{max} = \sigma_m + \tau_{max} = \frac{\sigma_x}{2} + \sqrt{\frac{1}{4}(\sigma_x)^2 + \tau^2} = 30 \text{ ksi}$$

From $\sigma_{max} = 30$ ksi, solve for τ: $\tau = 14.73$ ksi.
$$\tau = \frac{T\rho}{I_p} \Rightarrow 14.73 \text{ ksi} = \frac{T(3 \text{ in.})}{\frac{\pi}{2}\left[(3 \text{ in.})^4 - (2.5 \text{ in.})^4\right]} \Rightarrow T = 323(10^3) \text{ lb-in.}$$

As viewed from above, the stresses at the top of the pipe are

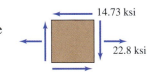

Points for Mohr's circle: A: $[22.8/2]$; B: $[22.8, -14.73]$; and C: $[0, 14.73]$.

$$2\theta_p = \tan^{-1}\left[\frac{\tau_{xy}}{(\sigma_x - \sigma_y)/2}\right] = \tan^{-1}\left[\frac{-14.73}{[22.8]/2}\right] = -52.26°$$

$$\tau_{max} = \sqrt{\left(\frac{22.8}{2}\right)^2 + (14.73)^2} = 18.63 \text{ ksi}$$

$$\sigma_{min} = \sigma_m - \tau_{max} = 11.4 - 18.63 = 7.23 \text{ ksi}$$

Principal Stress Element

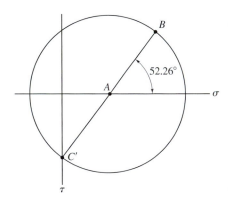

Failure occurs by separation on the plane on which $\sigma_{max} = 30$ ksi acts.

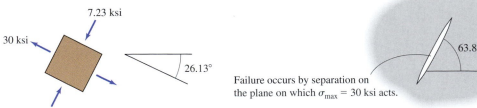

>>End Example Problem 7.11

A thermoplastic part is analyzed and, under a 100 N load, the stresses at the point considered critical are found to be as shown. A thermoplastic specimen of the same material was tested in tension and was found to yield plastically at a stress of 40 MPa. Using the maximum shear stress criterion, and assuming that the stresses are proportional to the load, determine the load at which yielding would occur at this point.

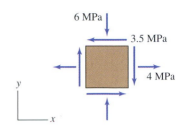

Solution

In the tension test, the tensile yield stress of 40 MPa corresponds to a maximum shear stress of 20 MPa. So yield will occur under a combination of stresses when the maximum shear stress also equals 20 MPa.

Under the applied load of 100 N, the maximum shear stress is

$$\tau_{max} = \sqrt{\frac{1}{4}(\sigma_y - \sigma_x)^2 + \tau_{xy}^2} = \sqrt{\frac{1}{4}(-6 - 4)^2 + (3.5)^2} = 6.10 \text{ MPa}$$

Since 6.10 MPa < 20 MPa, yield does not occur under the 100 N load. Imagine the load increases to a level $\Lambda(100 \text{ N})$. Since the stresses are all proportional to the load, we can express all stresses as a common factor, Λ, times the values they have when the load is 100 N.

Calculate the maximum shear stress under the increased stresses $\Lambda\sigma_x$, $\Lambda\sigma_y$, and $\Lambda\tau_{xy}$.

$$\tau_{max} = \sqrt{\frac{1}{4}(\Lambda\sigma_y - \Lambda\sigma_x)^2 + (\Lambda\tau_{xy})^2} = \Lambda\sqrt{\frac{1}{4}(\sigma_y - \sigma_x)^2 + (\tau_{xy})^2}$$

Notice that the maximum shear stress increases by the same factor Λ. So the maximum shear stress when the load is increased by factor Λ would be:

$$\tau_{max} = \Lambda\sqrt{\frac{1}{4}(\sigma_y - \sigma_x)^2 + (\tau_{xy})^2} = \Lambda(6.10 \text{ MPa})$$

Now, determine the factor Λ for the maximum shear stress to equal 20 MPa.

$$\tau_{max} = \Lambda(6.10 \text{ MPa}) = 20 \text{ MPa} \Rightarrow \Lambda = 20/6.10 = 3.28$$

Therefore, yielding will occur when the applied load equals $(3.28)(100 \text{ N}) = 328 \text{ N}$.

>>End Example Problem 7.12

Additional data on material properties needed to solve problems can be found in Appendix D or inside back cover.

7.83 Consider one loading to produce normal stresses on the x-y axes of $\sigma_x = 80$ MPa and $\sigma_y = -30$ MPa. An additional loading, producing only a shear stress, increases until a crack is observed to open up when the angle shown is $\alpha = 20°$. (a) Determine the shear stress τ_{xy} at which cracking occurs. (b) If this material were loaded in uniaxial tension, at what critical tensile stress would brittle fracture occur?

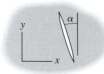

Prob. 7.83

7.84 Under purely uniaxial tension, brittle failure in a material is known to occur at 50 ksi. The normal stresses on the x-y axes are fixed at the level of $\sigma_x = 20$ ksi and $\sigma_y = 35$ ksi. Determine (a) the shear stress τ_{xy} at which cracking (brittle failure) occurs, and (b) the orientation of the plane on which the cracks appear.

7.85 Using a strain rosette to measure strains, cracks are found to initiate in a ceramic component at strains $\varepsilon_x = 6.2(10^{-4})$, $\varepsilon_y = 2.1(10^{-4})$, and $\gamma_{xy} = 6.7(10^{-4})$. The elastic moduli are $E = 400$ GPa, $\nu = 0.22$, and $G = 164$ GPa. Determine the critical tensile stress at which brittle fracture occurs and the anticipated orientation of the crack plane relative to the x-axis.

7.86 Cracking would occur in a material when stresses on the x-y axes are equal to $\sigma_x = 12$ ksi, $\sigma_y = 3$ ksi, and $\tau_{xy} = -7$ ksi. Determine the critical tensile stress at which brittle fracture occurs, and the anticipated orientation of the crack plane relative to the horizontal axis.

7.87 Elastic finite element analysis of a component is conducted. The loading is due to an applied force designated by F_0 applied at a particular point. When $F_0 = 100$ N, the highest stresses in the component are calculated to be $\sigma_x = -75$ MPa, $\sigma_y = 175$ MPa, and $\tau_{xy} = -60$ MPa. If the material cracks under a critical tensile stress of 350 MPa, determine the applied load F_0 at which cracking would occur.

7.88 Elastic finite element analysis of a component is conducted. The loading is due to an applied force designated by F_0 applied at a particular point. When $F_0 = 200$ lb, the highest stresses in the component are calculated to be $\sigma_x = -3760$ psi, $\sigma_y = 7800$ psi, and $\tau_{xy} = 4520$ ksi. If the material yields under a tensile stress of 55 ksi, determine the applied load F_0 at which yielding would occur.

7.89 The shear stress at a point of interest is known to be $\tau_{xy} = 40$ MPa. At that point, there is also a tensile stress of σ_y, but $\sigma_x = 0$. Slip lines due to plastic yielding are observed to first appear in the orientation shown with $\alpha = 15°$. (a) Determine the magnitude of the stress σ_y. (b) If σ_y alone acted, what would be the magnitude of σ_y at which slip lines on any plane would first be observed?

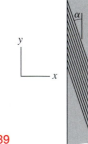

Prob. 7.89

7.90 Say that uniaxial tension at a level of 40 ksi causes ductile yielding. Yielding can also occur under a combination of compression and shear. (a) If the compressive stress in the x-direction is -25 ksi (and $\sigma_y = 0$), determine the shear stress τ_{xy}. (b) Determine the orientation of the slip lines.

7.91 A strain rosette is used to measure strains in a steel component. For a given load, the strains at a point are found to be $\varepsilon_x = 6.1(10^{-4})$, $\varepsilon_y = -3.8(10^{-4})$, and $\gamma_{xy} = 7.1(10^{-4})$. The elastic moduli are $E = 30(10^6)$ psi, $\nu = 0.3$, and $G = 11(10^6)$ psi. If the yield stress in tension is 36 ksi, by what percentage can the load be increased without yield occurring at that same point? Assume that the strains increase in proportion to the load.

7.92 A circular ceramic rod 35 mm in length and 8 mm in diameter is subjected to a twisting moment. Its elastic moduli are $E = 380$ GPa, $\nu = 0.2$, and $G = 158$ GPa. Determine the twist (relative angle of rotation) at which cracking would occur, if the same rod would have cracked under a tensile load of 6000 N.

Prob. 7.92

7.93 An aluminum bar would yield plastically at a tensile stress of 30 ksi. If the bar is 8 in. long, determine the acceptable range of diameters if it is to tolerate one end twisted by 1° relative to the other.

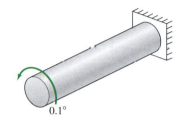

0.1°

Prob. 7.93

7.94 A twisting moment of 800 N-m first causes yielding in a circular steel tube that is 400 mm long with an outer diameter of 30 mm and an inner diameter of 24 mm. Determine the axial tensile force at which the same bar would yield.

Prob. 7.94

7.95 A bar of 20 mm in diameter and 300 mm in length is subjected to an axial compressive stress of 20 MPa. (a) If the material undergoes brittle failure under a tensile stress of 80 MPa, what twisting moment would just produce brittle failure while the compressive force acted? (b) If the material yields plastically at a yield strength of 80 MPa, what twisting moment would just produce yielding while the compressive force acted?

Prob. 7.95

7.96 A plastic to be used for a torsional snap fit is known to yield under a tensile stress of 8 ksi. Determine the allowable rotation of the tabs, if the snap fit is not to yield. The twisting member has a diameter of 0.3 in. and the tabs are 1 in. away from each of the fixed ends. (Take $G = 150$ ksi.)

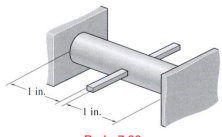

1 in.

1 in.

Prob. 7.96

7.97 A steel rod that is 40 mm long and has a diameter of 8 mm is composed of an alloy that yields plastically at a uniaxial tensile stress of 450 MPa. One end of the bar has been twisted relative to the other by 1.5°. Determine the amount by which the bar can be elongated, while it is already twisted, without yielding.

7.98 A bar that is 1.5 in. long and 0.25 in. in diameter is subjected to a tensile stress of 5000 psi. The bar is also inadvertently twisted. Cracks are observed to appear at an angle shown with $\theta_0 = 65°$. (a) Determine the torque that was applied. (b) If there had been no initial tension, what torque could have been applied without cracking?

Prob. 7.98

7.99 A circular aluminum bar is 200 mm long and has a diameter of 14 mm. Yielding would occur in this aluminum alloy under a tensile stress of 280 MPa. The elastic moduli are $E = 70$ GPa, $\nu = 0.33$, and $G = 26$ GPa. The bar is elongated by $\delta = 0.5$ mm and then twisted. Determine the relative rotation of one end of the rod with respect to the other at which yielding would occur. Determine the plane on which slip would be observed.

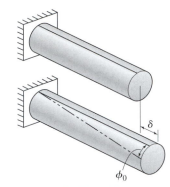

Prob. 7.99

7.100 A steel bar with rectangular cross-section is cantilevered at one end and the free end is subjected to a transverse force. The alloy yields at a tensile stress of 450 MPa. Use beam theory to predict the end deflection at which the bar would yield. Take the dimensions to be $a = 20$ mm, $b = 50$ mm, and $c = 400$ mm. Explain whether beam theory will give a good prediction of the end deflection.

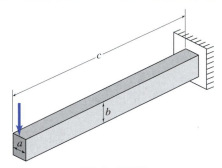

Prob. 7.100

7.101 A circular steel shaft which is 16 in. long and with a diameter of 1.2 in. is known to yield when subjected to a twisting moment of 16 kip-in. Instead, the rod is simply supported, and subjected to a transverse force in the center. Determine the center deflection at which yielding occurs. Explain whether beam theory will give a good prediction of the center deflection.

7.102 A circular rod is composed of a ceramic that cracks under a tensile stress of 370 MPa. The rod has a length of 60 mm and a diameter of 12 mm, and it is subjected to a bending moment of $M_0 = 40$ N-m. Determine the twisting moment T_0 that could be applied, in addition to the bending moment, before the rod cracks.

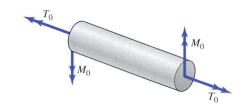

Prob. 7.102

7.103 A circular steel shaft of a length of 14 in. and a diameter of 0.75 in. is composed of an alloy that yields under a tensile stress of 60 ksi. The shaft is subjected to a twisting moment of $T_0 = 1500$ lb-in. Determine the bending moment M_0 that could be applied, in addition to the twisting moment, before the rod yields.

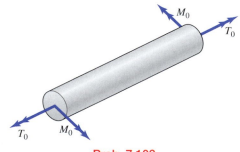

Prob. 7.103

7.104 The hollow steel rod with an outer diameter of 60 mm and an inner diameter of 55 mm is known to yield at a tensile stress of 325 MPa. Determine the magnitude of the force F_0 at which the member yields. Take the lengths to be $L_1 = 1.6$ m and $L_2 = 0.8$ m.

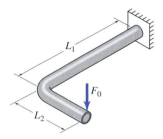

Prob. 7.104

7.105 A ceramic rod of a diameter of 7 mm is known to crack at a tensile stress of 200 MPa. Determine magnitude of the force F_0 at which the member cracks. Take the lengths to be $L_1 = 50$ mm and $L_2 = 40$ mm.

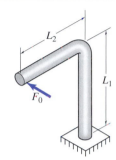

Prob. 7.105

7.106 An aluminum rod is subjected to a force of $F_0 = 20$ lb without failing. The rod has a diameter of 0.5 in. and a length of $L_1 = 14$ in. Determine the allowable range of the length L_2 if the rod is not to yield, given a uniaxial yield stress of 35 ksi.

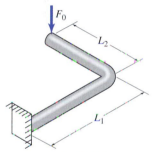

Prob. 7.106

Focused Application Problems

7.107 The allowable tensile force on drill pipe is often adjusted to account for the make-up torque that is applied while screwing neighboring segments together. Consider a drill pipe with a 5 in. outer diameter and 4.408 in. inner diameter. (a) If the yield stress of the pipe in tension is 100000 psi, what is the allowable tensile force on the pipe disregarding the make-up torque? (b) Say the make-up torque is 26000 lb-ft. What is the allowable tensile force when the make-up torque is accounted for?

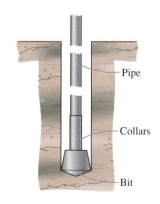

Prob. 7.107 (Appendix A3)

>>End Problems

7.7 Failure for Stresses in 3-D

We have studied the case of stresses acting in one plane. While this is a common situation, more generally all components of stress are present. For such a 3-D state of stress, we want to determine if failure will occur.

1. For a set of 3-D stress components, we could determine the stress acting on any inclined plane.

Here are the stresses on a 3-D element.

We can ask the same question that we asked for planar stresses: what stresses act across planes other than those parallel to the *x-y-z* axes?

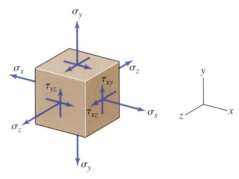

Here is a general plane through the 3-D element. The orientation of the plane must be described by two angles (rather than by the single angle θ, as was done in 2-D).

The force transmitted through this surface, per area, can be resolved into a force normal to the surface, σ (normal stress), and a force parallel to the surface, τ (shear stress). One can derive 3-D stress transformation formulas that give σ and τ in terms of the *x-y-z* stress components and the angles of the surface.

2. For 3-D stresses one can find planes on which the normal and shear stresses are maximum.

We are also interested in surfaces on which the normal stress or shear stress is maximum or minimum. The mathematics are complex, but the important aspects of what we found in 2-D hold also for 3-D.

In 3-D there are three surfaces, which are perpendicular to each other, on which the shear stress is zero. These planes are called the principal stress planes. The normal stresses on those planes are principal stresses.

Here are the principal stresses drawn on an element with surfaces aligned with the principal stress planes.

One uses matrix algebra to find the principal stresses, in terms of the *x-y-z* stresses.

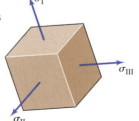

The largest of the three stresses, σ_I, is the maximum normal stress among all 3-D planes. The minimum of the three, σ_{III}, is the minimum normal stress among all 3-D planes. The stress on the third surface, σ_{II}, is the intermediate principal stress.

The absolute maximum shear stress from among all planes acts on the plane that is 45° between the principal planes of σ_I and σ_{III}.

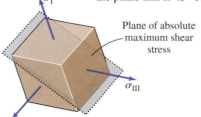

Plane of absolute maximum shear stress

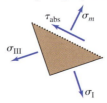

The shear stress has an absolute maximum value of $\tau_{abs} = (\sigma_I - \sigma_{III})/2$.

3. Track the absolute maximum shear stress even when the stresses are planar.

Even when the stresses are planar, the absolute maximum shear stress from among all 3-D planes, τ_{abs}, may be greater than the in-plane maximum shear stress, τ_{max}. We reconsider two cases of planar stresses and point out when the in-plane maximum shear stress is insufficient to predict failure.

Here we have bending and torsion.

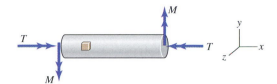

Here we have a cylindrical pressure vessel.

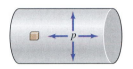

The stresses at the point in question are:

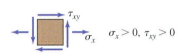

$\sigma_x > 0,\ \tau_{xy} > 0$

The stresses at the point in question are:

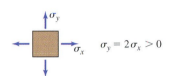

$\sigma_y = 2\sigma_x > 0$

From the 2-D analysis:

$\sigma_m = \sigma_x/2$	$\tau_{max} = \sqrt{\dfrac{1}{4}(\sigma_x)^2 + \tau_{xy}^2} > \sigma_m$
$\sigma_{max} = \sigma_m + \tau_{max} > 0$	$\sigma_{min} = \sigma_m - \tau_{max} < 0$

$\sigma_m = (\sigma_x + \sigma_y)/2$	$\tau_{max} = (\sigma_x - \sigma_x)/2 < \sigma_m$
$\sigma_{max} = \sigma_m + \tau_{max} > 0$	$\sigma_{min} = \sigma_m - \tau_{max} > 0$

In both cases the out-of-plane stresses are zero, so σ_z is a principal stress and $\sigma_z = 0$. Consider now the principal stresses in 3-D.

$\sigma_I = \sigma_m + \tau_{max} > 0, \qquad \sigma_{II} = 0, \sigma_{III} = \sigma_m - \tau_{max} < 0$

$\tau_{abs} = (\sigma_I - \sigma_{III})/2 = \tau_{max}$

$\sigma_I = \sigma_m + \tau_{max} > 0, \sigma_{II} = \sigma_m - \tau_{max} > 0, \sigma_{III} = 0$

$\tau_{abs} = (\sigma_I - \sigma_{III})/2 = (\sigma_m + \tau_{max})/2 > \tau_{max}$ (since $\sigma_m > \tau_{max}$)

Here τ_{abs} equals the in-plane maximum τ_{max}, so the planar analysis correctly gives the shear stress for use in the ductile failure criterion.

Here τ_{abs} exceeds the in-plane maximum τ_{max}, so the planar analysis gives an incorrect shear stress for use in the ductile failure criterion.

In summary: to predict ductile failure with 2-D stresses, compare as follows:

- If $\sigma_{max} > 0$ and $\sigma_{min} < 0$, then compare τ_{max} (in-plane maximum shear) with $\sigma_Y/2$
- If $\sigma_{max} > 0$ and $\sigma_{min} > 0$ or $\sigma_{max} < 0$ and $\sigma_{min} < 0$, then compare (the larger of $|\sigma_{max}|/2$ and $|\sigma_{min}|/2$) with $\sigma_Y/2$

4. Consider another means of predicting yield for combinations of stress: the von Mises stress.

A combination of stresses, σ_{VM}, named the von Mises stress after Richard von Mises, is an alternative measure of maximum shearing that is commonly used in ductile failure criteria.

$$\sigma_{VM} = \sqrt{\frac{(\sigma_x - \sigma_y)^2 + (\sigma_x - \sigma_z)^2 + (\sigma_y - \sigma_z)^2 + 6(\tau_{xy}^2 + \tau_{xz}^2 + \tau_{yz}^2)}{2}}$$

For uniaxial tension (say σ_x), σ_{VM} is equal to the tensile stress itself. So if yielding in uniaxial tension occurs at σ_Y, under a general combination of stresses we take yield to occur whenever $\sigma_{VM} = \sigma_Y$. One finds the x-y-z stress components, computes σ_{VM}, and compares with σ_Y.

σ_{VM} generally differs from twice the absolute maximum shear stress. Hence the predictions of ductile failure based on σ_{VM} and τ_{abs} are slightly different. Neither prediction is always in precise accord with experiments. Particularly in conjunction with finite element analysis of stresses, σ_{VM} is more commonly computed.

>>End 7.7

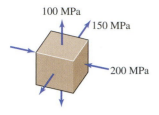

100 MPa

150 MPa

200 MPa

Finite element analysis of a part reveals that the principal stresses at a particular point are those shown in the figure. Say a specimen of the same material is tested in tension. Determine the factor of safety for the following cases. (a) Tensile specimen fractures in a brittle fashion at 400 MPa. (b) Tensile specimen yields in a ductile fashion at a stress of 400 MPa (use von Mises stress criterion). (c) Tensile specimen yields in a ductile fashion at a stress of 400 MPa (use absolute maximum shear stress criterion). For case (c) draw the planes on which the shear stress is maximum.

Solution

A brittle material fails when the maximum tensile stress reaches a critical value. In uniaxial tension, the uniaxial tension is itself the maximum tensile stress. Based on the tensile test, the material can withstand a maximum tensile stress of 400 MPa.

The largest of the three principal stresses, 150 MPa, is the maximum tensile stress.

So the factor of safety for this case is: Factor of safety $= \left(\dfrac{400 \text{ MPa}}{150 \text{ MPa}} \right) = 2.67.$

A ductile material fails due to shear. The von Mises stress is one way to capture the net shearing effect of all the stress components. The von Mises stress is defined so that the von Mises stress in uniaxial tension is the tensile stress itself. So the critical value of the von Mises would be 400 MPa. For this set of principal stresses,

$$\sigma_{VM} = \sqrt{\frac{1}{2}\left[(\sigma_I - \sigma_{II})^2 + (\sigma_I - \sigma_{III})^2 + (\sigma_{II} - \sigma_{III})^2\right]}$$

$$= \sqrt{\frac{1}{2}\left[(150 - 100)^2 + (150 + 200)^2 + (100 + 200)^2\right]} = 328 \text{ MPa}$$

So the factor of safety for this case is: Factor of safety $= \left(\dfrac{400 \text{ MPa}}{328 \text{ MPa}} \right) = 1.22.$

A ductile material fails due to shear. The absolute maximum shear stress is the maximum single shear stress among all planes through the point. The absolute maximum shear stress is half of the difference between the maximum and the minimum principal stresses. For uniaxial tension, the absolute maximum shear stress is half the tension, or 200 MPa.

Since the principal stresses are $\sigma_I = 150$ MPa, $\sigma_{II} = 100$ MPa, and $\sigma_{III} = -200$ MPa,

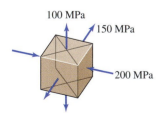

100 MPa

150 MPa

200 MPa

the absolute maximum shear stress is $\tau_{abs} = \left(\dfrac{\sigma_I - \sigma_{III}}{2} \right) = 175$ MPa.

So the factor of safety for this case is: Factor of safety $= \left(\dfrac{200 \text{ MPa}}{175 \text{ MPa}} \right) = 1.14.$

Absolute maximum shear stress is reached on the indicated planes at 45° between the planes on which σ_I and σ_{III} act.

>>End Example Problem 7.13

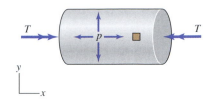

A pressure vessel with $p = 200$ psi internal pressure is subjected also to a twisting moment $T = 450(10^3)$ lb-in. The vessel has a mean diameter of 20 in. and a wall thickness of 0.25 in. (a) Determine Mohr's circle for an element on the front surface shown, and identify the three principal stresses. (b) Determine whether the in-plane maximum shear stress is equal to the absolute maximum shear stress. (c) Determine the tensile strength of a ductile material which has a factor of safety of 5 for this loading. Use the absolute maximum shear stress failure criterion.

Solution

Stresses due to the pressure vessel solution and torsion are:

$$\sigma_h = 2\sigma_a = \frac{pR}{t} = \frac{200\ \text{psi}\ (10\ \text{in.})}{(0.25\ \text{in.})} = 8000\ \text{psi} \qquad \tau = \frac{T\rho}{I_p} = \frac{\left(450(10^3)\ \text{lb-in.}\right)(20.125\ \text{in.})}{\dfrac{\pi}{2}\left[(20.125\ \text{in.})^4 - (19.875\ \text{in.})^4\right]} = 2900\ \text{psi}$$

So, the stresses on the x-y faces of the element are $\sigma_x = 4000$ psi, $\sigma_y = 8000$ psi, and $\tau_{xy} = 2900$ psi.

The in-plane maximum shear stress is:

$$\tau_{max} = \sqrt{\frac{1}{4}\left(\sigma_y - \sigma_x\right)^2 + \tau_{xy}^2} = \sqrt{\frac{1}{4}(8000 - 4000)^2 + (2900)^2} = 3520\ \text{psi}$$

$$\sigma_m = \frac{1}{2}\left(\sigma_x + \sigma_y\right) = \frac{1}{2}(4000 + 8000) = 6000\ \text{psi}$$

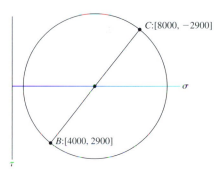

Draw Mohr's circle: $\sigma_{min} = 6000 - 3520 = 2480$ psi, $\sigma_{max} = 6000 + 3520 = 9520$ psi.

Note that the planar σ_{min} and σ_{max} are both positive.

If we think about 3-D principal stresses, then $\sigma_I = 9520$ psi, $\sigma_{II} = 2480$ psi, and $\sigma_{III} = 0$.

In this case, the absolute maximum shear stress is $\tau_{abs} = \left(\dfrac{\sigma_I - \sigma_{III}}{2}\right) = 4760$ psi.

Note the absolute maximum shear stress in this case is greater than the in-plane maximum shear stress.

A bar under uniaxial tension of $2(4760) = 9520$ psi would fail at an absolute maximum shear stress of 4760 psi.

With a factor of safety of 5, the bar must therefore have a tensile yield strength of 47.6 ksi.

>>End Example Problem 7.14

Additional data on material properties needed to solve problems can be found in Appendix D or inside back cover.

7.108 From a finite element analysis, the stresses at a critical point in an aluminum component are found to be as follows: $\sigma_x = 80$ MPa, $\sigma_y = 70$ MPa, $\sigma_z = -50$ MPa, $\tau_{xy} = -75$ MPa, $\tau_{xz} = 60$ MPa, and $\tau_{yz} = 30$ MPa. (a) Calculate the von Mises stress at this point. (b) If the material is ductile and has a uniaxial tensile yield strength of 270 MPa, determine the factor of safety using the von Mises stress criterion for yielding.

7.109 A steel alloy has a uniaxial yield strength of 580 MPa. Consider this steel to be subjected to only a shear stress τ. (a) Determine the value of τ at yield assuming a maximum shear stress yield criterion. (b) Determine the value of τ at yield assuming a von Mises yield criterion.

7.110 A plastic molding includes a handle that is loaded as shown. Yielding where the handle meets the support is of concern. The uniaxial yield strength of the plastic is 30 MPa. Take $L_1 = 50$ mm, $L_2 = 30$ mm, and $d = 8$ mm. (a) Determine the load F_0 at yield assuming a maximum shear stress yield criterion. (b) Determine the load F_0 at which yield occurs assuming a von Mises yield criterion.

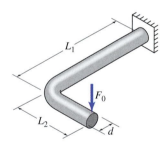

Prob. 7.110

7.111 An automotive component of complex shape experiences a primary load designated as F_0. Finite element analysis is carried out using a value $F_0 = 100$ N. From the results, the principal stresses at the point where failure is observed are found to be $\sigma_I = 20$ MPa, $\sigma_{II} = 10$ MPa, and $\sigma_{III} = -5$ MPa. Assume stresses increase in proportion to F_0. (a) Assuming the material experiences brittle failure at a uniaxial tensile stress of 140 MPa, determine the load F_0 at which the component would fail. (b) Assuming the material experiences ductile failure at a uniaxial tensile stress of 140 MPa, and that a maximum shear stress yield criterion applies, determine the load F_0 at which the component would fail. (c) Assuming the material experiences ductile failure at a uniaxial tensile stress of 140 MPa, and that a von Mises criterion applies, determine the load F_0 at which the component would fail.

7.112 The steel structure is composed of a hollow tube with an outer diameter of 10 in. and a wall thickness of 0.5 in. Take $L_1 = 12$ ft and $L_2 = 8$ ft. The steel has a uniaxial yield strength of 36 ksi. If the load $F_0 = 2500$ lb, determine the factor of safety for yielding at the base, assuming a von Mises yield criterion.

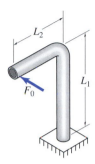

Prob. 7.112

7.113 A cylindrical vessel is pressurized to 1.7 MPa. The structure on which the vessel rests becomes damaged, resulting in uneven support that causes a twisting moment T_0 of $8(10^6)$ N-m on the vessel. The vessel has an outer diameter of 2.5 m and a wall thickness of 50 mm. The steel wall has a uniaxial yield strength of 600 MPa. Determine the factor of safety with respect to yielding (a) assuming a maximum shear stress yield criterion and (b) assuming a von Mises yield criterion.

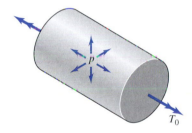

Prob. 7.113

>>End Problems

7.8 2-D Strain Transformations and Strain Rosettes _____

Just as a planar stress state is described by three components of stress, three strain components, ε_x, ε_y, and γ_{xy}, describe the strain in the plane at a point. From those components, one can calculate the strain in all other directions. A common experimental method of determining the strain state relies on these calculations.

1. Determine the normal strain of a line segment of general orientation in terms of the x-y strain components.

Say a planar element $ABCD$ at a point, of size 1×1, undergoes strains ε_x, ε_y, and γ_{xy}. Say we want the normal strain of a segment AE oriented at a general angle θ. Here we show the change in the segment EA when each of ε_x, ε_y, and γ_{xy} is non-zero.

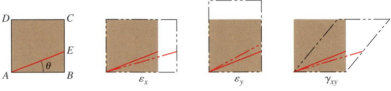

Segment length $|AE| = 1/\cos\theta$ (this holds for $\theta < 45°$, but the final formula is valid for all θ).

Since A does not move, we find the displacement of point E parallel to AE to determine the elongation of AE. We derive the strain only for one component, γ_{xy}; the others are similar.

$\delta_{EA} = u_{parallel} = u\cos\theta$, where $u = (\gamma_{xy})y$ and $y = \tan\theta$

$$\varepsilon(\theta) = \frac{\delta_{EA}}{|AE|} = \frac{\gamma_{xy}(\tan\theta)\cos\theta}{1/\cos} = \gamma_{xy}\sin\theta\cos\theta = \frac{1}{2}\gamma_{xy}\sin 2\theta$$

Note that the shear strain γ_{xy} produces no normal strain along the x-axis ($\theta = 0°$) or y-axis ($\theta = 90°$). But, for any other orientation, there is normal strain. This result is critical below.

2. Determine the shear strain of two initially perpendicular line segments of general orientation.

Now, say we want the shear strain of a pair of segments, AE and DF, which are mutually perpendicular before the strains ε_x, ε_y, and γ_{xy} act. The segments are oriented at a general angle θ.

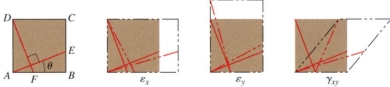

We find the rotation of each segment from the displacements perpendicular to that segment. Then, the shear strain corresponds to the change in angle between them, which is related to their respective rotations. Here we compute the effect of only ε_x, but ε_y and γ_{xy} are handled in the same way.

EA: $u_{perpendicular} = u\sin\theta$, where $u = (\varepsilon_x)x$ and $x = 1$

$$\omega_{EA} = \frac{u_{perpendicular}}{|EA|} = \frac{\varepsilon_x(1)\sin\theta}{1/\cos\theta} = \frac{1}{2}\varepsilon_x\sin 2\theta$$

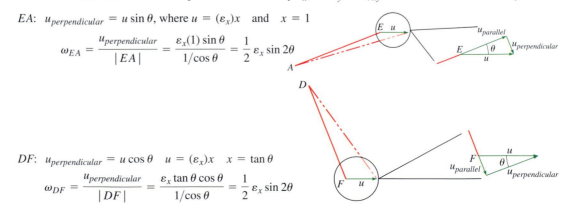

DF: $u_{perpendicular} = u\cos\theta$ $u = (\varepsilon_x)x$ $x = \tan\theta$

$$\omega_{DF} = \frac{u_{perpendicular}}{|DF|} = \frac{\varepsilon_x\tan\theta\cos\theta}{1/\cos\theta} = \frac{1}{2}\varepsilon_x\sin 2\theta$$

3. Calculate the contributions of all *x-y* strains to find the normal and shear on inclined axes.

Combine the results of similar derivations of $\varepsilon(\theta)$ for ε_x and ε_y and of $\gamma(\theta)$ for ε_y and γ_{xy} to find:

$$\varepsilon(\theta) = \frac{1}{2}(\varepsilon_x + \varepsilon_y) + \frac{1}{2}(\varepsilon_x - \varepsilon_y)\cos 2\theta + \frac{1}{2}\gamma_{xy}\sin 2\theta$$

$$\gamma(\theta) = \frac{1}{2}(\varepsilon_y - \varepsilon_x)\sin 2\theta + \gamma_{xy}\cos 2\theta$$

4. Recall that normal strains are measured with a strain gage, which can be oriented at any angle.

It is most common in practice to measure normal strain with a strain gage. A strain gage measures the strain parallel to the length of the wire in the foil. So normal strain can be measured in any direction with a suitably oriented strain gage. We can take advantage of this flexibility and measure normal strains in multiple directions.

Measures strain along x (ε_x)   Measures strain along y (ε_y)   Measures strain along θ ($\varepsilon(\theta)$)

5. Reconstruct the three *x-y* strains from the normal strains measured at three angles.

The first two orientations above each reveal only one *x-y* strain component. A strain gage oriented at angle θ captures contributions from all three components ε_x, ε_y, and γ_{xy}, although they cannot be separated with only that one measurement. From strains measured along three different directions, say along θ_A, θ_B, and θ_C, we can solve the following simultaneous equations to determine ε_x, ε_y, and γ_{xy}.

$$\varepsilon(\theta_A) = \frac{1}{2}(\varepsilon_x + \varepsilon_y) + \frac{1}{2}(\varepsilon_x - \varepsilon_y)\cos 2\theta_A + \frac{1}{2}\gamma_{xy}\sin 2\theta_A$$

$$\varepsilon(\theta_B) = \frac{1}{2}(\varepsilon_x + \varepsilon_y) + \frac{1}{2}(\varepsilon_x - \varepsilon_y)\cos 2\theta_B + \frac{1}{2}\gamma_{xy}\sin 2\theta_B$$

$$\varepsilon(\theta_C) = \frac{1}{2}(\varepsilon_x + \varepsilon_y) + \frac{1}{2}(\varepsilon_x - \varepsilon_y)\cos 2\theta_C + \frac{1}{2}\gamma_{xy}\sin 2\theta_C$$

6. Deduce the *x-y* strains by interpreting the measurements from a strain rosette, a combination of three strain gages oriented at different angles.

A strain rosette is a set of three strain gages on a single foil, from which three normal strains are measured.

Here is one of several common arrangements for a strain rosette, with two that are mutually perpendicular and a third that is at 45° to both.

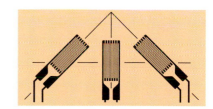

>>End 7.8

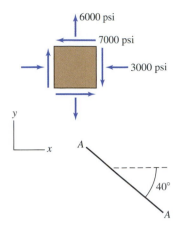

6000 psi
7000 psi
3000 psi

y

x

A

40°

A

For the stress state shown, determine the normal strain along the direction A-A in two ways. (a) Use the transformation formulas to determine the stresses perpendicular and parallel to A-A, and then use the stress–strain relation for the transformed stresses. (b) Determine the strains along the x-y axes, and then transform the strains to find the normal strain along A-A. Take $E = 30 \times 10^6$ psi, $G = 11.5 \times 10^6$ psi, and $\nu = 0.3$.

Solution

Stresses are: $\sigma_x = -3000$ psi, $\sigma_y = 6000$ psi, and $\tau_{xy} = -7000$ psi.

The stress parallel with A-A corresponds to $\theta = -40°$.

$$\sigma(-40°) = \frac{1}{2}(-3000 + 6000) + \frac{1}{2}(-3000 - 6000)\cos 2(-40°) - 7000 \sin 2(-40°) = 7610 \text{ psi}$$

The stress perpendicular to A-A corresponds to $\theta = 50°$.

$$\sigma(50°) = \frac{1}{2}(-3000 + 6000) + \frac{1}{2}(-3000 - 6000)\cos 2(50°) - 7000 \sin 2(50°) = -4610 \text{ psi}$$

The strain along A-A is computed from these stresses by:

$$\varepsilon_{A\text{-}A} = \frac{\sigma(-40°) - \nu\sigma(50°)}{E} = \frac{7610 \text{ psi} - 0.3(-4610 \text{ psi})}{30 \times 10^6 \text{ psi}} = 3.00 \times 10^{-4}$$

Alternatively, determine strains along the x-y axes:

$$\varepsilon_x = \frac{\sigma_x - \nu\sigma_y}{E} = \frac{-3000 \text{ psi} - 0.3(6000 \text{ psi})}{30 \times 10^6 \text{ psi}} = -1.60 \times 10^{-4}$$

$$\varepsilon_y = \frac{\sigma_y - \nu\sigma_x}{E} = \frac{6000 \text{ psi} - 0.3(-3000 \text{ psi})}{30 \times 10^6 \text{ psi}} = 2.30 \times 10^{-4}$$

$$\gamma_{xy} = \frac{\tau_{xy}}{G} = \frac{-7000 \text{ psi}}{11.5 \times 10^6 \text{ psi}} = -6.09 \times 10^{-4}$$

The strain along A-A is computed by transforming the strains ε_x, ε_y, and γ_{xy} to the normal strain ε along $\theta = -40°$.

$$\varepsilon(-40°) = \frac{1}{2}(-1.6 \times 10^{-5} + 2.3 \times 10^{-4}) + \frac{1}{2}(-1.6 \times 10^{-5} - 2.3 \times 10^{-4})\cos 2(-40°)$$

$$+ \frac{1}{2}(-6.09 \times 10^{-4})\sin 2(-40°) = 3.01 \times 10^{-4}$$

The two calculations of strain along A-A nearly agree with each other. They do not agree precisely, because the values given for G, E, and ν are not precisely consistent with that of an isotropic material, which should satisfy:

$$G = \frac{E}{2(1 + \nu)}$$

>>End Example Problem 7.15

A 45° strain rosette is mounted onto a steel structure in the orientation shown. The strain gages read $\varepsilon_a = 150(10^{-6})$, $\varepsilon_b = 240(10^{-6})$, $\varepsilon_c = -200(10^{-6})$. Determine the stresses on the x-y axes, as well as the principal stresses and maximum in-plane shear stress. Take the moduli to be $E = 200$ GPa, $G = 80$ GPa, and $\nu = 0.3$.

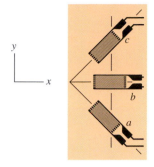

Solution

If the strains along the x-y axes were known, then the normal strain ε along a line oriented at a general angle θ is:

$$\varepsilon(\theta) = \frac{1}{2}(\varepsilon_x + \varepsilon_y) + \frac{1}{2}(\varepsilon_x - \varepsilon_y)\cos 2\theta + \frac{1}{2}\gamma_{xy}\sin 2\theta$$

Apply this equation to each of the three strain gages:

$$\varepsilon_a = \varepsilon(-45°) = \frac{1}{2}(\varepsilon_x + \varepsilon_y) + \frac{1}{2}(\varepsilon_x - \varepsilon_y)\cos -90° + \frac{1}{2}\gamma_{xy}\sin -90° = \frac{1}{2}(\varepsilon_x + \varepsilon_y) - \frac{1}{2}\gamma_{xy}$$

$$\varepsilon_b = \varepsilon(0°) = \frac{1}{2}(\varepsilon_x + \varepsilon_y) + \frac{1}{2}(\varepsilon_x - \varepsilon_y)\cos 0° + \frac{1}{2}\gamma_{xy}\sin 0° = \varepsilon_x$$

$$\varepsilon_c = \varepsilon(45°) = \frac{1}{2}(\varepsilon_x + \varepsilon_y) + \frac{1}{2}(\varepsilon_x - \varepsilon_y)\cos 90° + \frac{1}{2}\gamma_{xy}\sin 90° = \frac{1}{2}(\varepsilon_x + \varepsilon_y) + \frac{1}{2}\gamma_{xy}$$

Solve these equations for the given values of ε_a, ε_b, and ε_c to determine ε_x, ε_y, and γ_{xy}. To do so, first find ε_x from ε_b, and then ε_y and γ_{xy} by adding or subtracting ε_a and ε_c.

$$\varepsilon_x = 240(10^{-6}), \ \varepsilon_y = -290(10^{-6}), \text{ and } \gamma_{xy} = -350(10^{-6})$$

Given these strains, determine stresses by solving: $\quad \varepsilon_x = \dfrac{\sigma_x - \nu\sigma_y}{E} \qquad \varepsilon_y = \dfrac{\sigma_y - \nu\sigma_x}{E} \qquad \gamma_{xy} = \dfrac{\tau_{xy}}{G}$

$$\sigma_x = \frac{E(\varepsilon_x + \nu\varepsilon_y)}{(1 - \nu^2)} = 33.6 \text{ MPa} \qquad \sigma_y = \frac{E(\varepsilon_y + \nu\varepsilon_x)}{(1 - \nu^2)} = -47.9 \text{ MPa} \qquad \tau_{xy} = G\gamma_{xy} = -28.0 \text{ MPa}$$

Determine the maximum in-plane shear and principal stresses:

$$\tau_{max} = \sqrt{\frac{1}{4}(\sigma_y - \sigma_x)^2 + \tau_{xy}^2} = 49.5 \text{ MPa} \qquad \sigma_m = \frac{(\sigma_x + \sigma_y)}{2} = -7.14 \text{ MPa}$$

$$\sigma_{max} = \sigma_m + \tau_{max} = 42.3 \text{ MPa} \qquad \sigma_{min} = \sigma_m - \tau_{max} = -56.6 \text{ MPa}$$

>>End Example Problem 7.16

Additional data on material properties needed to solve problems can be found in Appendix D or inside back cover.

7.114 For the 45° rosette shown, determine the strains ε_x, ε_y, and γ_{xy}, if the strain gages read $\varepsilon_a = 550(10^{-6})$, $\varepsilon_b = 100(10^{-6})$, and $\varepsilon_c = -300(10^{-6})$.

7.115 A steel bridge girder is instrumented with a 45° rosette as shown. Under a test loading, the strain gages read $\varepsilon_a = 400(10^{-6})$, $\varepsilon_b = 250(10^{-6})$, and $\varepsilon_c = -150(10^{-6})$. Assuming a state of plane stress, determine the stress components in psi at the rosette.

7.116 One portion of an aluminum aircraft fuselage is load tested while instrumented with a 45° rosette as shown. The strain gages read $\varepsilon_a = 170(10^{-6})$, $\varepsilon_b = -200(10^{-6})$, and $\varepsilon_c = 250(10^{-6})$. Assuming a state of plane stress, determine the principal stresses and the maximum shear stress.

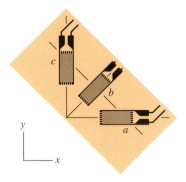

Probs. 7.114–116

7.117 For the 30° rosette shown, determine the strains ε_x, ε_y, and γ_{xy}, if the strain gages read $\varepsilon_a = 120(10^{-6})$, $\varepsilon_b = 350(10^{-6})$, and $\varepsilon_c = -200(10^{-6})$.

7.118 A steel structure in an earthquake prone region is instrumented with a 30° rosette as shown. During a minor quake, the strain gages read $\varepsilon_a = -270(10^{-6})$, $\varepsilon_b = -350(10^{-6})$, and $\varepsilon_c = 250(10^{-6})$. Assuming a state of plane stress, determine the stress components at the rosette.

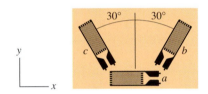

Probs. 7.117–119

7.119 An aluminum storage container on a truck is instrumented with a 30° rosette as shown while subjected to load. The strain gages read $\varepsilon_a = 400(10^{-6})$, $\varepsilon_b = -300(10^{-6})$, and $\varepsilon_c = 600(10^{-6})$. Assuming a state of plane stress, determine the principal stresses and the maximum shear stress in MPa.

7.120 Loads on a structure are expected to result in strains $\varepsilon_x = -370(10^{-6})$, $\varepsilon_y = 280(10^{-6})$, and $\gamma_{xy} = 450(10^{-6})$. If the structure is instrumented with a 45° rosette as shown, what should be the strains in the gages ε_a, ε_b, and ε_c?

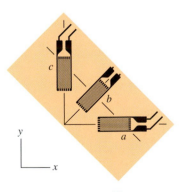

Prob. 7.120

7.121 Loads on a structure are expected to result in strains $\varepsilon_x = 230(10^{-6})$, $\varepsilon_y = 420(10^{-6})$, and $\gamma_{xy} = -180(10^{-6})$. If the structure is instrumented with a 30° rosette as shown, what should be the strains in the gages ε_a, ε_b, and ε_c?

7.122 FEA analysis predicts the stresses in a steel structure to be $\sigma_x = 56$ MPa, $\sigma_y = -40$ MPa, and $\tau_{xy} = -70$ MPa. If the structure is instrumented with a 30° rosette as shown, what should be the strains in the gages ε_a, ε_b, and ε_c?

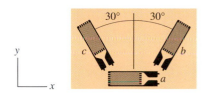

Probs. 7.121–122

7.123 A steel cylindrical pressure vessel has an outer diameter of 2.5 m and a wall thickness of 50 mm. Due to a rearrangement of supports, the vessel experiences a torque $T = 8(10^6)$ N-m. In addition, the pressure is raised by 1 MPa. If the structure is instrumented with a 30° rosette as shown, what should be the strains measured in the gages ε_a, ε_b, and ε_c due to the torque and pressure increase?

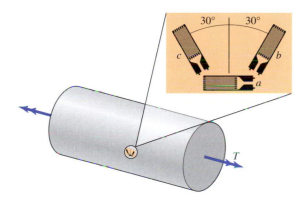

Prob. 7.123

>>End Problems

7.9 Fatigue

So far we have considered failures that would occur when an increasing load reaches a critical level. Also common, and more difficult to design for, are fatigue failures, which are due to fluctuating loads.

1. A load might not cause failure when applied once, but might cause failure if applied and removed multiple times.

You can bend a paper clip severely, and it won't break. But bend it back and forth repeatedly, and eventually it will break. This is the essence of fatigue failure: failure can occur due to a cyclically (repeatedly) applied load, even when a single loading of the same magnitude will not cause failure.

2. In engineering applications, it is common for loads to be applied multiple times.

There are many circumstances when parts are naturally subjected to cyclic loads. For example, a hip implant is loaded cyclically during walking, because weight is placed on and removed from the leg. The stress in a bicycle crank fluctuates with pedaling. Perhaps more detrimental to bicycle life are the more severe, though less frequent, impacts of the tires in potholes or bouncing off curbs.

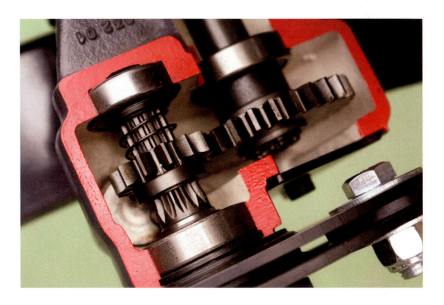

A rotating shaft is prone to fatigue. The forces on gears might lead to bending stresses in the shaft, with maximum tension at, say, the top of the shaft. Every time the shaft rotates, a particular element of material rotates from the top ($\sigma > 0$) to the bottom ($\sigma < 0$) and back. So each shaft rotation corresponds to a stress cycle, of which there can be thousands every minute, depending on the rotation speed.

3. Describe cyclic loads as load or stress vs. time.

Here we show the case of so-called fully reversed loading: the stress at a point fluctuates with time, with the maximum stress in tension equal to the maximum stress in compression. The cycling is described by the amplitude σ_a. The time between cycles, and the shape of the cycle with time, are usually much less important. We are interested in how σ_a affects the number of cycles that can be applied with failure.

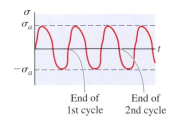

4. Plastic deformation is usually necessary for fatigue, which can occur locally even if the overall stress is below yield.

Plastic or permanent deformation is usually necessary for fatigue. The conventional yield stress corresponds to plastic deformation that occurs in a large region. But due to the microscopic mechanisms of plastic deformation, the movement of crystal defects or dislocations, plasticity can initiate locally, for example at an imperfectly ground surface, even if the overall stress is well below the yield stress.

Since fatigue failure can occur under cyclic loading due to local plastic deformation at imperfections, fatigue failure remains a concern even if we design to keep the overall stresses below the yield strength.

5. Fatigue damage in metals usually involves initiating a crack, which grows until catastrophic failure.

Fatigue failure often initiates because of enhanced stress near a particular feature (see stress concentrations in Section 7.10), or even at a rough spot on the surface, where local plastic deformation occurs. With enough stress cycles, this leads to the formation of a crack. Cracks grow with further cycles. Since the cracked area carries no load, the average stress on the uncracked portion increases as the crack grows. When the crack becomes too long, sudden failure occurs with the crack running across the part. Sometimes inspections can detect cracks before they lead to failure. Based on experience, engineers may decide to leave the crack as is, to repair it, or to replace the part.

Fatigue crack initiated and grew in this area

At failure, crack runs suddenly across this area

6. Plot the cyclic stress that can be tolerated vs. the number of cycles before failure (S-N curve).

Fatigue strength is depicted graphically with a so-called *S-N* curve: a plot of the cyclic stress amplitude (σ_a) to cause failure as a function of the number of cycles at which fatigue failure is observed to occur when cycling at that stress amplitude. If, for a chosen number of cycles, the cyclic stress is below the *S-N* curve, then the material will not fail in that number of cycles. Since fatigue strength is highly variable, a substantial factor of safety is generally employed.

Many materials, notably steels, when subjected to very low stress, can survive any number of cycles. The endurance limit is the stress below which infinite cycles can be tolerated. If many cycles are expected, parts are designed so stresses are well below the endurance limit. For materials that have no clear endurance limit, the maximum allowable stress might be that which safely allows for a large number of cycles, say 10^8.

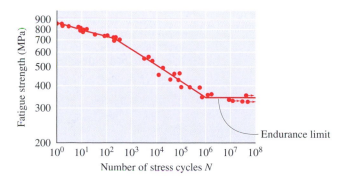

7.10 Stress Concentrations _____

In Chapter 2, we pointed out that simple formulas such as *P/A* for normal stress in axial loading may not accurately predict stress under some circumstances. Here we consider a set of common situations in which a better prediction of stress can be found.

1. Non-uniformities in a body's cross-section take on several common forms.

For design purposes, it is common for bars, shafts, and rods to have holes or sudden changes in cross-section. For example, shafts often have steps, sudden changes in cross-section that allow bearings, gears, and other elements to be located axially along the shaft. We have seen that shafts are subjected to twisting and bending moments. The stresses given by the simple formulas for axial, twisting, and bending must be altered to account for sudden changes in cross-section.

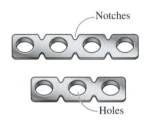

Reconstruction plates that are screwed to bone fragments to set fractures exhibit common features, notches and holes, at which the simple predictions of stress from Chapters 3–5 would also need to be revised.

2. Geometric details of steps, notches, and other features can sharply influence stresses.

A perfectly sharp step change in a shaft's cross-section is impossible to manufacture. Besides, the stresses would be unacceptably high where the smaller diameter segment meets the shoulder.

Instead, a fillet is added to make the transition from shoulder to smaller diameter more gradual. A larger radius fillet gives a more gradual transition and leads to lower stresses. But the stress at the fillet is still higher than in the neighboring portion.

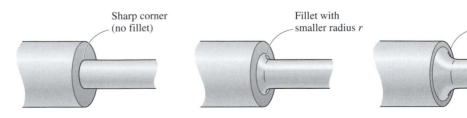

Sharp corner (no fillet)   Fillet with smaller radius *r*   Fillet with larger radius *r*

3. **At changes in cross-section, stresses redistribute so that the maximum stress is greater than would occur in a similarly loaded body of uniform cross-section.**

Near a change in cross-section, the stresses are redistributed relative to the nominal stress distribution that would hold in a straight bar. Since the maximum stress is higher, the stresses elsewhere are lower, so the net load is the same (e.g., axial force P or torque T).

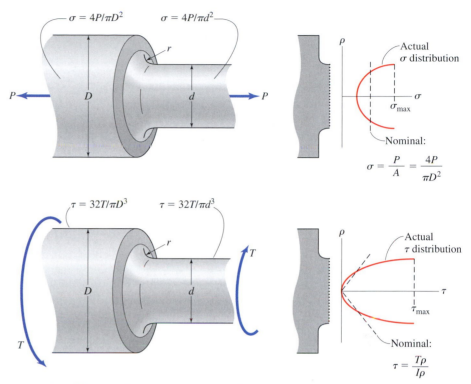

Features, such as fillets, notches, or holes, at which the maximum stresses are above those predicted by mechanics of materials, are called *stress concentrations*.

For design purposes, we want to predict the maximum stress, σ_{max} or τ_{max}.

4. **For relatively simple, standard features, we predict the maximum stress in a two-step process: calculate an approximate stress and then look up the correction to that approximation.**

For many standard features, one first estimates the stress with mechanics of materials, using the local cross-section, and ignoring the variation in the cross-section along the length. This stress estimate is called the nominal stress, σ_{nom}. The nominal stress would be found using the smaller cross-section. The actual stress maximum, σ_{max}, which accounts for the sudden change in cross-section, is higher than σ_{nom}.

The ratio of the maximum stress at the feature, σ_{max}, to the nominal stress, σ_{nom}, is termed the stress concentration factor, K:

$$K = \frac{\sigma_{max}}{\sigma_{nom}}$$

Since σ_{max} and σ_{nom} are both proportional to the load, the ratio $K = \sigma_{max}/\sigma_{nom}$ is independent of the load. K depends on the dimensions that describe the feature. For many standard features, such as holes and fillets, experiments or more complex methods have been used to determine σ_{max}. Values of the ratio, $K = \sigma_{max}/\sigma_{nom}$ have been recorded in tables or charts, for example those in Appendix H.

In brief, the procedure we will follow to compute the maximum stress σ_{max} will be: calculate σ_{nom} from mechanics of materials, look up K from a chart or table, and compute $\sigma_{max} = K\sigma_{nom}$.

5. **Stress concentrations affect failure primarily when the loading is cyclic or the material is brittle.**

Sometimes a stress concentration does not lower the strength of a body, particularly when the material is ductile and the load is not cyclic. This is because plastic deformation causes the actual stresses to be lower than those predicted by the stress concentration, which assumes elastic behavior. However, if the material is brittle, that is it remains elastic up to failure, or if the loading is cyclic as in fatigue, then the strength is often better predicted with a maximum stress that accounts for the stress concentration.

>>End 7.10

Find the maximum stress at the fillet of the stepped shaft subjected to torsion.

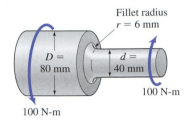

Solution

The stress concentration factor K is dimensionless or has no units. So in general, it depends on ratios of dimensions. In this problem, the ratios are D/d, r/d, and r/D. Only two of these ratios are independent, since for example $r/D = (r/d)/(D/d)$. So K depends on just two ratios, say D/d and r/d.

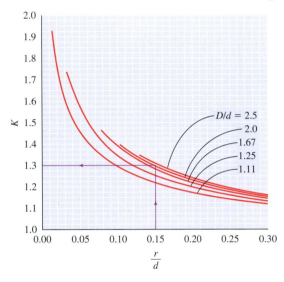

This chart displays the stress concentration factor K as a function of the ratios D/d and r/d, for a stepped shaft that is subjected to torsion.

This chart is taken from Appendix H, which includes charts that display stress concentration factors for other loadings and geometries. Additional purple marks have been added to explain the use of this chart for this particular problem.

Here are the general steps to be taken to determine a maximum stress due to a stress concentration and what that step means for this specific example.

General Step	Step for Particular Example
Find internal load (torque for torsion) at cross-section of fillet	Everywhere along shaft $T = 100$ N-m
Calculate nominal stress, the maximum stress at that cross-section if the bar had a uniform cross-section (no fillet)	$\tau_{nom} = 32T/\pi d^3 = 127.3$ MPa
Evaluate relevant ratios (D/d and r/d)	$D/d = 2$, $r/d = 0.15$
Look up K in chart	$K = 1.3$
Compute max stress $\tau_{max} = K\tau_{nom}$	$\tau_{max} = (1.3)(127$ MPa$) = 165$ MPa

>>End Example Problem 7.17

Several elements are mounted on the stepped shaft and exert the torques shown. The larger diameter of the shaft is 1.25 in., the smaller diameter is 1 in., and the fillet radius r is 0.05 in.

Determine the maximum shear stress in the shaft. Consider all cross-sections, and the stress concentration at the fillet.

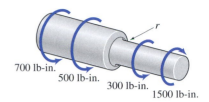

700 lb-in. 500 lb-in. 300 lb-in. 1500 lb-in.

Solution

See shaft redrawn with labels and the torques acting.

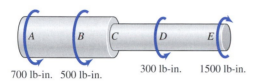

700 lb-in. 500 lb-in. 300 lb-in. 1500 lb-in.

Internal torque (magnitudes): $T_{AB} = 700$ lb-in., $T_{BC} = 1200$ lb-in., $T_{CD} = 1200$ lb-in., and $T_{DE} = 1500$ lb-in.

The step and fillet are at C, where the internal torque is 1200 lb-in.

Regions where we should look for maximum shear stress:

DE (where the internal torque is highest and the diameter is small)
C (where internal torque is not the highest, but there is a stress concentration)

In DE the maximum shear is on the outer surface:

$$\tau_{max} = \frac{2T}{\pi R^3} = \frac{2(1500 \text{ lb-in.})}{\pi(0.5 \text{ in.})^3} = 7640 \text{ psi}$$

At C, use the stress concentration relation $\tau_{max} = K\tau_{nom}$.

τ_{nom} is the maximum stress calculated in the narrow part of the shaft connected to the fillet, but with the stress concentration ignored.

$$\tau_{nom} = \frac{2T}{\pi R^3} = \frac{1200 \text{ lb-in.}}{\pi(0.5 \text{ in.})^3} = 6110 \text{ psi}$$

Find K: identify parameters $D/d = 1.25$ and $r/d = 0.05$. From table: $K = 1.575$.

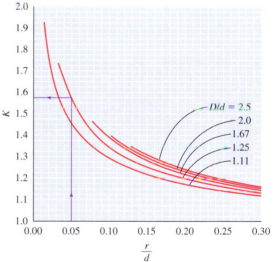

Therefore, at the fillet, $\tau_{max} = K\tau_{nom} = (1.575)(6110 \text{ psi}) = 9660 \text{ psi}$.

The maximum stress in the shaft is indeed at the fillet, even though the internal torque there (1200 lb-in.) is not at its maximum level (1500 lb-in.).

>>End Example Problem 7.18

Additional data on material properties needed to solve problems can be found in Appendix D or inside back cover.

Charts for Stress Concentration Factors can be found in Appendix H.

7.124 Determine the maximum stress in the bar shown if the axial force $P = 30$ kN. The plate widths are $a = 60$ mm and $b = 30$ mm, the thickness is $t = 4$ mm, and the fillet radius is $r = 10$ mm.

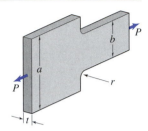

Prob. 7.124

7.125 Determine the maximum stress in the bar shown if the axial force $P = 25(10^3)$ lb. The plate has a width of $a = 4$ in. and a thickness of $t = 0.25$ in., and the hole diameter is 1.5 in.

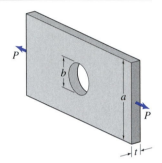

Prob. 7.125

7.126 The maximum stress in the aluminum bar is not to exceed 15 ksi. The plate has widths of $a = 4.5$ in. and $b = 3$ in., and a thickness of $t = 0.3$ in. The fillet radius is $r = 1$ in. Determine the maximum allowable axial force P.

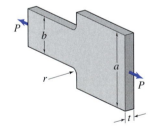

Prob. 7.126

7.127 The maximum stress in the nylon bar is not to exceed 40 MPa. The plate has a width of $a = 35$ mm and a thickness of $t = 3$ mm, and the hole diameter is 10 mm. Determine the maximum allowable axial force P.

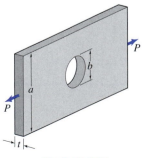

Prob. 7.127

7.128 The steel plate of thickness $t = 0.5$ in. is to be designed under the constraint that the maximum stress remains below 21 ksi. The hole must have a diameter of $b = 2$ in. Determine the allowable plate width a within the nearest 0.1 in. if it is to withstand an axial force of $P = 30$ kip.

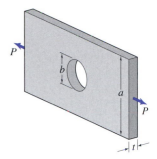

Prob. 7.128

7.129 The aluminum plate of thickness $t = 3.5$ mm is to be designed so the maximum stress remains below 100 MPa. The hole must have a diameter of $b = 25$ mm. Determine the allowable plate width a within the nearest 5 mm if it is to withstand an axial force of $P = 25$ kN.

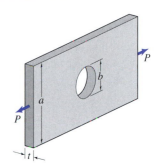

Prob. 7.129

7.130 Recommendations are sought regarding the fillet geometry in the steel bar. The widths are fixed at $a = 120$ mm and $b = 60$ mm, and the thickness is fixed at $t = 5$ mm. The bar is to withstand an axial force of 35 kN without the maximum stress exceeding 170 MPa. Specify the minimum allowable fillet radius.

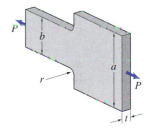

Prob. 7.130

7.131 The bar is subjected to the axial forces shown, with $F_1 = 3$ kN, $F_2 = 6$ kN, $F_3 = 11$ kN, and $F_4 = 20$ kN. Determine the maximum stress in the bar. The plate widths are $a = 40$ mm and $b = 25$ mm, the thickness is $t = 3$ mm, and the fillet radius is $r = 8$ mm.

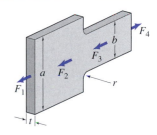

Prob. 7.131

7.132 The maximum stress in the bar shown must not exceed 80 MPa. The widths are $a = 60$ mm, $b = 120$ mm, and $c = 80$ mm, and the thickness is $t = 4$ mm. The fillet radii are $r_1 = 40$ mm and $r_2 = 30$ mm. Determine the maximum allowable axial force that can be applied to the bar. Assume that the stress concentrations due to the two steps can be analyzed independently.

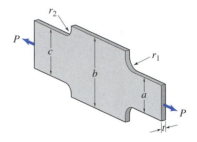

Prob. 7.132

7.133 The bar is subjected to an axial force of $P = 2$ kip. The plate has widths of $a = 3$ in. and $c = 1$ in., and a thickness of $t = 0.2$ in. The diameter of the hole is $b = 1.5$ in., and the fillet radius is $r = 0.25$ in. Assuming that the stress concentrations due to the hole and the fillet can be analyzed independently, determine the maximum stress in the bar.

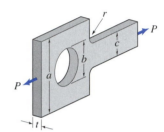

Prob. 7.133

7.134 The bar is subjected to an axial force of $P = 30$ kN. The widths are $a = 100$ mm, $b = 120$ mm, and $c = 90$ mm, and the thickness is $t = 3$ mm. The fillet radius $r_2 = 20$ mm. Determine the radius of fillet r_1 so that there is an equal likelihood of failure at the two fillets.

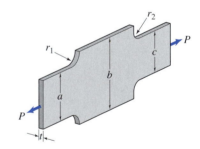

Prob. 7.134

7.135 The stepped shaft is subjected to torque of $T_1 = 300$ N-m. There is a fillet of radius $r = 3$ mm at the step. Determine the maximum stress in the shaft. ($L_1 = 400$ mm, $L_2 = 300$ mm, $d_1 = 30$ mm, and $d_2 = 50$ mm.)

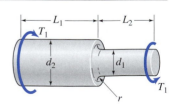

Prob. 7.135

7.136   The shaft is driven by a motor and has a step and fillet at B. The motor delivers 8 hp at 1000 rpm. The gear at C delivers 60% of the power and the gear at D delivers 40%. From A to B, the diameter is 1.5 in., from B to D the shaft is 0.75 in. The fillet radius is $r = 0.1$ in. Determine the maximum shear stress in the shaft.

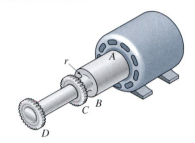

Prob. 7.136

7.137   The stepped shaft ($G = 27$ GPa) has a fillet ($r = 4$ mm) and is fixed at A. A torque is applied at C that produces a rotation at C of 1°. Determine the maximum shear stress in the shaft. The rotation must account for the differing radii of the two segments of the shaft. While the step and fillet produce a stress concentration, assume they have negligible effect on the rotation. ($L_1 = 350$ mm, $L_2 = 250$ mm, $d_1 = 20$ mm, and $d_2 = 25$ mm.)

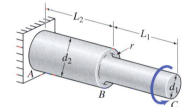

Prob. 7.137

7.138   The stepped shaft has a fillet of radius $r = 0.06$ in. and is subjected to the torques $T_1 = 600$ lb-in., $T_2 = 550$ lb-in., $T_3 = 60$ lb-in., and $T_4 = 10$ lb-in. Determine the maximum shear stress in the shaft. ($d_1 = 0.4$ in. and $d_2 = 0.8$ in.)

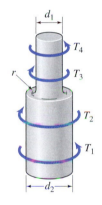

Prob. 7.138

7.139   The stepped shaft ($G = 80$ GPa) is subjected to torques $T_1 = 60$ N-m, $T_2 = 30$ N-m, $T_3 = 80$ N-m, and $T_4 = 10$ N-m. If the maximum shear stress is not to exceed 70 MPa, determine the minimum acceptable fillet radius r. ($d_1 = 20$ mm and $d_2 = 50$ mm.)

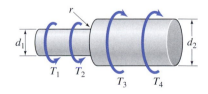

Prob. 7.139

7.140 The stepped shaft is fixed at A and is subjected to torques at C and D. The torque applied at D has a known value $T_D = 300$ lb-in. The step has a fillet with radius $r = 0.1$ in. If the maximum shear stress of 7 ksi is not to be exceeded, determine the maximum value for the torque T_C. ($L_1 = 10$ in., $L_2 = 2$ in., $L_3 = 8$ in., $d_1 = 0.8$ in., and $d_2 = 1$ in.)

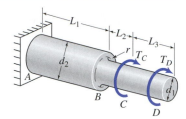

Prob. 7.140

7.141 Determine the maximum stress in the bar shown if the bending moment $M = 200$ N-m. The plate widths are $a = 50$ mm and $b = 150$ mm, the thickness is $t = 5$ mm, and the fillet radius is $r = 20$ mm.

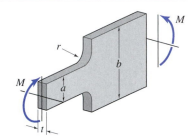

Prob. 7.141

7.142 Determine the maximum stress in the bar shown if the bending moment $M = 900$ lb-in. The plate widths are $a = 3$ in. and $b = 2$ in., the thickness is $t = 0.25$ in., and the fillet radius is $r = 0.8$ in.

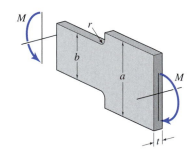

Prob. 7.142

7.143 If the maximum stress in the bar is not to exceed 14 ksi, determine the maximum allowable bending moment M. The plate widths are $a = 6$ in. and $b = 2$ in., the thickness is $t = 0.3$ in., and the fillet radius is $r = 1.2$ in.

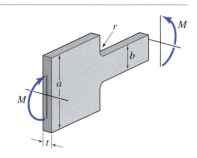

Prob. 7.143

7.144 If the maximum stress in the bar is not to exceed 100 MPa, determine the maximum allowable bending moment M. The plate widths are $a = 40$ mm and $b = 50$ mm, the thickness is $t = 3$ mm, and the fillet radius is $r = 12$ mm.

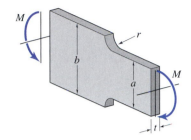

Prob. 7.144

7.145 Determine the maximum stress in the bar shown containing a semi-circular notch. The bending moment $M = 300$ N-m. The plate width is $a = 150$ mm, the thickness is $t = 5$ mm, and the notch radius is $r = 25$ mm.

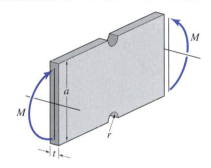

Prob. 7.145

7.146 The maximum stress in the bar with the semi-circular notch is to remain below 12 ksi. Determine the maximum allowable bending moment M. The plate width is $a = 2.5$ in., the thickness is $t = 0.125$ in., and the notch radius is $r = 0.175$ in.

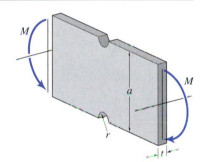

Prob. 7.146

7.147 A notch bend test on a plate with a semi-circular notch is conducted on the specimen shown. Determine the maximum stress when the load $P = 600$ lb. The specimen dimensions are $a = 1.5$ in., $b = 1$ in., $L = 6$ in., and $t = 0.2$ in., and the notch radius is $r = 0.175$ in.

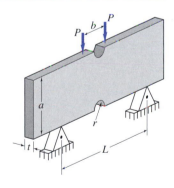

Prob. 7.147

>>End Problems

Chapter Summary

7.1–7.2 To detect failure, consider stresses on internal surfaces of all orientations.

Describe a surface of general orientation with θ, the angle from x-axis to the outward normal from the surface.

Denote normal and shear stresses on a surface of orientation θ by $\sigma(\theta)$ and $\tau(\theta)$. When positive, normal stress $\sigma(\theta)$ acts outward (tension), and shear stress $\tau(\theta)$ acts at 90° CCW from outward normal direction.

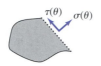

7.3 Use methods from Chapters 3–6 to find stress components along the x-y axes, σ_x, σ_y, and τ_{xy}, at a point of a body. From equilibrium of a portion of an element at that point, stresses $\sigma(\theta)$ and $\tau(\theta)$ on a surface of orientation θ can be related to the x-y components. The resulting **stress transformation formulas** are:

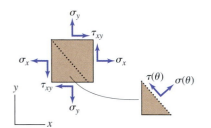

$$\sigma(\theta) = \frac{1}{2}(\sigma_x + \sigma_y) + \frac{1}{2}(\sigma_x - \sigma_y)\cos 2\theta + \tau_{xy}\sin 2\theta$$

$$\tau(\theta) = \tau_{xy}\cos 2\theta + \frac{1}{2}(\sigma_y - \sigma_x)\sin 2\theta$$

7.4 For given stresses σ_x, σ_y, and τ_{xy}, we can find the orientations θ of internal surfaces on which $\sigma(\theta)$ and $\tau(\theta)$ are maximized.

The **maximum shear stress** is: $\quad \tau_{max} = \sqrt{\frac{1}{4}(\sigma_y - \sigma_x)^2 + \tau_{xy}^2}$

$|\tau(\theta)| = \tau_{max}$ on the four surfaces of the element oriented at angle θ_s: $\quad \theta_s = \frac{1}{2}\tan^{-1}\left[\frac{(\sigma_y - \sigma_x)/2}{\tau_{xy}}\right]$

The maximum and minimum normal stresses, also called **principal stresses**, are σ_{max} and σ_{min}:

$$\sigma_{max} = \sigma_m + \tau_{max} \qquad \text{and} \qquad \sigma_{min} = \sigma_m - \tau_{max} \qquad \text{where} \qquad \sigma_m = \frac{1}{2}(\sigma_x + \sigma_y)$$

Maximum shear stresses and principal stresses can be drawn on elements as shown. The elements differ in orientation by 45°. On the maximum shear stress element, the normal stress is σ_m. On the principal stress element, the shear stress is zero.

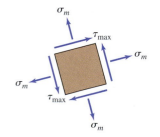

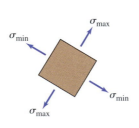

7.5 **Mohr's circle** graphically represents $\sigma(\theta)$ and $\tau(\theta)$, as they vary with θ.

On the circle, points A and B represent the stresses σ_x, σ_y, and τ_{xy}, and points A' and B' represent the stresses on an element oriented at angle θ from the x-axis. Points at the top, bottom, and sides of the circle represent stresses τ_{max}, σ_{max}, and σ_{min}.

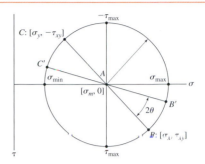

7.6 Failure for many materials falls into one of two categories, and depends on combined stresses as follows:

| **Ductile Failure** | | **Brittle Failure** | |

Material yields plastically and no longer springs back to initial shape starting at σ_Y.

Onset: Maximum shear stress τ_{max} reaches critical level $\sigma_Y/2$.

Material responds elastically until it breaks at σ_F.

Onset: Maximum normal stress σ_{max} reaches critical level σ_F.

7.7 For general 3-D stresses (σ_x, σ_y, σ_z, τ_{xy}, τ_{xz}, and τ_{yz}), there are three perpendicular planes on which shear stresses are zero. Normal stresses on these planes, σ_{I}, σ_{II}, and σ_{III}, are the principal stresses, with σ_{I} the maximum and σ_{III} the minimum normal stresses. The absolute maximum shear stress among all 3-D planes is $\tau_{abs} = (\sigma_I - \sigma_{III})/2$. The condition for brittle failure is $\sigma_I = \sigma_F$ if $\sigma_I > 0$. Alternative conditions for ductile failure are $\tau_{abs} = \sigma_Y/2$ or $\sigma_{VM} = \sigma_Y$, where the von Mises stress σ_{VM} is calculated from

$$\sigma_{VM} = \sqrt{\frac{\left(\sigma_x - \sigma_y\right)^2 + \left(\sigma_x - \sigma_z\right)^2 + \left(\sigma_y - \sigma_z\right)^2 + 6\left(\tau_{xy}^2 + \tau_{xz}^2 + \tau_{yz}^2\right)}{2}}$$

7.8 From strains on the x-y axes (ε_x, ε_y, and γ_{xy}), use definitions of normal and shear strain to find strains along lines, or on an element, inclined by θ with respect to the x-axis.

Strain Transformation Formulas are:

$$\varepsilon(\theta) = \frac{1}{2}(\varepsilon_x + \varepsilon_y) + \frac{1}{2}(\varepsilon_x - \varepsilon_y)\cos 2\theta + \frac{1}{2}\gamma_{xy}\sin 2\theta$$

$$\gamma(\theta) = \frac{1}{2}(\varepsilon_y - \varepsilon_x)\sin 2\theta + \gamma_{xy}\cos 2\theta$$

7.9 **Fatigue Failure:** Cyclic (repeated, on and off) loading can cause failure even if the maximum stress is significantly less than the yield stress.

7.10 **Stress Concentration:** Geometric features such as sudden changes in cross-section can cause higher stresses.

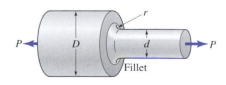

The stress at a change in cross-section is higher than the nominal stress, which is the maximum stress that would be calculated with the methods of Chapters 3, 4, and 5, assuming an unchanging cross-section.

Compute the actual maximum stress at the geometric feature from $\sigma_{max} = K\sigma_{nom}$, where the stress concentration factor $K(> 1)$ depends on geometric parameters (such as D, d, and r here). K is tabulated in handbooks. For ductile materials, stress concentrations are particularly important in fatigue loading.

>>End Chapter Summary

UNSTABLE DEFORMATIONS

Buckling is the tendency for a structure to deform in a secondary, undesirable mode when the load is too large.

Given its loading, we expect this beam simply to deflect in the plane (primary mode).

But, if the beam is not straight or the load does not act exactly through the center plane, the beam can also twist (secondary mode).

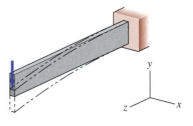
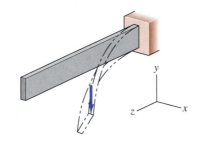

This is termed *lateral torsional buckling*. It is more likely to occur as the beam's stiffness in torsion decreases.

There are many other circumstances in which a structure buckles, that is, it deforms in a secondary mode, rather than in the desired, primary mode.

A long straight member subjected to axial forces, a column, has a primary mode of uniform compression. The column buckles – it bows – if the axial force is too large. In this chapter we study column buckling.

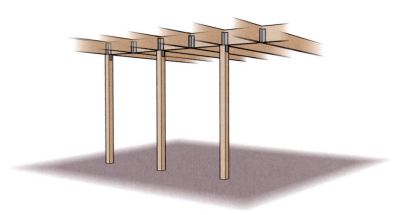

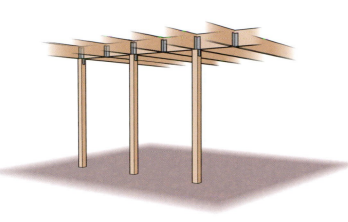

COLUMN BUCKLING

To see column buckling, squeeze a moderately stiff ruler lightly with two hands: it stays straight and compresses an undetectable amount.

Squeeze the same ruler with large enough forces and it buckles. The desired deformation pattern of simple axial compression is unstable: you cannot keep it straight even if you try.

As another example of column buckling, railroad tracks have been observed to buckle during a heat wave. Axial force had built up to excessive levels because of thermal expansion and axial constraint at some points.

Our goal is to discover how the buckling load, the axial force at which the primary mode becomes unstable and a secondary mode preferable, depends on the column's material and geometry.

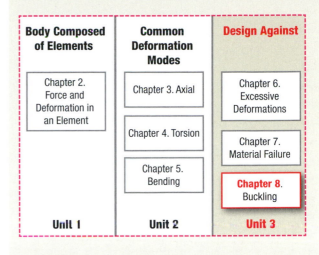

Body Composed of Elements	Common Deformation Modes	Design Against
Chapter 2. Force and Deformation in an Element	Chapter 3. Axial	Chapter 6. Excessive Deformations
	Chapter 4. Torsion	Chapter 7. Material Failure
	Chapter 5. Bending	**Chapter 8.** Buckling
Unit 1	Unit 2	Unit 3

Chapter Outline

A member can fail, not because the stresses cause the material to fail, but because it *buckles*, that is, the member's overall deformation pattern is not stable. In particular, members subjected to axial compression can bow out laterally (bending) if the axial force becomes too high. The critical axial force for buckling depends on the member's length, elastic modulus, and moment of inertia that controls bending (**8.1**). The critical force also depends on how the member is constrained against displacement and rotation at its ends (**8.2**). Failure under axial compression must also consider whether the material yields plastically, which becomes more likely than buckling as the member becomes shorter. Design formulas which capture the competition between buckling and material compressive failure have been devised for particular materials (**8.3**).

8.1 Buckling of Axially Loaded, Simply Supported Members

Some bodies under sufficiently high load can fail even if the stress nowhere reaches the material's limit. In a so-called buckling failure, the body sustains unstable, often large and unpredictable, deformations. The simple case of column buckling, namely a straight member under axial compression, is presented here.

1. A system is in an unstable equilibrium if even small disturbances cause large deflections from the equilibrium configuration.

In engineering, we usually require a system to be stable. A stable system, subjected to slight disturbances, will displace only slightly from its undisturbed state. Slight disturbances might include small additional loads or small deviations from nominal shape or dimensions. For example, a ball could be in equilibrium at the top or bottom of a round surface. It is stable at the bottom, but unstable at the top. If there is a horizontal force or the ball is not placed precisely on the top, it rolls off. But, the ball near the bottom would roll back to the bottom.

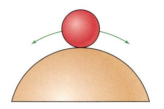

2. A straight member under axial compression is in equilibrium, but it will be unstable if the axial force is too high.

Squeeze a ruler axially with two hands. If the forces are small, the ruler stays straight. But, if the forces are large enough, it is hard to keep the ruler straight. The ruler will keep bowing one way or the other. We want to know the lowest force at which the ruler displays this unstable buckling behavior, and how that load depends on the ruler dimensions and material.

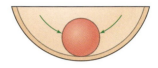

P (low) ———————————— P (low)

P (high) ⟶ ⟵ P (high)

3. Find the critical axial force by imposing equilibrium on the member in the buckled configuration.

We now determine the conditions under which a compressed axial member can be in equilibrium in a buckled configuration.

The result will depend on how the ends are restrained. We consider the basic simply supported case, corresponding to squeezing palms: the ends cannot move transversely to the beam, but they can rotate.

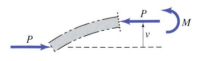

Buckling differs from all other analyses in mechanics of materials because we account for the change in geometry due to deflection (v) when imposing equilibrium.

A portion of the deflected beam in equilibrium must satisfy: $\sum M = 0 \Rightarrow M = -Pv$

Substitute the moment–curvature relation, $M = EI\kappa = EI\dfrac{d^2v}{dx^2},$ to obtain $EI\dfrac{d^2v}{dx^2} + Pv = 0.$

This differential equation for $v(x)$ over $0 < x < L$ depends on E, I, and P, which are constants.

For the simply supported case, the deflection at the ends must be zero: $v(0) = v(L) = 0$.

This differential equation is homogeneous: v appears in each term and the right hand side is zero. One solution, $v(x) = 0$, corresponds to the beam remaining straight (zero transverse deflection).

4. Look for the conditions under which non-zero solutions *v*(*x*) to the differential equation are possible.

The differential equation can be put in a standard form by dividing by EI.

$$\frac{d^2v}{dx^2} + \frac{P}{EI}v = 0$$

This second-order differential equation with constant coefficients has solutions of the form:

$$v(x) = A \sin \lambda x + B \cos \lambda x$$

Substitute this form for $v(x)$ into the differential equation and obtain

$$A\left[-\lambda^2 + \frac{P}{EI}\right]\sin(\lambda x) + B\left[-\lambda^2 + \frac{P}{EI}\right]\cos(\lambda x) = 0$$

The equation is satisfied at all points x, if $\lambda^2 = \dfrac{P}{EI}$

The end conditions, $v(0) = v(L) = 0$, must also be satisfied:

$$v(0) = A\sin\lambda(0) + B\cos\lambda(0) = B = 0. \qquad \text{Since } B = 0, \, \nu(L) = A\sin(\lambda L) = 0.$$

If $A = 0$, then $v(x) = 0$ (no buckling). So instead $\sin(\lambda L) = 0$. Values of λL that satisfy this equation are:

$$\lambda L = \pi, 2\pi, 3\pi, \ldots \quad \Rightarrow \lambda^2 = \frac{\pi^2}{L^2}, \frac{4\pi^2}{L^2}, \frac{9\pi^2}{L^2}, \ldots \quad \Rightarrow P = \frac{\pi^2 EI}{L^2}, \frac{4\pi^2 EI}{L^2}, \frac{9\pi^2 EI}{L^2}, \ldots$$

So, when the axial force P is less than any of these values, there is no solution to the differential equation other than $v(x) = 0$. But, non-zero (buckled) solutions for $v(x)$ are possible when P reaches the lowest of these values. So the critical or buckling load, P_{cr}, is given by

$$P_{cr} = \frac{\pi^2 EI}{L^2}$$

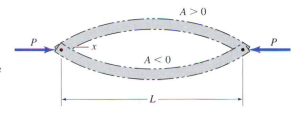

Because $\lambda = \pi/L$, the buckled shape is described by $v(x) = A\sin(x/L)$.

To avoid buckling, we design so that the axial force P is sufficiently less than the buckling load P_{cr}.

If we also constrained the deflection at the center to be zero, the buckled deflection would have this shape, and the critical load would correspond to the second critical value, $P_{cr} = \dfrac{4\pi^2 EI}{L^2}$. Such constraints are sometimes added to structures.

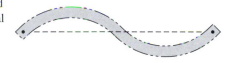

---

5. The critical buckling load, P_{cr}, increases with bending stiffness *EI* and decreases with length *L*.

P_{cr} increases with E: the stiffer the material, the higher the buckling load

P_{cr} increases with I: the more resistant the cross-section is to bending, the higher the buckling load

$$P_{cr} = \frac{\pi^2 EI}{L^2}$$

P_{cr} decreases with L: the shorter the member, the higher the buckling load

8.2 Buckling of Axially Loaded Members— Alternative Support Conditions

Members that are subjected to axial loading and prone to buckling may have end conditions that are not simply supported. The effect of support conditions can be found.

1. Find the critical load for buckling under alternative support conditions using the same differential equation analysis.

Members in compression which can buckle are not always simply supported. Different supports at the ends constrain the buckled shape in different ways. The corresponding values of v and v' at the ends lead to different values of λ (or buckling load) in $v(x) = A \sin \lambda x + B \cos \lambda x$.

Supports and Buckled Shape	End Conditions	Buckling Load
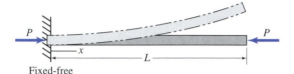 Fixed-free	$v(0) = 0, v'(0) = 0$	$P_{cr} = \dfrac{\pi^2 EI}{4L^2}$
 Fixed-fixed	$v(0) = 0, v'(0) = 0,$ $v(L) = 0, v'(L) = 0$	$P_{cr} = \dfrac{4\pi^2 EI}{L^2}$
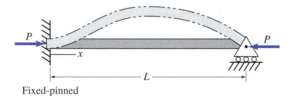 Fixed-pinned	$v(0) = 0, v'(0) = 0,$ $v(L) = 0$	$P_{cr} = \dfrac{2.046\pi^2 EI}{L^2}$

The critical buckling load still depends on the parameters: E, I, and L. Only the coefficient changes. In general, the more constrained the ends are, the greater is the buckling load.

2. Re-express the buckling load for different supports in terms of a simply supported member with a fictitious or "effective" length.

For the three new support cases above, the equations for the critical buckling load can be written:

$$\text{Fixed-free:} \quad P_{cr} = \frac{\pi^2 EI}{(2L)^2} \qquad \text{Fixed-fixed:} \quad P_{cr} = \frac{\pi^2 EI}{(0.5L)^2} \qquad \text{Fixed-pinned:} \quad P_{cr} = \frac{\pi^2 EI}{(0.7L)^2}$$

Recall that the buckling load for a beam of length L under simply supported conditions is: $\dfrac{\pi^2 EI}{L^2}$

Consider a simply supported column of length $2L$: from the equations, it buckles at the same load as a fixed-free column of length L. Likewise, simply supported columns of length $0.5L$ and $0.7L$ buckle at the same loads, respectively, as fixed-fixed and fixed-pinned columns of length L.

For a column of length L with general supports, we define its effective length, $L_e = KL$, as the length of a simply supported column that buckles at the same load. Handbooks tabulate the coefficient K for different support conditions; $K = 2, 0.5$, and 0.7 for fixed-free, fixed-fixed, and fixed-pinned columns, respectively.

3. The buckled shape of a generally supported beam of length L is the same as some portion of a simply supported beam of length equal to the effective length *KL*.

When a column buckles, the magnitude of the deflection is not determined. However, the deflection of each point relative to another, namely the shape, is determined. For the simply supported case, in $v(x) = A \sin \lambda x$, λ is determined, but the magnitude A is not. A generally supported column has the same buckled shape as some portion of a simply supported column with the corresponding effective length. That is why their buckling loads are equal.

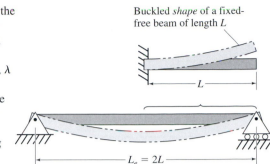

Buckled *shape* of a fixed-free beam of length L

$L_e = 2L$

The buckled shape of this fixed-free beam of length L is the same as the buckled shape of the right half of this simply supported beam of length 2L.

---

4. Define the radius of gyration to simplify the equation for the stress (rather than force) to cause buckling.

Later we will reconsider failure when either yielding or buckling could occur. This becomes simpler if the critical buckling force P_{cr} is converted to a critical buckling stress σ_{cr} by dividing by the area.

$$\sigma_{cr} = \frac{P_{cr}}{A} = \frac{\pi^2 EI}{(KL)^2 A}$$

The area A and the moment of inertia I are properties of the cross-section, and the ratio I/A has units of $(\text{length})^2$. We define the radius of gyration r by

$$I = Ar^2 \quad \text{or} \quad r = \sqrt{\frac{I}{A}}$$

While r is defined mathematically in terms of I and A, we can give physical meaning to the radius of gyration. Consider the polar moment of inertia I_p of a thin-walled circular cylinder of thickness t and radius R.

$$I_p = 2\pi R^3 t = (2\pi Rt)R^2 = AR^2$$

So $\sqrt{I_p/A}$ is the radius of the thin-walled cylinder, or how far the cross-sectional area is spread radially outward from the center. Likewise, $r = \sqrt{I/A}$ corresponds approximately to how far material of the cross-section is spread out from the center, and it is on the order of the cross-sectional dimensions.

Many cross-sections, for example I-beams, have different moments of inertia, I, in the two directions. The radius of gyration will have a different value depending on which I is being considered.

I_r, P_{cr} higher

I_r, P_{cr} lower

Using the radius of gyration, the critical stress for buckling, σ_{cr}, simplifies to $\sigma_{cr} = \dfrac{\pi^2 E}{(KL/r)^2}$.

So the buckling stress depends on the effective length KL relative to the radius of gyration r. KL/r is called the *slenderness ratio*, since it increases as the beam becomes longer relative to its thickness or cross-sectional dimensions.

>>End 8.2

8.3 Design Equations for Axial Compression

When designing a member under axial compression, one must consider both instability due to buckling and the compressive strength of the material.

1. In the simplest approach to axial loading of a column, failure corresponds to whichever critical stress is reached first—the yield stress or the buckling stress.

Here, failure stresses due to yielding and buckling are plotted as functions of KL/r, the slenderness of the compressed member. Yielding occurs at a constant stress σ_Y that depends on the material, but not on the length or cross-section. As we have seen, the buckling stress σ_{cr} decreases with increasing KL/r.

For a very stubby column (KL/r small), σ reaches σ_Y first $\Rightarrow$ yields.

For a very slender column (KL/r large), σ reaches σ_{cr} first $\Rightarrow$ buckles.

Failure is actually more complex. For columns that are neither very slender nor very stubby, buckling can occur when the material has partially yielded plastically.

Professional associations dedicated to the use of specific types of materials (e.g., steel, aluminum, or wood) publish design guidelines for predicting failure of columns.

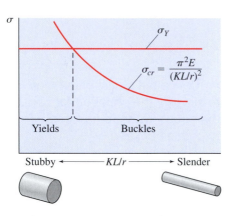

2. For steel columns, recommended design guidelines consider both yielding and buckling.

The Structural Stability Research Council has proposed formulas for designing steel columns under axial compression. Formulas, which give the maximum allowable stress, are based on research and empirical evidence and include appropriate factors of safety.

The first formula, for larger slenderness ratios (KL/r up to 200), predicts failure as due to buckling, with the recommended factor of safety of $23/12 = 1.92$.

$$\sigma_{allow} = \frac{12}{23}\frac{\pi^2 E}{(KL/r)^2} \qquad (KL/r)_c < (KL/r) < 200$$

The second formula, for smaller slenderness ratios, corresponds to plastic yielding with a factor of safety of 5/3 for very short columns ($KL/r = 0$) and joins smoothly with the first formula, if the so-called critical slenderness ratio $(KL/r)_c$ is chosen properly.

$$\sigma_{allow} = \frac{\left[1 - \dfrac{(KL/r)^2}{2(KL/r)_c^2}\right]\sigma_Y}{\left[\dfrac{5}{3} + \dfrac{3}{8}\dfrac{(KL/r)}{(KL/r)_c} - \dfrac{1}{8}\dfrac{(KL/r)^3}{(KL/r)_c^3}\right]} \qquad (KL/r) < (KL/r)_c$$

$(KL/r)_c$ is found by setting the buckling stress equal to $\sigma_Y/2$.

$$\frac{\pi^2 E}{(KL/r)_c^2} = \frac{\sigma_Y}{2} \Rightarrow (KL/r)_c = \pi\sqrt{\frac{2E}{\sigma_Y}}$$

This level for the buckling stress is rationalized as follows. There are already residual stresses due to manufacturing as high as $\sigma_Y/2$ in roll formed steel beams. With an applied stress of $\sigma_Y/2$, the total stress would be σ_Y.

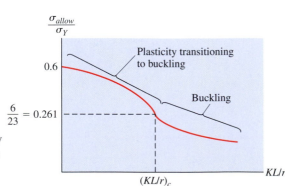

3. Recommended guidelines for aluminum columns consider both yielding and buckling.

The Aluminum Association specifies formulas for designing aluminum columns against buckling and plastic yielding. The formulas depend on the aluminum alloy and also reflect particular choices for factors of safety. For an alloy commonly used in building construction (2014-T6), the recommended formulas are:

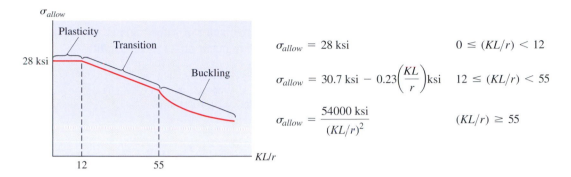

$$\sigma_{allow} = 28 \text{ ksi} \qquad 0 \le (KL/r) < 12$$

$$\sigma_{allow} = 30.7 \text{ ksi} - 0.23\left(\frac{KL}{r}\right)\text{ksi} \qquad 12 \le (KL/r) < 55$$

$$\sigma_{allow} = \frac{54000 \text{ ksi}}{(KL/r)^2} \qquad (KL/r) \ge 55$$

The formulas correspond to failure due to plastic yielding for short columns, failure due to buckling for long columns, and an interpolating formula for intermediate columns. The numerical coefficients (28 ksi, 30.7 ksi, etc.) and the values of KL/r that define the ranges (12 and 55) would change depending on the alloy and on the chosen factors of safety.

4. Recommended guidelines for wood columns consider both the compressive crushing strength parallel to the wood grain and buckling.

The American Forest and Paper Association publishes guidelines for designing structural wood members under axial compressive loading. In analogy with metals, the design formulas recognize that failure can occur due either to buckling or to axial crushing of wood fibers (analogous to yielding).

The formulas presented here are for columns of rectangular cross-section in particular, with the smaller cross-sectional dimension defined as d. The formulas correspond to: failure due to axial crushing for short columns, failure due to buckling for long columns, and an interpolating formula for intermediate columns.

$$\sigma_{allow} = F_c \qquad 0 \le (KL/d) < 11$$

$$\sigma_{allow} = F_c\left[1 - \frac{1}{3}\left(\frac{KL/d}{(KL/d)_c}\right)^4\right] \qquad 11 \le (KL/d) \le (KL/d)_c$$

$$\sigma_{allow} = \frac{0.3E}{(KL/d)^2} \qquad (KL/d) > (KL/d)_c$$

The formulas contain E and F_c, respectively the elastic modulus and the crushing stress (which reflects a factor of safety), which are chosen as appropriate to the wood and conditions of service. There is always a small discontinuity in σ_{allow} at the $KL/d = 11$. The transition value of the slender ratio, $(KL/d)_c$, is chosen so that σ_{allow} is continuous when $(KL/d) = (KL/d)_c$.

$$\frac{2F_c}{3} = \frac{0.3E}{(KL/d)_c^2} \Rightarrow (KL/d)_c = \sqrt{\frac{0.45E}{F_c}}$$

Typical values for E and F_c range over $1(10^6)$ psi to $1.8(10^6)$ psi and 700 psi to 2000 psi, respectively (or 7 GPa to 12 GPa, and 5 MPa to 14 MPa, respectively). This ratio $(KL/d)_c$ typically ranges over 18 to 30. The plot of σ_{allow} normalized by F_c here corresponds to a particular ratio $E/F_c = 1280$, and $(KL/d)_c = 24$.

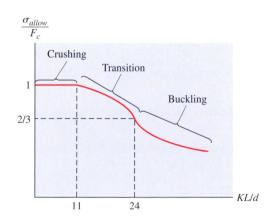

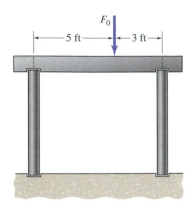

Two steel pipes which serve as posts are to be designed against buckling. The piping used has an outer diameter of 2.375 in. and a wall thickness of 0.109 in. At the top and bottom the pipes fit loosely into supports that prevent lateral motion, but permit rotation (pin supports). Determine the maximum allowable vertical force F_0 that can be tolerated if the factor of safety against buckling is to be 3. Show that yielding is highly unlikely to occur under these conditions.

Solution

The pipes are modeled as simply supported beams under axial loading. The critical buckling load is given by

$$P_{cr} = \frac{\pi^2 EI}{L^2}$$

The pipe has a circular cross-section with an outer diameter of 2.375 in. and a thickness of 0.109 in.

The outer radius is $\dfrac{2.375}{2}$ in. = 1.1875 in.

The inner radius is 1.1875 in. − 0.109 in. = 1.0785 in.

$$I = \frac{\pi}{4}\left[b^4 - a^4\right] = \frac{\pi}{4}\left[(1.1875 \text{ in.})^4 - (1.0785 \text{ in.})^4\right] = 0.499 \text{ in.}^4$$

The critical force for buckling of this pipe is found to be

$$P_{cr} = \frac{\pi^2 EI}{L^2} = \frac{\pi^2 (30 \times 10^6 \text{ psi})(0.499 \text{ in.}^4)}{[(18 \text{ ft})]^2} = 3170 \text{ lb}$$

For a factor of safety of 3, the maximum load on each pipe should not exceed

$$\frac{3170 \text{ lb}}{3} = 1056 \text{ lb}$$

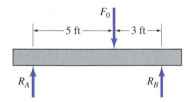

Under the loading shown, the forces in the two pipes are related to F_0 by

$$R_A = \frac{3}{(5 + 3)} F_0 = 0.375 F_0 \qquad R_B = \frac{5}{5 + 3} F_0 = 0.625 F_0$$

If $R_A < 1056$ lb, then $0.375 F_0 < 1056$ lb or $F_0 < \dfrac{1056 \text{ lb}}{0.375} = 2820 \text{ lb}$

If $R_B < 1056$ lb, then $0.625 F_0 < 1056$ lb or $F_0 < \dfrac{1056 \text{ lb}}{0.625} = 1690 \text{ lb}$

Since *neither* pipe may buckle, F_0 must be less than the *lower* of these two $(F_0)_{max} = 1690$ lb

The pipe area is $\pi[b^2 - a^2] = \pi[(1.1875 \text{ in.})^2 - (1.0785 \text{ in.})^2] = 0.776 \text{ in.}^2$

The maximum normal stress in the pipe is $\dfrac{1056 \text{ lb}}{0.776 \text{ in.}^2} = 1361$ psi, far less than the yield strength of any steel.

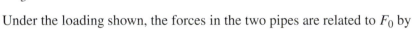

A 2 × 4 stud (finished cross-section: 1.5 in. × 3.5 in.) is 8 ft long and is modeled as simply supported. (a) Determine the critical load for buckling in the first mode in each of the two planes (take the end restraints to act as simple supports regardless of the plane buckling). (b) Assume a lateral restraint at the center point of the stud is added, which prevents deflection only in the softer direction. Determine the new buckling load assuming it goes into the second buckling mode shown and, by comparison with the result from (a), decide how it will buckle. Take Young's modulus to be $E = 1.7 \times 10^6$ psi.

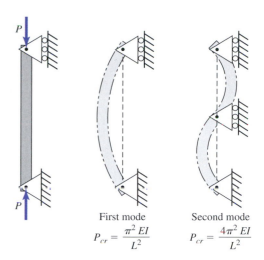

First mode
$$P_{cr} = \frac{\pi^2 EI}{L^2}$$

Second mode
$$P_{cr} = \frac{4\pi^2 EI}{L^2}$$

Solution

For buckling in the soft direction:

$$I = \frac{(3.5 \text{ in.})(1.5 \text{ in.})^3}{12} = 0.984 \text{ in.}^4$$

The critical load for buckling in the first mode in the soft direction is then:

$$P_{cr} = \frac{\pi^2 EI}{L^2} = \frac{\pi^2 (1.7 \times 10^6 \text{ psi})(0.984 \text{ in.}^4)}{(96 \text{ in.})^2} = 1792 \text{ lb}$$

For buckling in the stiff direction:

$$I = \frac{(1.5 \text{ in.})(3.5 \text{ in.})^3}{12} = 5.36 \text{ in.}^4$$

The critical load for buckling in the first mode in the stiff direction is

$$P_{cr} = \frac{\pi^2 EI}{L^2} = \frac{\pi^2 (1.7 \times 10^6 \text{ psi})(5.36 \text{ in.}^4)}{(96 \text{ in.})^2} = 9760 \text{ lb}$$

Consider now buckling in the second mode, with bending occurring in the soft direction ($I = 0.984$ in.^{4}):

$$P_{cr} = \frac{4\pi^2 EI}{L^2} = \frac{4\pi^2 (1.7 \times 10^6 \text{ psi})(0.984 \text{ in.}^4)}{(96 \text{ in.})^2} = 7170 \text{ lb}$$

Notice that this buckling load of 7170 lb is less than the load for first mode buckling in the stiff direction (9760 lb). So, the added lateral restraint at the midpoint compels the column to buckle in the second mode, in the softer direction ($I = 0.984$ in.^{4}).

>>End Example Problem 8.2

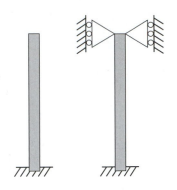

A wide-flange beam, designated as $W410 \times 85$, is built-in at its lower end. The beam must support a vertical force of 50 kN at its upper end without buckling. (a) Determine the maximum length which can be safely supported if the beam is free at the top. (b) In which direction should lateral motion be prevented at the top if it can be prevented in only one direction? What would the maximum allowable length be then? (See Appendix E.)

Solution

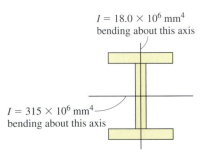

$I = 18.0 \times 10^6$ mm^4
bending about this axis

$I = 315 \times 10^6$ mm^4
bending about this axis

From Appendix E, the moments of inertia of wide-flange beam $W410 \times 85$ can be found. I is different for bending, and hence buckling, about each of the two axes.

Since the critical buckling load is proportional to I, the critical buckling load is less for the case with $I = 18.0 \times 10^6$ mm^4, assuming the supports are identical.

(a) For the case of a cantilevered-free column in general,

$$P_{cr} = \frac{\pi^2 EI}{4L^2}$$

Using $I = 18.0 \times 10^6$ mm^4 and solving for L we find:

$$L_{max} = \sqrt{\frac{\pi^2 EI}{4P_{cr}}} = \sqrt{\frac{\pi^2(200 \text{ GPa})(18 \times 10^{-6} \text{ m}^4)}{4(50000 \text{ N})}} = 13.32 \text{ m}$$

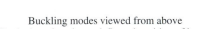

Buckling modes viewed from above
Dashed section shows deflected position of beam

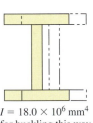
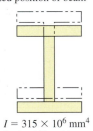

$I = 18.0 \times 10^6$ mm^4
for buckling this way

$I = 315 \times 10^6$ mm^4
for buckling this way

(b) If the column is to be prevented from deflecting at the top, it must be done to constrain the buckling mode corresponding to $I = 18.0 \times 10^6$ mm^4. If the column is so constrained, it becomes a cantilevered-pinned column, with critical load:

$$P_{cr} = \frac{2.046\pi^2 EI}{L^2}$$

The maximum length for this mode (with $I = 18.0 \times 10^6$ mm^4) is found to be:

$$L_{max} = \sqrt{\frac{2.046\pi^2 EI}{P_{cr}}} = \sqrt{\frac{2.046\pi^2(200 \text{ GPa})(18 \times 10^{-6} \text{ m}^4)}{(50000 \text{ N})}} = 38.13 \text{ m}$$

Since motion is unconstrained at the top in the other mode, we should check the maximum allowable length for this case also (with $I = 315 \times 10^6$ mm^4).

$$L_{max} = \sqrt{\frac{\pi^2 EI}{4P_{cr}}} = \sqrt{\frac{\pi^2(200 \text{ GPa})(315 \times 10^{-6} \text{ m}^4)}{4(50000 \text{ N})}} = 55.76 \text{ m}$$

We want neither buckling mode to occur, so we must take the minimum of the two: $L_{max} = 38.13$ m.

>>End Example Problem 8.3

Steel bracing members are formed using the cross-shaped cross-section shown. Assume the brace is 8 m long and is pinned at its ends. (a) Determine the radius of gyration for this cross-section and the slenderness ratio. (b) Determine the axial stress at which buckling occurs. (c) If the yield stress is 250 MPa, then what is the factor of safety against yielding when the buckling stress is reached?

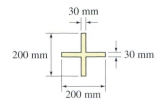

30 mm

200 mm

30 mm

200 mm

Solution

The second moment of inertia for the cross-section is:

$$I = \frac{(30 \text{ mm})(200 \text{ mm})^3}{12} + \frac{(200 \text{ mm} - 30 \text{ mm})(30 \text{ mm})^3}{12} = 2.038 \times 10^7 \text{ mm}^4$$

The area of the cross-section is:

$$A = (30 \text{ mm})(200 \text{ mm}) + (170 \text{ mm})(30 \text{ mm}) = 1.110 \times 10^4 \text{ mm}^2$$

The radius of gyration is:

$$r = \sqrt{\frac{I}{A}} = \sqrt{\frac{2.038 \times 10^7 \text{ mm}^4}{1.11 \times 10^4 \text{ mm}^2}} = 42.9 \text{ mm}$$

The slenderness ratio is:

$$L/r = \frac{8 \text{ m}}{42.9 \text{ mm}} = 187$$

The buckling stress is:

$$\sigma_{cr} = \frac{\pi^2 E}{(L/r)^2} = \frac{\pi^2 (200 \text{ GPa})}{(187)^2} = 56.6 \text{ MPa}$$

If the yield strength is 250 MPa, and if the maximum stress is kept below the buckling stress, then:

$$\text{Factor of safety against yielding} = \frac{250 \text{ MPa}}{56.6 \text{ MPa}} = 4.41$$

>>End Example Problem 8.4

Additional data on material properties needed to solve problems can be found in Appendix D or inside back cover.

8.1 A solid titanium implant 280 mm long has to withstand an axial compressive force of 1200 N without buckling. If the ends can pivot (and be modeled as simply supported), determine the minimum allowable diameter before buckling can occur.

Prob. 8.1

8.2 The steel post shown is cantilevered at its base and supports a weight at its upper end. Considering that buckling can occur about either plane, determine the maximum allowable weight it can support. The post length is $L = 4$ m, the other cross-sectional dimensions are $a = 120$ mm and $b = 250$ mm, and the uniform wall thickness is 15 mm.

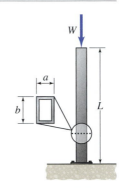

Prob. 8.2

8.3 Steel beams of wide-flange section W12 × 87 serve as a series of posts. Determine the maximum allowable distributed force per length that can be tolerated prior to buckling. Consider that buckling can occur in either transverse plane. Take the beams to be cantilevered at their base but pinned at the upper end (regardless of the plane of buckling). The spacing between posts is $w = 20$ ft, and the post length is $L = 14$ ft. (See Appendix E.)

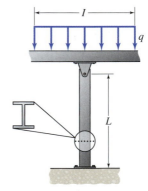

Prob. 8.3

8.4 The shelf shown is supported by two aluminum struts, one on each side. The struts are $h = 20$ mm wide, but the thickness is to be determined. Say the shelf is to support a maximum load at the end of $F_0 = 2000$ N. Determine the minimum allowable thickness of each of the struts, if it can buckle in either plane. Take $L_1 = 400$ mm, $L_2 = 600$ mm, and $L_3 = 300$ mm.

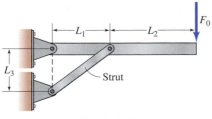

Prob. 8.4

8.5 A simple truss, which consists of steel members that are 0.125 in. × 0.75 in., must support a load of $F_0 = 300$ lb. Determine the maximum size of the truss (L) if the members are not to buckle either in or out of plane. Take $\theta_0 = 35°$.

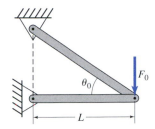

Prob. 8.5

8.6 The truss shown supporting the horizontal beam consists of three solid steel rods 14 mm in diameter. Determine the force per length q that can be supported without buckling of the members. Take $L_1 = 1.75$ m, $L_2 = 2.25$ m, $L_3 = 3.5$ m, and $L_4 = 3$ m.

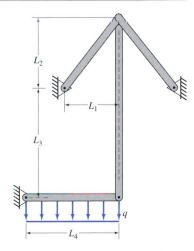

Prob. 8.6

8.7 The circular copper tube, 1300 mm long with an inner diameter of $d_i = 10$ mm and an outer diameter of $d_o = 12$ mm, is pinned at its ends to fixed bodies. At 25°C, there is no force in the bar. Determine the temperature at which the copper tube could buckle.

Prob. 8.7

8.8 The circular copper tube, $L_1 = 1300$ mm long with an inner diameter of $d_i = 10$ mm and an outer diameter of $d_o = 12$ mm, is pinned at each end to the center of diaphragms that are flexible. The diaphragms are copper strips that can pivot freely at their ends and have a width of 25 mm (into the paper), a thickness of 5.5 mm, and a length of $L_2 = 100$ mm. At 25°C, there is no force in the bar. Determine the temperature at which the copper tube could buckle.

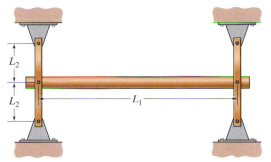

Prob. 8.8

8.9 A set of four wood members are lashed together at several points along the length with the goal of creating a bar with a larger cross-section that is less susceptible to buckling. Each member is 8 ft long and has a square cross-section of width $d = 0.875$ in. (so altogether the beam consisting of four members is 1.75 in. square). Using an elastic modulus of wood of $1.6(10^6)$ psi, and treating the members as simply supported, determine the buckling load under two assumptions: (a) the members are lashed together well enough that they function as a single solid beam and (b) the members bend individually, each taking one quarter of the load.

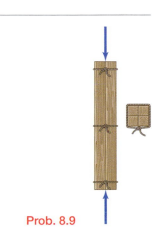

Prob. 8.9

8.10 A steel pipe with a 2 in. outer diameter and a wall thickness of 0.09 in. is built-in at its base and then stabilized with guy wires. If the resulting axial force in the pipe is to remain below 1/3 of the pipe's buckling load, determine the allowable tensions (assumed equal) in the two guy wires. Take $L_1 = 25$ ft and $L_2 = 20$ ft.

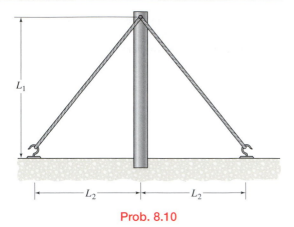

Prob. 8.10

8.11 An aluminum strut 30 in. in length has a solid 0.8 in. cross-section. Assume the end conditions of the strut can be modeled as simply supported. To reduce the rod weight by 25%, a rod with a hollow cross-section is considered. Determine (a) the critical buckling load of the hollow rod and (b) the percentage reduction in the critical load relative to the solid rod.

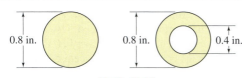

Prob. 8.11

8.12 A steel post is formed by welding two steel plates to a W360 × 64 wide-flange section. The plates are 15 mm thick, and their widths match the depth of the wide-flange section. If the post is 6 m long and modeled as simply supported, determine (a) the buckling load if A-A is the neutral plane of bending and (b) the buckling load if B-B is the neutral plane of bending. (See Appendix E.)

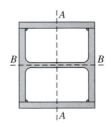

Prob. 8.12

8.13 A column is formed by nailing together three wood boards, each of which has a width to depth ratio of 1 to 4. Cross-sections of two arrangements (a) and (b) of the boards are shown. Determine the ratio of the critical buckling load for arrangement (a) to the critical buckling load for arrangement (b). Consider buckling in either plane for both arrangements.

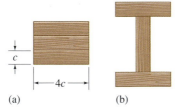

Prob. 8.13

8.14 The steel tube shown serves as a compression member. A hollow square cross-section with the same outer dimension, length, and material is to be used instead. (a) Determine the wall thickness, so that the square and circular cross-sections have the same weight per length (area). (b) Determine the ratio of the buckling strength of the square cross-section to that of the circular cross-section, assuming they have the same weight per length.

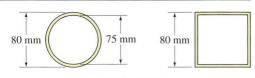

Prob. 8.14

8.15 The compression member with rectangular cross-section is supported as shown. The brackets and pins allow the member to pivot about the pin axis (deflections parallel to A-A), but prevent rotation against the brackets (deflections parallel to B-B). Determine the ratio a/b if the buckling load is to be the same for buckling in the two planes.

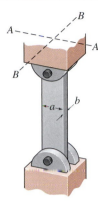

Prob. 8.15

8.16 The rigid horizontal member $ABCD$ is supported by two struts with ends that are modeled as pinned. The struts have the same cross-section, with area A, and moment of inertia I. Determine the load F_0 at which buckling in the plane first occurs in either strut. Take the length to be related by $L_1 = 1.5 L_2$. Specify the strut that buckles.

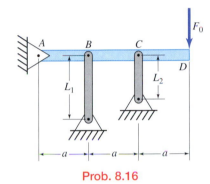

Prob. 8.16

8.17 A pipe is to be extracted from the ground via the cable that passes over a pulley which is pinned to the truss structure. The truss members are steel with hollow square cross-section that has a 3 in. outer dimension and a wall thickness of 0.125 in. Determine the maximum tension T that can be applied without either truss member buckling. Model the truss members as pinned at their ends, and assume the pulley can rotate freely about its pin.

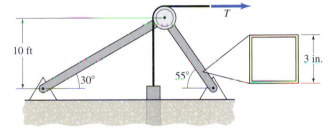

Prob. 8.17

8.18 A steel strip is built-in at one end and pinned at the other. The strip temperature is raised until it buckles. Determine the temperature rise ΔT_{crit} at which buckling occurs in terms of the given variables. Take $b < a$.

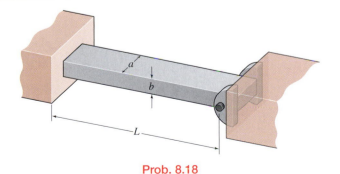

Prob. 8.18

8.19 The steel post *BC* supports the right end of the beam that has a uniform distributed force *q*. The post is a wide-flange section $W310 \times 74$, which is built-in at the base *C*, but free to pivot at *B*. Allowing for buckling in either plane, determine the critical distributed force *q* at which the post first buckles. Take $L_1 = 3$ m, $L_2 = 4$ m. (See Appendix E.)

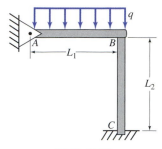

Prob. 8.19

8.20 A simply supported compression member consists of a standard structural steel cross-section. Because of the lower bending stiffness in one plane compared to the other for most standard cross-sections, such a member would tend to fail by buckling in the weaker direction. One strategy of raising the buckling load is to constrain the lateral motion at the center of the member—the beam then has a buckling load that corresponds to a member that is half as long. Say that the lateral motion is constrained only in the weaker direction. Consider how much this strategy raises the buckling load for two cross-sections: (a) a wide-flange section $W6 \times 25$ and (b) a standard channel section $C9 \times 15$. For each section, compute the ratio of the constrained buckling load to the unconstrained buckling load. Also, for each section, specify whether the plane of buckling changes when the constraint is added. (See Appendix E.)

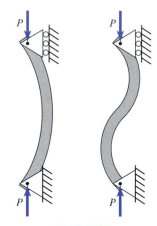

Prob. 8.20

Focused Application Problems

8.21 Each pylon of the cable stayed bridge has a cable that supports the deck and a backstay. Say the supporting cables carry 70% of the weight of the bridge which weighs 50 Mg. Even though the pylon is built-in at the bottom, the tensions in the backstays are chosen so that the net horizontal force on a pylon due to the two cables is zero. Say the steel pylon consists of two separate box sections, 800 mm $\times$ 1000 mm, 20 mm thick on the two sides of the deck that join at a certain point above the deck. Let the cables be attached to the pylons at height *h*, below which the two box sections are separated and equally share the loads of the cables. Determine the maximum allowable height, *h*, if the box sections are not to buckle in the weaker direction. Take the angles to be $\theta_1 = 30°$ and $\theta_2 = 15°$.

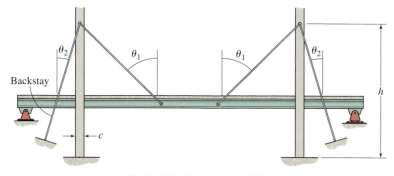

Prob. 8.21 (Appendix A2)

8.22 Because of the axial compression of the deck in a cable stayed bridge, buckling is a concern. In this simple bridge, each pylon carries one frontstay and one backstay as shown. There is no horizontal force between the pylon and the deck. Take all stays to have equal tensions. Let $I = 0.2$ m^4 for the reinforced concrete girder that forms the deck and $E = 40$ GPa. Determine the cable tension at which the segment of the girder between the cable anchorage points would buckle. Take the dimensions to be $w = 40$ m and $h = 60$ m.

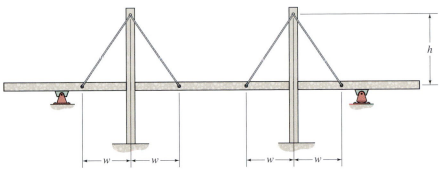

Prob. 8.22 (Appendix A2)

8.23 The problem of a column buckling simply under its own weight is important in a number of contexts. For example, to place enough force on a drill bit, relatively heavy drill collars are placed on the bottom of a drill string. However, if the drill collars are too heavy relative to their length, then the string can buckle under its own weight, something that is usually avoided. Even the problem of a uniform column buckling under its own weight is difficult to solve without more advanced methods. However, one can estimate the critical height at which a column of uniform properties buckles under its own weight as follows. Consider a column of mass density ρ, area A, and moment of inertia I. Let the column be fixed at its base but free at its top. Make the simplification, clearly only a rough approximation, that all the weight is applied at the top of the column. Show that the critical height, h_{crit1}, at which such a column would buckle is given by

Prob. 8.23 (Appendix A3)

$$h_{crit1} = \left[\frac{\pi^2 EI}{4\rho gA} \right]^{1/3}$$

Do you expect that the true critical length, found from treating the weight as uniformly distributed over the column length, to be greater or less than h_{crit1}? Explain.

8.24 The critical buckling length for a column under its own weight, with mass density ρ, area A, and moment of inertia I, is given by the formula

$$h_{crit} = 2.15 \left[\frac{EI}{\rho gA} \right]^{1/3}$$

Say this buckling mode limits the weight of collars that can be stacked on a drill bit in a drill string. Determine the maximum force that the weight of stacked collars can exert on a drill bit, without the string of collars buckling. Take the steel collars to have an outer diameter of 7 in. and an inner diameter of 2.81 in.

Chapter Summary

8.1 For many structures, the deformation becomes unstable if the load is too high. Such unstable deformations are termed **buckling**. For an axially compressed bar, also termed a column if buckling is of concern, the unstable deformation corresponds to bending.

One can compute the **buckling load**, the critical axial force at which buckling can occur P_{cr}.

For a simply supported column: $P_{cr} = \dfrac{\pi^2 EI}{L^2}$

The buckling load increases with bending stiffness EI and decreases with length L.

8.2 One can treat end conditions other than simple supports by viewing the member as having a longer or shorter **effective length** L_e.

$$P_{cr} = \frac{\pi^2 EI}{(KL)^2} = \frac{\pi^2 EI}{(L_e)^2}$$

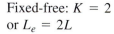

Fixed-free: $K = 2$
or $L_e = 2L$

Fixed-fixed: $K = 0.5$
or $L_e = 0.5L$

Fixed-pinned: $K = 0.7$
or $L_e = 0.7L$

8.3 To consider the possibility of plastic yielding as well as buckling, define a **buckling stress** in terms of the buckling load:

$$\sigma_{cr} = \frac{P_{cr}}{A} = \frac{\pi^2 EI}{(L_e)^2} \frac{1}{A}$$

Combine parameters I and A that describe the cross-section and define the radius of gyration r:

$$r = \sqrt{\frac{I}{A}}$$

Then, the buckling stress is found from:

$$\sigma_{cr} = \frac{\pi^2 E}{(L_e/r)^2}$$

To judge failure under axial compression, compare the stress σ with both the buckling stress σ_{cr} and the yield stress σ_Y. More detailed design guidelines for columns are available for specific materials.

>>End Chapter Summary

Focused Applications for Problems

CONTENTS

Bicycles

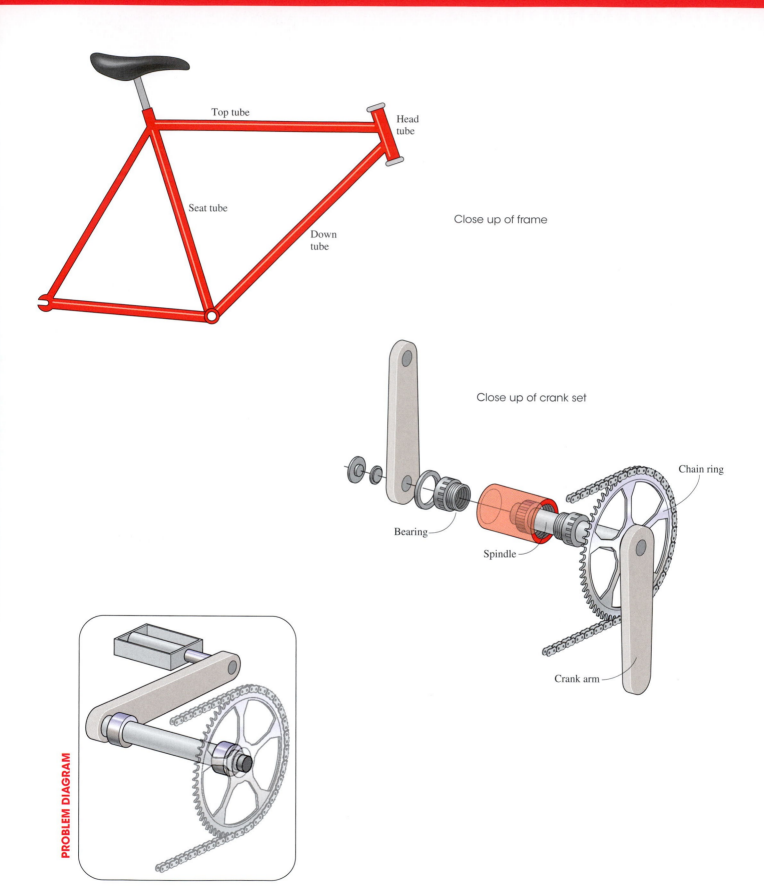

Top tube

Head tube

Close up of frame

Seat tube

Down tube

Close up of crank set

Chain ring

Bearing

Spindle

Crank arm

PROBLEM DIAGRAM

Close up of handle bars

Close up of fork

Cable-Stayed Bridges

In a cable-stayed bridge, the weight of the deck and roadway, which can be very long, is supported largely by tension cables that are attached to one or more pylons or towers.

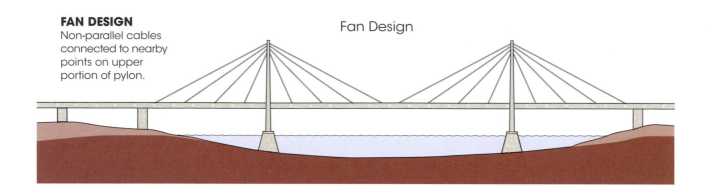

FAN DESIGN
Non-parallel cables connected to nearby points on upper portion of pylon.

Fan Design

Pylons can be of many different configurations. Here are three examples of configurations.

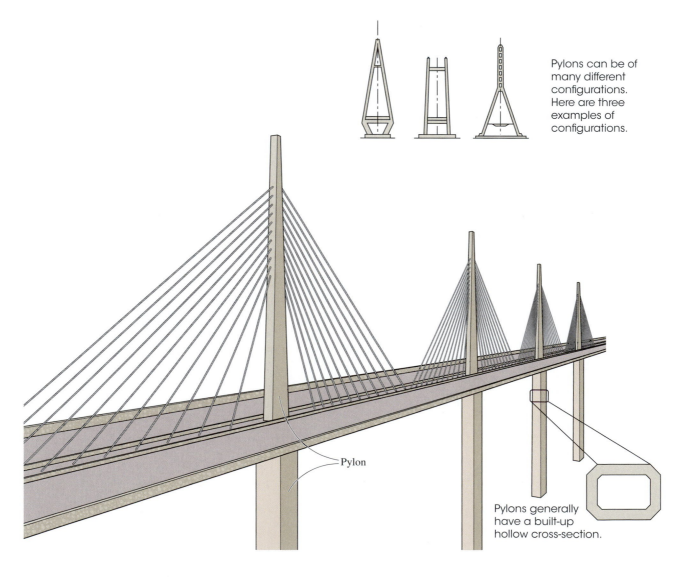

Pylon

Pylons generally have a built-up hollow cross-section.

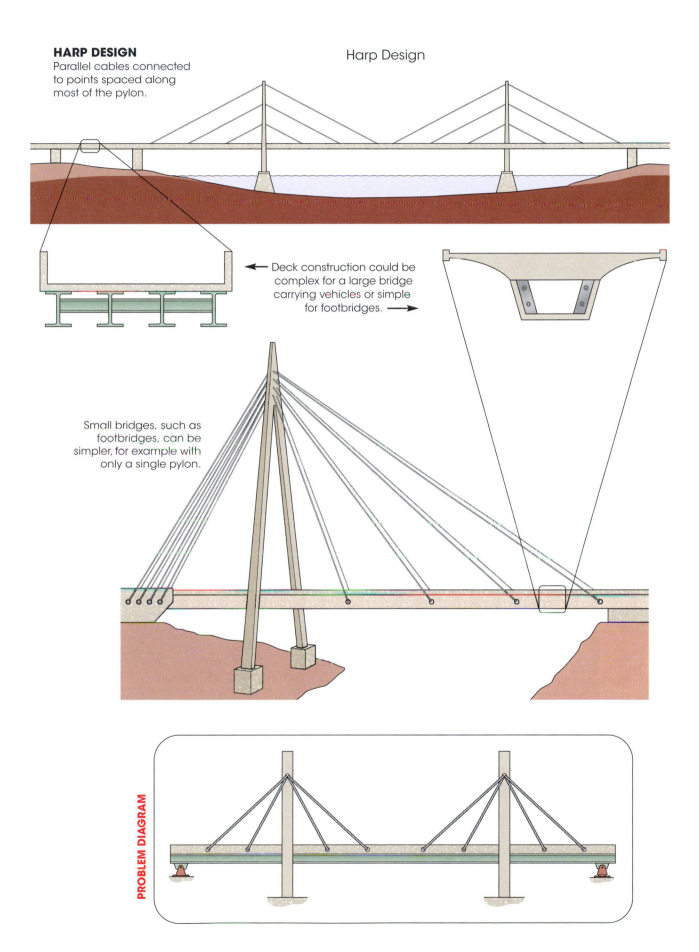

HARP DESIGN
Parallel cables connected to points spaced along most of the pylon.

Harp Design

← Deck construction could be complex for a large bridge carrying vehicles or simple for footbridges. →

Small bridges, such as footbridges, can be simpler, for example with only a single pylon.

PROBLEM DIAGRAM

Drilling

Drilling creates a deep hole from which oil or gas can be extracted. A drill string consists of a series of connected pipes and a bit. Drill strings can up to 10000 m (30000 ft) long.

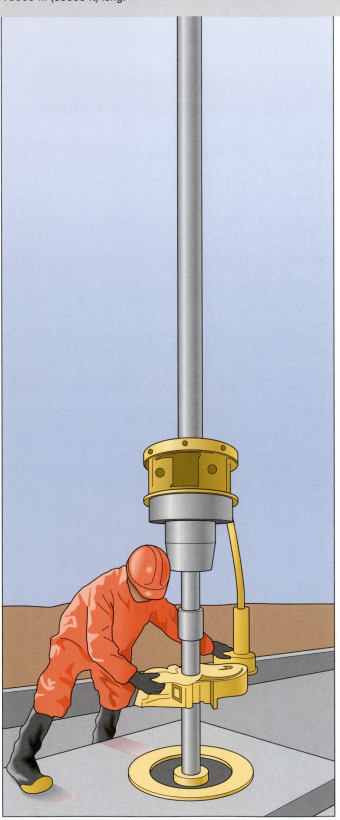

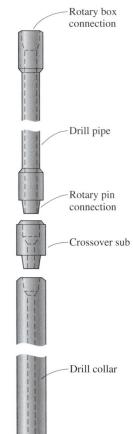

Rotary box connection

Each pipe has a box connection to mate with the pipe above and a pin connection at its lower end to mate with the pipe below.

Drill pipe

Rotary pin connection

Drill pipe occupies the upper portion of the drill string (~95% of total length).

Crossover sub

Collars are much heavier than drill pipe and occupy the lower portion of the drill string (~5% of total length).

Drill collar

Bit sub

Bit

Close up of connection between two drill pipes.

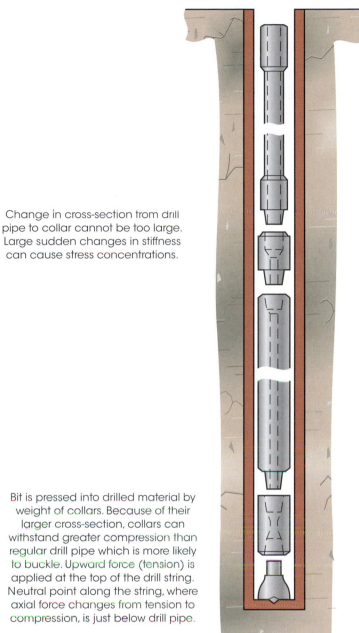

Mud is forced through the hollow pipe from top to bottom, and exits in gap between the pipe and ground, bringing drill debris up.

Gap between pipe and ground must be large enough so loss in mud pressure is small.

Change in cross-section from drill pipe to collar cannot be too large. Large sudden changes in stiffness can cause stress concentrations.

Torque applied during assembly of the drill string must produce sufficient initial compression between faces of adjacent pipe so no opening and leakage can occur (which could occur if pipe bends while passing through curved hole).

Bit is pressed into drilled material by weight of collars. Because of their larger cross-section, collars can withstand greater compression than regular drill pipe which is more likely to buckle. Upward force (tension) is applied at the top of the drill string. Neutral point along the string, where axial force changes from tension to compression, is just below drill pipe.

Main loads on a drill string: combination of torsion and axial loading (some additional bending).

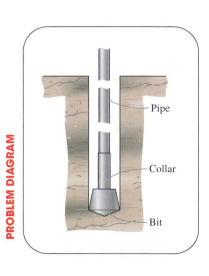

Pipe

Collar

Bit

PECTORAL FLY MACHINE

Cord disk

Cord disk

Cord disk

Pivot axis

Pivoting arm

Pivoting arm

Pivot axis

Swinging arm

Side peg

Handle

Weight stack

Exerciser's hand pulls handle causing swinging and pivoting arms and cord disk to rotate about pivot axis, which draws in cord and lifts weight stack.

PROBLEM DIAGRAM

Pivoting arm

Swinging arm

LEG CURL MACHINE

Front View

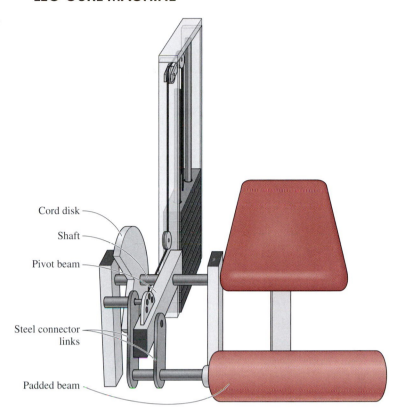

Cord disk

Shaft

Pivot beam

Steel connector links

Padded beam

Top View

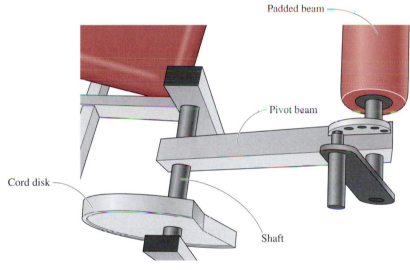

Padded beam

Pivot beam

Cord disk

Shaft

Exerciser's legs lift padded beam, which causes pivot beam, shaft, and cord disk to rotate about shaft axis, which draws in cord and lifts weight stack.

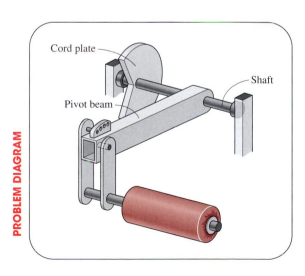

PROBLEM DIAGRAM

Cord plate

Shaft

Pivot beam

Fracture Fixation

To promote healing of bone fractures, we seek to:

- Bring fracture faces of two bone fragments closer together and in the correct orientations so they can heal (by physician).
- Prevent subsequent relative motion of bone fragments (by hardware).
- If possible, compress the fractured faces together (by hardware).

FRACTURE SURFACE COMPRESSION

Unthreaded portion of screw passes through one fragment and screws into other fragment. When tightened, the screw goes into tension and pulls two bone fragments together, compressing fractured faces together.

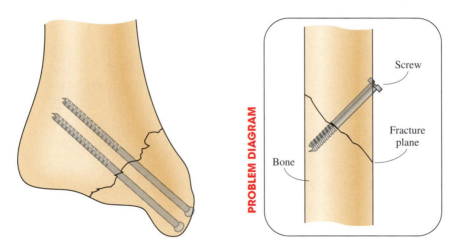

Plates with screws can be used to induce compression at the fracture surface. In the mechanism shown here, given the points where they initially penetrate the bone fragments, the screws cause the fragments to move toward one another as the screw heads seat into the plate.

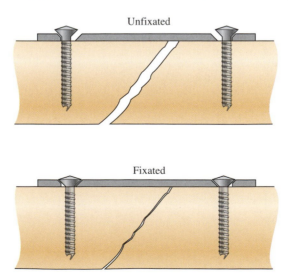

NO FRACTURE SURFACE COMPRESSION

Intramedullary Nail
Long rod (nail) runs parallel to bone length and occupies intramedullary space at center of bone. Rod spans fracture plane and is connected to each bone fragment by screws.

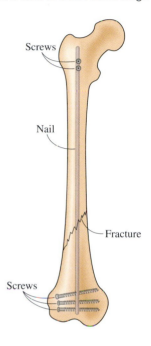

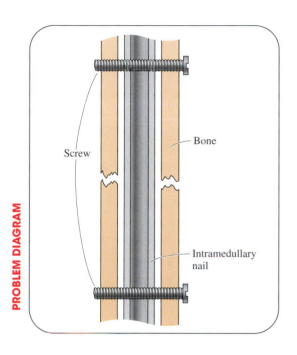

External Fixation System
Long threaded pins are inserted into bone fragments (on both sides of fracture) and the pins are rigidly mounted to an external rod.

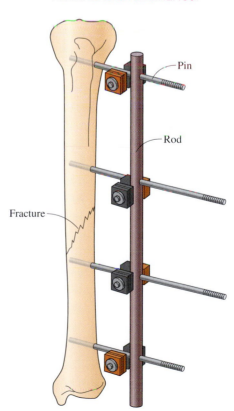

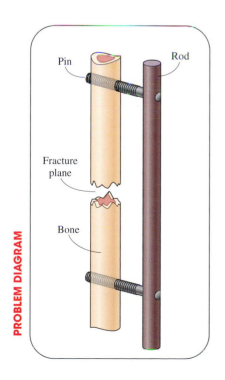

Wind Turbines

Winds cause airfoil-shaped blades of a wind turbine to rotate. A gearbox is used to convert the lower speed of the shaft from the blades to the higher shaft speed needed to drive the electric generator. The turbine is located on top of a tall tower, which must withstand significant loads.

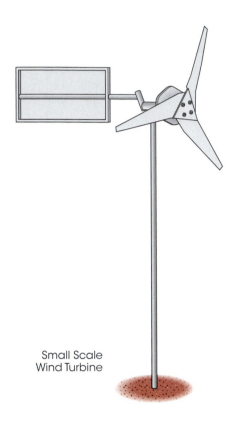

Small Scale
Wind Turbine

Large Scale
Wind Turbine

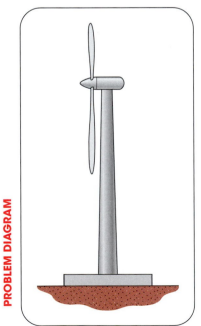

PROBLEM DIAGRAM

Wind turbine towers
typically have hollow
cross-sections.

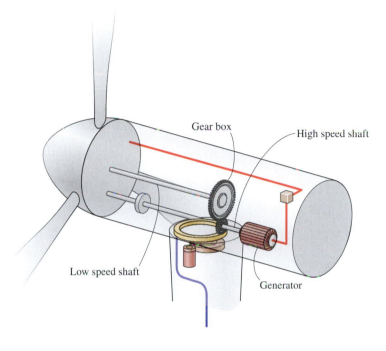

Close-up of Gear box and Generator

Gear box

High speed shaft

Low speed shaft

Generator

Gear box converts low-speed of shaft from blades to
higher speed needed to generate electricity.

Blades also have hollow
cross-section, but with
airfoil shape.

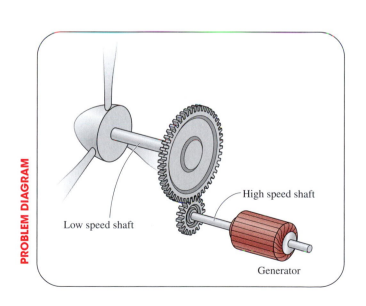

PROBLEM DIAGRAM

Low speed shaft

High speed shaft

Generator

Theory of Properties of Areas

B-1 Centroid and Second Moment of Inertia

Here we define and explain geometric properties of areas that are relevant to *Mechanics of Materials*. The properties defined are tabulated for simple shapes in Appendix C.

1. Define the centroid.

The centroid C corresponds to the geometric center of an area A. For simple shapes like rectangles and circles, C is at the center. In general, the x- and y-coordinates of the centroid, $\bar{x}$ and $\bar{y}$, are equal to values of the x- and y-coordinates *averaged* over the area A. Since the average is over an infinite number of points, it is found, in general, by integration.

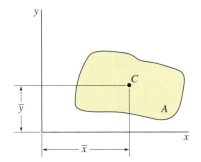

$$\bar{x} = \frac{1}{A} \int_A x \, dA$$

$$\bar{y} = \frac{1}{A} \int_A y \, dA$$

The centroid has a fixed position relative to the area. But the values $\bar{x}$ and $\bar{y}$ depend on the location of the area relative to the x-y origin.

2. Recognize that the centroid lies on a symmetry line, when present.

An area with a line of symmetry contains corresponding pairs of points, each of which has an average coordinate on the symmetry line. So $\bar{x}$ of the centroid is located on the symmetry line.

Whether an area is symmetric or not, if the x or y origin is located at the centroid, then the respective integral $\int_A x \, dA$ or $\int_A y \, dA$ is zero.

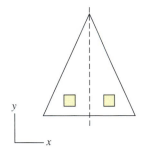

3. Define the second moment of inertia.

The second moment of inertia (sometimes referred to as a second moment of area) quantifies how spread out the area is from a given plane, often from planes that pass through the centroid. I_x captures the spread about the x plane or in the y-direction. I_y captures the spread about the y plane or in the x-direction. By integrating (averaging) the square of x (or y), the second moment of inertia captures the spread in both directions from the plane.

$$I_x = \int_A y^2 \, dA$$

$$I_y = \int_A x^2 \, dA$$

4. Observe the properties of the second moment of inertia.

As an example, here are the second moments of inertia of this ellipse about its center (centroid). Note the units are length to the fourth power.

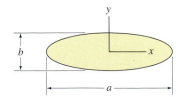

$$I_x = \frac{\pi a b^3}{4}$$

$$I_y = \frac{\pi a^3 b}{4}$$

Since $a > b$ in the case drawn, the ellipse is spread more in the x-direction than the y. Correspondingly, $I_y > I_x$ since the larger dimension, a, is cubed.

5. Distinguish area from the moment of inertia.

Two areas can be equal in area but have very different moments of inertia. The left figure has the same area as the right, but it has a much larger I_x and I_y since the area is spread from the center. The bending and twisting resistance of a rod with the left cross-section would be greater because its moments of inertia are greater.

6. Calculate the second moments of inertia about other axes using the parallel axis theorem.

Sometimes one knows the second moment of inertia about a plane through the centroid, but wants the second moment of inertia about a different parallel plane. Take the origin of the axes x'-y' to be centered at the centroid. Say we want the moment of inertia about different axes, x-y, located at distances d_x and d_y away.

$$I_x = \int_A (y' + d_y)^2 dA = \int_A y'^2 dA + 2d_y \int_A y' dA + d_y^2 \int_A dA$$

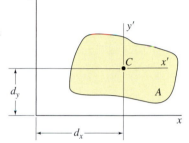

The integral over y' is zero because y' is at the centroid. The integral over y'^2 is designated as I_{cx} because it is the second moment of inertia about the centroid. The last integral equals simply A. A similar formula can be found for the shift from x' to x.

The pair of formulas for finding moments of inertia for axes away from the centroid is called the *Parallel Axis Theorem.*

$$I_x = I_{cx} + Ad_y^2 \qquad I_y = I_{cy} + Ad_x^2$$

7. Recognize composite areas and use their individual centroids and moments of inertia appropriately to find those of the whole area.

Sometimes an area is a combination, or composite, of simpler areas. If the centroids and moments of inertia of the simpler areas are known, then those of the composite can be found.

This figure  can be recognized as a combination of two rectangles, each having an obvious centroid. 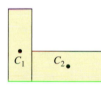 The centroid of the composite is the average of centroids, each weighted by its area.

$$\bar{x} = \frac{1}{A}\int_A x\,dA = \frac{1}{A_1 + A_2}\left[\int_{A_1} x\,dA + \int_{A_2} x\,dA\right] = \frac{A_1\bar{x}_1 + A_2\bar{x}_2}{A_1 + A_2}s \qquad \bar{y} = \frac{1}{A}\int_A y\,dA = \frac{A_1\bar{y}_1 + A_2\bar{y}_2}{A_1 + A_2}$$

Often one wants the moments of inertia about the centroid of the composite. It is the sum of the moments of inertia of the two areas.

$$I_x = \int_A y^2 dA = \int_{A_1} y^2 dA + \int_{A_2} y^2 dA = I_{x1} + I_{x2}$$

But, the moments of inertia tabulated for simple areas (e.g., in Appendix C) are about the simple area's centroid. We can use the parallel axis theorem to convert the moments of inertia so they are about the composite centroid C:

$$I_{x1} = I_{cx1} + A_1 d_{y1}^2 \qquad I_{x2} = I_{cx2} + A_2 d_{y2}^2 \qquad I_{y1} = I_{cy1} + A_1 d_{x1}^2 \qquad I_{y2} = I_{cy2} + A_2 d_{x2}^2$$

$$I_x = I_{cx1} + A_1 d_{y1}^2 + I_{cx2} + A_1 d_{y2}^2$$

$$I_y = I_{cy1} + A_1 d_{x1}^2 + I_{cy2} + A_1 d_{x2}^2$$

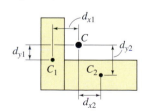

As seen in Example B-1.1, composite areas can have holes of a simple shape. The contributions of the holes are subtracted from the summations for the centroids and moments of inertia.

>>End B-1

B-2 Products of Inertia and Principal Axes of Inertia _____

Sometimes areas of interest are not symmetric about the *x*- or *y*-axis.
The most useful moments of inertia for such areas may not be those
about the *x-y* axes.

1. Define the product
of inertia.

Another moment of inertia that is useful to non-symmetric areas is the product of inertia, I_{xy}.

$$I_{xy} = \int_A xy\, dA$$

I_{xy} can be positive, negative, or zero.

Consider these ellipses, tilted at 45°

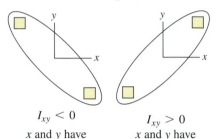

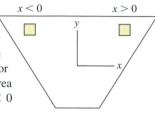

$I_{xy} = 0$ when a body has at least one plane of symmetry. Here, for every *y*, there is an area element for both $x < 0$ and $x > 0$.

$I_{xy} < 0$
x and *y* have
different signs.

$I_{xy} > 0$
x and *y* have
same signs.

This tilted ellipse is symmetric about axes *x'-y'* rotated by 45° from *x-y*, in which case $I_{x'y'} = 0$.

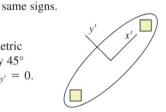

While many areas have no symmetry planes about any axes, they will always have $I_{x'y'} = 0$ about some axes *x'-y'*, as we next see.

The parallel axis theorem also applies to I_{xy}. Since I_{xy} can be > 0 or < 0, be careful with the signs of d_x and d_y when shifting from axes at centroid *x'-y'* to new axes *x-y*.

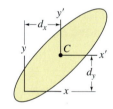

$$I_{xy} = I_{cxy} + A d_x d_y$$

2. Compute how the
moments of inertia change
for a general tilting of axes.

To study the effect of tilting axes on the moments of inertia, we define *x'-y'* as tilted by angle θ relative to the *x-y* axes.

$$x' = x \cos\theta + y \sin\theta \qquad y' = -x \sin\theta + y \cos\theta$$

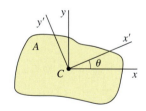

Define the moments of inertia about the new *x'-y'* axes.

$$I_{x'} = \int_A y'^2 dA = \int_A [-x \sin\theta + y \cos\theta]^2 dA \qquad I_{y'} = \int_A x'^2 dA = \int_A [x \cos\theta + y \sin\theta]^2 dA$$

$$I_{x'y'} = \int_A x'y'\, dA = \int_A [x \cos\theta + y \sin\theta][-x \sin\theta + y \cos\theta]\, dA$$

Simplify these expressions using trig identities: $\sin 2\theta = 2 \sin\theta \cos\theta$ and $\cos 2\theta = \cos^2\theta - \sin^2\theta$

$$I_{x'} = \frac{I_x + I_y}{2} + \frac{I_x - I_y}{2} \cos 2\theta - I_{xy} \sin 2\theta \qquad I_{y'} = \frac{I_x + I_y}{2} - \frac{I_x - I_y}{2} \cos 2\theta + I_{xy} \sin 2\theta$$

$$I_{x'y'} = \frac{I_x - I_y}{2} \sin 2\theta + I_{xy} \cos 2\theta$$

3. Recall moments of inertia for the tilted ellipse.

For the tilted ellipse, when x'-y' axes are parallel with the ellipse axes, $I_{x'y'} = 0$.

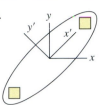

Inspect the figure and consider the spread about the x'- and y'-axes. The area appears to be maximally spread along x' ($I_{y'}$ maximum) and along the y'-axis, the area appears to minimally spread ($I_{x'}$ minimum). We next show this observation holds for all areas: the axes on which $I_{x'y'} = 0$ are those on which $I_{x'}$ and $I_{y'}$ are maximum or minimum.

---

4. Find the angle of the axes at which the moment of inertia is maximized.

Let $I_{x'}$ be maximized or minimized at $\theta = \theta_P$: $\left.\dfrac{\partial I_{x'}}{\partial \theta}\right|_{\theta=\theta_P} = -(I_x - I_y)\sin 2\theta_P - 2I_{xy}\cos 2\theta_P = 0$

The angle θ_P therefore satisfies:

$$\theta_P = \frac{1}{2}\arctan\left[\frac{-I_{xy}}{(I_x - I_y)/2}\right]$$

This triangle is consistent with the formula for $2\theta_P$.

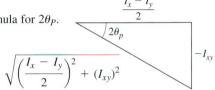

Use this triangle to evaluate $\sin 2\theta_P$ and $\cos 2\theta_P$.

$$\sin 2\theta_P = \frac{-I_{x'y'}}{\sqrt{\left(\dfrac{I_x - I_x}{2}\right)^2 + I_{xy}^2}} \qquad \cos 2\theta_P = \frac{\dfrac{I_x - I_x}{2}}{\sqrt{\left(\dfrac{I_x - I_x}{2}\right)^2 + I_{xy}^2}}$$

With these expressions for $\sin 2\theta_P$ and $\cos 2\theta_P$, we find $I_{x'y'} = 0$.

Likewise evaluate $I_{x'}$ at the angle θ_P and at $\theta_P + 90°$ to find the maximum and minimum $I_{x'}$.

$$I_{max} = \frac{I_x + I_y}{2} + \sqrt{\left(\frac{I_x - I_y}{2}\right)^2 + (I_{xy})^2} \qquad I_{min} = \frac{I_x + I_y}{2} - \sqrt{\left(\frac{I_x - I_y}{2}\right)^2 + (I_{xy})^2}$$

The x'-y' axes, which lie parallel and perpendicular to θ_p, are called the principal axes. The maximum and minimum I, I_{max} and I_{min}, are called the principal moments of inertia. On the principal axes, $I_{x'y'} = 0$.

---

5. Analyze a general cross-section as follows.

For a general cross-section, find I_x, I_y, and I_{xy}, treating it as a composite, if necessary.

From I_x, I_y, and I_{xy}, find the principal angle, θ_P, I_{max}, and I_{min}.

Try to determine which of the axes (at θ_p or $\theta_P + 90°$) corresponds to I_{max} and I_{min} by inspecting the spread of the shape. If it is not obvious by inspection, evaluate $I_{x'}$ using the formula in terms of $\sin 2\theta_P$ and $\cos 2\theta_P$. You should obtain either I_{max} or I_{min}.

>>End B-2

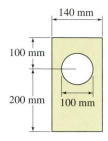

140 mm

100 mm

200 mm   100 mm

Determine the centroid of the figure shown and the moment of inertia about the horizontal axis through the centroid.

Solution

The figure can be viewed as the difference between two areas: a rectangle (1) and a circle (2).

Each individual area has its own centroid at its center. To determine the centroid of the composite, we define $\bar{y}_1$ and $\bar{y}_2$ as the positions of the centroids relative to a common baseline: the base of rectangle (1).

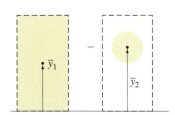

$$A_1 = (300 \text{ mm})(140 \text{ mm}) = 42(10^3) \text{ mm}^2 \quad \bar{y}_1 = 150 \text{ mm}$$

$$A_2 = \frac{\pi}{4}(100 \text{ mm})^2 = 3.927(10^3) \text{ mm}^2 \quad \bar{y}_2 = 200 \text{ mm}$$

Since the circle is a hole, its contribution is subtracted, both from the numerator and the denominator (which contains the true area).

$$\bar{y} = \frac{A_1\bar{y}_1 - A_2\bar{y}_2}{A_1 - A_2} = 138.5 \text{ mm}$$

The moments of inertia for the rectangle and the circle are tabulated about their centers. Since these do not coincide with the centroid of the whole figure $\bar{y}$, we need to apply the parallel axis theorem.

From the parallel axis theorem, the moments of inertia of the rectangle and circle about the centroid $\bar{y}$ are:

$$I_{x1} = I_{cx1} + A_1 d_{y1}^2 = \frac{1}{12}(140 \text{ mm})(300 \text{ mm})^3 + 42(10)^3 \text{ mm}^2(150 \text{ mm} - 138.5 \text{ mm})^2 = 321(10^6) \text{ mm}^4$$

$$I_{x2} = I_{cx2} + A_2 d_{y2}^2 = \frac{\pi}{4}(50 \text{ mm})^4 + \pi(50)^2 \text{ mm}^2(200 \text{ mm} - 138.5 \text{ mm})^2 = 39.5(10^6) \text{ mm}^4$$

Notice that the term Ad_y^2 is added for both the circle and the rectangle; the moments of inertia of each area about its own centroid is less than the moment of inertia about any point above or below.

Now recognize that the circle represents material that is removed (a circular *hole*). Its moment of inertia must be subtracted from the moment of inertia of the rectangle.

$$I_x = I_{x1} - I_{x2} = 321(10^6) \text{ mm}^4 - 39.5(10^6) \text{ mm}^4 = 281(10^6) \text{ mm}^4$$

>>End Example Problem B-1

Determine the centroid of the figure shown and the moment of inertia about the horizontal axis through the centroid.

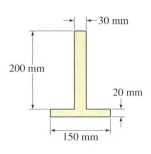

Solution

The figure can be viewed as adding two rectangles.

The individual areas have their own centroid, each at its center. To determine the centroid of the composite, we define $\bar{y}_1$ and $\bar{y}_2$ as the positions of the centroid relative to a common baseline: the base of rectangle (1).

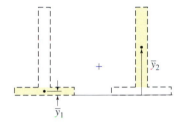

$$A_1 = (150 \text{ mm})(20 \text{ mm}) = 3(10^3) \text{ mm}^2 \qquad \bar{y}_1 = 10 \text{ mm}$$
$$A_2 = (30 \text{ mm})(200 \text{ mm}) = 6(10^3) \text{ mm}^2 \qquad \bar{y}_2 = 120 \text{ mm}$$

Find the centroid: $\quad \bar{y} = \dfrac{A_1\bar{y}_1 + A_2\bar{y}_2}{A_1 + A_2} = 83.3 \text{ mm}$

The second moments of inertia of the individual rectangular areas are tabulated about their centers, which do not coincide with the centroid of the whole figure $\bar{y}$.

From the parallel axis theorem, the moments of inertia of the rectangles about the centroid $\bar{y}$ are:

$$I_{x1} = I_{cx1} + A_1 d_{y1}^2 = \frac{1}{12}(150 \text{ mm})(20 \text{ mm})^3 + 3(10)^3 \text{ mm}^2(10 \text{ mm} - 83.3 \text{ mm})^2 = 16.23(10^6) \text{ mm}^4$$

$$I_{x2} = I_{cx2} + A_2 d_{y2}^2 = \frac{1}{12}(30 \text{ mm})(200 \text{ mm})^3 + 6(10)^3 \text{ mm}^2(120 \text{ mm} - 83.3 \text{ mm})^2 = 28.1(10^6) \text{ mm}^4$$

Add the moments of inertia to get that for the whole area:

$$I_x = I_{x1} + I_{x2} = 16.23(10^6) \text{ mm}^4 + 28.1(10^6) \text{ mm}^4 = 44.3(10^6) \text{ mm}^4$$

An alternative approach is to subtract two rectangular holes from a rectangle covering the entire region.

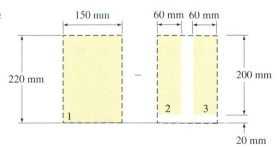

$$b_1 = (150 \text{ mm}) \qquad h_1 = (220 \text{ mm})$$
$$b_2 = b_3 = (60 \text{ mm}) \qquad h_2 = h_3 = (200 \text{ mm})$$

Notice the bottom of the holes are above the bottom of region 1, so:

$$\bar{y}_1 = 110 \text{ mm} \qquad \bar{y}_2 = \bar{y}_3 = 20 \text{ mm} + 100 \text{ mm} = 120 \text{ mm}$$

Once the parallel axis theorem is applied for the individual areas, this approach leads, of course, to the same $\bar{y}$ and I_x.

>>End Example Problem B-2

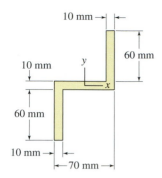

Determine the moments of inertia I_x and I_y of the figure shown.

Solution

Because the right and left legs are the same length, the figure obviously has its centroid at the center of the horizontal portion.

The figure is divided into areas as shown. (Alternatively, the vertical legs could be 60 mm + 10 mm = 70 mm, and then the horizontal piece 70 mm − 20 mm = 50 mm.)

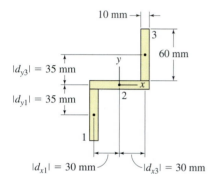

$$A_1 = (10 \text{ mm})(60 \text{ mm}) = 600 \text{ mm}^2$$
$$\bar{x}_1 = -30 \text{ mm} \qquad \bar{y}_1 = -35 \text{ mm}$$

$$A_2 = (70 \text{ mm})(10 \text{ mm}) = 700 \text{ mm}^2$$
$$\bar{x}_2 = 0 \qquad \bar{y}_2 = 0$$

$$A_3 = (10 \text{ mm})(60 \text{ mm}) = 600 \text{ mm}^2$$
$$\bar{x}_3 = 30 \text{ mm} \qquad \bar{y}_3 = 35 \text{ mm}$$

$$I_{cx1} = I_{cx3} = \frac{1}{12}(10 \text{ mm})(60 \text{ mm})^3 = 180(10^3) \text{ mm}^4$$

$$I_{cx2} = \frac{1}{12}(70 \text{ mm})(10 \text{ mm})^3 = 5.83(10^3) \text{ mm}^4$$

$$I_x = \left[I_{cx1} + A_1(d_{y1})^2\right] + \left[I_{cx2} + A_2(d_{y2})^2\right] + \left[I_{cx3} + A_3(d_{y3})^2\right]$$
$$I_x = \left[180(10^3) \text{ mm}^4 + 600 \text{ mm}^2(35 \text{ mm})^2\right] + \left[5.83(10^3) \text{ mm}^4 + 700 \text{ mm}^2(0)^2\right]$$
$$+ \left[180(10^3) \text{ mm}^4 + 600 \text{ mm}^2(35 \text{ mm})^2\right] = 1.836(10^6) \text{ mm}^4$$

$$I_{cy1} = I_{cy3} = \frac{1}{12}(60 \text{ mm})(10 \text{ mm})^3 = 5(10^3) \text{ mm}^4$$

$$I_{cy2} = \frac{1}{12}(10 \text{ mm})(70 \text{ mm})^3 = 286(10^5) \text{ mm}^4$$

$$I_y = \left[I_{cy1} + A_1(d_{x1})^2\right] + \left[I_{cy2} + A_2(d_{x2})^2\right] + \left[I_{cy3} + A_3(d_{x3})^2\right]$$
$$I_y = \left[5(10^3) \text{ mm}^4 + 600 \text{ mm}^2(30 \text{ mm})^2\right] + \left[286(10^5) \text{ mm}^4 + 700 \text{ mm}^2(0)^2\right]$$
$$+ \left[5(10^3) \text{ mm}^4 + 600 \text{ mm}^2(30 \text{ mm})^2\right] = 1.376(10^6) \text{ mm}^4$$

>>End Example Problem B-3

Determine the product of inertia of the figure shown and then the principal axes and the moments of inertia. Use the moments of inertia I_x and I_y found in Example Problem B-3.

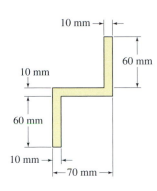

Solution

The figure is divided into areas as shown. (Alternatively, the vertical legs could be 60 mm + 10 mm = 70 mm, and then the horizontal piece 70 mm − 20 mm = 50 mm.)

Since each of the rectangles individually is symmetric with respect to be x- and y-axes, the product of inertia of each (about its own centroid) is zero.

$$I_{cxy1} = I_{cxy2} = I_{cxy3} = 0$$

Apply the parallel axis theorem for each area to find I_{xy} of the figure.

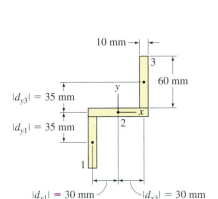

$$I_{xy} = \left[I_{cxy1} + A_1(d_{x1})(d_{y1})\right] + \left[I_{cxy2} + A_2(d_{x2})(d_{y2})\right] + \left[I_{cxy3} + A_3(d_{x3})(d_{y3})\right]$$

$$I_{xy} = \left[0 + 600 \text{ mm}^2(-30 \text{ mm})(-35 \text{ mm})\right] + \left[0 + 700 \text{ mm}^2(0)(0)\right]$$

$$+ \left[0 + 600 \text{ mm}^2(30 \text{ mm})(35 \text{ mm})\right] = 1.260(10^6) \text{ mm}^4$$

Determine the angle of the principal axes.

$$\theta_P = \frac{1}{2}\arctan\left[\frac{-I_{xy}}{(I_x - I_y)/2}\right] = \frac{1}{2}\arctan\left[\frac{-1.260 \text{ mm}^4}{(1.836 \text{ mm}^4 - 1.376 \text{ mm}^4)/2}\right] = -39.83°$$

Compute the principal moments of inertia:

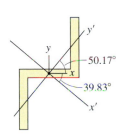

$$I_{max} = \frac{I_x + I_y}{2} + \sqrt{\left(\frac{I_x - I_y}{2}\right)^2 + (I_{xy})^2} = 2.887(10^6) \text{ mm}^4$$

$$I_{min} = \frac{I_x + I_y}{2} - \sqrt{\left(\frac{I_x - I_y}{2}\right)^2 + (I_{xy})^2} = 0.325(10^6) \text{ mm}^4$$

Since the figure is longer in the y' direction, one expects $I_{x'}$ to correspond to I_{max} and $I_{y'}$ to I_{min}. But, this can be verified by finding $I_{x'}(\theta_p)$.

$$I_{x'} = \frac{I_x + I_y}{2} + \frac{I_x - I_y}{2}\cos 2\theta_p - I_{xy}\sin 2\theta_p = \frac{1.836(10^6) \text{ mm}^4 + 1.376(10^6) \text{ mm}^4}{2}$$

$$+ \frac{1.836(10^6) \text{ mm}^4 - 1.376(10^6) \text{ mm}^4}{2}\cos 2(39.83°) - 1.26(10^6) \text{ mm}^4 \sin 2(39.83°) = 2.887(10^6) \text{ mm}^4$$

>>End Example Problem B-4

Tabulated Properties of Areas

Notation: A = area

$\bar{x}, \bar{y}$ = distances to centroid C

I_x, I_y = moments of inertia with respect to the x- and y-axes, respectively

I_{xy} = product of inertia with respect to the x- and y-axes

$I_p = I_x + I_y$ = polar moment of inertia with respect to the origin of the x- and y-axes

I_{BB} = moment of inertia with respect to axis B-B

1.

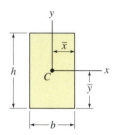

Rectangle   (Origin of axes at centroid)

$$A = bh \qquad \bar{x} = \frac{b}{2} \qquad \bar{y} = \frac{h}{2}$$

$$I_x = \frac{bh^3}{12} \qquad I_y = \frac{hb^3}{12} \qquad I_{xy} = 0 \qquad I_p = \frac{bh}{12}(h^2 + b^2)$$

2.

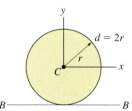

Circle   (Origin of axes at center)

$$A = \pi r^2 = \frac{\pi d^2}{4} \qquad I_x = I_y = \frac{\pi r^4}{4} = \frac{\pi d^4}{64}$$

$$I_{xy} = 0 \qquad I_p = \frac{\pi r^4}{2} = \frac{\pi d^4}{32} \qquad I_{BB} = \frac{5\pi r^4}{4} = \frac{5\pi d^4}{64}$$

3.

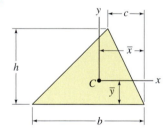

Triangle   (Origin of axes at centroid)

$$A = \frac{bh}{2} \qquad \bar{x} = \frac{b + c}{3} \qquad \bar{y} = \frac{h}{3}$$

$$I_x = \frac{bh^3}{36} \qquad I_y = \frac{bh}{36}(b^2 - bc + c^2)$$

$$I_{xy} = \frac{bh^2}{72}(b - 2c) \qquad I_p = \frac{bh}{36}(h^2 + b^2 - bc + c^2)$$

4.

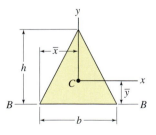

Isosceles triangle   (Origin of axes at centroid)

$$A = \frac{bh}{2} \qquad \bar{x} = \frac{b}{2} \qquad \bar{y} = \frac{h}{3}$$

$$I_x = \frac{bh^3}{36} \qquad I_y = \frac{hb^3}{48} \qquad I_{xy} = 0$$

$$I_p = \frac{bh}{144}(4h^2 + 3b^2) \qquad I_{BB} = \frac{bh^3}{12}$$

(*Note:* For an equilateral triangle, $h = \sqrt{3}\,b/2$.)

5.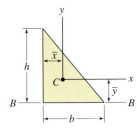

Right triangle (Origin of axes at centroid)

$$A = \frac{bh}{2} \qquad \bar{x} = \frac{b}{3} \qquad \bar{y} = \frac{h}{3}$$

$$I_x = \frac{bh^3}{36} \qquad I_y = \frac{hb^3}{36} \qquad I_{xy} = -\frac{b^2h^2}{72}$$

$$I_p = \frac{bh}{36}(h^2 + b^2) \qquad I_{BB} = \frac{bh^3}{12}$$

6.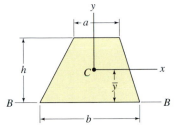

Trapezoid (Origin of axes at centroid)

$$A = \frac{h(a + b)}{2} \qquad \bar{y} - \frac{h(2a + b)}{3(a + b)}$$

$$I_x = \frac{h^3(a^2 + 4ab + b^2)}{36(a + b)} \qquad I_{BB} = \frac{h^3(3a + b)}{12}$$

7.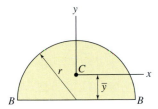

Semicircle (Origin of axes at centroid)

$$A = \frac{\pi r^2}{2} \qquad \bar{y} = \frac{4r}{3\pi}$$

$$I_x = \frac{(9\pi^2 - 64)r^4}{72\pi} \approx 0.1098r^4 \qquad I_y = \frac{\pi r^4}{8} \qquad I_{xy} = 0 \qquad I_{BB} = \frac{\pi r^4}{8}$$

8.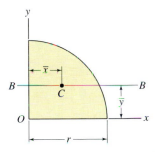

Quarter circle (Origin of axes at center of circle)

$$A = \frac{\pi r^2}{4} \qquad \bar{x} = \bar{y} = \frac{4r}{3\pi}$$

$$I_x = I_y = \frac{\pi r^4}{16} \qquad I_{xy} = \frac{r^4}{8} \qquad I_{BB} = \frac{(9\pi^2 - 64)r^4}{144\pi} \approx 0.05488r^4$$

9.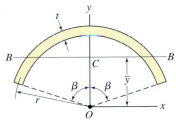

Thin circular arc (Origin of axes at center of circle)
Approximate formulas for case when t is small

β = angle in radians   (For a semicircular arc, $\beta = \pi/2$.)

$$A = 2\beta rt \qquad \bar{y} = \frac{r \sin \beta}{\beta}$$

$$I_x = r^3t(\beta + \sin \beta \cos \beta) \qquad I_y = r^3t(\beta - \sin \beta \cos \beta)$$

$$I_{xy} = 0 \qquad I_{BB} = r^3t\left(\frac{2\beta + \sin 2\beta}{2} - \frac{1 - \cos 2\beta}{\beta}\right)$$

10.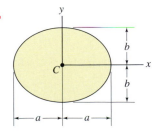

Ellipse (Origin of axes at centroid)

$$A = \pi ab \qquad I_x = \frac{\pi ab^3}{4} \qquad I_y = \frac{\pi ba^3}{4}$$

$$I_{xy} = 0 \qquad I_p = \frac{\pi ab}{4}(b^2 + a^2)$$

Material Properties

These tables contain typical values of material properties useful in *Mechanics of Materials*. Properties in these tables should be used when properties are needed for problems in this book, but are not specified. Each material listed refers to a whole class of materials. There can be wide variation even within a class. These tables also help the student see at a glance how distinct classes of materials differ from one another with respect to mechanical properties. For design purposes more exact values, which can depend sensitively on chemistry and processing conditions, should be sought from material suppliers.

International System (SI)

Material	Mass Density ρ (kg/m^3)	Young's Modulus E (GPa)	Shear Modulus G (GPa)	Poisson Ratio ν	Yield Stress σ_Y (MPa)	Thermal Expansion α (10^{-6}/°C)
Steel	7850	200	80	0.3	350	12
Aluminum	2700	70	27	0.33	270	23
Copper	8900	115	45	0.33	70	17
Glass	2600	70	27	0.2	—	9
Wood	600	12	—	—	—	—
Plastics	1000	2.5	1	0.4	40	100
Rubber	1100	0.002	0.0007	0.45	—	150

U.S. Customary System (USCS)

Material	Weight Density γ (lb/ft^3)	Young's Modulus E (10^6 psi)	Shear Modulus G (10^6 psi)	Poisson Ratio ν	Yield Stress σ_Y (ksi)	Thermal Expansion α (10^{-6}/°F)
Steel	490	30	11.5	0.3	50	6.5
Aluminum	170	10	3.8	0.33	40	13
Copper	550	17	6.5	0.33	10	9.5
Glass	170	10	4	0.2	—	5
Wood	35	1.7	—	—	—	—
Plastics	70	0.35	0.13	0.4	6	55
Rubber	75	0.003	0.001	0.45	—	80

Comments:

Yield stress given is a nominal value; for any particular material, yield stress depends on chemistry and processing. In the case of some materials yield stress is not well defined.

Wood: modulus is in bending; since wood is highly anisotropic, other properties are not tabulated.

Geometric Properties
of Structural Shapes

Wide-Flange Sections or W Shapes   USCS Units

Designation	Area A	Depth d	Web Thickness t_w	Flange		1-1 Axis			2-2 Axis		
				Width b_f	Thickness t_f	I	S	r	I	S	r
in. × lb/ft	in.^{2}	in.	in.	in.	in.	in.^{4}	in.^{3}	in.	in.^{4}	in.^{3}	in.
W24 × 104	30.6	24.06	0.500	12.750	0.750	3100	258	10.1	259	40.7	2.91
W24 × 94	27.7	24.31	0.515	9.065	0.875	2700	222	9.87	109	24.0	1.98
W24 × 84	24.7	24.10	0.470	9.020	0.770	2370	196	9.79	94.4	20.9	1.95
W24 × 76	22.4	23.92	0.440	8.990	0.680	2100	176	9.69	82.5	18.4	1.92
W24 × 68	20.1	23.73	0.415	8.965	0.585	1830	154	9.55	70.4	15.7	1.87
W24 × 62	18.2	23.74	0.430	7.040	0.590	1550	131	9.23	34.5	9.80	1.38
W24 × 55	16.2	23.57	0.395	7.005	0.505	1350	114	9.11	29.1	8.30	1.34
W18 × 71	20.8	18.47	0.495	7.635	0.810	1170	127	7.50	60.3	15.8	1.7
W18 × 65	19.1	18.35	0.450	7.590	0.750	1070	117	7.49	54.8	14.4	1.69
W18 × 60	17.6	18.24	0.415	7.555	0.695	984	108	7.47	50.1	13.3	1.69
W18 × 55	16.2	18.11	0.390	7.530	0.630	890	98.3	7.41	44.9	11.9	1.67
W18 × 50	14.7	17.99	0.355	7.495	0.570	800	88.9	7.38	40.1	10.7	1.65
W18 × 46	13.5	18.06	0.360	6.060	0.605	712	78.8	7.25	22.5	7.43	1.29
W18 × 40	11.8	17.90	0.315	6.015	0.525	612	68.4	7.21	19.1	6.35	1.27
W18 × 35	10.3	17.70	0.300	6.000	0.425	510	57.6	7.04	15.3	5.12	1.22
W16 × 100	29.4	16.97	0.585	10.425	0.985	1490	175	7.10	186	35.7	2.51
W16 × 77	22.6	16.52	0.455	10.295	0.760	1110	134	7.00	138	26.9	2.47
W16 × 57	16.8	16.43	0.430	7.120	0.715	758	92.2	6.72	43.1	12.1	1.60
W16 × 50	14.7	16.26	0.380	7.070	0.630	659	81.0	6.68	37.2	10.5	1.59
W16 × 45	13.3	16.13	0.345	7.035	0.565	586	72.7	6.65	32.8	9.34	1.57
W16 × 36	10.6	15.86	0.295	6.985	0.430	448	56.5	6.51	24.5	7.00	1.52
W16 × 31	9.12	15.88	0.275	5.525	0.440	375	47.2	6.41	12.4	4.49	1.17
W16 × 26	7.68	15.69	0.250	5.500	0.345	301	38.4	6.26	9.59	3.49	1.12
W14 × 53	15.6	13.92	0.370	8.060	0.660	541	77.8	5.89	57.7	14.3	1.92
W14 × 43	12.6	13.66	0.305	7.995	0.530	428	62.7	5.82	45.2	11.3	1.89
W14 × 38	11.2	14.10	0.310	6.770	0.515	385	54.6	5.87	26.7	7.88	1.55
W14 × 34	10.0	13.98	0.285	6.745	0.455	340	48.6	5.83	23.3	6.91	1.53
W14 × 30	8.85	13.84	0.270	6.730	0.385	291	42.0	5.73	19.6	5.82	1.49
W14 × 26	7.69	13.91	0.255	5.025	0.420	245	35.3	5.65	8.91	3.54	1.08
W14 × 22	6.49	13.74	0.230	5.000	0.335	199	29.0	5.54	7.00	2.80	1.04

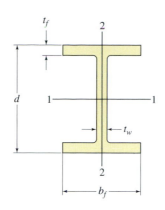

Wide-Flange Sections or W Shapes   USCS Units

Designation	Area A	Depth d	Web Thickness t_w	Flange		1-1 Axis			2-2 Axis		
				Width b_f	Thickness t_f	I	S	r	I	S	r
in. × lb/ft	in.^{2}	in.	in.	in.	in.	in.^{4}	in.^{3}	in.	in.^{4}	in.^{3}	in.
W12 × 87	25.6	12.53	0.515	12.125	0.810	740	118	5.38	241	39.7	3.07
W12 × 50	14.7	12.19	0.370	8.080	0.640	394	64.7	5.18	56.3	13.9	1.96
W12 × 45	13.2	12.06	0.335	8.045	0.575	350	58.1	5.15	50.0	12.4	1.94
W12 × 26	7.65	12.22	0.230	6.490	0.380	204	33.4	5.17	17.3	5.34	1.51
W12 × 22	6.48	12.31	0.260	4.030	0.425	156	25.4	4.91	4.66	2.31	0.847
W12 × 16	4.71	11.99	0.220	3.990	0.265	103	17.1	4.67	2.82	1.41	0.773
W12 × 14	4.16	11.91	0.200	3.970	0.225	88.6	14.9	4.62	2.36	1.19	0.753
W10 × 100	29.4	11.10	0.680	10.340	1.120	623	112	4.60	207	40.0	2.65
W10 × 54	15.8	10.09	0.370	10.030	0.615	303	60.0	4.37	103	20.6	2.56
W10 × 45	13.3	10.10	0.350	8.020	0.620	248	49.1	4.32	53.4	13.3	2.01
W10 × 39	11.5	9.92	0.315	7.985	0.530	209	42.1	4.27	45.0	11.3	1.98
W10 × 30	8.84	10.47	0.300	5.810	0.510	170	32.4	4.38	16.7	5.75	1.37
W10 × 19	5.62	10.24	0.250	4.020	0.395	96.3	18.8	4.14	4.29	2.14	0.874
W10 × 15	4.41	9.99	0.230	4.000	0.270	68.9	13.8	3.95	2.89	1.45	0.810
W10 × 12	3.54	9.87	0.190	3.960	0.210	53.8	10.9	3.90	2.18	1.10	0.785
W8 × 67	19.7	9.00	0.570	8.280	0.935	272	60.4	3.72	88.6	21.4	2.12
W8 × 58	17.1	8.75	0.510	8.220	0.810	228	52.0	3.65	75.1	18.3	2.10
W8 × 48	14.1	8.50	0.400	8.110	0.685	184	43.3	3.61	60.9	15.0	2.08
W8 × 40	11.7	8.25	0.360	8.070	0.560	146	35.5	3.53	49.1	12.2	2.04
W8 × 31	9.13	8.00	0.285	7.995	0.435	110	27.5	3.47	37.1	9.27	2.02
W8 × 24	7.08	7.93	0.245	6.495	0.400	82.8	20.9	3.42	18.3	5.63	1.61
W8 × 15	4.44	8.11	0.245	4.015	0.315	48.0	11.8	3.29	3.41	1.70	0.876
W6 × 25	7.34	6.38	0.320	6.080	0.455	53.4	16.7	2.70	17.1	5.61	1.52
W6 × 20	5.87	6.20	0.260	6.020	0.365	41.4	13.4	2.66	13.3	4.41	1.50
W6 × 16	4.74	6.28	0.260	4.030	0.405	32.1	10.2	2.60	4.43	2.20	0.966
W6 × 15	4.43	5.99	0.230	5.990	0.260	29.1	9.72	2.56	9.32	3.11	1.46
W6 × 12	3.55	6.03	0.230	4.000	0.280	22.1	7.31	2.49	2.99	1.50	0.918
W6 × 9	2.68	5.90	0.170	3.940	0.215	16.4	5.56	2.47	2.19	1.11	0.905

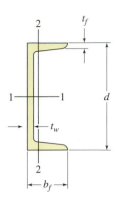

American Standard Channels or C Shapes   USCS Units

Designation	Area A	Depth d	Web Thickness t_w		Flange Width b_f		Flange Thickness t_f		1-1 Axis I	1-1 Axis S	1-1 Axis r	2-2 Axis I	2-2 Axis S	2-2 Axis r
in. × lb/ft	in.^{2}	in.	in.		in.		in.		in.^{4}	in.^{3}	in.	in.^{4}	in.^{3}	in.
C15 × 50	14.7	15.00	0.716	11/16	3.716	$3\frac{3}{4}$	0.650	5/8	404	53.8	5.24	11.0	3.78	0.867
C15 × 40	11.8	15.00	0.520	1/2	3.520	$3\frac{1}{2}$	0.650	5/8	349	46.5	5.44	9.23	3.37	0.886
C15 × 33.9	9.96	15.00	0.400	3/8	3.400	$3\frac{3}{8}$	0.650	5/8	315	42.0	5.62	8.13	3.11	0.904
C12 × 30	8.82	12.00	0.510	1/2	3.170	$3\frac{1}{8}$	0.501	1/2	162	27.0	4.29	5.14	2.06	0.763
C12 × 25	7.35	12.00	0.387	3/8	3.047	3	0.501	1/2	144	24.1	4.43	4.47	1.88	0.780
C12 × 20.7	6.09	12.00	0.282	5/16	2.942	3	0.501	1/2	129	21.5	4.61	3.88	1.73	0.799
C10 × 30	8.82	10.00	0.673	11/16	3.033	3	0.436	7/16	103	20.7	3.42	3.94	1.65	0.669
C10 × 25	7.35	10.00	0.526	1/2	2.886	$2\frac{7}{8}$	0.436	7/16	91.2	18.2	3.52	3.36	1.48	0.676
C10 × 20	5.88	10.00	0.379	3/8	2.739	$2\frac{3}{4}$	0.436	7/16	78.9	15.8	3.66	2.81	1.32	0.692
C10 × 15.3	4.49	10.00	0.240	1/4	2.600	$2\frac{5}{8}$	0.436	7/16	67.4	13.5	3.87	2.28	1.16	0.713
C9 × 20	5.88	9.00	0.448	7/16	2.648	$2\frac{5}{8}$	0.413	7/16	60.9	13.5	3.22	2.42	1.17	0.642
C9 × 15	4.41	9.00	0.285	5/16	2.485	$2\frac{1}{2}$	0.413	7/16	51.0	11.3	3.40	1.93	1.01	0.661
C9 × 13.4	3.94	9.00	0.233	1/4	2.433	$2\frac{3}{8}$	0.413	7/16	47.9	10.6	3.48	1.76	0.962	0.669
C8 × 18.75	5.51	8.00	0.487	1/2	2.527	$2\frac{1}{2}$	0.390	3/8	44.0	11.0	2.82	1.98	1.01	0.599
C8 × 13.75	4.04	8.00	0.303	5/16	2.343	$2\frac{3}{8}$	0.390	3/8	36.1	9.03	2.99	1.53	0.854	0.615
C8 × 11.5	3.98	8.00	0.220	1/4	2.260	$2\frac{1}{4}$	0.390	3/8	32.6	8.14	3.11	1.32	0.781	0.625
C7 × 14.75	4.33	7.00	0.419	1/2	2.299	$2\frac{1}{4}$	0.366	3/8	27.2	7.78	2.51	1.38	0.779	0.564
C7 × 12.25	3.60	7.00	0.314	5/16	2.194	$2\frac{1}{4}$	0.366	3/8	24:2	6.93	2.60	1.17	0.703	0.571
C7 × 9.8	2.87	7.00	0.210	3/16	2.090	$2\frac{1}{8}$	0.366	3/8	21.3	6.08	2.72	0.968	0.625	0.581
C6 × 13	3.83	6.00	0.437	7/16	2.157	$2\frac{1}{8}$	0.343	5/16	17.4	5.80	2.13	1.05	0.642	0.525
C6 × 10.5	3.09	6.00	0.314	5/16	2.034	2	0.343	5/16	15.2	5.06	2.22	0.866	0.564	0.529
C6 × 8.2	2.40	6.00	0.200	3/16	1.920	$1\frac{7}{8}$	0.343	5/16	13.1	4.38	2.34	0.693	0.492	0.537
C5 × 9	2.64	5.00	0.325	5/16	1.885	$1\frac{7}{8}$	0.320	5/16	8.90	3.56	1.83	0.632	0.450	0.489
C5 × 6.7	1.97	5.00	0.190	3/16	1.750	$1\frac{3}{4}$	0.320	5/16	7.49	3.00	1.95	0.479	0.378	0.493
C4 × 7.25	2.13	4.00	0.321	5/16	1.721	$1\frac{3}{4}$	0.296	5/16	4.59	2.29	1.47	0.433	0.343	0.450
C4 × 5.4	1.59	4.00	0.184	3/16	1.584	$1\frac{5}{8}$	0.296	5/16	3.85	1.93	1.56	0:319	0.283	0.449
C3 × 6	1.76	3.00	0.356	3/8	1.596	$1\frac{5}{8}$	0.273	1/4	2.07	1.38	1.08	0.305	0.268	0.416
C3 × 5	1.47	3.00	0.258	1/4	1.498	$1\frac{1}{2}$	0.273	1/4	1.85	1.24	1.12	0.247	0.233	0.410
C3 × 4.1	1.21	3.00	0.170	3/16	1.410	$1\frac{3}{8}$	0.273	1/4	1.66	1.10	1.17	0.197	0.202	0.404

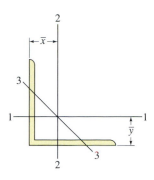

Angles Having Equal Legs   USCS Units

Size and Thickness	Weight per Foot	Area A	1-1 Axis				2-2 Axis				3-3 Axis
			I	S	r	$\bar{y}$	I	S	r	$\bar{x}$	r
in.	lb	in.^{2}	in.^{4}	in.^{3}	in.	in.	in.^{4}	in.^{3}	in.	in.	in.
L8 × 8 × 1	51.0	15.0	89.0	15.8	2.44	2.37	89.0	15.8	2.44	2.37	1.56
L8 × 8 × $\frac{3}{4}$	38.9	11.4	69.7	12.2	2.47	2.28	69.7	12.2	2.17	2.28	1.58
L8 × 8 × $\frac{1}{2}$	26.4	7.75	48.6	8.36	2.50	2.19	48.6	8.36	2.50	2.19	1.59
L6 × 6 × 1	37.4	11.0	35.5	8.57	1.80	1.86	35.5	8.57	1.80	1.86	1.17
L6 × 6 × $\frac{3}{4}$	28.7	8.44	28.2	6.66	1.83	1.78	28.2	6.66	1.83	1.78	1.17
L6 × 6 × $\frac{1}{2}$	19.6	5.75	19.9	4.61	1.86	1.68	19.9	4.61	1.86	1.68	1.18
L6 × 6 × $\frac{3}{8}$	14.9	4.36	15.4	3.53	1.88	1.64	15.4	3.53	1.88	1.64	1.19
L5 × 5 × $\frac{3}{4}$	23.6	6.94	15.7	4.53	1.51	1.52	15.7	4.53	1.51	1.52	0.975
L5 × 5 × $\frac{1}{2}$	16.2	4.75	11.3	3.16	1.54	1.43	11.3	3.16	1.54	1.43	0.983
L5 × 5 × $\frac{3}{8}$	12.3	3.61	8.74	2.42	1.56	1.39	8.74	2.42	1.56	1.39	0.990
L4 × 4 × $\frac{3}{4}$	18.5	5.44	7.67	2.81	1.19	1.27	7.67	2.81	1.19	1.27	0.778
L4 × 4 × $\frac{1}{2}$	12.8	3.75	5.56	1.97	1.22	1.18	5.56	1.97	1.22	1.18	0.782
L4 × 4 × $\frac{3}{8}$	9.8	2.86	4.36	1.52	1.23	1.14	4.36	1.52	1.23	1.14	0.788
L4 × 4 × $\frac{1}{4}$	6.6	1.94	3.04	1.05	1.25	1.09	3.04	1.05	1.25	1.09	0.795
L3$\frac{1}{2}$ × 3$\frac{1}{2}$ × $\frac{1}{2}$	11.1	3.25	3.64	1.49	1.06	1.06	3.64	1.49	1.06	1.06	0.683
L3$\frac{1}{2}$ × 3$\frac{1}{2}$ × $\frac{1}{2}$	8.5	2.48	2.87	1.15	1.07	1.01	2.87	1.15	1.07	1.01	0.687
L3$\frac{1}{2}$ × 3$\frac{1}{2}$ × $\frac{1}{4}$	5.8	1.69	2.01	0.794	1.09	0.968	2.01	0.794	1.09	0.968	0.694
L3 × 3 × $\frac{1}{2}$	9.4	2.75	2.22	1.07	0.898	0.932	2.22	1.07	0.898	0.932	0.584
L3 × 3 × $\frac{3}{8}$	7.2	2.11	1.76	0.833	0.913	0.888	1.76	0.833	0.913	0.888	0.587
L3 × 3 × $\frac{1}{4}$	4.9	1.44	1.24	0.577	0.930	0.842	1.24	0.577	0.930	0.842	0.592
L2$\frac{1}{2}$ × 2$\frac{1}{2}$ × $\frac{1}{2}$	7.7	2.25	1.23	0.724	0.739	0.806	1.23	0.724	0.739	0.806	0.487
L2$\frac{1}{2}$ × 2$\frac{1}{2}$ × $\frac{3}{8}$	5.9	1.73	0.984	0.566	0.753	0.762	0.984	0.566	0.753	0.762	0.487
L2$\frac{1}{2}$ × 2$\frac{1}{2}$ × $\frac{1}{4}$	4.1	1.19	0.703	0.394	0.769	0.717	0.703	0.394	0.769	0.717	0.491
L2 × 2 × $\frac{3}{8}$	4.7	1.36	0.479	0.351	0.594	0.636	0.479	0.351	0.594	0.636	0.389
L2 × 2 × $\frac{1}{4}$	3.19	0.938	0.348	0.247	0.609	0.592	0.348	0.247	0.609	0.592	0.391
L2 × 2 × $\frac{1}{8}$	1.65	0.484	0.190	0.131	0.626	0.546	0.190	0.131	0.626	0.546	0.398

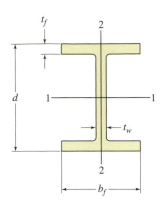

Wide-Flange Sections or W Shapes   SI Units

Designation	Area A	Depth d	Web Thickness t_w	Flange Width b_f	Flange Thickness t_f	1-1 Axis I	1-1 Axis S	1-1 Axis r	2-2 Axis I	2-2 Axis S	2-2 Axis r
mm × kg/m	mm²	mm	mm	mm	mm	10⁶ mm⁴	10³ mm³	mm	10⁶ mm⁴	10³ mm³	mm
W610 × 155	19 800	611	12.70	324.0	19.0	1 290	4 220	255	108	667	73.9
W610 × 140	17 900	617	13.10	230.0	22.2	1 120	3 630	250	45.1	392	50.2
W610 × 125	15 900	612	11.90	229.0	19.6	985	3 220	249	39.3	343	49.7
W610 × 113	14 400	608	11.20	228.0	17.3	875	2 880	247	34.3	301	48.8
W610 × 101	12 900	603	10.50	228.0	14.9	764	2 530	243	29.5	259	47.8
W610 × 92	11 800	603	10.90	179.0	15.0	646	2 140	234	14.4	161	34.9
W610 × 82	10 500	599	10.00	178.0	12.8	560	1 870	231	12.1	136	33.9
W460 × 97	12 300	466	11.40	193.0	19.0	445	1 910	190	22.8	236	43.1
W460 × 89	11 400	463	10.50	192.0	17.7	410	1 770	190	20.9	218	42.8
W460 × 82	10 400	460	9.91	191.0	16.0	370	1 610	189	18.6	195	42.3
W460 × 74	9 460	457	9.02	190.0	14.5	333	1 460	188	16.6	175	41.9
W460 × 68	8 730	459	9.14	154.0	15.4	297	1 290	184	9.41	122	32.8
W460 × 60	7 590	455	8.00	153.0	13.3	255	1 120	183	7.96	104	32.4
W460 × 52	6 640	450	7.62	152.0	10.8	212	942	179	6.34	83.4	30.9
W410 × 85	10 800	417	10.90	181.0	18.2	315	1 510	171	18.0	199	40.8
W410 × 74	9 510	413	9.65	180.0	16.0	275	1 330	170	15.6	173	40.5
W410 × 67	8 560	410	8.76	179.0	14.4	245	1 200	169	13.8	154	40.2
W410 × 53	6 820	403	7.49	177.0	10.9	186	923	165	10.1	114	38.5
W410 × 46	5 890	403	6.99	140.0	11.2	156	774	163	5.14	73.4	29.5
W410 × 39	4 960	399	6.35	140.0	8.8	126	632	159	4.02	57.4	28.5
W360 × 79	10 100	354	9.40	205.0	16.8	227	1 280	150	24.2	236	48.9
W360 × 64	8 150	347	7.75	203.0	13.5	179	1 030	148	18.8	185	48.0
W360 × 57	7 200	358	7.87	172.0	13.1	160	894	149	11.1	129	39.3
W360 × 51	6 450	355	7.24	171.0	11.6	141	794	148	9.68	113	38.7
W360 × 45	5 710	352	6.86	171.0	9.8	121	688	146	8.16	95.4	37.8
W360 × 39	4 960	353	6.48	128.0	10.7	102	578	143	3.75	58.6	27.5
W360 × 33	4 190	349	5.84	127.0	8.5	82.9	475	141	2.91	45.8	26.4

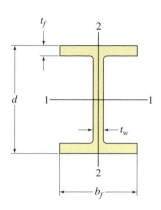

Wide-Flange Sections or W Shapes   SI Units

| | | | Web Thickness | Flange | | 1-1 Axis | | | 2-2 Axis | | |
Designation	Area A	Depth d	t_w	Width b_f	Thickness t_f	I	S	r	I	S	r
mm × kg/m	mm²	mm	mm	mm	mm	10^6 mm⁴	10^3 mm³	mm	10^6 mm⁴	10^3 mm³	mm
W310 × 129	16 500	318	13.10	308.0	20.6	308	1940	137	100	649	77.8
W310 × 74	9 480	310	9.40	205.0	16.3	165	1060	132	23.4	228	49.7
W310 × 67	8 530	306	8.51	204.0	14.6	145	948	130	20.7	203	49.3
W310 × 39	4 930	310	5.84	165.0	9.7	84.8	547	131	7.23	87.6	38.3
W310 × 33	4 180	313	6.60	102.0	10.8	65.0	415	125	1.92	37.6	21.4
W310 × 24	3 040	305	5.59	101.0	6.7	42.8	281	119	1.16	23.0	19.5
W310 × 21	2 680	303	5.08	101.0	5.7	37.0	244	117	0.986	19.5	19.2
W250 × 149	19 000	282	17.30	263.0	28.4	259	1840	117	86.2	656	67.4
W250 × 80	10 200	256	9.40	255.0	15.6	126	984	111	43.1	338	65.0
W250 × 67	8 560	257	8.89	204.0	15.7	104	809	110	22.2	218	50.9
W250 × 58	7 400	252	8.00	203.0	13.5	87.3	693	109	18.8	185	50.4
W250 × 45	5 700	266	7.62	148.0	13.0	71.1	535	112	7.03	95	35.1
W250 × 28	3 620	260	6.35	102.0	10.0	39.9	307	105	1.78	34.9	22.2
W250 × 22	2 850	254	5.84	102.0	6.9	28.8	227	101	1.22	23.9	20.7
W250 × 18	2 280	251	4.83	101.0	5.3	22.5	179	99.3	0.919	18.2	20.1
W200 × 100	12 700	229	14.50	210.0	23.7	113	987	94.3	36.6	349	53.7
W200 × 86	11 000	222	13.00	209.0	20.6	94.7	853	92.8	31.4	300	53.4
W200 × 71	9 100	216	10.20	206.0	17.4	76.6	709	91.7	25.4	247	52.8
W200 × 59	7 580	210	9.14	205.0	14.2	61.2	583	89.9	20.4	199	51.9
W200 × 46	5 890	203	7.24	203.0	11.0	45.5	448	87.9	15.3	151	51.0
W200 × 36	4 570	201	6.22	165.0	10.2	34.4	342	86.8	7.64	92.6	40.9
W200 × 22	2 860	206	6.22	102.0	8.0	20.0	194	83.6	1.42	27.8	22.3
W150 × 37	4 730	162	8.13	154.0	11.6	22.2	274	68.5	7.07	91.8	38.7
W150 × 30	3 790	157	6.60	153.0	9.3	17.1	218	67.2	5.54	72.4	38.2
W150 × 22	2 860	152	5.84	152.0	6.6	12.1	159	65.0	3.87	50.9	36.8
W150 × 24	3 060	160	6.60	102.0	10.3	13.4	168	66.2	1.83	35.9	24.5
W150 × 18	2 290	153	5.84	102.0	7.1	9.19	120	63.3	1.26	24.7	23.5
W150 × 14	1 730	150	4.32	100.0	5.5	6.84	91.2	62.9	0.912	18.2	23.0

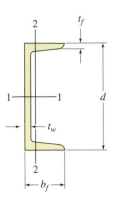

American Standard Channels or C Shapes   SI Units

Designation	Area A	Depth d	Web Thickness t_w	Flange Width b_f	Flange Thickness t_f	1-1 Axis I	1-1 Axis S	1-1 Axis r	2-2 Axis I	2-2 Axis S	2-2 Axis r
mm × kg/m	mm²	mm	mm	mm	mm	10⁶ mm⁴	10³ mm³	mm	10⁶ mm⁴	10³ mm³	mm
C380 × 74	9 480	381.0	18.20	94.4	16.50	168	882	133	4.58	61.8	22.0
C380 × 60	7 610	381.0	13.20	89.4	16.50	145	761	138	3.84	55.1	22.5
C380 × 50	6 430	381.0	10.20	86.4	16.50	131	688	143	3.38	50.9	22.9
C310 × 45	5 690	305.0	13.00	80.5	12.70	67.4	442	109	2.14	33.8	19.4
C310 × 37	4 740	305.0	9.83	77.4	12.70	59.9	393	112	1.86	30.9	19.8
C310 × 31	3 930	305.0	7.16	74.7	12.70	53.7	352	117	1.61	28.3	20.2
C250 × 45	5 690	254.0	17.10	77.0	11.10	42.9	338	86.8	1.61	27.1	17.0
C250 × 37	4 740	254.0	13.40	73.3	11.10	38.0	299	89.5	1.40	24.3	17.2
C250 × 30	3 790	254.0	9.63	69.6	11.10	32.8	258	93.0	1.17	21.6	17.6
C250× 23	2 900	254.0	6.10	66.0	11.10	28.1	221	98.4	0.949	19.0	18.1
C230 × 30	3 790	229.0	11.40	67.3	10.50	25.3	221	81.7	1.01	19 2	16.3
C230 × 22	2 850	229.0	7.24	63.1	10.50	21.2	185	86.2	0.803	16.7	16.8
C230 × 20	2 540	229.0	5.92	61.8	10.50	19.9	174	88.5	0.733	15.8	17.0
C200 × 28	3 550	203.0	12.40	64.2	9.90	18.3	180	71.8	0.824	16.5	15.2
C200 × 20	2 610	203.0	7.70	59.5	9.90	15.0	148	75.8	0.637	14.0	15.6
C200 × 17	2 180	203.0	5.59	57.4	9.90	13.6	134	79.0	0.549	12.8	15.9
C180 × 22	2 790	178.0	10.60	58.4	9.30	11.3	127	63.6	0.574	12.8	14.3
C180 × 18	2 320	178.0	7.98	55.7	9.30	10.1	113	66.0	0.487	11.5	14.5
C180 × 15	1 850	178.0	5.33	53.1	9.30	8.87	99.7	69.2	0.403	10.2	14.8
C150 × 19	2 470	152.0	11.10	54.8	8.70	7.24	95.3	54.1	0.437	10.5	13.3
C150 × 16	1 990	152.0	7.98	51.7	8.70	6.33	83.3	56.4	0.360	9.22	13.5
C150 × 12	1 550	152.0	5.08	48.8	8.70	5.45	71.7	59.3	0.288	8.04	13.6
C130 × 13	1 700	127.0	8.25	47.9	8.10	3.70	58.3	46.7	0.263	7.35	12.4
C130 × 10	1 270	127.0	4.83	44.5	8.10	3.12	49.1	49.6	0.199	6.18	12.5
C100 × 11	1 370	102.0	8.15	43.7	7.50	1.91	37.5	37.3	0.180	5.62	11.5
C100 × 8	1 030	102.0	4.67	40.2	7.50	1.60	31.4	39.4	0.133	4.65	11.4
C75 × 9	1 140	76.2	9.04	40.5	6.90	0.862	22.6	27.5	0.127	4.39	10.6
C75 × 7	948	76.2	6.55	38.0	6.90	0.770	20.2	28.5	0.103	3.83	10.4
C75 × 6	781	76.2	4.32	35.8	6.90	0.691	18.1	29.8	0.082	3.32	10.2

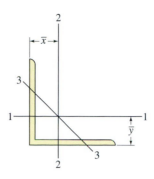

Angles Having Equal Legs   SI Units

| Size and Thickness | Mass per Meter | Area | 1-1 Axis | | | | 2-2 Axis | | | | 3-3 Axis |
| | | | I | S | r | y | I | S | r | x | r |
mm	kg	mm²	10^6 mm⁴	10^6 mm³	mm	mm	10^6 mm⁴	10^6 mm³	mm	mm	mm
L203 × 203 × 25.4	75.9	9 680	36.9	258	61.7	60.1	36.9	258	61.7	60.1	39.6
L203 × 203 × 19.0	57.9	7 380	28.9	199	62.6	57.8	28.9	199	62.6	57.8	40.1
L203 × 203 × 12.7	39.3	5 000	20.2	137	63.6	55.5	20.2	137	63.6	55.5	40.4
L152 × 152 × 25.4	55.7	7 100	14.6	139	45.3	47.2	14.6	139	45.3	47.2	29.7
L152 × 152 × 19.0	42.7	5 440	11.6	108	46.2	45.0	11.6	108	46.2	45.0	29.7
L152 × 152 × 12.7	29.2	3 710	8.22	75.1	47.1	42.7	8.22	75.1	47.1	42.7	30.0
L152 × 152 × 9.5	22.2	2 810	6.35	57.4	47.5	41.5	6.35	57.4	47.5	41.5	30.2
L127 × 127 × 19.0	35.1	4 480	6.54	73.9	38.2	38.7	6.54	73.9	38.2	38.7	24.8
L127 × 127 × 12.7	24.1	3 060	4.68	51.7	39.1	36.4	4.68	51.7	39.1	36.4	25.0
L127 × 127 × 9.5	18.3	2 330	3.64	39.7	39.5	35.3	3.64	39.7	39.5	35.3	25.1
L102 × 102 × 19.0	27.5	3 510	3.23	46.4	30.3	32.4	3.23	46.4	30.3	32.4	19.8
L102 × 102 × 12.7	19.0	2 420	2.34	32.6	31.1	30.2	2.34	32.6	31.1	30.2	19.9
L102 × 102 × 9.5	14.6	1 840	1.84	25.3	31.6	29.0	1.84	25.3	31.6	29.0	20.0
L102 × 102 × 6.4	9.8	1 250	1.28	17.3	32.0	27.9	1.28	17.3	32.0	27.9	20.2
L89 × 89 × 12.7	16.5	2 100	1.52	24.5	26.9	26.9	1.52	24.5	26.9	26.9	17.3
L89 × 89 × 9.5	12.6	1 600	1.20	19.0	27.4	25.8	1.20	19.0	27.4	25.8	17.4
L89 × 89 × 6.4	8.6	1 090	0.840	13.0	27.8	24.6	0.840	13.0	27.8	24.6	17.6
L76 × 76 × 12.7	14.0	1 770	0.915	17.5	22.7	23.6	0.915	17.5	22.7	23.6	14.8
L76 × 76 × 9.5	10.7	1 360	0.726	13.6	23.1	22.5	0.726	13.6	23.1	22.5	14.9
L76 × 76 × 6.4	7.3	927	0.514	9.39	23.5	21.3	0.514	9.39	23.5	21.3	15.0
L64 × 64 × 12.7	11.5	1 450	0.524	12.1	19.0	20.6	0.524	12.1	19.0	20.6	12.4
L64 × 64 × 9.5	8.8	1 120	0.420	9.46	19.4	19.5	0.420	9.46	19.4	19.5	12.4
L64 × 64 × 6.4	6.1	766	0.300	6.59	19.8	18.2	0.300	6.59	19.8	18.2	12.5
L51 × 51 × 9.5	7.0	877	0.202	5.82	15.2	16.2	0.202	5.82	15.2	16.2	9.88
L51 × 51 × 6.4	4.7	605	0.146	4.09	15.6	15.1	0.146	4.09	15.6	15.1	9.93
L51 × 51 × 3.2	2.5	312	0.080	2.16	16.0	13.9	0.080	2.16	16.0	13.9	10.1

Wood Structural Member Properties

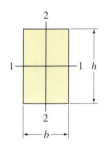

Properties of Surfaced Lumber

Nominal Dimensions $b \times h$	Net Dimensions $b \times h$	Area $A = bh$	Axis 1-1		Axis 2-2		Weight per Linear Foot (Weight Density = 35 lb/ft^3)
			Moment of Inertia $I_1 = \dfrac{bh^3}{12}$	Section Modulus $S_1 = \dfrac{bh^2}{6}$	Moment of Inertia $I_2 = \dfrac{hb^3}{12}$	Section Modulus $S_2 = \dfrac{hb^2}{6}$	
in.	in.	in.^{2}	in.^{4}	in.^{3}	in.^{4}	in.^{3}	lb
2 × 4	1.5 × 3.5	5.25	5.36	3.06	0.98	1.31	1.3
2 × 6	1.5 × 5.5	8.25	20.80	7.56	1.55	2.06	2.0
2 × 8	1.5 × 7.25	10.88	47.63	13.14	2.04	2.72	2.6
2 × 10	1.5 × 9.25	13.88	98.93	21.39	2.60	3.47	3.4
2 × 12	1.5 × 11.25	16.88	177.98	31.64	3.16	4.22	4.1
3 × 4	2.5 × 3.5	8.75	8.93	5.10	4.56	3.65	2.1
3 × 6	2.5 × 5.5	13.75	34.66	12.60	7.16	5.73	3.3
3 × 8	2.5 × 7.25	18.13	79.39	21.90	9.44	7.55	4.4
3 × 10	2.5 × 9.25	23.13	164.89	35.65	12.04	9.64	5.6
3 × 12	2.5 × 11.25	28.13	296.63	52.73	14.65	11.72	6.8
4 × 4	3.5 × 3.5	12.25	12.51	7.15	12.51	7.15	3.0
4 × 6	3.5 × 5.5	19.25	48.53	17.65	19.65	11.23	4.7
4 × 8	3.5 × 7.25	25.38	111.15	30.66	25.90	14.80	6.2
4 × 10	3.5 × 9.25	32.38	230.84	49.91	33.05	18.89	7.9
4 × 12	3.5 × 11.25	39.38	415.28	73.83	40.20	22.97	9.6
6 × 6	5.5 × 5.5	30.25	76.3	27.7	76.3	27.7	7.4
6 × 8	5.5 × 7.5	41.25	193.4	51.6	104.0	37.8	10.0
6 × 10	5.5 × 9.5	52.25	393.0	82.7	131.7	47.9	12.7
6 × 12	5.5 × 11.5	63.25	697.1	121.2	159.4	58.0	15.4
8 × 8	7.5 × 7.5	56.25	263.7	70.3	263.7	70.3	13.7
8 × 10	7.5 × 9.5	71.25	535.9	112.8	334.0	89.1	17.3
8 × 12	7.5 × 11.5	86.25	950.5	165.3	404.3	107.8	21.0

APPENDIX G
Tabulated Beam Deflections

G-1   Deflections and Slopes of Cantilever Beams

$v =$ deflection in the y direction (positive upward)
$v' = dv/dx =$ slope of the deflection curve
$\delta_B = -v(L) =$ deflection at end B of the beam (positive downward)
$\theta_B = -v'(L) =$ angle of rotation at end B of the beam (positive clockwise)
$EI =$ constant

G-1.1

$$v = -\frac{Px^2}{6EI}(3L - x) \qquad v' = -\frac{Px}{2EI}(2L - x)$$

$$\delta_B = \frac{PL^3}{3EI} \qquad \theta_B = \frac{PL^2}{2EI}$$

G-1.2

$$v = -\frac{M_0 x^2}{2EI} \qquad v' = -\frac{M_0 x}{EI}$$

$$\delta_B = \frac{M_0 L^2}{2EI} \qquad \theta_B = \frac{M_0 L}{EI}$$

G-1.3

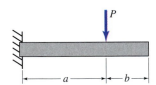

$$v = -\frac{Px^2}{6EI}(3a - x) \qquad v' = -\frac{Px}{2EI}(2a - x) \qquad (0 \le x \le a)$$

$$v = -\frac{Pa^2}{6EI}(3x - a) \qquad v' = -\frac{Pa^2}{2EI} \qquad (a \le x \le L)$$

$$\text{At } x = a: \qquad v = -\frac{Pa^3}{3EI} \qquad v' = -\frac{Pa^2}{2EI}$$

$$\delta_B = \frac{Pa^2}{6EI}(3L - a) \qquad \theta_B = \frac{Pa^2}{2EI}$$

G-1.4

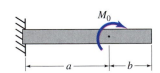

$$v = -\frac{M_0 x^2}{2EI} \qquad v' = -\frac{M_0 x}{EI} \qquad (0 \le x \le a)$$

$$v = -\frac{M_0 a}{2EI}(2x - a) \qquad v' = -\frac{M_0 a}{EI} \qquad (a \le x \le L)$$

$$\text{At } x = a: \qquad v = -\frac{M_0 a^2}{2EI} \qquad v' = -\frac{M_0 a}{EI}$$

$$\delta_B = \frac{M_0 a}{2EI}(2L - a) \qquad \theta_B = \frac{M_0 a}{EI}$$

G-1.5

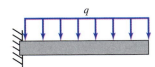

$$v = -\frac{qx^2}{24EI}(6L^2 - 4Lx + x^2) \qquad v' = -\frac{qx}{6EI}(3L^2 - 3Lx + x^2)$$

$$\delta_B = \frac{qL^4}{8EI} \qquad \theta_B = \frac{qL^3}{6EI}$$

G-1.6

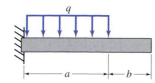

$$v = -\frac{qx^2}{24EI}(6a^2 - 4ax + x^2) \qquad (0 \le x \le a)$$

$$v' = -\frac{qx}{6EI}(3a^2 - 3ax + x^2) \qquad (0 \le x \le a)$$

$$v = -\frac{qa^3}{24EI}(4x - a) \qquad v' = -\frac{qa^3}{6EI} \qquad (a \le x \le L)$$

At $x = a$: $\quad v = -\frac{qa^4}{8EI} \qquad v' = -\frac{qa^3}{6EI}$

$$\delta_B = \frac{qa^3}{24EI}(4L - a) \qquad \theta_B = \frac{qa^3}{6EI}$$

G-1.7

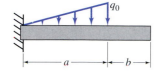

$$v(x) = \frac{q_0 x^2}{120EI}\left[10ax - 20a^2 - x^3/a\right] \qquad (0 < x < a)$$

$$v'(x) = \frac{q_0 x}{24EI}\left[6ax - 8a^2 - x^3/a\right] \qquad (0 < x < a)$$

$$v(x) = \frac{-q_0 a^3}{120EI}\left[\frac{11}{120}a + \frac{1}{8}(x - a)\right] \qquad v'(x) = \frac{-q_0 a^3}{8EI} \qquad (a < x < L)$$

At $x = a$: $\quad v = \frac{-11q_0 a^4}{120EI} \qquad v' = \frac{-q_0 a^3}{8EI}$

$$\delta_B = \frac{q_0 a^3}{EI}\left[\frac{11}{120}a + \frac{1}{8}(L - a)\right] \qquad \theta_B = \frac{q_0 a^3}{8EI}$$

G-2   Deflections and Slopes of Simply Supported Beams

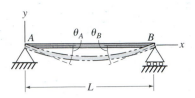

v = deflection in the y direction (positive upward)
$v' = dv/dx$ = slope of the deflection curve
$\delta_C = -v(L/2)$ = deflection at midpoint of the beam (positive downward)
x_1 = distance from support A to point of maximum deflection
$\delta_{max} = -v_{max}$ = maximum deflection (positive downward)
$\theta_A = -v'(0)$ = angle of rotation at left-hand end of the beam (positive clockwise)
$\theta_B = v'(L)$ = angle of rotation at right-hand end of the beam (positive counterclockwise)
EI = constant

G-2.1

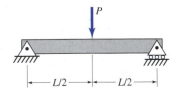

$$v = -\frac{Px}{48EI}(3L^2 - 4x^2) \qquad v' = -\frac{P}{16EI}(L^2 - 4x^2) \qquad \left(0 \le x \le \frac{L}{2}\right)$$

$$\delta_C = \delta_{max} = \frac{PL^3}{48EI} \qquad \theta_A = \theta_B = \frac{PL^2}{16EI}$$

G-2.2

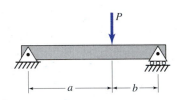

$$v = -\frac{Pbx}{6LEI}(L^2 - b^2 - x^2) \qquad v' = -\frac{Pb}{6LEI}(L^2 - b^2 - 3x^2) \qquad (0 \le x \le a)$$

$$\theta_A = \frac{Pab(L + b)}{6LEI} \qquad \theta_B = \frac{Pab(L + a)}{6LEI}$$

If $a \ge b$, $\delta_C = \dfrac{Pb(3L^2 - 4b^2)}{48EI}$ If $a \le b$, $\delta_C = \dfrac{Pa(3L^2 - 4a^2)}{48EI}$

If $a \ge b$, $x_1 = \sqrt{\dfrac{L^2 - b^2}{3}}$ and $\delta_{max} = \dfrac{Pb(L^2 - b^2)^{3/2}}{9\sqrt{3}LEI}$

If $a < b$, $x_1 = L - \sqrt{\dfrac{L^2 - a^2}{3}}$ and $\delta_{max} = \dfrac{Pa(L^2 - a^2)^{3/2}}{9\sqrt{3}LEI}$

G-2.3

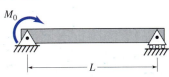

$$v = -\frac{M_0 x}{6LEI}(2L^2 - 3Lx + x^2) \qquad v' = -\frac{M_0}{6LEI}(2L^2 - 6Lx + 3x^2)$$

$$\delta_C = \frac{M_0 L^2}{16EI} \qquad \theta_A = \frac{M_0 L}{3EI} \qquad \theta_B = \frac{M_0 L}{6EI}$$

$$x_1 = L\left(1 - \frac{\sqrt{3}}{3}\right) \quad \text{and} \quad \delta_{max} = \frac{M_0 L^2}{9\sqrt{3}EI}$$

G-2.4

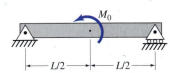

$$v = -\frac{M_0 x}{24LEI}(L^2 - 4x^2) \qquad v' = -\frac{M_0}{24LEI}(L^2 - 12x^2) \qquad \left(0 \le x \le \frac{L}{2}\right)$$

$$\delta_C = 0 \qquad \theta_A = \frac{M_0 L}{24EI} \qquad \theta_B = -\frac{M_0 L}{24EI}$$

G-2.5

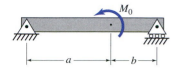

$$v = -\frac{M_0 x}{6LEI}(6aL - 3a^2 - 2L^2 - x^2) \qquad (0 \le x \le a)$$

$$v' = -\frac{M_0}{6LEI}(6aL - 3a^2 - 2L^2 - 3x^2) \qquad (0 \le x \le a)$$

At $x = a$: $\qquad v = -\frac{M_0 ab}{3LEI}(2a - L) \qquad v' = -\frac{M_0}{3LEI}(3aL - 3a^2 - L^2)$

$$\theta_A = \frac{M_0}{6LEI}(6aL - 3a^2 - 2L^2) \qquad \theta_B = \frac{M_0}{6LEI}(3a^2 - L^2)$$

G-2.6

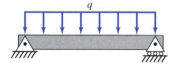

$$v = -\frac{qx}{24EI}(L^3 - 2Lx^2 + x^3)$$

$$v' = -\frac{q}{24EI}(L^3 - 6Lx^2 - 4x^3)$$

$$\delta_C = \delta_{max} = \frac{5qL^4}{384EI} \qquad \theta_A = \theta_B = \frac{qL^3}{24EI}$$

G-2.7

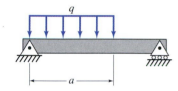

$$v = -\frac{qx}{24LEI}(a^4 - 4a^3L + 4a^2L^2 + 2a^2x^2 - 4aLx^2 + Lx^3) \qquad (0 \le x \le a)$$

$$v' = -\frac{q}{24LEI}(a^4 - 4a^3L + 4a^2L^2 + 6a^2x^2 - 12aLx^2 + 4Lx^3) \qquad (0 \le x \le a)$$

$$v = -\frac{qa^2}{24LEI}(-a^2L + 4L^2x + a^2x - 6Lx^2 + 2x^3) \qquad (a \le x \le L)$$

$$v' = -\frac{qa^2}{24LEI}(4L^2 + a^2 - 12Lx + 6x^2) \qquad (a \le x \le L)$$

$$\theta_A = \frac{qa^2}{24LEI}(2L - a)^2 \qquad \theta_B = \frac{qa^2}{24LEI}(2L^2 - a^2)$$

G-2.8

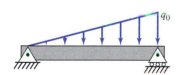

$$v = -\frac{q_0 x}{360LEI}(7L^4 - 10L^2x^2 + 3x^4)$$

$$v' = -\frac{q_0}{360LEI}(7L^4 - 30L^2x^2 + 15x^4)$$

$$\delta_C = \frac{5q_0L^4}{768EI} \qquad \theta_A = \frac{7q_0L^3}{360EI} \qquad \theta_B = \frac{q_0L^3}{45EI}$$

$$x_1 = 0.5193L \qquad \delta_{max} = 0.00652\frac{q_0L^4}{EI}$$

Stress Concentration Factors

H.1

Axial Loading of Flat Bars with Shoulder Fillets

$$\sigma_{nom} = \frac{P}{dt}$$

$$\sigma_{max} = K\sigma_{nom}$$

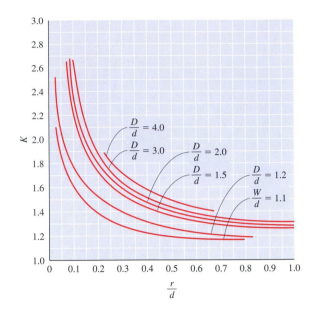

H.2

Axial Loading of Flat Bars with Circular Holes

$$\sigma_{nom} = \frac{P}{(D - 2r)t}$$

$$\sigma_{max} = K\sigma_{nom}$$

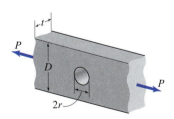

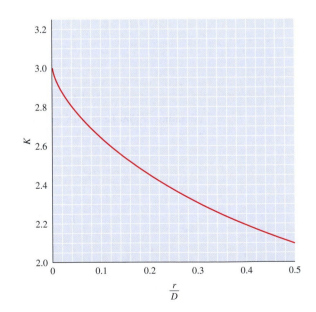

H.3

Torsional Loading of Circular Bars with Shoulder Fillets

$$\tau_{nom} = \frac{16T}{\pi d^3}$$

$$\tau_{max} = K\tau_{nom}$$

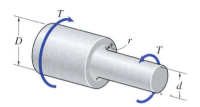

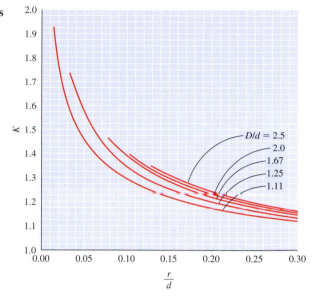

H.4

Bending of Flat Bars with Shoulder Fillets

$$\sigma_{nom} = \frac{6M}{td^2}$$

$$\sigma_{max} = K\sigma_{nom}$$

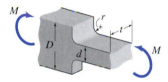

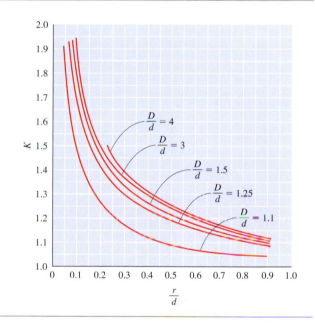

H.5

Bending of Flat Bars with Notches

$$\sigma_{nom} = \frac{6M}{td^2}$$

$$\sigma_{max} = K\sigma_{nom}$$

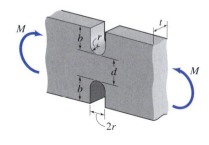

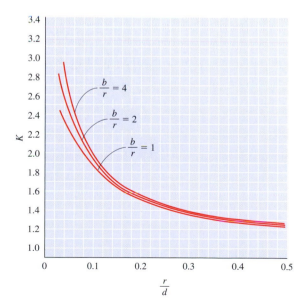

APPENDIX I
Advanced Methods and Derivations

I-1 Shear Stress and Twist in Thin-Walled Shaft Subjected to Torsion

We determine the shear stress in a thin-walled shaft subjected to a twisting moment as follows.

The force, dF, due to the shear stress, τ, acting on a portion of the wall of thickness t and length ds, acts tangentially and is equal to

$$dF = \tau t\, ds$$

T, the total moment about the point O, is found by integrating contributions to the moment around the closed curve defining the cross-section

$$T = \oint r\tau t\, ds$$

where r is the perpendicular distance from O to the line of action of dF.

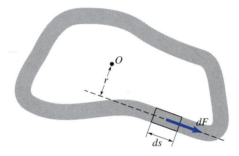

As shown in Section 4.9, the product τt is constant around the wall, so the integral simplifies to

$$T = \tau t \oint r\, ds$$

The integral depends only on the geometry of the cross-section, but it is challenging to evaluate analytically. However, it can be interpreted from the following drawing. The quantity $r\, ds$ is equal to twice the area of the shaded triangle. The full set of triangles, for all line segments ds along the perimeter, define the area of cross-section, A_m, contained by the midsurface.

$$T = \tau t(2A_m) \quad \Rightarrow \quad \tau = \frac{T}{2t A_m}$$

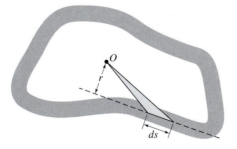
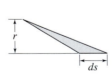
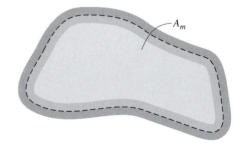

The twist of a thin-walled shaft of length, L, subjected to twisting moment, T, is found by applying conservation of energy (see Sections 6.6–6.7). Let the rotation at one end of the shaft be zero. Then the rotation at the other end, $\Delta\phi$, is related to the external work by

$$External\ Work = \frac{1}{2}T\Delta\phi$$

The shear stress for a given torque has just been determined. The stored energy can be found by integrating the strain energy density due to shear over the volume:

$$Stored\ Energy = \int_V \left(\frac{\tau^2}{2G}\right)dV = \int_{s_m}\left[\left(\frac{1}{2G}\right)\left(\frac{T}{2tA_m}\right)^2\right]Lt\,ds = \frac{T^2Lts_m}{8Gt^2A_m^2} = \frac{T^2Ls_m}{8GtA_m^2}$$

where s_m is the arc length along the mid-surface.

We can equate External Work and Stored Energy to determine the twist, $\Delta\phi$, in terms of the torque and shaft parameters:

$$\Delta\phi = \frac{TLs_m}{4GtA_m^2}$$

I-2 Method of Singularity Functions

The method of singularity functions seeks to solve the following equations that interrelate the loads on a beam, the internal loads (shear force and bending moment), and the deflection.

$$\frac{dV}{dx} = -q \qquad \frac{dM}{dx} = V \qquad \frac{d^2v}{dx^2} = \frac{M}{EI}$$

The customary method of solving for deflections via integration, presented in Section 5.16, becomes difficult when: (i) distributed loads act over only part of the beam, and (ii) rather than distributed loads, there are concentrated forces and moments. The method of singularity functions addresses both of those challenges.

Definitions of Discontinuity and Singularity Functions

First, let us define special discontinuity functions that help represent uniform and triangular distributions. The distributions act over the portion of the beam extending from a general point on the beam, $x = a$, to the right end of the beam ($x = L$):

uniform: $\quad q(x) = q_0\langle x - a\rangle^0 = \begin{cases} 0 & x < a \\ q_0 & x > a \end{cases}$

triangular: $\quad q(x) = q_1\langle x - a\rangle^1 = \begin{cases} 0 & x < a \\ q_1(x - a) & x > a \end{cases}$

Both of these discontinuity functions are zero to the left of a, and non-zero to the right of a.

In fact, one can define more general power-law discontinuity functions:

$$q(x) = q_n\langle x - a\rangle^n = \begin{cases} 0 & x < a \\ q_n(x - a)^n & x > a \end{cases}$$

We need to integrate these discontinuity functions once to obtain $V(x)$ and then a second time to obtain $M(x)$. We define the integral of these discontinuity functions following the usual rule for integrating x to a power:

$$\int \langle x - a\rangle^n dx = \frac{\langle x - a\rangle^{n+1}}{n + 1}$$

In order to represent concentrated forces and moments, we introduce additional special functions, which are called singularity functions.

Point force of magnitude P (positive, downward):

$$q(x) = P\langle x - a\rangle^{-1} = \begin{cases} 0 & x \neq a \\ P & x = a \end{cases}$$

Point moment of magnitude M_0 (positive, counter-clockwise):

$$q(x) = M_0\langle x - a\rangle^{-2} = \begin{cases} 0 & x \neq a \\ M_0 & x = a \end{cases}$$

We define the integrals of these two singularity functions as follows:

$$\int \langle x - a\rangle^{-1} dx = \langle x - a\rangle^0 \qquad\qquad \int \langle x - a\rangle^{-2} dx = \langle x - a\rangle^{-1}$$

Note that these integrals differ from those for $n > -1$, in that there is no term in the denominator. Note that a can be any point along the beam including the left end ($a = 0$).

Procedure for Analyzing Beams with Discontinuity and Singularity Functions

The procedure will consist of four steps:

1. Represent external loads on the beam with one loading function $q(x)$.
2. Integrate $q(x)$ once to find $V(x)$ and again to find $M(x)$, with no integration constants.
3. Integrate $M(x)/EI$ once to find the slope v' and again to find v, generating two integration constants.
4. Evaluate integration constants in v' and v using deflections and/or slopes at supports.

Step 1

We assume that the loads and beam supports are specified. We first determine the support reactions using equilibrium. Then, the beam can be viewed as having just external loads. (The actual types of supports are invoked again later when integration constants are evaluated.)

As we have done so far, let $x = 0$ correspond to the left end of the beam, with the positive x-axis extending along the beam.

We write down the loads on the beam as a loading function $q(x)$, which is the sum of discontinuity and singularity functions, each of which captures one load. As an example, this beam had a cantilever support at the left, along with a concentrated force and a distributed force.

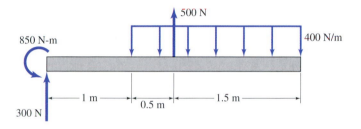

The loading function for this case is:

$$q(x) = 850 \text{ N-m}\langle x - 0\rangle^{-2} - 300 \text{ N}\langle x - 0\rangle^{-1} + 400 \text{ N/m}\langle x - 1\rangle^{0} - 500 \text{ N}\langle x - 1.5\rangle^{-1}$$

Notice that the sign of the forces at $x = 0$ and $x = 1.5$ m are negative, to reflect the fact that the forces act upward.

We could also represent a distributed force that extends over a portion of the beam that does not include either end, such as in this beam.

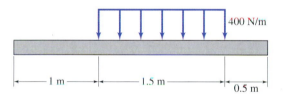

This loading could be represented as a downward uniformly distributed force that acts over $1 \text{ m} < x < 3$ m, plus an upward uniformly distributed force that acts over $2.5 \text{ m} < x < 3$ m. Therefore, the portion of the loading function that represents this loading is

$$q(x) = \cdots + 400 \text{ N/m}\langle x - 1\rangle^{0} - 400 \text{ N/m}\langle x - 2.5\rangle^{0} + \cdots$$

When a concentrated force or moment acts at the right end of the beam, it is not included in the loading function. That is because the discontinuity and singularity functions would be non-zero only to the right of the point $x = a$ and there are no points to the right of the right end.

Step 2

Next, we integrate the loading function to obtain $V(x)$. Notice that the integral of each discontinuity or singularity functions is another discontinuity or singularity function.

$$V(x) = -\int q(x)dx$$

For the beam above

$$V(x) = -\int q(x)dx = -850 \text{ N-m}\langle x - 0\rangle^{-1} + 300 \text{ N}\langle x - 0\rangle^0 - 400 \text{ N/m}\langle x - 1\rangle^1 + 500 \text{ N}\langle x - 1.5\rangle^0$$

We integrate a second time to obtain $M(x)$:

$$M(x) = \int V(x)dx$$

For the beam above

$$M(x) = \int V(x)dx = -850 \text{ N-m}\langle x - 0\rangle^0 + 300 \text{ N}\langle x - 0\rangle^1 - \frac{400}{2}\text{N/m}\langle x - 1\rangle^2 + 500 \text{ N}\langle x - 1.5\rangle^1$$

You can double check the values of shear force and bending moment at the ends. The left end corresponds to x just to the right of 0, and the right corresponds to x just to the left of 3.5 m. For the beam above

$$V(0^+) = 300 \text{ N and } M(0^+) = -850 \text{ N-m}$$

since no terms beyond $\langle x - 0\rangle$ contribute.

$$V(3\text{m}^-) = 300 \text{ N} - (400 \text{ N/m})(2 \text{ m}) + 500 \text{ N} = 0$$

$$M(3\text{m}^-) = -850 \text{ N-m} + 300 \text{ N}(3 \text{ m}) - \frac{400}{2}\text{N/m}(2 \text{ m})^2 + 500 \text{ N}(1.5 \text{ m}) = 0$$

since all terms contribute.

Step 3
The next step is to integrate $M(x)/EI$ once to obtain the slope, v', plus an integration constant. For the beam above

$$v'(x) = \frac{1}{EI}\left[-850 \text{ N-m }\langle x - 0\rangle^1 + \frac{300 \text{ N}}{2}\langle x - 0\rangle^2 - \frac{400}{6}\text{N/m }\langle x - 1\rangle^3 + \frac{500 \text{ N}}{2}\langle x - 1.5\rangle^2 + c_1\right]$$

Now, integrate v' to obtain $v(x)$, plus a second integration constant. For the same beam

$$v(x) = \frac{1}{EI}\left[-\frac{850}{2}\text{N-m}\langle x - 0\rangle^2 + \frac{300 \text{ N}}{6}\langle x - 0\rangle^3 - \frac{400}{24}\text{N/m}\langle x - 1\rangle^4 + \frac{500 \text{ N}}{6}\langle x - 1.5\rangle^3 + c_1 x + c_2\right]$$

Step 4
The next step is to evaluate the integration constants using deflection or slope at supports. This step is done exactly as presented in Section 5.16, except one needs to be careful evaluating the singular and discontinuity functions.

For the beam above, it was assumed that the left end was cantilevered, in which case the beam must satisfy zero slope and deflection there, that is, $v'(0) = 0$ and $v(0) = 0$. Furthermore, all the terms, except for the constants, are equal to zero at $x = 0$, and therefore $c_1 = 0$ and $c_2 = 0$.

Example Problem Using Singularity Functions

Determine the distributions of shear force, bending moment, and deflection for this beam.

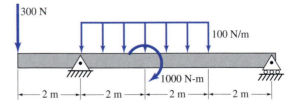

The support reactions are found from equilibrium and are shown here.

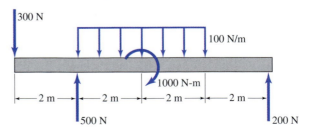

The loading function for this beam is given by:

$$q(x) = 300 \text{ N}\langle x - 0 \rangle^{-1} - 500 \text{ N}\langle x - 2 \rangle^{-1}$$
$$+ 100 \text{ N/m}\langle x - 2 \rangle^{0} - 100 \text{ N/m}\langle x - 6 \rangle^{0} - 1000 \text{ N-m}\langle x - 4 \rangle^{-2}$$

Integrate once to obtain the shear force:

$$V(x) = -\int q(x)dx = -300 \text{ N}\langle x - 0 \rangle^{0} + 500 \text{ N}\langle x - 2 \rangle^{0}$$
$$- 100 \text{ N/m}\langle x - 2 \rangle^{1} + 100 \text{ N/m}\langle x - 6 \rangle^{1} + 1000 \text{ N-m}\langle x - 4 \rangle^{-1}$$

Integrate again to obtain the bending moment:

$$M(x) = \int V(x)dx = -300 \text{ N}\langle x - 0 \rangle^{1} + 500 \text{ N}\langle x - 2 \rangle^{1}$$
$$- \frac{100}{2} \text{ N/m}\langle x - 2 \rangle^{2} + \frac{100}{2} \text{ N/m}\langle x - 6 \rangle^{2} + 1000 \text{ N-m}\langle x - 4 \rangle^{0}$$

If we evaluate the shear force and bending moments at the ends, we can see that they all have the correct values.

At the left end

$$V(0^{+}) = -300 \text{ N} \qquad M(0^{+}) = 0$$

At the right end

$$V(8^{-}) = -300 \text{ N} + 500 \text{ N} - (100 \text{ N/m})(8 - 2) \text{ m} + (100 \text{ N/m})(8 - 6) \text{ m} = -200 \text{ N}$$
$$M(8^{-}) = -(300 \text{ N})(8 - 0)\text{m} + (500 \text{ N})(8 - 2)\text{m}$$
$$- \left(\frac{100}{2} \text{ N/m} \right)[(8 - 2)\text{m}]^{2} + \left(\frac{100}{2} \text{ N/m} \right)[(8 - 6)\text{m}]^{2} + 1000 \text{ N-m} = 0$$

Now integrate $M(x)/EI$ to find the slope:

$$v'(x) = \frac{1}{EI}\left[-\frac{300}{2} \text{ N}\langle x - 0 \rangle^{2} + \frac{500}{2} \text{ N}\langle x - 2 \rangle^{2} - \frac{100}{6} \text{ N/m}\langle x - 2 \rangle^{3} + \frac{100}{6} \text{ N/m}\langle x - 6 \rangle^{3} + 1000 \text{ N-m}\langle x - 4 \rangle^{1} + c_1 \right]$$

Integrate one more time to find the deflection:

$$v(x) = \frac{1}{EI}\left[-\frac{300}{6} \text{ N}\langle x - 0 \rangle^{3} + \frac{500}{6} \text{ N}\langle x - 2 \rangle^{3} - \frac{100}{24} \text{ N/m}\langle x - 2 \rangle^{4} + \frac{100}{24} \text{ N/m}\langle x - 6 \rangle^{4} + \frac{1000}{2} \text{ N-m}\langle x - 4 \rangle^{2} + c_1 x + c_2 \right]$$

Choose the integration constants to satisfy $v(2) = 0$ and $v(8) = 0$.

$$v(2) = \frac{1}{EI}\left[-\frac{300}{6} \text{ N}[(2 - 0)\text{m}]^{3} + c_1(2 \text{ m}) + c_2 \right] = 0$$

$$v(8) = \frac{1}{EI}\left[-\frac{300}{6} \text{ N}[(8 - 0)\text{m}]^{3} + \frac{500}{6} \text{ N}[(8 - 2)\text{m}]^{3} - \left(\frac{100}{24} \text{ N/m} \right)[(8 - 2)\text{m}]^{4} \right.$$
$$\left. + \left(\frac{100}{24} \text{ N/m} \right)[(8 - 6)\text{m}]^{4} + \left(\frac{1000}{2} \text{ N-m} \right)[(8 - 4)\text{m}]^{2} + c_1(8 \text{ m}) + c_2 \right] = 0$$

These equations simplify to

$$-400 \text{ N-m}^{3} + c_1(2 \text{ m}) + c_2 = 0 \qquad -\frac{118400}{24} \text{ N-m}^{3} + c_1(8 \text{ m}) + c_2 = 0$$

Solve these equations to obtain the constants $c_1 = 756 \text{ N-m}^{2}$ and $c_2 = -1111 \text{ N-m}^{3}$.

I-3 Derivation of Stress Transformation Formulas

In Section 7.3, we showed that the equilibrium of the triangular element could be expressed as

$$\sum F_x = -\sigma_x[th\cos\theta] - \tau_{xy}[th\sin\theta] + \sigma(\theta)\cos\theta[th] - \tau(\theta)\sin\theta[th] = 0$$

$$\sum F_y = -\tau_{xy}[th\cos\theta] - \sigma_y[th\sin\theta] + \sigma(\theta)\sin\theta[th] + \tau(\theta)\cos\theta[th] = 0$$

Notice that th can be canceled in all terms. Combine the force summation equations as follows:

$$\left(\sum F_x\right)\cos\theta + \left(\sum F_y\right)\sin\theta = 0$$

which leads to

$$\sigma(\theta) = \sigma_x\cos^2\theta + \sigma_y\sin^2\theta + 2\tau_{xy}\sin\theta\cos\theta$$

Apply the following standard trigonometric identities:

$$\cos^2\theta = \frac{1}{2}[1 + \cos 2\theta] \qquad \sin^2\theta = \frac{1}{2}[1 - \cos 2\theta] \qquad 2\sin\theta\cos\theta = \sin 2\theta$$

to simplify $\sigma(\theta)$ to the form:

$$\sigma(\theta) = \frac{1}{2}(\sigma_x + \sigma_y) + \frac{1}{2}(\sigma_x - \sigma_y)\cos 2\theta + \tau_{xy}\sin 2\theta$$

Now, combine the force summation equations instead as follows:

$$-\left(\sum F_x\right)\sin\theta + \left(\sum F_y\right)\cos\theta = 0$$

which leads to

$$\tau(\theta) = \tau_{xy}(\cos^2\theta - \sin^2\theta) + (\sigma_y - \sigma_x)\sin\theta\cos\theta$$

By applying the trigonometric identities, one can now simplify $\tau(\theta)$ to the form:

$$\tau(\theta) = \tau_{xy}\cos 2\theta + \frac{1}{2}(\sigma_y - \sigma_x)\sin 2\theta$$

I-4 Derivation of Equations for Maximum Normal and Shear Stress

Equations for maximum shear and normal stresses, which were given in Section 7.4, can be derived by starting from the stress transformation formulas.

$$\sigma(\theta) = \frac{1}{2}(\sigma_x + \sigma_y) + \frac{1}{2}(\sigma_x - \sigma_y)\cos 2\theta + \tau_{xy}\sin 2\theta$$

$$\tau(\theta) = \tau_{xy}\cos 2\theta + \frac{1}{2}(\sigma_y - \sigma_x)\sin 2\theta$$

We set the first derivative of $\tau(\theta)$ with respect to θ equal to zero to find the maximum and minimum shear stress among all angles θ:

$$\frac{d\tau(\theta)}{d\theta} = \frac{d\left[\tau_{xy}\cos 2\theta + \frac{1}{2}(\sigma_y - \sigma_x)\sin 2\theta\right]}{d\theta} = -2\tau_{xy}\sin 2\theta + (\sigma_y - \sigma_x)\cos 2\theta = 0$$

The equality holds if the angle has the value θ_s, which satisfies:

$$\frac{\sin 2\theta_s}{\cos 2\theta_s} = \tan 2\theta_s = \frac{(\sigma_y - \sigma_x)/2}{\tau_{xy}}$$

We can draw two triangles that are consistent with this value for $\tan 2\theta_s$:

Notice that the hypotenuse, H, is given by

$$H = \sqrt{\left(\frac{\sigma_y - \sigma_x}{2}\right)^2 + \tau_{xy}^2}$$

We now determine the shear stress $\tau(\theta)$ at the two solutions for θ_s. One solution corresponds to the left triangle

$$\cos 2\theta_s = \frac{\tau_{xy}}{H} \qquad \sin 2\theta_s = \frac{(\sigma_y - \sigma_x)/2}{H}$$

For this solution we find

$$\tau(\theta_s) = \tau_{xy}\cos 2\theta_s + \frac{1}{2}(\sigma_y - \sigma_x)\sin 2\theta_s = \frac{[\tau_{xy}]^2}{H} + \frac{[(\sigma_y - \sigma_x)/2]^2}{H} = \sqrt{\left(\frac{\sigma_y - \sigma_x}{2}\right)^2 + \tau_{xy}^2}$$

The second solution corresponds to the right triangle:

$$\cos 2\theta_s = \frac{-\tau_{xy}}{H} \qquad \sin 2\theta_s = \frac{-(\sigma_y - \sigma_x)/2}{H}$$

For this solution we find

$$\tau(\theta_s) = \tau_{xy}\cos 2\theta_s + \frac{1}{2}(\sigma_y - \sigma_x)\sin 2\theta_s = \frac{-[\tau_{xy}]^2}{H} + \frac{-[(\sigma_y - \sigma_x)/2]^2}{H} = -\sqrt{\left(\frac{\sigma_y - \sigma_x}{2}\right)^2 + \tau_{xy}^2}$$

Thus, the shear stress has maximum and minimum values of $\pm\tau_{max}$, where

$$\tau_{max} = \sqrt{\left(\frac{\sigma_y - \sigma_x}{2}\right)^2 + \tau_{xy}^2}$$

Note: since $\cos\alpha$ and $\sin\alpha$ both change sign if α increases by π (or 180°), the maximum shear stress changes sign when θ_s increases by $\pi/2$ (or 90°). So, the maximum and minimum shear stress is reached on the four faces of a square.

Now, we set the first derivative of $\sigma(\theta)$ with respect to θ to find the maximum and minimum normal stress among all angles θ:

$$\frac{d\sigma(\theta)}{d\theta} = \frac{d\left[\frac{1}{2}(\sigma_x + \sigma_y) + \frac{1}{2}(\sigma_x - \sigma_y)\cos 2\theta + \tau_{xy}\sin 2\theta\right]}{d\theta} = -(\sigma_x - \sigma_y)\sin 2\theta + 2\tau_{xy}\cos 2\theta = 0$$

The equality holds if the angle has the value θ_p, which satisfies:

$$\frac{\sin 2\theta_p}{\cos 2\theta_p} = \tan 2\theta_p = \frac{\tau_{xy}}{(\sigma_x - \sigma_y)/2}$$

We can derive a relation between θ_s and θ_p by first noting that they involve a ratio of the same two quantities and a difference in sign:

$$\tan 2\theta_s = \frac{(\sigma_y - \sigma_x)/2}{\tau_{xy}} = -\left[\frac{1}{\dfrac{\tau_{xy}}{(\sigma_x - \sigma_y)/2}}\right] = -\frac{1}{\tan 2\theta_p}$$

The following trigonometric identity is now useful:

$$\tan\left(\frac{\pi}{2} - \beta\right) = \frac{\sin\left(\frac{\pi}{2} - \beta\right)}{\cos\left(\frac{\pi}{2} - \beta\right)} = \frac{\cos \beta}{\sin \beta} = \frac{1}{\tan \beta}$$

If we set $\beta = 2\theta_p$, then this trigonometric identity becomes

$$\tan\left(\frac{\pi}{2} - 2\theta_p\right) = \frac{1}{\tan 2\theta_p}$$

Use this form of the identity to substitute for $1/\tan 2\theta_p$ above:

$$\tan 2\theta_s = -\frac{1}{\tan 2\theta_p} = -\tan\left(\frac{\pi}{2} - 2\theta_p\right) = \tan\left(2\theta_p - \frac{\pi}{2}\right)$$

Therefore, the angles of maximum shear and maximum normal stress, θ_s and θ_p, are related by

$$\theta_s = \theta_p - \frac{\pi}{4} \quad \text{or} \quad \theta_p = \theta_s + \frac{\pi}{4}$$

So the angles differ by 45°.

Consider now the maximum or minimum values that the normal stress has at the angle θ_p. We can again draw two triangles that are consistent with the formula for $\tan 2\theta_p$:

Notice the hypotenuse, H, is the same as above, namely equal to $\tau_{\max}$.

For the left triangle

$$\cos 2\theta_p = \frac{(\sigma_x - \sigma_y)/2}{H} \qquad \sin 2\theta_p = \frac{\tau_{xy}}{H}$$

and the corresponding normal stress is

$$\frac{1}{2}(\sigma_x + \sigma_y) + \frac{1}{2}(\sigma_x - \sigma_y)\cos 2\theta_p + \tau_{xy}\sin 2\theta_p$$

$$= \frac{1}{2}(\sigma_x + \sigma_y) + \frac{1}{\tau_{\max}}\left[\left[\frac{(\sigma_x - \sigma_y)}{2}\right]^2 + [\tau_{xy}]^2\right] = \frac{1}{2}(\sigma_x + \sigma_y) + \tau_{\max}$$

For the right triangle

$$\cos 2\theta_p = \frac{-(\sigma_x - \sigma_y)/2}{H} \qquad \sin 2\theta_p = \frac{-\tau_{xy}}{H}$$

and the corresponding normal stress is

$$\frac{1}{2}(\sigma_x + \sigma_y) + \frac{1}{2}(\sigma_x - \sigma_y)\cos 2\theta_p + \tau_{xy}\sin 2\theta_p$$

$$= \frac{1}{2}(\sigma_x + \sigma_y) + \frac{1}{\tau_{max}}\left[-\left[\frac{(\sigma_x - \sigma_y)}{2}\right]^2 - [\tau_{xy}]^2 \right] = \frac{1}{2}(\sigma_x + \sigma_y) - \tau_{max}$$

Since τ_{max} is defined as positive, the first solution to $\dfrac{d\sigma(\theta)}{d\theta} = 0$ corresponds to the maximum normal stress and the second to the minimum normal stress. We can write the maximum and minimum values of normal stress as

$$\sigma_{max} = \sigma_m + \tau_{max} \qquad \sigma_{min} = \sigma_m - \tau_{max}$$

in terms of the mean stress, $\sigma_m = (\sigma_x + \sigma_y)/2$, and the maximum shear stress τ_{max}.

We complete the picture of stresses by noting that the normal stress equals σ_m on all four planes on which the shear is maximum or minimum because

$$-2\tau_{xy}\sin 2\theta_s + (\sigma_y - \sigma_x)\cos 2\theta_s = 0$$

and so

$$\sigma(\theta_s) = \frac{1}{2}(\sigma_x + \sigma_y) + \frac{1}{2}(\sigma_x - \sigma_y)\cos 2\theta_s + \tau_{xy}\sin 2\theta_s = \frac{1}{2}(\sigma_x + \sigma_y) = \sigma_m$$

Correspondingly, the shear stress is zero on the four planes on which the normal stress is maximum or minimum because

$$-(\sigma_x - \sigma_y)\sin 2\theta_p + 2\tau_{xy}\cos 2\theta_p = 0$$

Drawing of the maximum shear stress and principal stress elements as described in Section 7.4 are consistent with all the relations derived here. Furthermore, θ for the plane on which the normal stress is σ_{max} is 45° greater than θ for the plane on which the shear stress is $+\tau_{max}$. Recall from Section 7.4 that once a particular value of θ_s has been extracted from

$$\theta_s = \frac{1}{2}\tan^{-1}\left[\frac{(\sigma_y - \sigma_x)/2}{\tau_{xy}}\right]$$

One must evaluate $\tau(\theta_s)$ to determine if τ at that θ_s is $+\tau_{max}$ or $-\tau_{max}$. Thereafter, the maximum shear stress and principal stress elements can be found unambiguously.

Answers to Selected Problems

Chapter 2

2.2 $F = 200$ N, $M = 82$ N-m, $T = 18$ N-m

2.4 $F = 2$ N, $M = 0.21$ N-m

2.6 $V = 241$ lb, $N = 64.7$ lb, $M = 2660$ lb-in.

2.8 $F = 60$ kN, $M = 3900$ kN-m

2.10 $\sigma_{AB} = 60.1$ MPa (tensile),
$\sigma_{BC} = 26.7$ MPa (compressive)

2.12 (a) $A = 0.001866$ m^2,
(b) $A = 0.002$ m^2

2.14 $A = 0.00811$ m^2, 1032 strands

2.16 (a) $P_{additional} = 3.84\ (10^6)$ lb,
(b) $P_{additional} = 4.41\ (10^6)$ lb

2.18 (a) $\sigma = 248$ psi,
(b) $\sigma = 364$ psi

2.20 (a) $p_{avg} = 64$ kPa,
(b) $p_{max} = 21$ kPa,
(c) Will not lift off

2.22 $\varepsilon = 0.4\ (10^{-3})$

2.24 $h > 2.78$ in.

2.26 (a) $\varepsilon = 0.0345$,
(b) $L = 2.17$ in.

2.28 (a) Strain ratio = 0.00943,
(b) $\delta = -2.11$ mm

2.30 $\varepsilon_{AC} = 8.34\ (10^{-3})$

2.32 (a) $E = 10.7$ GPa,
(b) $\nu = 0.295$

2.36 (a) $d_2 = 6.71$ mm,
(b) $\delta_t = 0.00559$ mm

2.38 $L = 0.505$ μm

2.40 (a) $\delta_{AB} = 0.1743$ mm,
(b) $\delta_{BC} = -0.422$ mm,
(c) $\delta_{AC} = 0.248$ mm

2.42 $F_0 = 3820$ lb, $\delta = 0.1254$ in.

2.44 $\delta = 3.86$ mm

2.46 $P = 43.2$ kN

2.48 $E = 3.9$ GPa

2.50 (b) $E = 34\ (10^3)$ ksi,
(c) $\sigma_Y = 71$ ksi

2.52 (a) $E = 3.1$ GPa,
(b) $\sigma_Y = 17$ MPa

2.54 (a) $P_{max} = 5430$ N,
(b) $\delta = 0.337$ mm

2.58 Displacement ranges from 0.826 in. (down)
to 0.326 in. (up)

2.60 $\tau_A = 2.5$ psi, $\tau_B = 5$ psi, $\tau_C = 1.25$ psi

2.62 $v_1 = 0.00625$ in. (up), $v_3 = 0.025$ in. (up),
$v_4 = 0.0219$ in. (up)

2.64 $v = 0.8$ mm

2.66 $F_0 = 4320$ N

2.68 $\tau = 26.0$ kPa

2.70 $T = \left[\dfrac{2\pi h R^3 G}{t}\right]\Delta\phi$

2.72 (a) $\sigma_{bearing} = 4800$ psi,
(b) $\tau = 3060$ psi

2.74 $P = 880$ N

2.76 $\tau = 180.4$ MPa

2.78 $\tau_A = 10.61$ MPa, $\tau_B = 15.03$ MPa

2.80 (a) $\tau = 23.9$ MPa,
(b) $\sigma_{bearing} = 45.8$ MPa

2.82 $d = 0.252$ in.

2.84 $\tau = 768$ psi

Chapter 3

3.2 (a) $T = 73.0$ N,
(b) 0.7 turns

3.4 $M = 576$ kN-m

3.6 $v = 1.011$ mm

3.8 $\delta_{BC} = 0.0716$ mm (separate)

3.10 $\sigma = 13.65$ ksi (tensile)

3.12 $v = 0.0699$ in.

3.14 $\Delta\varepsilon_{BC} = 0.001019$

3.16 $\sigma_{AB} = -14.3$ ksi, $\sigma_{BC} = -17.6$ ksi

3.18 $F_1 = 42.0$ kip, $F_2 = 60.0$ kip

3.20 $v = 0.01883$ mm (up)

3.22 $\Delta\varepsilon = 0.837\,(10^{-3})$

3.24 (a) $\varepsilon_2 = 0.764\,(10^{-3})$,
(b) $\varepsilon_1 = 0.001042$

3.28 $\sigma = 25.3$ MPa

3.30 Cannot be used (σ would be 51.9 ksi)

3.32 $\sigma = 32.9$ MPa

3.34 $\dfrac{L_3}{L_4} = \dfrac{L_1}{L_2}$

3.36 $v_D = 1.078$ mm

3.38 $v_D = \dfrac{4}{3}\left[\dfrac{0.6\text{ m}}{\cos 40° \cos 50°}\right]\varepsilon$

3.40 $u_B = 0.25$ mm (left), $v_B = 0.1443$ mm (down)

3.42 $(F_0)_x = \dfrac{E_2 A_2 u_0}{L_2} + \dfrac{E_1 A_1 (\sin 30°)^2 u_0}{L_1}$,

$(F_0)_y = \dfrac{E_1 A_1 (\sin 30°)(\sin 60°)u_0}{L_1}$

3.44 $P_1 = \dfrac{2}{3}P, P_2 = -\dfrac{1}{3}P, u_E = \dfrac{5}{3}\dfrac{P}{k}$

3.46 $F_1 = 3300$ lb, $F_2 = 6000$ lb

3.48 $F_1 = 825$ N, $F_2 = 4050$ N

3.50 $f = 671$ Hz

3.52 Rotation of deck $= 4.71°$

3.54 Restoring force $= 1.380$ kN

3.56 $P = 856$ kN

3.58 $v_A = 3.06$ mm, $v_B = 3.23$ mm, $v_C = 3.40$ mm

3.60 $u = 0.00992$ mm

3.62 $v_B = \dfrac{F_0}{E_b A_b}\left[L_2 + \dfrac{L_1}{1 + 2\dfrac{E_p A_p L_1}{E_b A_b L_3}}\right]$

3.64 $\dfrac{F_0}{v_B} = \dfrac{E_b A_b}{L_2}$

3.66 (a) $\sigma_{AB} = 14.01$ ksi, $\sigma_{DE} = 2.28$ ksi,
(b) $v_C = 0.01892$ in.

3.70 Torsional stiffness $= 10580$ N-m/rad

3.72 (a) $v_0 = 0.024$ mm,
(b) fractional contribution of
reorientation $= 0.424\,(10^{-3})$

3.74 (a) $\Delta T = 3110$ N,
(b) $\Delta T = 3020$ N

3.76 $\sigma_{max} = 24.4$ MPa, $\sigma_{min} = 17.91$ MPa

3.78 $R_1 = \dfrac{k_0}{k_1}R_0, R_2 = \dfrac{k_0}{k_2}R_0$

3.82 $\Delta T_{upper} = 138.6$ lb, $\Delta T_{lower} = 135.1$ lb

3.84 (a) $\sigma_s = 128.9$ MPa, $\sigma_b = 80.6$ MPa,
(b) $u = 0.01045$ mm (left)

3.88 (a) $P_1 = \dfrac{\alpha\Delta T L_1}{\dfrac{L_1}{E_1 A_1} + \dfrac{4L_2}{E_2 A_2}}$ (tensile),

$P_2 = \dfrac{2\alpha\Delta T L_1}{\dfrac{L_1}{E_1 A_1} + \dfrac{4L_2}{E_2 A_2}}$ (tensile),

(b) $u_Q = \dfrac{4\alpha\Delta T L_1 L_2}{L_1\dfrac{E_2 A_2}{E_1 A_1} + 4L_2}$ (left)

3.90 $u = 1.426$ in.

3.92 $\Delta P_{spring} = 0.0588$ lb (decreases)

3.94 $\alpha = 37.6\,(10^{-6})\,(°C)^{-1}$

3.96 $T \sim 220$ °C

3.98 (a) $\sigma_m = \dfrac{E_m(\alpha_f - \alpha_m)\Delta T}{1 + \dfrac{E_m A_m}{E_f A_f}}$,

(b) $\alpha_{comp} = \alpha_f + \dfrac{(\alpha_m - \alpha_f)}{1 + \dfrac{E_f A_f}{E_m A_m}}$

3.100 $N = 3.79$ N

3.102 $N_{max} = 113.1$ N

Chapter 4

4.2 $\phi = 0.158°$ about $+z$-axis

4.4 (a) $\phi_C = 0.5°$ about $-z$-axis,
(b) $\phi_E = 0.2°$ about $+z$-axis

4.6 $\gamma = 0.1309 \ (10^{-3})$

4.8 $\gamma = 0.333 \ (10^{-3})$

4.10 $\gamma = 0.623 \ (10^{-3})$

4.12 $\Delta\phi = 3.14°$

4.14 $\phi = 3.37°$

4.16 $\Delta u = 6.37$ mm

4.18 $\Delta\phi = 17.19°$

4.20 $\phi = 8.54°$ about $+x$-axis

4.22 $\tau = 2410$ psi

4.24 $F_1 = 3.93$ lb

4.26 $\tau = 9600$ psi

4.28 (a) $F_1 = 18.41$ lb,
(b) $\phi = 68.8°$ about $+y$-axis

4.30 $p = 0.0142$ psi

4.34 Force per length $= 842$ lb/in.

4.36 $P = 2880$ N

4.38 $\tau = 2070$ psi, $\Delta\phi = 0.869°$

4.40 $T = 26.8 \ (10^3)$ lb-ft

4.42 $\tau = 30.5$ MPa

4.44 $L = 0.371$ m

4.46 $d = 16.58$ mm

4.48 $d_{inner} = 1.75$ in., $\Delta T = 6810$ lb-in.

4.50 Torsional stiffness $= \dfrac{\pi G d^4}{16 w_1^2 w_2}$

4.52 Hollow shaft stiffness is 0.973 of solid shaft stiffness

4.54 $d = 39.6$ mm

4.56 Fraction of yield stress $= 0.1907$

4.58 $\phi = 3.23°$

4.60 Hollow-to-solid ratios (a) Weight: 0.72,
(b) Stiffness: 2.11,
(c) Strength: 1.56

4.62 Torsional stiffness $= 0.219 \ (10^6)$ lb-in./rad

4.64 $\tau_{max} = 81.3$ MPa, $\Delta\phi = 3.75°$

4.66 Stiffness $= 0.1862 \ (10^6)$ N/rad

4.68 $\Delta T = 80.9$ N-m

4.70 For width w_1: Stiffness $= 466$ N-m/rad,
for width w_2: Stiffness $= 998$ N-m/rad

4.72 $\tau_{max} = 267$ psi, $\Delta\phi = 0.201°$

4.74 Hollow square tube of outer dimensions
2 in. $\times$ 2 in.

4.76 $\Delta\phi = 9.96°$

4.78 $\tau = 1371$ psi, $\Delta\phi = 0.1874°$

4.80 Stiffness ratio of tubular nail to slotted
nail $= 272$

4.82 (a) $\gamma = 0.00307$,
(b) $\gamma = 0.001650$

4.84 $\phi = 92.68°$, $\tau_{max} = 71.7$ MPa

4.86 $\phi_C - \phi_B = 1.009°$, $\phi_C - \phi_A = 3.17°$

4.88 -300 N-m $< T_A < -100$ N-m

4.90 $\tau_{max} = 30.2$ MPa, $\phi = 58.4°$

4.92 $T = 0.251 \ (10^6)$ N-m, $d_{shaft} = 172.4$ mm

4.96 $\tau_{max} = 5.47$ MPa (at top of tower)

4.98 $\tau_{max} = 38.2$ MPa, $\phi = 0.547°$

4.100 Stiffness $= 45.8$ lb/in.

4.102 $\phi_B = T_0 \left(\dfrac{a}{a + 2L} \right) \dfrac{32L}{\pi G d^4}$,

$\tau_{max} = T_0 \left(\dfrac{2L}{a + 2L} \right) \dfrac{16}{\pi d^3}$

4.104 $u = 1.679$ mm

4.106 $\dfrac{T}{\Delta\phi/L} = \dfrac{\pi}{32} G_0 d^4 \left(\dfrac{76}{81} \right)$

4.108 $(\tau_{max})_{bone} = 1.045$ MPa,
$(\tau_{max})_{nail} = 10.47$ MPa

4.110 $d > 20.1$ mm

4.112 $\phi_A - \phi_B = 0.279°$, $\phi_B - \phi_{motor} = 0.871°$

4.114 (a) $d_{min} = 4.10$ in.,
(b) $P = 323$ hp

4.116 $P = 356$ W, $\tau_{max} = 13.88$ MPa

4.118 $\Delta\phi - 21.7°$

Chapter 5

5.2 $V = -8000$ lb, $M = -24\,(10^3)$ lb-ft

5.4 $V = -17.5$ kN, $M = -12.5$ kN-m

5.6 $M_0 = 80$ kN-m, $V = 22.5$ kN, $M = 25$ kN-m

5.8 $V_{max} = 300$ N, $M_{min} = -1100$ N-m

5.10 $V_{max} = 184$ lb, $V_{min} = -200$ lb,
$M_{min} = -1600$ lb-in.

5.12 $V_{max} = 45$ N, $M_{min} = -16.5$ N-m

5.14 $V_{max} = 828$ N, $V_{min} = -672$ N,
$M_{max} = 2890$ N-m

5.16 $q = 15.38$ lb/ft, $V_{min} = -461$ lb,
$M_{min} = -300$ lb-ft

5.18 $M_0 = 1280$ lb-in., $V_{max} = 53.3$ lb,
$M_{max} = 800$ lb-in., $M_{min} = -480$ lb-in.

5.20 $0 < P < 233$ lb

5.22 $a = 0.207\,L$

5.24 $V_{max} = 224$ N, $V_{min} = -224$ N,
$M_{max} = 500$ N-m

5.26 $V_{max} = 600$ N, $V_{min} = -600$ N,
$M_{min} = -600$ N-m

5.32 $V_{min} = -400$ N, $M_{max} = 5.2$ N-m,
$M_{min} = -10$ N-m

5.34 $V_{min} = -176.2$ N, $V_{max} = 176.2$ N,
$M_{min} = -61.8$ N-m

5.36 $V_{max} = 1064$ N, $M_{min} = -74.5$ N-m

5.38 $T = 3.59\,(10^6)$ N, $M_{max} = 10.52\,(10^6)$ N-m

5.40 $V_{max} = 343$ lb, $V_{min} = -100$ lb,
$M_{max} = 1200$ lb-in.

5.42 $V_{max} = 12.34$ kN, $M_{min} = -123.4$ kN-m

5.44 Strain along y; $\varepsilon_A > 0$, $\varepsilon_B = 0$, $\varepsilon_C < 0$,
$\varepsilon_D < 0$, $\varepsilon_E < 0$

5.46 Strain along x; $\varepsilon_A > 0$, $\varepsilon_B > 0$, $\varepsilon_C > 0$,
$\varepsilon_D = 0$, $\varepsilon_E < 0$

5.48 Strain along x; $\varepsilon_A < 0$, $\varepsilon_B = 0$, $\varepsilon_C > 0$,
$\varepsilon_D > 0$, $\varepsilon_E > 0$

5.50 Strain along x; $\varepsilon_A > 0$, $\varepsilon_B = 0$, $\varepsilon_C < 0$,
$\varepsilon_D < 0$, $\varepsilon_E < 0$

5.52 Strain along z; $\varepsilon_A < 0$, $\varepsilon_B < 0$, $\varepsilon_C < 0$,
$\varepsilon_D = 0$, $\varepsilon_E > 0$

5.54 Strain along x; $\varepsilon_A < 0$, $\varepsilon_B < 0$, $\varepsilon_C < 0$,
$\varepsilon_D = 0$, $\varepsilon_E > 0$

5.56 $\sigma_A = -20$ MPa, $\sigma_B = -20$ MPa, $\sigma_C = -20$ MPa,
$\sigma_D = 0$, $\sigma_E = 20$ MPa

5.58 $\sigma_A = -30$ MPa, $\sigma_B = -30$ MPa, $\sigma_C = -30$ MPa,
$\sigma_D = 0$, $\sigma_E = 30$ MPa

5.60 $\sigma_A = 20$ MPa, $\sigma_B = 0$, $\sigma_C = -20$ MPa,
$\sigma_D = -20$ MPa, $\sigma_E = -20$ MPa

5.62 $\sigma_A = 30$ MPa, $\sigma_B = 30$ MPa, $\sigma_C = 30$ MPa,
$\sigma_D = 0$, $\sigma_E = -30$ MPa

5.64 $\sigma_A = 30$ MPa, $\sigma_B = 30$ MPa, $\sigma_C = 30$ MPa,
$\sigma_D = 0$, $\sigma_E = -30$ MPa

5.66 $\sigma_A = -20$ MPa, $\sigma_B = 0$, $\sigma_C = 20$ MPa,
$\sigma_D = 20$ MPa, $\sigma_E = 20$ MPa

5.68 $\sigma_{max} = 110.5$ MPa, at B

5.70 $\sigma_{max} = 56.5$ MPa, at A

5.72 (a) $\varepsilon = 0.612\,(10^{-3})$,
(b) $\varepsilon = 0.1029\,(10^{-3})$

5.74 (a) $M_{max} = 700$ N-m,
(b) $M_{max} = 350$ N-m

5.76 (a) $M_{max} = 4.06$ N-m,
(b) $M_{max} = 3.25$ N-m

5.78 $E = 25.6$ MPa, $\varepsilon = 0.00208$ (compression)

5.80 $M_2 = 29.7\,(10^3)$ lb-in., compression at top

5.82 (a) $\sigma_{max} = 9530$ psi,
(b) $\sigma_{max} = 8540$ psi

5.84 (a) $\sigma = 10.8$ MPa (tension/top),
$\sigma = 34.2$ MPa (compression/bottom),
(b) $\sigma = 47.9$ MPa (tension/bottom),
$\sigma = 15.12$ MPa (compression/top)

5.86 $\bar{y} = 0.521$ in., $I = 0.0274$ in.^{4}

5.88 $\sigma_{max} = 7420$ psi

5.90 $M_{max} = 460 \, (10^6)$ N-m

5.92 $\sigma_{max} = 179.6$ MPa (at top of tower)

5.94 (a) $\varepsilon = 0.1980 \, (10^{-3})$,
(b) $M = 451$ lb-in.

5.96 (a) $\sigma = 20.9$ ksi (tension/top),
$\sigma = 10.35$ ksi (compression/bottom),
(b) $M = 2830$ lb-in.

5.98 $\sigma = 978$ psi (tension/top),
$\sigma = 317$ psi (compression/bottom)

5.100 $M_{max} = 728$ lb-in. (M_B, compression/bottom)

5.102 (a) $M_{max} = 1355$ lb-in.,
(b) $M_{max} = 2740$ lb-in.

5.104 $\Delta\theta = 0.599°$

5.106 $E = 26.3$ MPa

5.108 $M_{max} = 5530$ lb-in.

5.110 (a) $\sigma = 1043$ psi (tension),
(b) $\sigma = 2090$ psi (compression)

5.112 (a) $\sigma_{max} = 218$ psi (tension and compression),
(b) $\sigma_{max} = 145.4$ psi (tension and compression)

5.114 (a) $\sigma = 48.6$ MPa (compression),
(b) $\sigma = 20.8$ MPa (compression)

5.116 $\sigma_{max} = 13.06$ ksi (tension/bottom, $x = 7.5$ ft),
$\sigma_{max} = 10.45$ ksi (compression/bottom, $x = 3$ ft)

5.118 (a) $\sigma_{max} = 14.49$ ksi (at change in cross-section),
(b) $\sigma_{max} = 9390$ psi (at support)

5.120 $q = 392$ lb/ft

5.122 $\sigma_{max} = 209$ MPa

5.124 $F_{max} = 8220$ N

5.126 Factor of safety $= 7.5$

5.128 $\sigma_{max} = 92.6$ MPa

5.130 $(F_0)_{max} = 295$ lb

5.132 (a) $\sigma = 176.0$ MPa,
(b) $\sigma = 440$ MPa

5.134 $\sigma = 47.9$ MPa (bottom)

5.136 $\sigma = 34.6$ MPa

5.138 (a) $S = \dfrac{\pi}{4}\left(\dfrac{d}{2}\right)^3$,
(b) $S = 2.65 \, (10^{-6})$ m^3

5.140 $M_{max} = 37.0$ N-m, B

5.142 (a) $S = 1425$ mm^3,
(b) $S = 1206$ mm^3

5.144 $EI = 627 \, (10^6)$ lb-in.^{2}

5.146 (a) $S = 41.7 \, (10^{-9})$ m^3,
(b) $S = 39.8 \, (10^{-9})$ m^3 (bottom),
$S = 32.2 \, (10^{-9})$ m^3 (top)

5.148 $d_{inner} = 21.9$ mm, $d_{outer} = 31.9$ mm

5.150 $S_{collar}/S_{pipe} = 4.26$

5.152 (a) safe,
(b) safe,
(c) unsafe

5.154 $M_{max} = 27.0 \, (10^3)$ lb-in.

5.156 $(EI)_{clad}/(EI)_{alum} = 2.37$

5.158 (a) $(EI)_{comp} = 24.7 \, (10^6)$ N-m^2,
(b) $(EI)_{steel\ plates} = 21.8 \, (10^6)$ N-m^2,
(c) $(EI)_{wood\ separated} = 2.87 \, (10^6)$ N-m^2,
(d) $(EI)_{wood\ together} = 0.410 \, (10^6)$ N-m^2

5.160 $q = 23.3 \, (10^3)$ N/m

5.162 $q = 4710$ N/m

5.166 $\tau_{max}/\sigma_{max} = 0.0075$

5.168 (a) $\tau_{max} = 421$ psi,
(b) $\sigma_{max} = 8160$ psi

5.170 $\tau_{max} = \left(\dfrac{h}{12L}\right)\sigma_{max}$

5.172 $V = 16.07\,(10^3)$ lb

5.174 $V = 15.50\,(10^3)$ lb

5.176 $V = 8030$ N

5.178 $V = 10.81$ kN

5.180 $V = 92.7\,(10^3)$ lb

5.182 (a) $P = 522\,(10^3)$ lb,
    (b) $P = 232\,(10^3)$ lb

5.184 $V = 1267$ N

5.186 $V_{bolt} = 2860$ lb

5.188 $s = 75.6$ mm

5.190 $v(x) = \dfrac{q}{360LEI}\left[10L^2x^3 - 3x^5 - 7L^4x\right]$

5.192 In $(0 < x < a)$:

$v(x) = \dfrac{-M_0}{LEI}\left[\dfrac{x^3}{6} + \left(\dfrac{a^2}{2} - La + \dfrac{L^2}{3}\right)x\right]$,

$v'(x) = \dfrac{-M_0}{LEI}\left[\dfrac{x^2}{2} + \left(\dfrac{a^2}{2} - La + \dfrac{L^2}{3}\right)\right]$

In $(a < x < L)$:

$v(x) = \dfrac{M_0}{LEI}\left[\dfrac{-x^3}{6} + \dfrac{Lx^2}{2} - \left(\dfrac{a^2}{2} + \dfrac{L^2}{3}\right)x + \dfrac{La^2}{2}\right]$,

$v'(x) = \dfrac{M_0}{LEI}\left[\dfrac{-x^2}{2} + Lx - \left(\dfrac{a^2}{2} + \dfrac{L^2}{3}\right)\right]$

5.194 In $(0 < x < a)$:

$v(x) = \dfrac{M_0}{6EI}\left[3x^2 - (6a + 2b)x + 3a^2 + 2ab\right]$,

$v'(x) = \dfrac{M_0}{6EI}\left[6x - (6a + 2b)\right]$

In $(a < x < a + b)$:

$v(x) = \dfrac{M_0}{6bEI}\left[(a + b - x)^3 + b^2x - b^2(a + b)\right]$,

$v'(x) = \dfrac{M_0}{6bEI}\left[-3(a + b - x)^2 + b^2\right]$

5.196 In $(0 < x < a)$:

$v(x) = \dfrac{-q}{EI}\left[\dfrac{x^4}{12} + \left(\dfrac{b^3}{12} - \dfrac{ba^2}{3} - \dfrac{a^3}{3}\right)x + \dfrac{a^4}{4} + \dfrac{ba^3}{3} + \dfrac{b^3a}{12}\right]$,

$v'(x) = \dfrac{-q}{EI}\left[\dfrac{x^3}{3} + \dfrac{b^3}{12} - \dfrac{ba^2}{3} - \dfrac{a^3}{3}\right]$

In $(a < x < a + b)$:

$v(x) = \dfrac{q}{EI}\left[-\dfrac{(a + b - x)^4}{24} + \left(\dfrac{b^2 - a^2}{2b}\right)\dfrac{(a + b - x)^3}{6}\right.$
$\left. + \left(\dfrac{b^3}{24} - \dfrac{ba^2}{12}\right)x + (a + b)\left(\dfrac{ba^2}{12} - \dfrac{b^3}{24}\right)\right]$,

$v'(x) = \dfrac{q}{EI}\left[\dfrac{(a + b - x)^3}{6} - \left(\dfrac{b^2 - a^2}{2b}\right)\dfrac{(a + b - x)^2}{2}\right.$
$\left. + \left(\dfrac{b^3}{24} - \dfrac{ba^2}{12}\right)\right]$

5.198 $c = \sqrt{L^2 - \dfrac{4EI}{Dw_0}},\ L_{min} = \sqrt{\dfrac{4EI}{Dw_0}}$

5.202 $v' = \dfrac{17.97\,(10^3)\ \text{lb-in.}^2}{EI}$ (CW)

5.204 $q = 95.2\,(10^3)$ lb/ft

5.206 Spring constant $= 14.73$ N/mm

5.208 $v = 0.0709$ in. (up)

5.210 $v = \dfrac{0.01406\ \text{N-m}^3}{EI}$ (up), $v' = \dfrac{0.0625\ \text{N-m}^2}{EI}$ (CCW)

5.212 $P = \dfrac{5}{4}qa$

5.214 $M_0 = 450$ lb-ft

5.216 $\delta_B = \dfrac{qL^4}{30EI}$

5.218 $\theta_A = \dfrac{-qL^3}{648EI},\ \theta_B = \dfrac{-5qL^3}{648EI}$

5.220 $v = \dfrac{7}{128}\dfrac{M_0L^2}{EI}$ (up), $v' = \dfrac{11}{96}\dfrac{M_0L}{EI}$ (CW)

5.222 $P_3 = 0.842$ lb

5.224 8.55 N/m $<$ Stiffness < 58.5 N/m

5.226 $v = 0.01074$ in.

5.228 $\Delta T = 93.2\,(10^3)$ N

5.230 $v = 0.1196$ mm/kN

5.232 (a) $u = 15.08$ mm,
    (b) $\theta = 5.04°$

5.234 $\theta_A = \dfrac{-Pab}{3EI}$

5.236 $u_B = 0,\ v_B = \dfrac{Pa^3}{3EI} - \dfrac{Pca^2}{2EI}$ (down),

$\theta_B = \dfrac{Pa^2}{2EI} - \dfrac{Pac}{EI}$ (CW)

5.238 $u_B = 0$, $v_B = \dfrac{qca^3}{3EI} + \dfrac{qc^2a^2}{4EI}$ (down),

$\theta_B = \dfrac{qca^2}{2EI} + \dfrac{qc^2a}{2EI}$ (CW)

5.240 $u_C = 0$, $v_C = \dfrac{qa^4}{8EI} - \theta_0 a$ (down)

5.242 $u_C = u_0$ (right), $v_C = \dfrac{Pb^3}{3EI} + \theta_0 b$ (down)

5.244 $u_C = 0$, $v_C = \dfrac{Pb^3}{3EI} + \dfrac{Pb^2a}{3EI}$ (down)

5.246 $u_D = \dfrac{qa^2b^2}{4EI}$ (right), $v_D = \dfrac{qb^4}{8EI} + \dfrac{7}{24}\dfrac{qb^3a}{EI}$ (down)

5.248 (a) $v_{\text{center}} = \dfrac{1}{24}\dfrac{qa^4}{EI}$ (up),

(b) $v_C = \dfrac{7qa^4}{24EI}$ (down)

5.250 $u_D = \dfrac{Pcb^2}{2EI} + \dfrac{Pa^2b}{2EI} + \dfrac{Pcab}{EI}$ (right),

$v_D = \dfrac{Pc^3}{3EI} + \dfrac{Pa^2c}{2EI} + \dfrac{Pc^2a}{EI} + \dfrac{Pc^2b}{EI}$ (down)

5.252 $\theta = T\left[\dfrac{2c}{Ebh^3} + \dfrac{2a}{G\pi R^4}\right]$

5.254 $v = 0.420$ in.

5.256 $P = 20.0$ N

5.258 $u = 67.6$ mm, $v = 9.96$ mm, $\theta = 17.07°$

5.260 $R_A = \dfrac{7}{32}qL$, $R_B = \dfrac{5}{16}qL$, $R_C = \dfrac{-1}{32}qL$

5.262 $R_B = P\left[\dfrac{6\dfrac{EI}{k} - 3a^3}{6\dfrac{EI}{k} + 2a^3}\right]$,

$M_B = Pa\left[\dfrac{12\dfrac{EI}{k} - a^3}{6\dfrac{EI}{k} + 2a^3}\right]$,

$v_A = \dfrac{Pa^3}{EI}\left[\dfrac{16\dfrac{EI}{k} + \dfrac{7}{32}a^3}{6\dfrac{EI}{k} + 2a^3}\right]$

5.264 $R_A = R_D = \dfrac{2}{15}qL$, $R_B = R_C = \dfrac{11}{30}qL$

5.266 $R_C = R_D = \dfrac{P}{2}\left[\dfrac{1}{1 + \dfrac{1}{16}\dfrac{E_1 I_1}{E_2 I_2}\left(\dfrac{L_2}{L_1}\right)^3}\right]$,

$R_A = P\left[\dfrac{\dfrac{1}{16}\dfrac{E_1 I_1}{E_2 I_2}\left(\dfrac{L_2}{L_1}\right)^3}{1 + \dfrac{1}{16}\dfrac{E_1 I_1}{E_2 I_2}\left(\dfrac{L_2}{L_1}\right)^3}\right]$, $M_A = R_A L_1$

5.268 $v_B = \alpha_2 \Delta T L_2 \left[\dfrac{1}{1 + \dfrac{3E_1 I_1}{L_1^3}\dfrac{L_2}{E_2 A_2}}\right]$ (down),

$\sigma = \dfrac{E_2 \alpha_2 \Delta T}{1 + \dfrac{L_1^3}{3E_1 I_1}\dfrac{E_2 A_2}{L_2}}$

5.272 (a) $M_{B\text{-}B} = \left[E_t\dfrac{\pi}{2}R^3 t + 4E_r A_r R^2\right]\kappa$,

(b) $M_{C\text{-}C} = \left[E_t\dfrac{\pi}{2}R^3 t + 4E_r A_r R^2\right]\kappa$

5.274 (a) $\kappa = \dfrac{4M}{\pi\left[E_A\left(R_2^4 - R_1^4\right) + E_B\left(R_3^4 - R_2^4\right)\right]}$,

(b) $(\sigma_A)_{\text{max}} = \dfrac{4MR_2}{\pi\left[\left(R_2^4 - R_1^4\right) + \dfrac{E_B}{E_A}\left(R_3^4 - R_2^4\right)\right]}$,

$(\sigma_B)_{\text{max}} = \dfrac{4MR_3}{\pi\left[\dfrac{E_A}{E_B}\left(R_2^4 - R_1^4\right) + \left(R_3^4 - R_2^4\right)\right]}$

Chapter 6

6.2 $F_y = 750$ lb (normal), $F_z = 500$ lb (shear), $M_x = 500$ lb-ft (bending), $M_y = 3500$ lb-ft (twisting), $M_z = 5250$ lb-ft (bending)

6.4 $F_y = 6$ kN (shear), $F_z = 4$ kN (normal), $M_x = 1$ kN-m (bending), $M_y = 2$ kN-m (bending), $M_z = 3$ kN-m (twisting)

6.6 (a) $F_1 = 0$, $F_2 = 25.6$ kN, $F_3 = 16$ kN, (b) $F_x = 16$ kN (shear), $F_y = 25.6$ kN (normal), $M_z = 16$ kN-m (bending)

6.8 $F_y = 1000$ lb (normal), $M_x = 5000$ lb-ft (bending), $M_y = 1333$ lb-ft (bending)

6.12 $F_y = 1600$ N (shear), $M_x = 240$ N-m (twisting), $M_z = 272$ N-m (bending)

6.14 $F_y = 1600$ N (shear), $F_z = 524$ N (shear),
$M_x = 240$ N-m (twisting), $M_y = 174$ N-m (bending),
$M_z = 272$ N-m (bending)

6.16 (a) $\sigma = 11.64$ ksi, $\tau = 1294$ psi,
(b) $\sigma = 32.3$ psi,
(c) $\tau = 17.97$ psi

6.18 (a) $\sigma = 65.7$ MPa, $\tau = 8.96$ MPa,
(b) $\sigma = 0.1643$ MPa,
(c) $\tau = 0.1095$ MPa

6.20 $c = 16.67$ mm

6.22 $F_0 = 8.77$ lb

6.24 (a) $T = 132.6$ lb,
(b) $F = 2250$ lb

6.26 $\sigma = 60.8$ MPa, $\tau = 69.4$ MPa

6.28 $\sigma = 1113$ psi, $\tau = 2190$ psi

6.30 $\sigma = 36.1$ MPa, fraction of stress due to axial $= 0.0256$

6.32 $p = 104.2$ psi

6.34 (a) $p_{max} = 240$ psi,
(b) $p_{max} = 200$ psi

6.36 $p = 0.733$ MPa, $\sigma_{max} = 22.0$ MPa

6.38 $\Delta\sigma_h = 13.91$ MPa

6.40 $t = 130$ mm

6.42 $\sigma_h = p_{atm}\left[\dfrac{D_o - D_i}{2t}\right]$, $\sigma_a = p_{atm}\left[\dfrac{D_o - D_i}{4t}\right]$

6.44 $t = 2.38$ in.

6.46 $\Delta p = 187.5$ psi

6.48 $p = 7.50$ MPa

6.50 top: $\varepsilon_h = 0.328\,(10^{-3})$, $\varepsilon_a = -0.0660\,(10^{-3})$
side: $\varepsilon_h = 0.288\,(10^{-3})$, $\varepsilon_a = 0.0676\,(10^{-3})$

6.52 $p = 0.0682$ psi

6.54 (a) $\sigma = -4.73$ psi,
(b) $\delta_z = -0.00474$ in.

6.56 (a) $p = 4650$ Pa,
(b) $p = 0.539$ MPa

6.58 $\theta_y = 6.05°$, $\theta_z = 2.70°$

6.60 $\theta_y = 16.80°$

6.62 $\theta_0 = 21.1°$

6.64 $\theta = 0.1959°$

6.66 $v_C = \dfrac{F_0}{k}\left[\dfrac{L_1 + L_2}{L_1}\right]^2$

6.68 $\phi_B = \dfrac{T_0(L_1 + L_2)}{4GI_p}$

6.70 $\theta_B = \dfrac{M_0 L}{3EI}$

6.72 $\theta_{center} = \dfrac{M_0 L}{12EI}$

6.74 $v_0 = \dfrac{F_0}{3E}\left[\dfrac{(L_1 + L_2)^3 - L_2^3}{I_1} + \dfrac{L_2^3}{I_2}\right]$

6.76 $v_D = \dfrac{P}{EI}\left[\dfrac{a^3}{3} + \dfrac{a^2 b}{2}\right]$

6.78 $u_0 = \dfrac{\pi}{2}\dfrac{F_0 R^3}{EI}$

6.80 $w_0 = F_0 R^3\left[\dfrac{\pi}{2EI} + \dfrac{3\pi}{2GI_p}\right]$

Chapter 7

7.2 $\theta = 200°$

7.8 $\sigma > 0$, $\tau > 0$

7.12 (a) $\sigma > 0$, $\tau > 0$,
(b) $\sigma > 0$, $\tau < 0$,
(c) $\sigma < 0$, $\tau > 0$,
(d) $\sigma > 0$, $\tau < 0$

7.14 $\theta = 245°$, $\sigma > 0$, $\tau < 0$

7.18 (a) $\theta = 15°$, $\sigma > 0$, $\tau > 0$,
(b) $\theta = 125°$, $\sigma < 0$, $\tau > 0$,
(c) $\theta = 230°$, $\sigma > 0$, $\tau < 0$,
(d) $\theta = -65°$, $\sigma < 0$, $\tau > 0$

7.20 $\sigma = -3270$ psi, $\tau = -5810$ psi

7.22 $\sigma = 4470$ psi, $\tau = -897$ psi

7.24 $\sigma = -34.5$ MPa, $\tau = 47.5$ MPa, $\sigma = -85.4$ MPa

7.26 $\sigma = 170.0$ MPa, $\tau = 33.3$ MPa, $\sigma = 80.1$ MPa

7.28 $\sigma_x = 5.71$ MPa, $\sigma_y = -40.7$ MPa, $\tau_{xy} = -79.8$ MPa

7.30 $\sigma_x = 100$ MPa, $\sigma_y = -167.5$ MPa, $\tau_{xy} = -70$ MPa,
$\tau = 150.8$ MPa

7.36 $\varepsilon = 0.00342$

7.38   Angle while deformed = 90.01298°

7.40 $E = 200$ GPa, $\nu = 0.3$

7.42 $G = 26.5$ GPa

7.44 $\sigma_{screw} = 9.58$ MPa

7.46 $\alpha > 73.3°$

7.48 $\tau_{max} = 112.4$ MPa, $\sigma_m = -15$ MPa, $\theta_s = -28.86°$

7.50 $\sigma_{max} = 6760$ psi, $\sigma_{min} = -5760$ psi, $\theta_p = 75.69°$,
$\tau_{max} = 6260$ psi, $\sigma_m = 500$ psi, $\theta_s = 30.69°$

7.52   Line element is 31.7° CCW from vertical,
$\varepsilon = 0.863 \, (10^{-3})$

7.54   Line element is 39.3° CCW from vertical,
$\varepsilon = -0.202 \, (10^{-3})$

7.56   Element at 69.4°, length of one side = 1.000414 in.,
length of other side = 0.999493 in.

7.58   Element at −35.8°, $\gamma = 0.025$,
both sides = 0.998286 in.

7.60 $\tau_{max} = 8070$ psi, $\sigma_m = -5040$ psi, $\theta_s = 19.33°$

7.62 $\sigma_{max} = 2195$ psi, $\sigma_{min} = -195$ psi, $\theta_p = 16.60°$

7.64 $\sigma_{max} = 0.268$ MPa, $\sigma_{min} = -0.088$ MPa, $\theta_p = 29.8°$

7.66 $\tau_{max} = 5025$ psi, $\sigma_m = 4180$ psi, $\theta_s = -28.15°$,
$\sigma_{max} = 9200$ psi, $\sigma_{min} = -845$ psi, $\theta_p = 16.85°$

7.68 $\tau_{max} = 1025$ psi, $\sigma_m = 3075$ psi, $\theta_s = 45°$,
$\sigma_{max} = 4100$ psi, $\sigma_{min} = 2025$ psi, $\theta_p = 0°$

7.70 $\sigma_{max} = 6480$ psi, $\tau_{max} = 3720$ psi

7.72   (a) $\tau_{max} = 8860$ psi,
(b) $\tau_{max} = 9540$ psi

7.74 $\sigma = 5600$ psi, $\tau = -1964$ psi

7.76 $\tau_{max} = 5385$ psi, $\sigma_m = 10$ ksi

7.78 $\tau_{max} = 30$ MPa, $\sigma_m = 24$ MPa

7.80 $\sigma_{max} = 7580$ psi, $\sigma_{min} = 3790$ psi

7.82 $\tau_{max} = 9340$ psi, $\sigma_m = 4950$ psi

7.84   (a) $\tau_{xy} = 21.2$ ksi,
(b) Crack on plane $\theta = 54.7°$

7.86 $\sigma = 15.82$ ksi, crack on plane $\theta = -28.6°$

7.88 $F_0 = 750$ lb

7.90   (a) $\tau_{xy} = 15.61$ ksi,
(b) Slip lines on $\theta = 19.34°$, $\theta = -70.7°$

7.92 $\Delta\phi = 0.379°$

7.94 $P = 130.1$ kN

7.96 $\phi = 10.19°$

7.98   (a) $T = 9.14$ lb-in.,
(b) $T = 19.60$ lb-in.

7.100 $v = 4.80$ mm,
$v \ll 400$ mm, so beam theory gives
a good prediction

7.102 $T_0 = 75.6$ N-m

7.104 $F_0 = 1132$ N

7.106 $L_2 < 16.3$ in.

7.108   (a) $\sigma_{VM} = 215$ MPa,
(b) Factor of safety = 1.26

7.110   (a) $F_0 = 25.9$ N,
(b) $F_0 = 26.8$ N

7.112   Factor of safety = 2.92

7.114 $\varepsilon_x = 550 \, (10^{-6})$, $\varepsilon_y = -300 \, (10^{-6})$, $\gamma_{xy} = -50 \, (10^{-6})$

7.116 $\sigma_{max} = 6200$ psi, $\sigma_{min} = 10.7$ psi, $\tau_{max} = 3095$ psi

7.118 $\sigma_x = -57.8$ MPa, $\sigma_y = -12.67$ MPa, $\tau_{xy} = -55.4$ MPa

7.120 $\varepsilon_a = -370 \, (10^{-6})$, $\varepsilon_b = 180 \, (10^{-6})$, $\varepsilon_c = 280 \, (10^{-6})$

7.122 $\varepsilon_a = 340 \, (10^{-6})$, $\varepsilon_b = -561 \, (10^{-6})$, $\varepsilon_c = 305 \, (10^{-6})$

7.124 $\sigma_{max} = 385$ MPa

7.126 $P = 9060$ lb

7.128 $a = 9.6$ in.

7.130 $r = 26.4$ mm

7.132 $P_{max} = 14.44$ kN

7.134 $r_1 = 33$ mm

7.136 $\tau_{max} = 8220$ psi

7.138 $\tau_{max} = 5970$ psi

7.140 $T_c = 252$ lb-in.

7.142 $\sigma_{max} = 6800$ psi

7.144 $M_{max} = 61.5$ N-m

7.146 $M_{max} = 578$ lb-in.

Chapter 8

8.2 $W = 698$ kN

8.4 $t = 9.67$ mm

8.6 $q = 482$ N/m

8.8 $\Delta T = 139.2$ °C

8.10 $T = 43.3$ lb

8.12 (a) $P_{cr} = 15.54 \ (10^6)$ N,
(b) $P_{cr} = 7.82 \ (10^6)$ N

8.14 (a) $t = 1.950$ mm,
(b) Ratio of buckling strength $= 1.352$

8.16 $F_0 = \dfrac{7}{9} \left(\dfrac{\pi^2 EI}{L_2^2} \right)$

8.18 $\Delta T = \dfrac{2.041}{12} \dfrac{\pi^2 b^2}{L^2 \alpha}$

8.20 (a) $\dfrac{P_{constrained}}{P_{unconstrained}} = 3.12$,

(b) $\dfrac{P_{constrained}}{P_{unconstrained}} = 4$

8.22 $T = 111.2 \ (10^6)$ N

8.24 $W = 14210$ lb

Key Terms

Chapter 2

Elements (2.1.2)
We view a body as composed of many small (theoretically infinitesimal) cubes of material. An element is defined as one such cube of material in the body.

Internal Force (2.2.4)
If we view a body as composed of two adjacent parts that meet at an imaginary plane or line passing through the body, then the internal force (moment) is defined as the force (moment) exerted by the two adjacent parts on each other.

Normal Stress (2.3.3)
Normal stress, σ, describes the element-level intensity of internal normal force, and it is defined as the normal force exerted by two adjacent elements on each other, divided by the area on which the force acts.

Normal Strain (2.4.4)
Normal strain, ε, describes the element-level intensity of deformation due to elongation, and it is defined as the increase in length of an element, due to deformation, divided by the element's original length.

Young's Modulus (2.6.3)
Young's modulus (or Elastic modulus), E, captures the intrinsic stiffness of a material when it elongates elastically, and it is defined as the proportionality between normal stress and normal strain when the material is subjected to normal stress in one direction.

Poisson Ratio (2.6.8)
Poisson ratio, ν, captures the contraction in one direction when a material is elongated in a perpendicular direction, and it is defined as the proportionality between the magnitudes of transverse and longitudinal normal strains when the material is subjected to only normal stress in the longitudinal direction.

Yield Stress (2.7.3)
Yield stress, σ_y, is defined as the uniaxial tensile stress at which a ductile material begins to exhibit noticeable plastic deformation.

Shear Strain (2.9.2)
Shear strain, γ, describes the element-level intensity of deformation due to shape change, and it is defined as the tangent of the angle change, due to deformation, between two lines that are originally perpendicular.

Shear Stress (2.9.5)
Shear stress, τ, describes the element-level intensity of internal shear force, and it is defined as the shear force exerted by two adjacent elements on each other, divided by the area on which the force acts.

Shear Modulus (2.9.7)
Shear modulus, G, captures the intrinsic stiffness of a material when it shears elastically, and it is defined as the proportionality between shear stress and shear strain.

Chapter 3

Displacement (3.1.6)
Displacement, u or v, is defined as the distance through which a point on a body moves in some direction, due to applied loads.

Thermal Strain (3.5.2)
Thermal strain, ε_{th}, is defined as the normal strain that an element, unconstrained by any neighboring elements, would undergo if the temperature is altered.

Coefficient of Thermal Expansion (3.5.2)
Coefficient of thermal expansion, α, captures the intrinsic tendency of a material to expand or contract due to temperature changes, and it is defined as the proportionality between the thermal strain and the change in temperature.

Chapter 4

Twist (4.1.2)
Twist, $\Delta\phi$, is defined as the difference in rotation angle of one cross-section of a shaft relative to another, due to torsional deformation.

Twist per Length (4.1.5)
Twist per length captures the intensity of the torsional deformation, and it is defined as the difference in rotation angle of one cross-section of a shaft relative to another, divided by the distance between the cross-sections.

Polar Moment of Inertia (4.4.5)
Polar moment of inertia, I_p, captures the contribution of a circular shaft's cross-section to its resistance to twisting, and it is defined mathematically as the integral over the cross-sectional area of the square of the radial position from the center.

Strength (4.5.1)
Strength is a general term that captures the maximum load that a body can safely carry, but it is defined in each circumstance with regard to the type of loading and how the load is defined.

Stiffness (4.5.1)
Stiffness is a general term that captures the proportionality between the load and the deformation when a body deforms elastically, but it is defined in each circumstance with regard to the type of loading and how the load and deformation are defined.

Chapter 5

Radius of Curvature (5.1.4)
Radius of curvature, ρ, captures the intensity with which a segment of a beam has deformed in bending, and ρ is defined as the radius of the circle into which a longitudinal line of the initially straight beam has deformed. The reciprocal, $\kappa = 1/\rho$, is referred to as the curvature.

Shear Force and Bending Moment (5.3.4)

View the beam as composed of two adjacent portions that meet at an imaginary plane or line passing transversely through the beam. The shear force, V, is defined as the internal force that the two portions exert on each other in the direction transverse to the beam. The bending moment, M, is defined as the internal moment or couple that the two portions exert on each other about the axis perpendicular to the plane of bending.

Neutral Plane and Neutral Axis (5.6.4)

The neutral plane is defined as the plane that experiences no strain when a beam is bent. The neutral axis is the intersection between the plane of bending and the neutral plane.

Moment of Inertia (5.8.2)

Moment of inertia, I, captures the contribution of a beam's cross-section to its resistance to bending, and it is defined mathematically as the integral over the cross-sectional area of the square of the distance away from the neutral plane.

Section Modulus (5.11.4)

Section modulus, S, captures the relation between the bending moment and the maximum resulting bending stress in the beam. S depends on the shape of the cross-section, in particular, on the moment of inertia, I, and on the distance from the neutral plane to the most remote points in the cross-section.

Shear Flow (5.15.2)

In transverse shear loading of a beam, we consider the loads exerted between adjacent portions that meet at planes parallel to the longitudinal axis. The shear flow, q, is defined as the longitudinal shear force per unit longitudinal length.

Chapter 6

Isotropic (6.4.2)

A material is defined as isotropic if it has the same mechanical properties, regardless of the orientation of the material when tested mechanically. For an isotropic, elastic material, E, G, and ν have unique values regardless of orientation.

Strain Energy Density (6.6.3)

Strain energy density is defined as the elastic energy stored in a deformed element divided by the element volume.

Chapter 7

Maximum Shear Stress (7.4.1)

When considering the stresses acting on elemental surfaces of different orientations passing through the same point, the maximum shear stress, τ_{max}, is defined as the magnitude of the shear stress that is maximum from among all possible surface orientations.

Principal Stresses (7.4.1)

When considering the stresses acting on elemental surfaces of different orientations passing through the same point, the principal stresses, σ_{max} and σ_{min}, are defined as the maximum and minimum normal stresses from among all possible surface orientations.

Index

failure, 482
lateral torsional, 480
load, 498
stress, 485, 486, 491, 499
Built-up cross-sections
shear flow, 310–317

C

Cables
axial loading and, 128–133
Cable-stayed bridge, 504–505
Cantilever beams
deflections and slopes of, 536–537
variables for tabulated, 328
Centroid, 262, 263
calculations, 515, 518–519
of composite, 272, 273, 363
defined, 514
line of symmetry, 514
Circle, 522
moment of inertia calculation for, 263, 265
Circular shafts
shear strain in, 140–147
shear stress in, 150–161
vs. non-circular, 176
Circumferential (hoop) stress, 381
Coefficient of thermal expansion, defined,
120, 135
Column buckling, 481; *See also* Buckling
Combined loading, 364–411
conservation of energy, 400–409
deflections by, 392–397, 411
elastic stress–strain relationship, 386–391, 411
internal loads; *See* Internal loads
of pressure vessels, 380–385
strain energy, 398–399
stresses with, 372–379
Composite areas
calculations, 515
Composite beam; *See also* Beam(s)
bending of, 296–297, 300–303
Composite cross-sections
bending of, 272–279
centroid of, 272, 273, 363
Compression, tension *vs.*, 93
Compressive stress, 33
Concentration factors
stress, 540–541
Conditions of equilibrium, 12

Conservation of energy, 398, 400–409, 411
Constraints
beams, 222
symbols, 222
Cross-sections, 517
beams, 250–251
and bending stiffness, 290–295
and bending strength, 290–295
composite, bending of, 272–279
3-D deflection and rotation of, 392
elastic deformation and, 86
internal force at, 92
internal loads at, 366–367
non-symmetric, bending of, 298–303
stresses on, 372–373
thin-walled and built-up (shear flow), 310–317
Cross-section, shaft; *See also* Shafts
circular *vs.* noncircular, 176
equal area, stiffness of, 167
stiffness/strength, 164, 217
Curvature, 362; *See also* Radius of curvature (ρ)
defined, 318
deflections and, 318–319, 363
Cyclic loads, 466, 479
Cyclic stress, 467
Cylindrical pressure vessel, 380, 381, 410

D

Deflections, 162, 220, 363; *See also* Bending
beams, 393
bending moments, 318–361
of cantilever beams, 536–537
by combined internal loads, 392–397, 411
curvature, 318–319, 363
and failure, 7, 20
related to internal loads, 318–327
of simply supported beams, 538–539
using tabulated solutions, 328–353, 363
Deformation; *See also* Displacement
axial and shear, 392
axially loaded members, 100
in bending, 219, 220–221
intensity of, 40
multiple axial forces and, 92
plastic, 467
radius of curvature, 221
in statically indeterminate problems, 108
strains and, 415
stress and, 86

SI Prefixes

Prefix	Symbol	Multiplication Factor
tera	T	10^{12}
giga	G	10^{9}
mega	M	10^{6}
kilo	k	10^{3}
milli	m	10^{-3}
micro	μ	10^{-6}
nano	n	10^{-9}
pico	p	10^{-12}

USCS Abbreviations

1 kip = 1000 lb	1 ksi = 1000 psi

Conversions

Quantity	USCS to SI	SI to USCS
Length	1 in. = 25.4 mm 1 ft = 0.3048 m	1 m = 39.37 in. 1 m = 3.281 ft
Force	1 lb = 4.448 N	1 N = 0.2248 lb
Power	1 hp = 745.7 W	1 kW = 1.341 hp
Pressure; Stress	1 psi = 6.895 kPa	1 MPa = 145.0 psi

Properties of Common Areas
(see Appendix C for more)

Rectangle

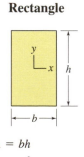

$$A = bh$$

$$I_x = \int_A y^2 \, dA = \frac{1}{12}bh^3$$

$$I_y = \int_A x^2 \, dA = \frac{1}{12}b^3h$$

Circle

$$A = \pi r^2$$

$$I_p = \int_A \rho^2 \, dA = \frac{\pi}{2}r^4$$

$$I_x = I_y = \frac{\pi}{4}r^4$$